AF527785

Introduction to Infectious Viral Diseases of Animals

Dr. Manoranjan Rout is presently working as Senior Scientist at ICAR-National Institute on Foot and Mouth Disease, International Centre for Foot and Mouth Disease, Bhubaneswar, Odisha. He has moved through a brilliant career-corridor with incredible academic achievements throughout his school, college and university days. He was the topper in his school career receiving merit scholarships in 5th, 7th and 10th class securing best 14th position in Orissa State 10th H.S.C. Board Examination. He was the "Borasambar-Lewis Gold Medal" winner for being the highest scorer in George High School, Bargarh, Orissa in High School Certificate Examination in the year 1994 conducted by the Board of Secondary Education, Orissa. He graduated from the College of Veterinary Science and Animal Husbandry, OUAT, Bhubaneswar, Odisha in 2003 securing the second position throughout his BVSc career. He received MVSc and PhD degree in the discipline of Veterinary Pathology from Indian Veterinary Research Institute (IVRI), Izatnagar, Uttar Pradesh in the year 2006 and 2009, respectively securing the first position. He is the recipient of the "Best Student Award" for the year 2008-2009 with gold medal in his PhD career at IVRI being the institute/ university topper. He is the first position holder in all India ARS/NET Examination - 2007, conducted by ASRB in the combined discipline of Veterinary Pathology and Veterinary Parasitology. He joined ARS in November 2009 as trainee scientist at ICAR-NAARM, Hyderabad and subsequently got his first posting as scientist at ICAR-Project Directorate on Foot and Mouth Disease, Mukteswar serving for over a decade at the place. He is the recipient of many prestigious awards during his career. He holds the life membership of 15 different esteemed scientific professional/ scholarly societies. He has published more than 85 research papers in prestigious national and international journals, 45 review/popular articles, 12 book/manual chapters to his credit. He is currently acting as the editorial member/reviewer in different journals of national and international repute.

Dr. Raj Pal Diwakar is currently working as Assistant Professor in the Department of Veterinary Microbiology, College of Veterinary Science & A.H, Acharya Narendra Deva University of Agriculture & Technology, Kumarganj, Ayodhya, Uttar Pradesh. He obtained undergraduate degree from the GBPUA&T, Pantnagar, UK in the year 2004 and master's degree in Veterinary Microbiology from the College of Veterinary Science and Animal Husbandry, NDUA&T, Kumarganj, Faizabad in 2006. Presently, he is pursing PhD from the same institute. His research interests focus mainly on Enterobacteria associated with calf diarrhoea especially *E. coli*. He has prepared a total of 07 manuals for the students of different courses of Veterinary Microbiology and published more than 46 Research articles, 51 review articles, 08 short communications, 12 lead papers, 32 popular articles and 06 case study reports. He has authored 09 books in different courses of Veterinary Microbiology for BVSc & AH and MVSc students and 02 books for animal owners and paravets. He is currently working as Editorial member/Reviewer in different reputed national and international journals.

Introduction to Infectious Viral Diseases of Animals

– Editors –

Dr. Manoranjan Rout

M.V.Sc., Ph.D. (Veterinary Pathology)

Senior Scientist,
ICAR-National Institute on Foot and Mouth Disease,
International Centre for Foot and Mouth Disease,
Bhubaneswar, Odisha

Dr. Raj Pal Diwakar

M.V.Sc. (Veterinary Microbiology)

Assistant Professor,
Department of Veterinary Microbiology,
College of Veterinary Science and Animal Husbandry,
ANDUA&T, Ayodhya, Uttar Pradesh

2025

Daya Publishing House®

A Division of

Astral International Pvt. Ltd.

New Delhi – 110 002

ISBN 978-93-5919-302-1

Published by : **Daya Publishing House®**
A Division of
Astral International Pvt. Ltd.
– ISO 9001:2015 Certified Company –
4736/23, Ground Floor, Ansari Road, Darya Ganj
New Delhi-110 002
Ph. 011-43549197, 8130496929
E-mail: info@astralint.com
Website: www.astralint.com

Printed at : **Replika Press Pvt. Ltd.**

To thine own self be true.

William Shakespeare

Reading maketh a full man; conference a ready man; and writing an exact man.

Francis Bacon

To our parents, family members, friends and well-wishers, in grateful appreciation of all they have done to inspire, support and encourage us in our every endeavor.

Manoranjan Rout

Raj Pal Diwakar

Foreword

Viruses are constantly evolving and at much faster rate now than ever before. Accordingly, the dynamics of diseases caused by the viruses are also constantly changing. These changes in recent times are across the species in both animal and plant kingdom including livestock and other domestic animals. Many new and novel viral pathogens have been detected in the last two to three decades and even the existing ones have undergone mutations to become more pathogenic with many acquiring zoonotic potential. Livestock is an integral component of India's growth story, particularly in rural economy by providing for economic, nutritional and food security. However, the presence of diseases in livestock, particularly of viral origin, is a major impediment in the growth of this sector in India. Huge resources in terms of financial, human and material are being spent in control and containment of these viral diseases of livestock so as to minimize the loss and finally overcome the menace. In order to do so, complete and the latest knowledge on the disease and the pathogen is of utmost importance.

In this context, I find the new book on "Introduction to Infectious Viral Diseases of Animals" is relevant and a timely compilation of latest knowledge on the most important viral diseases, written by eminent experts in specific diseases. All the contributing authors of the book are among the best experts in the country having contributed through their experience gained over many years of dedicated work in renowned research and educational institutions. The knowledge available in each of the chapters on specific diseases in the book are latest and exhaustive. I believe this book is going to be immensely useful as a textbook for both undergraduate students of veterinary science as well as for the postgraduate students of veterinary virology/microbiology. This book could also be a ready reckoner for those involved in planning, management and execution of control programme of viral diseases of livestock in India. The book has thoughtfully included a few major diseases of pet and companion animals, which could be useful not just for the students, but also for veterinary practitioners serving this sector.

The book in this edition covered most of the important viral diseases of livestock. For readers desiring more, inclusion of a few additional chapters on important viral diseases of aviary, equidae and wildlife species, as well as on some emerging and exotic diseases of animals particularly of zoonotic importance could be considered by the editors for their next edition of the book. Overall, this book contains plethora of useful knowledge on important viral diseases of livestock and the measures for their control and containment.

Santanu Kumar Bandyopadhyay

BVSc&AH, MVSc, PhD (Cambridge), DSc (Honoris Causa), FNAAS

Ex-Member, Agricultural Scientist Recruitment Board, ICAR

Ex-Animal Husbandry Commissioner to Government of India,

Ex-Sr Technical Coordinator, FAO of UN, Vietnam,

Ex-RSU-SAARC Coordinator, FAO of UN, Nepal

Formerly, Head, Division of Virology, IVRI, Mukteswar

Preface

It is truly said that 'nature is the best teacher'. After the unprecedented COVID pandemic, common man has gained a fair knowledge on what is a virus, what it causes and what precautions one should take to prevent the viral infection. It is also true that amongst all pathogens, viruses continue to account for a large chunk of animal infections. In recent years there has been emergence and re-emergence of many viral infections in animals. "Introduction to Infectious Viral Diseases of Animals" is the fruit of all painstaking efforts and dedication of many researchers in their respective fields of veterinary science who have sincerely diverted some hours out of their precious time-pool. Though this excellent book is entitled "Introduction to Infectious Viral Diseases of Animals", it is of course beyond and far more than just an 'Introduction'. The book attempts to present the essential information on virus, the disease, pathogenesis, epidemiology, pathology, diagnosis and control with every effort being made to incorporate recent advances in the subject. It is true to say that the tireless and painstaking efforts and expertise of the contributing authors and editors could make it possible to bring out the compilation reflecting their immense enthusiasm and dedication.

When we first dreamt of editing a book on viral diseases of animals, we did so because of our awareness of its importance and necessity for the students and practitioners of the veterinary profession. At some points of time, we became aware of an incredible text book on infectious diseases involving different etiologies that inspired us to focus on the viral agents causing diseases in different species as a single source summary. Over the past one year, our rationale for producing this book has extended and it made us believe that the demand for such a text is even more pressing. This is because after the COVID pandemic, the awareness on virus diseases has been increased among the public.

Our main objective in compiling this book has been to provide a ready guide to the infectious viral diseases of livestock for veterinarians, students and others who are involved with animals.

The authors and publisher feel happy to present the book on "Introduction to Infectious Viral Diseases of Animals". No veterinarian can ever practice veterinary medicine without confronting different faces of animal diseases in much detail. This is nowhere probably more vividly explained than in the "Introduction to Infectious Viral Diseases of Animals". The germ of the idea for this book first infected our minds few years before. We have been the eyewitnesses to the severe outbreaks of foot and mouth disease, classical swine fever, PPR, lumpy skin disease, contagious ecthyma, African swine fever *etc.* at different parts of the country. We have watched the economic hardships the farmers and animal keepers face as well as the struggles of large elite farms during such outbreaks even during our student days. In every instance, we have sought to find appropriate answers to the questions. However, the answers were scattered throughout the veterinary literature that was prompting the veterinarians to access to the state-of-the-art computer databases. The readership can rapidly gain access to a comprehensive body of knowledge with the significance of the most recent advances in the field of virology in the carefully crafted and edited chapters covered here. As envisioned by the editors, this textbook will be of utmost utility not only for the undergraduate students but also for the students pursuing post-graduation in veterinary virology and pathology.

Our understanding and knowledge of animal diseases has derived immense benefits from newer inventions and advances made in the field of veterinary science especially in diagnostic virology. We feel it is badly necessary to conceptualize and compile all such advancements in one place in the form of a book for our budding talents who are pursuing veterinary and allied courses that will definitely add the existing resources of information on viral diseases. Each chapter of the book has been designed with different viruses causing different diseases in different species of animals including farm and companion animals that will hopefully help the students to understand the diseases in detail and keep abreast with the latest developments made in the particular diseases in the field of virology and laboratory diagnostics followed by details of the virus, disease epidemiology and pathogenesis. Finally, this valuable compilation will act as a guide and help the readers to update their knowledge pool with newer information on different aspects.

Without the hard work with sincere efforts and inputs of many intellects, this book would never have kissed the line of completion. Researchers from diverse scientific backgrounds have contributed topics of their choice and long involvement as chapters in this book. We sincerely thank each and every author who worked hard to provide us with different chapters of their expertise and for being patient with the waiting time and sometimes editorial changes that were requested at different points of time. The constant support and input of our authors have had a significant impact on the final structure of this text. We gratefully acknowledge the dedication of the authors for bringing their expertise and fresh perspectives to the text. We hope that each contributing author will be as proud to be associated with this book as we are. We again thank all the authors who unselfishly shared the photographs in their respective chapters. The staffs at the publisher have been extremely supportive from the beginning that needs great appreciation. We hope that this book will be of value and service to the veterinarians and students of undergraduate and postgraduate. If any reader learns just a small fraction of the amount that we learnt while writing and editing, then our purpose will be served.

Writing or editing any book is never easy or painless and we are therefore indebted to many people who have assisted us and contributed in different ways. In this line, it was truly an honour to receive inspiring words from Dr. S.K. Bandyopadhyay, the doyen in the discipline of veterinary virology. I am eternally grateful to sir for his critical comments to improve the book. Sir has expressed his invaluable views in the 'foreword' section with unwavering optimism. In truth, he has inspired us a lot for the work showering his blessings in the form of criticisms, suggestions and constant encouragements while editing this book. Sir, we fall in short of lexicon while expressing our heartfelt gratitude for your thoughtful remarks and constant motivation that provided us with a stimulus to keep pushing through the tough times. Special acknowledgements are also due to our publisher. Our families have always given us support and encouragement during the entire journey. Our parents, to whom we owe so much, are the subject of the dedication at the beginning of the book. The beautiful colorful butterflies feeding on the sweet nectar in the interior of flowers, marching of ants together at a consistent speed to gather food showing their incredible sense of discipline and unity, and lyrics from the honey bees of the fragile nature of our planet, coupled with the kindness of friends and the inspiring words from the well-wishers, have made our task far more pleasurable than ever expected. We are truly grateful. We welcome constructive comments and suggestions from all students and teachers who use our book as part of their studies, teachings and academics. We are grateful to many academicians for their comments and valuable criticisms during the preparation of this book. We look forward to receiving similar input in the future times too. We wish to express our sincere gratitude to all the authors and the publisher. With pride and privilege, we acknowledge the contribution of all our authors who are involved in teaching and research in various renowned universities and institutes from various states of the country. This book will decide its own fate while its journey through different brains and will find a place in the hearts of the readers. Our endeavor will be successful, if the book is found to be useful for students, faculties as well as for the veterinarians.

Manoranjan Rout

Raj Pal Diwakar

Contents

List of Contributors

Dr. A. Anand Kumar, M.V.Sc., Ph.D. (Veterinary Pathology)

Professor and Head, Department of Veterinary Pathology, College of Veterinary Science, Sri Venkateswara Veterinary University, Tirupati - 517502, Andhra Pradesh, India

7aakumar@gmail.com

Mr. Ankit Prasad Kelwan, M.Sc. (Microbiology)

Young Professional – II , Pox Virus Laboratory, Division of Virology, Indian Veterinary Research Institute, Mukteswar Campus – 263138, Nainital, Uttarakhand, India

prasadankit840@gmail.com

Dr. A.P. Madhusoodan, M.V.Sc., Ph.D. (Veterinary Physiology)

Scientist, ICAR-Indian Veterinary Research Institute, Mukteswar Campus - 263138, Nainital, Uttarakhand, India

madhu.jnvet.918@gmail.com

Dr. Abhishek M. Todkari,

M.V.Sc. Scholar, Division of Virology, ICAR-Indian Veterinary Research Institute, Mukteswar Campus - 263138, Nainital, Uttarakhand, India

Dr. Ambuj Shrivastava, M.Sc. (Botany)

Scientist D, Virology Division, Defence Research and Development Establishment, Jhansi Road, Gwalior - 474002, Madhya Pradesh, India

ambujshrivastava@hotmail.com

Dr. Amit Kumar, M.V.Sc., Ph.D. (Veterinary Virology)

Scientist, Pox Virus Laboratory, Division of Virology, Indian Veterinary Research Institute, Mukteswar Campus - 263138, Nainital, Uttarakhand, India

amitvet87@gmail.com

Dr. Anita M. Shete, M.Sc., Ph.D. (Microbiology)

Scientist E, Maximum Containment Facility, ICMR-National Institute of Virology, Pune, Maharashtra - 411021, India

anitaaich2008@gmail.com

Mr. Anshuman Behera, M.Sc.

CSIR-JRF, Virology Division, Defence Research and Development Establishment, Jhansi Road, Gwalior - 474002, Madhya Pradesh, India

anshuman.behera2601@gmail.com

Dr. Aswin P. Kumar

M.V.Sc. Scholar, Division of Virology, ICAR-Indian Veterinary Research Institute, Mukteswar Campus - 263138, Nainital, Uttarakhand, India

Dr. B. Mondal, MVSc, Ph.D. (Veterinary Virology)

Principal Scientist, ICAR-Indian Veterinary Research Institute, Mukteswar Campus - 263138, Nainital, Uttarakhand, India

bimalendu.m@gmail.com

Dr. Balamurugan Vinayagamurthy, M.V.Sc., Ph.D. (Veterinary Virology)

Post Doc (Germany), FNAAS, FIVS, FISVIB, AFNAVS, Principal Scientist, ICAR-National Institute of Veterinary Epidemiology and Disease Informatics, Yelahanka, Bengaluru - 560064, Karnataka, India

balavirol@gmail.com

Dr. C. Ravishankar, M.V.Sc., Ph.D. (Veterinary Virology)

Professor, Department of Veterinary Microbiology, College of Veterinary and Animal Sciences, Kerala Veterinary and Animal Sciences University, Pookode, Lakkidi, Kerala - 673 576, India

chinturavi@rediffmail.com

Dr. Deepak Y. Patil, M.Sc., Ph.D. (Applied Biology)

Scientist B, Maximum Containment Facility, ICMR-National Institute of Virology, Pune, Maharashtra - 411021, India

patildeepak8@gmail.com

Dr. Deepika Bisht, M.V.Sc., Ph.D. (Veterinary Biotechnology)

ICAR-Indian Veterinary Research Institute, Mukteswar Campus - 263138, Nainital, Uttarakhand, India,

dpbisht4n@gmail.com

Ms. Ekta Gupta, M.Sc. (Zoology)

UGC-SRF, Virology Division, Defence Research and Development Establishment, Jhansi Road, Gwalior - 474002, Madhya Pradesh, India

ektagupta1912@gmail.com

Dr. G.B. Manjunatha Reddy, M.V.Sc., Ph.D. (Veterinary Pathology)

Senior Scientist, ICAR-National Institute of Veterinary Epidemiology and Disease Informatics, Yelahanka, Bengaluru - 560064, Karnataka, India

gbmpatho@gmail.com

Dr. Geeta Devi Leishangthem, M.V.Sc., Ph.D. (Veterinary Pathology)

Assistant Professor, Department of Veterinary Pathology, Guru Angad Dev Veterinary and Animal Sciences University, Ludhiana, Punjab, India

drgeetapatho@gmail.com

Dr. Gnanavel Venkatesan, M.V.Sc., Ph.D. (Veterinary Virology)

Senior Scientist, FMD Research Centre, IVRI, Bengaluru campus, Karnataka - 560024, India

gnanamvirol@gmail.com

Dr. H.B. Chethan Kumar, M.V.Sc., Ph.D. (Veterinary Public Health)

ICAR-National Institute of Veterinary Epidemiology and Disease Informatics, Yelahanka, Bengaluru - 560064, Karnataka, India

chethuhb@gmail.com

Dr. Jyoti S. Kumar, M.Sc., Ph.D. (Microbiology)

Scientist F, Virology Division, Defence Research and Development Establishment, Jhansi Road, Gwalior - 474002, Madhya Pradesh, India

Dr. K. Nagarajan, M.V.Sc., Ph.D. (Veterinary Pathology)

Professor, Department of Veterinary Pathology, Madras Veterinary College, Vepery, Chennai – 600007, Tamil Nadu, India

nagavet@gmail.com

Dr. K. Sumod, M.V.Sc., Ph.D. (Veterinary Immunology and Life Sciences)

Assistant Professor, Department of Veterinary Microbiology, College of Veterinary and Animal Sciences, Kerala Veterinary and Animal Sciences University, Pookode, Lakkidi, Kerala - 673576, India,

sumodk@kvasu.ac.in

Dr. K.K. Rajak, M.V.Sc., Ph.D. (Veterinary Virology)

Principal Scientist, Biological Products Division, ICAR-Indian Veterinary Research Institute, Izatnagar, Bareilly - 243122, Uttar Pradesh, India

kaushalvirol@gmail.com

Dr. K.P. Singh, M.V.Sc., Ph.D. (Veterinary Pathology)

Diplomate ICVP, FNAVSc, FIAVP, Joint Director, Centre for Animal Disease Research and Diagnosis (CADRAD), ICAR-Indian Veterinary Research Institute, Izatnagar, Bareilly - 243122, Uttar Pradesh, India

karam.singh@rediffmail.com

Dr. K.V. Jayanth

M.V.Sc. Scholar, Department of Veterinary Microbiology, College of Veterinary and Animal Sciences, Kerala Veterinary and Animal Sciences University, Pookode, Lakkidi, Kerala - 673 576, India

Jayanthkv7760@gmail.com

Dr. Karam Chand, M.V.Sc., Ph.D. (Veterinary Virology)

Senior Scientist, ICAR-Indian Veterinary Research Institute,Mukteswar Campus - 263138, Nainital, Uttarakhand, India

virusshield@gmail.com

Mr. L. Dilip, M.F.Sc.

Technical Manager, KVAFSU-CVA Rabies Diagnostic Laboratory, WOAH (Founded as OIE) Reference Laboratory for Rabies, Department of Microbiology, Veterinary College, KVAFSU, Hebbal, Bangalore - 560024, Karnataka, India

Dr. Manoranjan Rout, M.V.Sc., Ph.D. (Veterinary Pathology)

Senior Scientist, ICAR-National Institute on Foot and Mouth Disease, Bhubaneswar, Odisha - 752050, India

drmrout@gmail.com

Dr. Saravanan Subramaniam, M.V.Sc., Ph.D. (Veterinary Virology)

Principal Scientist, ICAR-National Institute on Foot and Mouth Disease, Bhubaneswar, Odisha - 752050, India

saranvirol@gmail.com

Dr. Jajati Keshari Mohapatra, M.V.Sc., Ph.D. (Veterinary Virology)

Principal Scientist, ICAR-National Institute on Foot and Mouth Disease, Bhubaneswar, Odisha - 752050, India

jajati1@gmail.com

Dr. Rabindra Prasad Singh, M.V.Sc., Ph.D. (Veterinary Virology)

Director, ICAR-National Institute on Foot and Mouth Disease, Bhubaneswar, Odisha - 752050, India

Rabindra.Singh@icar.gov.in

Dr. N. Shivasharanappa, M.V.Sc., Ph.D. (Veterinary Pathology)

Senior Scientist, ICAR-National Institute of Veterinary Epidemiology and Disease Informatics, Yelahanka, Bengaluru - 560064, Karnataka, India,

drshivasharan@gmail.com

Dr. Niranjan Mishra, M.V.Sc., Ph.D. (Veterinary Virology)

Principal Scientist, ICAR-National Institute of High Security Animal Diseases, Anand Nagar, Bhopal, Madhya Pradesh - 462022, India

mishranir@rediffmail.com

Dr. Nitish S. Kharayat, M.V.Sc., Ph.D. (Animal Reproduction)

ICAR-Indian Veterinary Research Institute, Mukteswar Campus - 263138, Nainital, Uttarakhand, India

nkharayat03@gmail.com

Dr. P. Sai Mounica, M.V.Sc.

Junior Research Fellow, ICAR-National Institute of Veterinary Epidemiology and Disease Informatics, Yelahanka, Bengaluru - 560064, Karnataka, India,

mounica20sai@gmail.com

Dr. Paban Kumar Dash, M.V.Sc., Ph.D. (Veterinary Microbiology)

Scientist G and Head, Virology Division, Defence Research and Development Establishment, Jhansi Road, Gwalior - 474002, Madhya Pradesh, India

pabandash@rediffmail.com

Dr. Pankaj Deka, M.V.Sc., Ph.D. (Veterinary Microbiology)

Assistant Professor cum Assistant Virologist, Regional Research Centre, AICRP on FMD, Department of Microbiology, College of Veterinary Science, AAU, Khanapara, Guwahati - 22, Assam, India

drpankajaau@gmail.com

Dr. Pragya D. Yadav, M.Sc., Ph.D. (Biotechnology)

Scientist F and Group leader, Maximum Containment Facility, ICMR-National Institute of Virology, Pune, Maharashtra - 411021, India

hellopragya22@gmail.com

Dr. R. Rajasekhar, M.V.Sc., Ph.D. (Veterinary Microbiology)

Assistant Professor, Department of Veterinary Microbiology, College of Veterinary and Animal Sciences, Kerala Veterinary and Animal Sciences University, Pookode, Lakkidi, Kerala - 673576, India,

rajasekhar@kvasu.ac.in

Dr. R. Sharada, M.V.Sc., Ph.D. (Veterinary Microbiology)

Professor and Quality Manager, KVAFSU-CVA Rabies Diagnostic Laboratory, WOAH (Founded as OIE) Reference Laboratory for Rabies, Dept. of Microbiology, Veterinary College, KVAFSU, Hebbal, Bangalore-560024, India

Dr. Rahul Singh Arya, M.V.Sc., Ph.D. (Veterinary Pathology)

Associate Professor, Department of Veterinary Pathology, DUVASU, Mathura, Uttar Pradesh, India

aryarahulsingh@gmail.com

Dr. Ramesh Somvanshi, M.V.Sc., Ph.D. (Veterinary Pathology)

FRCVS (Uppsala, Sweden), DICVP, Former Acting Joint Director, CADRAD, Former Head, Division of Pathology, Former Head and Ex-Emeritus Scientist (ICAR), ICAR-Indian Veterinary Research Institute, Izatnagar, Bareilly - 243122, Uttar Pradesh, India

dr.rsomvanshi@gmail.com

Dr. Rima R. Sahay, MD. (Community Medicine)

Scientist C, Maximum Containment Facility, ICMR-National Institute of Virology, Pune, Maharashtra - 411021, India

dr.rima.sahay@gmail.com

Dr. S. Isloor, M.V.Sc., Ph.D. (Veterinary Virology)

Professor and Laboratory Director, KVAFSU-CVA Rabies Diagnostic Laboratory, WOAH (Founded as OIE) Reference Laboratory for Rabies, Dept. of Microbiology, Veterinary College, KVAFSU, Hebbal, Bangalore - 560024, Karnataka, India

kisloor@gmail.com

Dr. S. Kalaiyarasu, M.V.Sc., Ph.D. (Veterinary Immunology)

Senior Scientist, ICAR-National Institute of High Security Animal Diseases, Anand Nagar, Bhopal, Madhya Pradesh - 462022, India

kalai82vetmic@gmail.com

Dr. S. Nandi, M.V.Sc., Ph.D. (Veterinary Virology)

Principal Scientist, CADRAD, ICAR-Indian Veterinary Research Institute, Izatnagar, Bareilly - 243122, Uttar Pradesh, India

sukdebnandi@yahoo.in

Dr. G.K. Sharma, M.V.Sc., Ph.D. (Veterinary Virology)

Senior Scientist, CADRAD, ICAR-Indian Veterinary Research Institute, Izatnagar, Bareilly,Uttar Pradesh - 243 122, India

gaurvet@gmail.com

Dr. S.K. Biswas, M.V.Sc., Ph.D. (Veterinary Virology)

Principal Scientist, CADRAD, ICAR-Indian Veterinary Research Institute, Izatnagar, Bareilly - 243122, Uttar Pradesh, India

sanchaybiswas@gmail.com

Dr. R. Mishra, M.V.Sc., Ph.D. (Veterinary Microbiology)

Student completed Ph.D. from ICAR-Indian Veterinary Research Institute, Izatnagar, Bareilly - 243122, Uttar Pradesh, India

Dr. S. Nautiyal, M.V.Sc., Ph.D. (Veterinary Microbiology)

Student completed Ph.D. from ICAR-Indian Veterinary Research Institute, Izatnagar, Bareilly - 243122, Uttar Pradesh, India

Dr. K.G. Sai Balaji, M.V.Sc,

Ph.D. Scholar, ICAR-Indian Veterinary Research Institute, Izatnagar, Bareilly - 243122, Uttar Pradesh, India

Dr. S.Shashank

M.V.Sc. Scholar, Department of Veterinary Microbiology, College of Veterinary and Animal Sciences, Kerala Veterinary and Animal Sciences University, Pookode, Lakkidi, Kerala - 673576, India

shashi.as98@gmail.com

Dr. Shashi Bhushan Sudhakar, M.V.Sc., Ph.D. (Veterinary Microbiology)

Senior Scientist, ICAR-National Institute of High Security Animal Diseases, Anand Nagar, Bhopal, Madhya Pradesh - 462022, India

sudha.vet@gmail.com

Dr. Sachin Kumar, M.V.Sc., Ph.D. (Molecular Virology)

1N-103, Department of Biosciences and Bioengineering, Indian Institute of Technology Guwahati, Assam - 781039, India

sachinku@iitg.ac.in

Dr. Sandeep Bhatia, M.V.Sc., Ph.D. (Veterinary Microbiology and Immunology)

Principal Scientist, ICAR-National Institute of High Security Animal Diseases, Anand Nagar, Bhopal, Madhya Pradesh - 462022, India

sbhatia1967@gmail.com

Dr. Sangeeta Das, M.V.Sc. (Veterinary Microbiology)

Veterinary Officer, Institute of Veterinary Biologicals, Khanapara - 22, Guwahati, Assam, India

sangeetadas31280@gmail.com

Mr. Satyendu Nandy, B.D.S., M.Sc., M.Tech.

Department of Biosciences and Bioengineering, Indian Institute of Technology, Guwahati - 781039, Assam, India

Satyendu.nandy@iitg.ac.in

Dr. Shashi Sharma, M.Sc., Ph.D. (Applied Microbiology and Biotechnology)

Scientist E, Virology Division, Defence Research and Development Establishment, Jhansi Road, Gwalior - 474002, Madhya Pradesh, India

Dr. Shubham Saini, M.V.Sc. (Veterinary Public Health)

Ph.D. Scholar, Division of Veterinary Public Health, ICAR-Indian Veterinary Research Institute, Izatnagar, Bareilly - 243122, Uttar Pradesh

indiashubhs.bandyopadhyaya@gmail.com

Dr. Shriya Rawat, M.V.Sc., Ph.D. (Veterinary Public Health)

College of Veterinary and Animal Science, Sardar Vallabh Bhai Patel University of Agriculture and Technology, Meerut - 250110, Uttar Pradesh, India

shriyarawat@gmail.com

Dr. Siddharth Gautam, M.V.Sc., Ph.D. (Veterinary Pathology)

Scientist, ICAR-Indian Veterinary Research Institute, Mukteswar Campus - 263138, Nainital, Uttarakhand, India

gautam.ivri.m@gmail.com

Dr. Sonam Sarita Bal, M.V.Sc.

Ph.D. Scholar, Department of Veterinary Pathology, College of Veterinary Science, Guru Angad Dev Veterinary and Animal Sciences University, Ludhiana, Punjab, India

sonamsarita1996@gmail.com

Dr. Sreelekshmy Mohandas, M.V.Sc., Ph.D. (Veterinary Pathology)

Scientist C, Maximum Containment Facility, ICMR-National Institute of Virology, Pune, Maharashtra - 411021, India

sreelekshmy88@gmail.com

Dr. Sushil Kumar Sharma, M.Sc., Ph.D. (Microbiology)

Scientist E, Virology Division, Defence Research and Development Establishment, Jhansi Road, Gwalior - 474002, Madhya Pradesh, India

sushilrdbc@rediffmail.com

Dr. Z.B. Dubal, M.V.Sc., Ph.D. (Veterinary Public Health)

Principal Scientist, ICAR-National Institute of Veterinary Epidemiology and Disease Informatics, Yelahanka, Bengaluru - 560064, Karnataka, India,

drzunjar@yahoo.co.in

Dr. H.R. Jogi, M.V.Sc.

Ph.D. Scholar, Division of Biological Products, ICAR-Indian Veterinary Research Institute, Izatnagar, Bareilly - 243122, Uttar Pradesh, India

harshjogi6@gmail.com

Dr. A.K. Yadav, M.V.Sc., Ph.D. (Veterinary Virology)

Scientist, Division of Biological Products, ICAR-Indian Veterinary Research Institute, Izatnagar, Bareilly - 243122, Uttar Pradesh, India

drajayyadav07@gmail.com

Dr. M. Bhatt, M.V.Sc., Ph.D. (Veterinary Virology)

Scientist, CADRAD, ICAR-Indian Veterinary Research Institute, Izatnagar, Bareilly - 243122, Uttar Pradesh, India

bhatt.mukesh20@gmail.com

Dr. C. Einstein, M.V.Sc.

Ph.D. Research Scholar, Division of Biological Products, ICAR-Indian Veterinary Research Institute, Izatnagar, Bareilly - 243122, Uttar Pradesh, India

908chris@gmail.com

Dr. Revathi, M.V.Sc.

Ph.D. Research Scholar, Division of Biological Products, ICAR-Indian Veterinary Research Institute, Izatnagar, Bareilly - 243122, Uttar Pradesh, India

revathiannepu@gmail.com

Mr. R.K. Thakur, M.Sc.

Senior Research Fellow, Division of Biological Products, ICAR-Indian Veterinary Research Institute, Izatnagar, Bareilly - 243122, Uttar Pradesh, India

ravithakuratarra786@gmail.com

Dr. V. Rai, M.V.Sc., Ph.D. (Veterinary Virology)

Scientist, ICAR-National Research Centre on Pig, Rani, Guwahati, Assam, India

vishal.rai@icar.gov.in

Chapter 1

Foot and Mouth Disease: An Immemorial Menace of Global Livestock

Manoranjan Rout, Saravanan Subramaniam, Jajati Keshari Mohapatra and Rabindra Prasad Singh*

ICAR-National Institute on Foot and Mouth Disease, International Centre for Foot and Mouth Disease, Bhubaneswar – 752 050, Odisha
**e-mail: drmrout@gmail.com, Manoranjan.Rout@icar.gov.in*

ABSTRACT

Foot and mouth disease (FMD) is incredibly contagious, devastating, debilitating and fast-spreading acute viral disease of higher mammalian cloven-hoofed animals including cattle, buffalo, sheep, goat, pig, mithun, yak as well as more than 70 wildlife species. Despite significant efforts, FMD continues to afflict over 100 countries across the world, while some 70 nations have achieved official recognition as FMD-free with or without vaccination by World Organization for Animal Health (WOAH). The disease is caused by FMD virus (FMDV), which belongs to the genus Aphthovirus within the Picornaviridae family, and exists as seven serotypes O, A, C, Asia 1, SAT1, SAT2 and SAT3. The virus spreads rapidly among susceptible hosts through various routes, including aerosols, making it difficult to contain once introduced into a population. The signature symptoms of the disease are pyrexia, anorexia, salivation and vesicle formation on the foot and mouth. Besides typical clinico-pathological attributes, diagnosis of FMD is accomplished by various techniques such as sandwich ELISA, multiplex RT-PCR, real-time PCR, virus isolation etc. Indirect nonstructural protein (NSP) ELISA, LPB/SPC ELISA, VNT can detect antibodies against FMDV NSPs and SPs. Vaccination remains a cornerstone in the prevention and control of FMD, complemented by stringent biosecurity measures. This chapter serves as a valuable resource, offering foundational knowledge on FMD for students and readers alike, facilitating a better understanding of the disease's complexities and the strategies employed to combat it.

Keywords: *FMD, Serotype, Species, Symptoms, ELISA, PCR, Prevention, Vaccine.*

Introduction with Historical Backdrop

Globalization has unlocked numerous economic opportunities through international livestock trade, but it has also heightened the risk of emergence, re-emergence, and spread of various infectious agents (VanderWaal and Deen, 2018). Among the pathogen-induced epizootics affecting animals, foot and mouth disease (FMD) stands out as a significant barrier to agricultural trade, causing substantial production losses and disrupting agricultural activities worldwide. FMD remains a major threat to livestock globally, resulting in estimated annual economic losses ranging from US$6.5 to US$21 billion in endemic countries, with approximately US$5 billion attributed to production losses and vaccination alone (Capozzo *et al.*, 2021). FMD incursion in an FMD-free country can lead to an annual economic loss of more than US$1.5 billion (Hardham *et al.*, 2020). The disease is widely recognized as a primary concern for many countries in terms of economic impact and is known by various synonyms such as aphthous fever (derived from the Greek word "aphtha", meaning vesicles of the mouth), epizootic aphthae, and infectious aphthous stomatitis (Arzt *et al.*, 2010; VanderWaal and Deen, 2018).

The documented history of FMD dates back 1546 when an Italian monk Hieronymous Fracasorius described an epidemic in cattle near Venice/Verona, Italy. The threat of FMD persisted in cattle throughout subsequent centuries, but the etiological agent remained undiscovered until the late 19th century when Friedrich Loeffler and Paul Frosch, students of Robert Koch, made significant breakthrough. They conducted an experiment where they passed the lymph fluid derived from vesicles of FMD affected cattle through bacteria-proof filter candles. To their surprise, characteristic disease symptoms appeared when the filtered fluid was inoculated into calves, supporting their hypothesis that an unseen agent had passed through the filter candles and caused the disease in the animals. This experiment led to the identification of the first genetically diverse virus associated with FMD, which has been a long-standing challenge in the agricultural sector, causing devastation in animals (Loeffler and Frosch, 1898).

Geographical Distribution

Currently, FMD is absent in North America, Central America, Australia, New Zealand, Japan, and until recently, in most countries in continental Western Europe. However, it remains endemic in Africa, Asia, the Middle East, and in most parts of South America. Sporadic outbreaks of FMD have occurred in disease-free countries, with the notable exceptions of New Zealand, Greenland,

Iceland, and the smaller islands of Oceania. FMD endemic areas have been categorized into seven geographical FMDV pools (1-7), characterized by the presence of similar viruses. These pools often result from ecological similarities, common livestock exchange, and shared cultural traditions (Rweyemamu *et al.*, 2008). Movement of FMDV from one pool to another has been reported, highlighting its transboundary nature. Most recently, during 2023, outbreaks of the SAT-2 serotype of FMDV were reported for the first time in the Middle East, specifically in Iraq and Jordan, as well as in Turkey. Additionally, in May 2023, South Korea reported its first outbreak of FMD since 2019.

Discovery of Virus Serotypes

FMD Virus (FMDV) exists in seven immunologically distinct forms called serotypes namely O, A, C, Asia 1, SAT (Southern African Territories)-1, -2 and -3) (Grubman and Baxt, 2004). The former four are also known as Euro-Asiatic serotypes circulating mainly in Europe, Asia and South America, while the latter three are collectively known as Southern African Territories (SAT) serotypes reported circulating mainly in African continent. Serotypes O and A were initially discovered by Vallee and Carre (1922). They found that cattle recovered from clinical disease due to an FMDV originated in France became re-infected immediately when mingled with animals infected with FMDV originated in Germany. They decided to name these serotypes after their place of origin, O for the department of Oise in France and A for Allemagne (the French word for Germany). Their work was subsequently extended by Waldmann and Trautwein (1926) with the discovery of a third entity named as serotype C. Three additional serotypes were detected in samples originating from South Africa and hence named as Southern African Territories 1, 2 and 3 (SAT1, SAT2, SAT3) (Brooksby, 1958). The seventh serotype, Asia 1 was initially detected in a sample in 1954 collected from a water buffalo at Okara, Punjab, Pakistan (Brooksby and Rogers, 1957).

The serotypes have been further divided into topotypes (the geographically isolated genotypes), genotypes, lineages and sub-lineages, usually restricted to specific geographical regions. Serotypes O, A and C have been distributed across the globe with serotype O being globally attributed to the majority of the outbreaks. Despite being attributed to a lesser number of FMD outbreaks compared to serotype O, serotype A has a widespread distribution. In contrast, serotype Asia 1 and the SAT viruses are normally restricted to Asia and sub-Saharan Africa, respectively (Rweyemamu *et al.*, 2008). The FMDV serotype C is now apparently extinct (Jamal and Belsham, 2018).

Structural and Genomic Organization of FMDV

FMDV is the type species of the *Aphthovirus* genus in the *Picornaviridae* family. When observed under electron microscopy, the FMD virion (140S particle) appears as a roughly spherical particle with a smooth surface, measuring about 25-30 nm in diameter. The virus consists of a single-stranded, positive-sense, non-segmented RNA genome of approximately 8,500 bases surrounded by four structural proteins, forming an icosahedral capsid that constitute approximately 69 percent of the virion mass. The capsid is composed of 60 copies of the capsomers (Rueckert, 1996), with each capsomer consisting of four structural polypeptides referred to as viral protein (VP1 [1D], VP2 [1B], VP3 [1C], and VP4 [1A]). While VP1, VP2 and VP3 are exposed on the surface of the virus, VP4 is located internally. The nucleotide sequences of the VP1 coding region are of particular significance due to their role in virus attachment, entry, protective immunity and serotype specificity, have been used for genetic characterization of FMDV strains.

FMDV genome contains a single open reading frame (ORF) that codes for a large polyprotein, which after translation, is processed into several mature proteins. The ORF can be divided into four regions based on the function of the mature proteins they encode, namely L, P1, P2 and P3. The P1 region codes for four structural proteins VP1-VP4, which are involved in capsid formation, while the L, P2 and P3 regions code for several nonstructural proteins (NSPs) that play roles in virus genome replication and protein cleavage activities. The FMDV polyprotein includes two separate proteases: L^{pro} and $3C^{pro}$. Following the P1 region is the P2 region, which encodes three viral NSPs: 2A, 2B and 2C, while the P3 region encodes NSP 3A, three copies of VPg, $3C^{pro}$ and $3D^{pol}$. Notably, $3D^{pol}$ is the viral RNA-dependent RNA polymerase. At the 5′ end of the genome, the L region encodes the N-terminal component of the polyprotein, known as the L protein, a papain-like protease (L^{pro}) activity, which also acts as an inhibitor of host protein synthesis and has been identified as a viral virulence factor. Additionally, an IRES (Internal Ribosomal Entry Site) element has been identified within the 5′ untranslated region (5′UTR) of the viral RNA, facilitating efficient translation of the viral RNA into protein.

Quasi-Species Concept

The quasi-species theory of molecular evolution was first proposed to elucidate the dynamics of error-prone replication, self-organization and adaptability observed in the primitive replicons around 4000 million years before (Eigen, 1992). Virologists have adopted the term "viral quasi-species" to refer to these populations, highlighting their genetic diversity and the ongoing processes of genetic variation, competition, and natural selection that shape them (Domingo, 2006). Thus, a quasi-species is defined as a population of similar genomes (Nowak, 2006). Quasi-species dynamics is obviously manifested in RNA viruses that display short duplication times, generally high fecundity and error-prone replication; traits that have been maintained despite having probable ancient origin.

RNA viruses in general, including FMDV, exhibits very high mutation rates, typically ranging from 10^{-3} to 10^{-5} mutations per nucleotide site per genome replication due to the lack of error correction mechanisms during RNA replication (Domingo and Holland, 1988). The high mutation rate of FMDV and other RNA viruses results in differences, or mutations, between replicated genomes and the original parental genome. These mutations can range from 0.1 to 10 base positions per replicated genome.

Stability of the Virus

In general, FMDV is relatively prone to heat and insensitive to cold. Sunlight quickly destroys the virus but it may persist on pasture for long periods at low temperatures. Virus free of tissue gets effectively destroyed by boiling, but autoclaving is the safest way for heat disinfection. The virus gets inactivated at pH <7.0, but can survive at high temperatures in presence of proteins therefore enhancing virus persistence in the environment (Racaniello, 2011). Sodium hydroxide (NaOH) or 1-2 percent formalin, 4 percent Sodium carbonate (Na_2CO_3) can destroy the virus in few minutes. The virus remains viable at 4°C for up to one year, while loses its infectivity at 56°C. It can survive for 14 days in dry feces, >6 months in slurry and for 39 days in winter. The virus can persist for 10-12 weeks on clothing and feed, up to one month on hair, while in the infected premises, it may persist for >1 year. Bull semen frozen at -79°C can favor survival of the virus for >2 months.

Susceptible Host Species

FMD is a disease of higher mammalian cloven-hoofed animals including cattle, buffalo, sheep, goat, pig, mithun, yak as well as more than 70 wildlife species (Thomson *et al.*, 2003). There are reports of natural infection in various non-cloven-hoofed wildlife species including Asiatic elephant (*Elephas maximus*) and African elephant (*Loxodonta africana*) (Hedger, 1981; Alexandersen and Mowat, 2005); eastern gray kangaroo (*Macropus giganteus*) (Bhattacharya *et al.*, 2003); Brazilian tapir (*Tapirus terrestris*) and Asiatic tapir (*Tapirus indicus*) (Weaver *et al.*, 2013) and brown bear (*Ursus arctos*) (Grosso, 1957) as well. Certain unconventional species have been reported to have contracted FMD in recent years, thereby broadening the heteroclite list of susceptible species. Officer *et al.* (2014) reported FMD in Asiatic black bears (*Ursus thibetanus*) in Vietnam. FMDV infection in the domestic dog (*Canis lupus familiaris*) has also been reported in Iran (Waters *et al.*, 2021). These instances highlight the broad range of species that can be affected by FMDV, underscoring the importance of vigilance and control measures to prevent the spread of the disease across different animal populations.

Epizootiology

FMD is a disease of high morbidity and low mortality. The disease is transmitted to other animals by three main routes: (a) direct contact between infected and healthy susceptible animal; (b) indirect contact between animals through various means *e.g.*, contaminated person or vehicle, inanimate objects contaminated with virus or excretions from infected animals, contaminated hay or straw, milk, clothing of animal handlers, shoes or equipment/vehicle tires contaminated with virus, ingestion of contaminated animal products/feed, vaccination using contaminated vaccines, contaminated semen used during artificial insemination; (c) droplet inhalation through aerosol route.

The inhalation is the main route of infection in ruminants. Pigs can exhale a significant amount of virus in their breath, making them potentially infectious to other animals nearby. Pigs are considerably less susceptible to aerosol infection than ruminants. Pigs usually become infected with the virus by oral route eating virus contaminated products, direct contact with infected animal, or when exposure to a heavily contaminated environment. A pig may require up to 6000 50 percent tissue culture infective dose ($TCID_{50}$) that is 600 times more than the exposure to aerosol virus required by a cattle or sheep to cause infection (Kitching and Alexandersen, 2002).

Transmissibility

Transmissibility is one of the significant characteristics of FMDV that makes it a 'fast moving disease'. The transmissibility of FMD is influenced by factors such as the viral load, environmental conditions, host susceptibility, and management practices.

The nasopharynx and epithelium of the tongue, nose, muzzle and coronary bands of susceptible animals are the sites of choice where the virus typically replicates producing characteristic lesions. The name "foot-and-mouth disease" reflects the characteristic lesions that occur not only in the mouth but also on the feet and other parts of the body. Virus replication in the epithelia of these sites leads to formation of blisters or vesicles heavily packed with FMDV. When these blisters rupture, vesicular fluid with high concentrations of viral particles contaminates the nearby areas facilitating its spread. Humans can also play an important role in spreading FMDV to new regions.

Host Reactivity

The degree of clinical severity of FMD depends on the infecting virus strain and the host species (Arzt *et al.*, 2011). Acute clinical FMD tends to be more severe in pigs compared to ruminant species (Alexandersen *et al.*, 2003). However, pigs also clear the virus more efficiently than the ruminants, and there is no observable subclinical carrier status for FMDV in porcine species (Stenfeldt *et al.*, 2014; 2016b). Though pigs are the producers of huge infectious aerosol virus, they are less susceptible to airborne infection as compared to ruminants (Donaldson and Alexandersen, 2001). Certain FMDV strains may exhibit differences in virulence between cattle and pigs, indicating host-specific variations in the molecular pathways of FMDV infection (Stenfeldt *et al.*, 2016a). Among all of the susceptible livestock species, sheep played a major role in the 2001 United Kingdom FMD outbreak. Clinical diagnosis of FMD in sheep can be challenging, leading to potential spread of the disease to other livestock before detection. Sheep are highly susceptible to virus infection via aerosol and can excrete airborne virus, making them a significant source of transmission during outbreaks (Kitching and Hughes, 2002). It is essential to comprehend the variations in susceptibility and transmission dynamics among different host species in order to develop efficient control strategies and lessen the expansion of FMD.

Mode of Transmission

Affected animals in acute phase shed the virus in various bodily secretions and excretions, including exhaled air, saliva, naso-lachrymal fluid, milk, urine, feces, semen. Irrespective of the entry of virus through skin or gastrointestinal tract, pharyngeal mucosa acts as the primary predilection and replication site., Infected animals can start shedding the virus as early as four days before the visible symptoms appear. Some animals can remain as the carrier and continue to excrete the virus for years even after recovery. The vesicles in oral mucosa (tongue and dental pad) and pedal mucosa (heel bulbs and inter-digital space) normally rupture within one day, releasing highly infectious vesicular fluid containing up to 10^8 infectious virus units per ml (Woodbury, 1995). Materials that come into contact with the vesicular fluid become contaminated and can transmit the disease to other susceptible animals. This emphasizes how crucial it is to follow stringent biosecurity rules, such as good sanitation and disinfection techniques, to stop the spread of FMD.

Airborne Transmission: Fast Moving Disease

Airborne transmission of FMDV has been blamed to have been the cause behind several historic FMD outbreaks. For example, outbreaks of FMD in cattle in Jersey during 1974 and 1981, as well as on the Isle of Wight during 1981, were traced back to outbreaks in Brittany, France (Donaldson *et al.*, 1982). The epidemiological evidence gathered during these outbreaks suggested that airborne transmission was the most likely route of infection for the animals. During these outbreaks virus would have been transported over distances up to 500 km (Donaldson *et al.*, 1982). Similarly, the 1967-1968 epidemic of FMD in the UK, which covered an approximate distance of 60-150 km, was also attributed to airborne transmission (Hugh-Jones and Wright, 1970). The 2001 epidemic of FMD in the UK was also believed to have been caused by airborne transmission (Gloster *et al.*, 2003).These incidents demonstrate the possibility that FMDV could travel great distances in the air and spread the disease between countries and regions.

Carrier Status

A 'Carrier' is defined as an animal from which live virus can be isolated at 28 days, or later after infection (Sutmoller *et al.*, 1968). Van Bekkum and associates (1959) were among the first to demonstrate that live FMDV could be recovered from oesophageal-pharyngeal fluids (OPF) of cattle during the convalescent phase of FMD. Currently, virus isolation from OPF is considered to be the most sensitive tool to identify carriers, although reverse transcription-PCR (RT-PCR) assays have increased sensitivity. Infected ruminant species such as cattle, buffalo, sheep and goats, after recovery, may become carriers carrying the virus in the pharynx. African buffalo, cattle, sheep and goat can carry the virus for 5 years, 3.5 years, 9 months, 6 months, respectively (Salt, 1993; Stenfeldt *et al.*, 2014).The virus may persist in a herd of African buffalo for 24 years or longer (Condy *et al.*, 1985). Unlike ruminants, domestic pigs (*Sus scrofa*) typically do not carry FMDV; however, Mohamed *et al.* (2011) reported isolation of virus from feral pigs after 33-35 days of infection. Similarly, Mezencio *et al.* (1999) found evidence of the FMDV genome in sera collected from recovered pigs, although no live virus has been isolated without any epidemiological support for the persistence of live virus in recovered pigs. The role of carrier animals in the spread of virus in natural field situation is still debatable/controversial. Water buffalo (*Bubalus bubalis*) may carry virus for 1-2 year (Barros *et al.*, 2007) and other wildlife species like Greater kudu (*Tragelaphus strepsiceros*) for up to 160 days (Hedger *et al.*, 1972), Eland (*Taurotragus oryx*) for 32 days, wildebeest (*Connochaetes taurinus*) for up to 45 days (Hedger *et al.*, 1972), and sable (*Hippotragus niger*) remain viremic up to 28 days (Ferris *et al.*, 1989). Cervine species like fallow deer, sika deer and white-tailed deer can also become carriers. The role of carrier animals in the spread of the virus in natural field situations remains debatable and controversial.

FMD: A Concern for Zoonosis?

While FMD primarily affects cloven-hoofed animals, it is indeed a very rare zoonotic entity. Most people who come into contact with the virus may become mechanical carriers of the virus for up to 36 hours without becoming viremic. However, about 40 reported cases exist where humans were infected with FMDV exhibiting clinical signs such as fever, malaise and oral blisters that resolved in a week. These cases occurred in individuals who had close contact with sick or infected livestock. Thus, as human infection is rare, there is no major public health concern for FMD, hence is not generally considered to have significant zoonotic potential (Nieldbalski *et al.*, 2006).

Pathogenesis

The disease pathogenesis has been extensively studied in cattle and pigs. Epithelial cells of the mucosa of oropharynx or nasopharynx are the sites of primary replication of FMDV followed by the spread into the circulation. From there virus reaches cornified, stratified squamous epithelia through the lymphoid system. As the virus disseminates throughout the body, the virus gets amplified at these sites especially the epithelia of the skin and mouth. However, recent research suggests that the respiratory epithelium itself may not be a primary site of infection (Alexandersen *et al.*, 2002 a, b). In cases where animals get infected through the respiratory tract, the initial replication of the virus occurs in pharyngeal area and lungs followed by viremic spread to other tissues and organs.

Several studies have suggested that the lung or pharyngeal areas are the sites of initial virus replication, with subsequent rapid dissemination into other areas such as oral and pedal epithelial areas (Sutmoller and McVicar, 1976), probably mediated by cells of monocyte/macrophage origin. FMDV is then distributed throughout the body to reach the epithelium of oro-pharynx, oral cavity, feet, the udder and heart. Vesicles develop on the feet and tongue. The nasopharyngeal epithelium in cattle has been attributed to the site of primary infection, while epithelial crypts of the oropharyngeal and laryngo-pharyngeal tonsils are the sites of primary infection in pigs receiving infection through oral exposure. Primary infection in sheep occurs in either epithelial crypts of the oropharyngeal or laryngopharyngeal tonsils (similar to pig) or within the nasopharyngeal mucosa (similar to cattle). Infection of cattle generally occurs via the respiratory route, though can also occur through abrasions on the skin or mucous membranes that requires 10000 times more virus (Donaldson *et al.*, 1987).

Pigs are less susceptible to aerosol infection than cattle, still excrete more aerosolized virus than cattle or sheep. Pigs usually become infected either by eating FMDV-contaminated food, by direct contact with infected animals, or being exposed to areas having FMDV-infected animals. Foot lesions are the most common findings in pigs, while lesions at other sites occur less frequently. Tongue lesions are relatively less noticeable than in cattle. In young piglets or kids, the disease remains fatal due to myocardial involvement. Recent findings indicate that in pig, huge viral replication occurs in the nasal mucosa than in lungs (Oleksiewicz *et al.*, 2001).

Clinical Symptoms

The incubation period for FMD in natural scenario takes a range of 1-14 days (Arzt *et al.*, 2010). When susceptible animals come in contact with clinically infected animals, the symptoms develop in 3 to 5 days. The severity of clinical signs of the disease varies with the host species, age and its immunity, strain and dose of the virus. The signs can range from a mild or in apparent in sheep and goats to a severe disease occurring in cattle and pig. Vesicles on feet may cause lameness. Young calves, lamps, kids and piglets succumb without any premonitory signs due to necrotizing myocarditis. The calves show prominent signs of myocarditis, whereas piglets manifest gastroenteritis.

The incubation period for FMD is 3-8 days in small ruminants (Mahmoud *et al.*, 2019). In sheep and goat, the most frequent clinical sign is lameness and other clinical signs tend to be very mild including dullness showing reluctance to walk segregating itself from the rest of the flock, decrease in food consumption, fever and small vesicles or erosions on the dental pad, lips, gums may make the detection of the disease difficult. Development of vesicles on teats of milking sheep and goats affect milk yield. Ewes may abort while lambs and kids are the victims of heart failure. Infected pigs become lethargic and remain huddled together with little interest in food, initially show mild signs of lameness, blanching of the skin around the coronary bands and may develop a fever of up to 42°C.

Macro- and Microscopic Pathology

The lesions of FMD comprise of vesicles and erosions in mouth, feet and udder. The erosions upon secondary bacterial infection become ulcers. Progressive swelling, necrosis and lysis of keratinocytes in the deeper layers of the epidermis and accumulation of fluid in the space (Radostits *et al.*, 2007). The pathology of FMD in pigs is similar to that in cattle and sheep (Kitching, 2002; Kitching and Hughes, 2002). The disease is characterized by formation of vesicles and erosions inside the mouth (tongue, dental pad and gums) and the skin of the muzzle, interdigital spaces, udder, teats and coronary band (Blowey and Weaver, 2003), in the teats and udders of lactating cows. In pig, vesicles develop on the coronary band and heel of the feet, snout, lower jaw and tongue. Lesions around the coronary bands are the most consistent findings in pigs. Vesicles on tongue of pigs are mostly found far back on the tongue or as tiny vesicles/erosions near the tip of the tongue.

The first histopathological changes in the cornified, stratified squamous epithelium are observed as ballooning degeneration and increased cytoplasmic eosinophilic staining of the cells in the stratum spinosum and intercellular edema. This is followed by necrosis and mononuclear cell and granulocyte infiltration. Vesicles develop by separation of the epithelium from the underlying connective tissue and filling of the cavity with vesicular fluid. Interestingly, no gross or microscopic lesions are observed on the soft palate and the dorsal part of pharynx even though this area contains significant amounts of virus. The epithelia covering these areas are highly specialized and consist of non-cornified, stratified squamous epithelia.

Fatal Myocarditis in Young Animals

The heart muscle cells (cardiomyocytes) of the young animals are affected by lympho-histiocytic myocarditis with necrosis of myocytes and infiltration of mononuclear cells. The terms 'tiger stripes' or 'tiger heart' or 'tigroid heart' are often used to depict gross pathological changes associated with FMDV myocarditis. Grossly, the ventricular walls appear streaked with patches of yellow tissue interspersed with apparently normal myocardium giving the typical 'tigroid heart' (looking like skin of tiger) appearance. Signature histological findings suggestive of FMDV-induced myocarditis may be predominantly acute and necrotizing or subacute to chronic with various hallmarks of inflammation. Inflammation typically includes lympho-histiocytic infiltration and edema, whereas cardiomyocyte degeneration and necrosis may occur as individual cells, small clusters, or may be regionally extensive (Stenfeldt *et al.*, 2016a).

Diagnosis

Diagnosis of FMD depends on clinical signs, pathological lesions together with laboratory examination to establish the detection of viral antigen, viral genome or specific antibodies in samples by different serological and nucleic acid recognition assays.

Clinical picture and pathology: The clinical signs exhibited by the diseased animals, such as vesicles and erosions of the mucous membranes of the oral mucosa, gum, tongue, teats and interdigital spaces (Dorea *et al.*, 2017), are characteristic features of FMD. These clinical manifestations can provide valuable clues for a prompt diagnosis of the disease. Tissues to be submitted for histopathology should include oral mucosa and skin containing vesicles or fresh erosions, heart, udder and pancreas. Most animals infected with FMD will not die and since it is important to make prompt diagnosis from clinical cases, histopathology of necropsy of material is often secondary.

Virus isolation: Isolation of the virus is considered the gold standard for the diagnosis of viral diseases. Primary cell culture of bovine (Snowdon, 1966), ovine and porcine (Bachrach *et al.*, 1955) origin as well as cell lines are considered to be generally suitable for isolation of FMDV (OIE, 2012). Some stable cell lines like IBRS-2 (Nair, 1987), MVPK-1 clone 7 (Dinka *et al.*, 1977; Nair, 1987), LFBK cell line, ZZ-R 127 (Brehm *et al.*, 2009) and baby hamster kidney (BHK21) are known to be susceptible to FMDV. Among these, BHK21 is particularly common in laboratory and vaccine industry settings.

In the diagnostic process, the clinical suspensions are inoculated into these cell culture systems and then incubated at 37°C. The cultures are examined for cytopathic effect (CPE) typically within 24 to 48 hours post infection. The most sensitive cell culture for FMDV is the primary bovine thyroid (BTY), and primary lamb kidney (LK) cells, with LFBK cells being among the most sensitive cell lines available.

Virus neutralization test (VNT): VNT is considered the gold standard tool for detection of neutralizing antibodies to FMDV structural proteins and is an approved test for the certification trade of animals and animal products, despite being time-consuming, liable for contamination and requiring special facilities.

Serological tests: The disease can be diagnosed by CFT, VNT, solid phase competitive ELISA (SPCE), liquid phase blocking ELISA (LPBE) and non-structural protein antibody detection ELISAs, enzyme linked immune electro transfer blot assay.

Nonstructural protein (NSP) ELISAs: NSP ELISA successfully differentiates between infected and vaccinated animals (DIVA) (Paton *et al.*, 2006). In India, battery of competitive and indirect ELISAs using 3AB3 (Mohapatra *et al.*,

2011a), 3ABC (Sharma *et al.*, 2012), truncated 2C (2 Ct) (Mahajan *et al.*, 2013), 3B (Mohapatra *et al.*, 2014), 2B (Biswal *et al.*, 2014), 3D (Mahajan *et al.*, 2015) have been developed and utilized for DIVA purpose at ICAR-National Institute on Foot and Mouth Disease (ICAR-NIFMD). The 3AB3 indirect ELISA is routinely applied in India for serosurveillance of bovines since more than a decade with consistent results. The development and implementation of DIVA assays, represent significant advancements in FMD diagnosis and control. These assays allow for the discrimination between animals that have been infected with FMDV and those that have been vaccinated, facilitating more targeted and effective disease management strategies.

Sandwich ELISA: The detection of FMDV antigen and identification of virus serotype is achieved by sandwich ELISA. In India, a double antibody sandwich ELISA (Bhattacharya *et al.*, 1996) is routinely used as a frontline test for identification of FMDV serotypes.

Nucleic acid recognition methods: The polymerase chain reaction (PCR) techniques are used for rapid identification of FMDV. The reverse-transcription PCR (RT-PCR) helps to detect the FMDV genome in clinical samples.

RT-PCR assay: RT-PCR for FMDV detection was first reported by Meyer *et al.* (1991), where a conserved genomic region encoding the RNA polymerase was amplified and analyzed by agarose gel electrophoresis followed by further characterization through restriction enzyme digestion or Southern blotting. Five FMDV serotypes were distinguished by conventional RT-PCR technique targeting magnification of the VP1 gene (Ahmed *et al.*, 2012). RT-PCR has proven to be useful tool for fast and sensitive diagnosis of FMD. Höfner *et al.* (1993) also demonstrated the detection of FMDV in clinical samples targeting conserved regions of 1A and 2A/2B. Furthermore, Dill *et al.* (2017) developed a rapid and cost-effective universal RT-PCR for virus genotyping and phylogenetic analysis of FMDV. These advancements in RT-PCR technology have greatly enhanced our ability to detect and characterize FMDV, facilitating more effective disease control and surveillance efforts.

Lineage differentiating RT-PCR: Genotype and lineage differentiating RT-PCR for serotype A (Mohapatra *et al.*, 2007), Asia 1 (Mohapatra *et al.*, 2006), O (Subramaniam *et al.*, 2018) and multiplex PCR (Mohapatra *et al.*, 2011b) for specific detection VP3^{59} deletion group circulating in India were developed. These assays play a crucial role in the detection and identification of circulating Indian FMDV strains, allowing for the differentiation of genotypes and lineages. They serve as valuable tools for preliminary screening before proceeding to sequencing and phylogenetic analyses, which are fundamental approaches for further characterization and understanding of FMDV strains

RT-LAMP: It is an isothermal nucleic acid amplification technique carried out at a constant temperature obviating the need of a thermocycler. The assay is based on the principle of DNA amplification by an autocycling strand displacement reaction. RT-LAMP is extremely sensitive for the simple and rapid detection of FMDV with a comparable sensitivity and specificity to RT-PCR and real-time PCR (Farooq *et al.*, 2015). The assay is performed using a set of two specially designed inner primers and two outer primers and a DNA polymerase with high strand displacement activity (Notomi *et al.*, 2000). The primers recognize 6 independent target sequences in the initial stage and 4 independent sequences during the later stages of the LAMP reaction. The reaction is carried out in less than an hour using a standard water bath or heating block and the results can be visualized with the naked eye. RT-LAMP was first employed for FMDV diagnosis by Dukes *et al.* (2006) targeting the FMDV 3D RNA polymerase gene. Li *et al.* (2009) also used RT-LAMP for detection of FMDV by targeting FMDV 3D gene.

Multiplex PCR: The technique is more rapid and sensitive. Earlier, multiplex PCR targeting VP1 region, which allows for the differentiation of FMDV serotypes, was developed (House and House, 1989). In India, the RT-mPCR developed by Giridharan *et al.* (2005) is utilized on ELISA negative clinical samples to further confirm the presence of FMDV. This technique enhances diagnostic capabilities by providing an additional check on samples that initially test negative by ELISA. Furthermore, Lung *et al.* (2011) demonstrated the use of multiplex RT-PCR for simultaneous detection and differentiation of FMDV serotypes, vesicular stomatitis virus (VSV), swine vesicular disease virus (SVDV) and vesicular exanthema of swine virus (VESV).

Real-time RT-PCR: Real-time RT-PCR assays targeting the 1D region in a multiplex mode have been utilized for concurrent detection and identification of FMDV serotypes in samples. Owing to its higher sensitivity and specificity, it was used for detection of FMDV in carrier animals (Zhang and Alexandersen, 2003). Additionally, a TaqMan assay has proven to be robust and effective for the detection of FMDV in parallel with antigen ELISA (Reid *et al.*, 2003). Currently, two different rRT-PCR TaqMan assays are commonly employed; one targeting the IRES within the 5′ UTR and the other targeting the 3D region of the FMDV genome (Callahan *et al.*, 2002). With the quick and precise diagnosis of FMDV offered by these molecular diagnostic assays, disease surveillance, outbreak containment, and carrier animal monitoring are all greatly aided.

Point of care or pen-side diagnostic tests/Chromatographic strip test: A monoclonal antibody (Mab)-based chromatographic strip test or lateral flow device (LFD) for FMD diagnosis has been developed, which reacts against FMDV of all seven serotypes (Reid *et al.*, 2001). These lateral flow assays (LFAs) are highly sensitive and easy to use at the site of the animal, making them suitable for use by unskilled personnel (Subramaniam *et al.*, 2013). The LFA based on FMDV antigen detection is easy to use and can be utilized on the farm to reduce the time required for transport and laboratory diagnosis of FMDV (Oem *et al.*, 2009). These quick diagnostic tests provide a practical and effective way to identify FMDV infections in animal populations, allowing for the quick implementation of control and intervention measures to the illness from spreading.

Biosensors for detection of FMDV: An immunobiosensor using a piezo-electric (PZ) crystal was developed and standardized for FMD diagnosis and virus typing (Gajendragad *et al.*, 2001).

Infrared thermography detection: Thermal imaging has been explored as a potential tool for the early detection of clinical cases. However, the assessment of hoof temperature by thermal imaging had limited accuracy for detection of FMD (Knight-Jones *et al.*, 2016). In a color scale, the warmest areas are depicted as white or red, while the coolest areas appear blue or black. While thermal imaging may not be highly accurate for FMD detection on its own, combining it with other standard methods of lameness detection on-farm could improve its effectiveness (Alsaaod *et al.*, 2015).

Microarray for FMDV: Microarray technology provides greater screening capabilities for FMDV detection (Subramaniam *et al.*, 2012). This technology employs a strategy of microarray hybridization, which is essentially the reverse of a standard dot-blot technique (Schrenzel *et al.*, 2009). A microarray-based test using FMD DNA chip containing 155 oligonucleotide probes which were designed to be serotype-specific and targeted regions within the VP3 and VP1-2A regions of the FMDV genome. Each probe was approximately 35-45 base pairs long, allowing for specific and sensitive detection of FMDV strains (Baxi *et al.*, 2006). High throughput screening capabilities, quick detection of various serotypes, and automation possibilities are just a few benefits of using microarray technology for FMDV detection.

Antigen detection by histopathology and immunohistochemistry: Immunohistochemistry (IHC) is a valuable method for detecting FMDV antigen in tissue sections (Mahmuda *et al.*, 2017). This technique involves fixing tissues samples in 10 percent neutral phosphate buffered formalin and embedding them in paraffin. The tissue sections are then stained by hematoxylin and eosin (H and E) for examination (Yamada *et al.*, 2018). In the case of FMDV detection, the Universal Immuno-enzyme Polymer method with a HISTFINE simple stain Max PO (M) kit (Nichirei, Tokyo, Japan) is used according to the manufacturer's instructions. The specificity of the staining is confirmed using monoclonal antibody specific for FMDV serotype, and counterstained with hematoxylin (Yamada *et al.*, 2018).

Differential Diagnosis

Distinguishing between FMD and other vesicular diseases such as swine vesicular disease (SVD), vesicular stomatitis (VS), and vesicular exanthema of swine (VES) is crucial for accurate diagnosis. VS can affect multiple species and may cause fever, lameness, and salivation, while VES primarily affects swine and can cause vesicular lesions on the skin. FMDV and SVDV are both picornaviruses, but they belong to different genera. FMDV is classified in the *Aphthovirus* genus, while SVDV belongs to the *Enterovirus* genus. SVDV is highly resistant to both low and high pH conditions, whereas FMDV is stable in a narrower pH range, typically between pH 7 and 7.8

Diagnosing FMD in sheep and goats can be challenging due to similarities with other diseases. FMD can be distinguished by the absence of respiratory and gastrointestinal symptoms from Peste des petits ruminants (PPR). Facial edema and nasal ulceration are characteristic signs of Bluetongue. Presence of pock lesions is a distinguishing feature of Capripox. Orf typically presents with scabby lesions around the mouth, which may be absent in FMD. Pneumonic Pasteurellosis and Contagious Caprine Pleuropneumonia (CCPP) primarily involve respiratory symptoms.

Disinfectants and Antiviral Compounds

FMDV is susceptible to wide variety of disinfectants such as sodium hydroxide, sodium carbonate, citric acid and Virkon-S26. The virus is labile to excess pH. Therefore, both acids (*e.g.*, citric acid) and bases (*e.g.*, caustic soda or sodium hydroxide) have the ability to destroy the virus.

Favipiravir (T-705, 6-fluoro-3-hydroxy-pyrazine-2-carboxamide), has shown promise as a potential treatment for FMD and recommended by the Japanese Government for treatment against FMDV. Favipiravir was developed for use against influenza viruses (de Avila *et al.*, 2017). This agent hinders ribonucleic acid (RNA) dependent RNA polymerase within the virus (Furuta *et al.*, 2013). Other chemical compounds that have shown potential against FMDV include ribavirin, BQR, mizorbine, VX497 and GS9620. These compounds may offer alternative treatment options in the future.

In addition to antiviral drugs, newly developed disinfectants have also shown efficacy against FMDV. These include CAC-717, accelerated hydrogen peroxide (AHP), and Virkon S. AHP, in particular, has demonstrated significant virucidal activity against FMDV, resulting in a $>5\log_{10}$ reduction in virus titre at a 1/40 dilution. Sodium carbonate solution (4 percent) and various commercial products are also widely used for disinfection of FMDV in endemic countries.

Containment Laboratories for FMDV Research

Leakage of infectious FMDV from the laboratory can cause devastating outbreaks, as was realized in several instances. Therefore, the necessity of a containment laboratory for safe handling of FMDV was realized, leading to establishment of such laboratories at several countries across the world. Improvements include passage of exhaust air through HEPA filters, heat inactivation of all liquid wastes, incineration of animal tissues and carcasses, and decontamination of all other materials leaving the containment facilities by steam under pressure, ethylene oxide, formaldehyde gas, and other suitable disinfectants. Mandatory change of clothes in/shower out requirements for all personnel and mandatory quarantine of all personnel and visitors from contact with cloven-hoofed animals for seven days after leaving the laboratories are also followed.

The Insel Reims in Germany in 1909, the Pirbright laboratory (World Reference Laboratory, WRL) in the United Kingdom in 1924, Lindholm Island in Denmark in 1925, the Centro Panamericano de Fie´breAfto´sa (PanAftosa) in Brazil in 1951, the Plum Island Animal Disease Center (PIADC) located off the eastern tip of Long Island, New York, the United States in 1953 and ICAR-National Institute on Foot and Mouth Disease (ICAR-NIFMD)-International Centre for FMD (ICFMD) in India in 2017 were established specifically to study and conduct cutting edge research works on FMDV in containment facility. These facilities are essential to our comprehension of the virus, to the development of vaccines, diagnostic tools, and control strategies, as well as to our international cooperation in the fight against FMD.

Prevention and Control

The efforts of Waldmann and colleagues (1937) in Germany led to the development of an inactivated FMD vaccine. By replicating the FMDV in BHK21 cells, they enabled commercially viable large-scale production of virus in suspension cultures, which could then be inactivated by treatment with aziridines such as binary ethylenemine and mixed with adjuvant in order to formulate the final vaccine to be used in the field. Ensuring the quality of vaccine and implementing effective vaccination practice are crucial for prevention and control of FMD. Besides symmetrical vaccination of all susceptible hosts, control of movement of animals is also important in controlling FMD. Real-time disease reporting is also vital to promptly characterize circulating virus strains, allowing for the matching of vaccines to the strains in use. Maintenance of proper biosecurity measures is essential to prevent the introduction and spread of the disease. Many countries including India are embarking on the stepwise progressive control pathway (PCP) approach to improve their FMD control capacity in a sustainable manner.

Conclusions

The outbreaks of FMD in a number of FMD-free countries such as Taiwan in 1997 and the United Kingdom in 2001, have underscored the significant threat posed by this highly infectious disease. Sustained public awareness is crucial to translate disease control efforts into effective action. FMD is no doubt a 'difficult to control disease' due to factors such as diverse viral strains within serotypes, multiple susceptible hosts, varying host reactivity to the virus, short-lived immunity from inactivated vaccines, inadequate movement control of animals, vaccine quality concerns, improper vaccination practices, under-reporting of disease incidences, ignorance to investigate each and every outbreak *etc.* When all these loopholes could be addressed with a coordinated effort, then only the disease can be controlled. Other challenges faced in FMD eradication are the economic considerations and social and political challenges. India is having a will to prevent and control FMD as an economically important viral disease of livestock which causes huge annual losses of about Rs 20000 crore. Government aided livestock health and disease control programme (LHDCP) is instrumental in such effort. In the past, India has successfully eradicated Rinderpest and African Horse Sickness. The country has a mechanism and expertise in place to prevent and control FMD and eliminate it in near future. As FMD is clinically indistinguishable from infections resulting from other similar vesicular disease-causing viruses, early diagnosis is critical for efficient disease control. Various diagnostic methods ranging from conventional virus isolation and competitive- and blocking-antigen ELISA to molecular-based methods such as RT-PCR have been developed over the years. However, there is a need to bridge the gap between laboratory-based assays and their practical field application. A multimodal strategy that incorporates new tactics and technologies into control initiatives is needed to eradicate FMD. Effective health management strategies should incorporate outbreak prevention techniques and interventions designed to address the unique issues presented by FMD. Achieving steady progress towards the eradication of FMD requires cooperation amongst stakeholders, including governments, veterinary authorities, research institutes, and the livestock sector.

REFERENCES

Ahmed, H.A., Salem, S.A., Habashi, A.R., Arafa, A.A., Aggour, M.G., Salem, G.H., Gaber, A.S., Selem, O., Abdelkader, S.H., Knowles, N.J., Madi, M., Valdazo-González, B., Wadsworth, J., Hutchings, G.H., Mioulet, V., Hammond, J.M. and King, D.P. (2012). Emergence of foot-and-mouth disease virus SAT 2 in Egypt during 2012. Transbound Emerg Dis, 59(6): 476-481. doi: 10.1111/tbed.12015.

Alexandersen, S. and Mowat, N. (2005). Foot-and-mouth disease: Host range and pathogenesis. in foot and mouth disease virus. Springer-verlag Berlin, Berlin, Current Topics in Microbiology and Immunology, no. 288, pp. 9-42.

Alexandersen, S., Brotherhood, I. and Donaldson, A.I. (2002a). Natural aerosol transmission of foot-and-mouth disease virus to pigs: minimal infectious dose for strain O1 Lausanne. Epidemiol Infect, 128(2): 301-312. doi: 10.1017/s095026880100646x.

Alexandersen, S., Zhang, Z. and Donaldson. A.I. (2002b). Aspects of the persistence of foot-and-mouth disease virus in animals - the carrier problem. Microbes Infect, 4(10): 1099-1110. doi: 10.1016/s1286-4579(02)01634-9.

Alexandersen, S., Zhang, Z., Donaldson, A.I. and Garland, A.J.M. (2003). The pathogenesis and diagnosis of foot-and-mouth disease. J Comp Pathol, 129(1): 1-36. doi:10.1016/S0021-9975(03)00041-0.

Alsaaod, M., Schaefer, A.L., Büscher, W. and Steiner, A. (2015). The Role of Infrared Thermography as a non-invasive tool for the detection of lameness in cattle. Sensors (Basel), 15(6): 14513-14525. doi: 10.3390/s150614513.

Arzt, J., Baxt, B., Grubman, M.J., Jackson, T., Juleff, N., Rhyan, J., Rieder, E., Waters, R. and Rodriguez, L.L. (2011). The pathogenesis of foot-and-mouth disease II: viral pathways in swine, small ruminants, and wildlife; myotropism, chronic syndromes, and molecular virus-host interactions. Transbound Emerg Dis, 58(4): 305-326. doi: 10.1111/j.1865-1682.2011.01236.x.

Arzt, J., Pacheco, J.M. and Rodriguez, L.L. (2010). The early pathogenesis of foot-and-mouth disease in cattle after aerosol inoculation. Identification of the nasopharynx as the primary site of infection. Vet Pathol, 47(6): 1048-1063. doi:10.1177/0300985810372509.

Bachrach, H.L., Hess, W.R. and Callis, J.J. (1955). Foot-and-mouth disease virus: its growth and cytopathogenicity in tissue culture. Science, 122(3183): 1269-1270. doi: 10.1126/science.122.3183.1269.

Barros, J.J., Malirat, V., Rebello, M.A., Costa, E.V. and Bergmann, I.E. (2007). Genetic variation of foot-and-mouth disease virus isolates recovered from persistently infected water buffalo (*Bubalus bubalis*). Vet Microbiol, 120(1-2): 50-62. doi: 10.1016/j.vetmic.2006.10.023.

Baxi, M.K., Baxi, S., Clavijo, A., Burton, K.M. and Deregt, D. (2006). Microarray-based detection and typing of foot-and-mouth disease virus. Vet J, 172(3): 473-481. doi: 10.1016/j.tvjl.2005.07.007.

Bhattacharya, S., Banerjee, R., Ghosh, R., Biswas, A. and Chatterjee, A. (2003). Identification of foot-and mouth disease from a captive kangaroo in a zoological garden in India. Vet Rec, 153(16): 504-505. doi: 10.1136/vr.153.16.504.

Bhattacharya, S., Pattnaik, B. and Venkataramanan, R. (1996). Development and application of a sandwich enzyme-linked immunosorbent assay (ELISA) for type identification of foot-and-mouth disease (FMD) virus in direct field materials. Indian J Anim Sci, 66: 1201-1209.

Biswal, J.K., Jena, S., Mohapatra, J.K., Bisht, P. and Pattnaik, B. (2014). Detection of antibodies specific for foot-and-mouth disease virus infection using indirect ELISA based on recombinant nonstructural protein 2B. Arch Virol, 2014 159(7): 1641-1650. doi: 10.1007/s00705-013-1973-3.

Blowey, R.W. and Weaver, A.D. (2003). Color atlas of diseases and disorders of cattle, 2nd ed. Mosby Publisher, pp: 187-189.

Brehm, K.E., Ferris, N.P., Lenk, M., Riebe, R., Haas, B. (2009). Highly sensitive fetal goat tongue cell line for detection and isolation of foot-and-mouth disease virus. J Clin Microbiol, 47(10): 3156-3160. doi: 10.1128/JCM.00510-09.

Brooksby, J.B. (1958). The virus of foot-and-mouth disease. Adv Virus Res, 5: 1-37.

Brooksby, J.B. and Rogers, J. (1957). Methods used in typing the virus of Foot-and-mouth disease at Pirbright, 1950–55. In Methods of typing and cultivation of foot-and-mouth disease virus: Project No. 208. Paris: European Productivity Agency of the Organization of European Cooperation (OEEC); 1957.

Callahan, J.D., Brown, F., Osorio, F.A., Sur, J.H., Kramer, E., Long, G.W., Lubroth, J., Ellis, S.J., Shoulars, K.S., Gaffney, K.L., Rock, D.L. and Nelson, W.M. (2002). Use of a portable real-time reverse transcriptase-polymerase chain reaction assay for rapid detection of foot-and-mouth disease virus. Am Vet Med Assoc, 220: 1636-1642.

Capozzo, A.V., Pérez-Filgueira, M., Vosloo, W. and Gay, C.G. (2021). Editorial: FMD Research: Bridging the Gaps With Novel Tools. Front. Vet. Sci, 8: 686141. doi: 10.3389/fvets.2021.686141.

Condy, J.B., Hedger, R.S., Hamblin, C. and Barnett, I.T. (1985). The duration of the foot-and-mouth disease virus carrier state in African buffalo (i) in the individual animal and (ii) in a free-living herd. Comp Immunol Microbiol Infect Dis, 8: 259-265.

de Avila, A.I., Moreno, E., Perales, C. and Domingo, E. (2017). Favipiravir can evoke lethal mutagenesis and extinction of foot-and-mouth disease virus. Virus Res, 233: 105112. doi: 10.1016/j.virusres.2017.03.014.

Dill, V., Beer, M. and Hoffmann, B. (2017). Simple, quick and cost-efficient: a universal RT-PCR and sequencing strategy for genomic characterisation of foot-and-mouth disease viruses. J Virol Methods, 246: 58-64. doi: 10.1016/j.jviromet.2017.04.007.

Dinka, S.K., Swaney, L.M. and Vicar, J.W. (1977). Selection of a stable clone of the MVPK-1 fetal porcine kidney cell for assays of foot-and-month disease virus. Can J Microbiol, 23(3): 295-299. doi: 10.1139/m77-043. PMID: 192426.

Domingo, E., Martin, V., Perales, C., Grande-Pérez, A., García-Arriaza, J. and Arias, A. (2006). Viruses as quasispecies: biological implications. Curr Top Microbiol Immunol, 299: 51-82. doi: 10.1007/3-540-26397-7_3. PMID: 16568896; PMCID: PMC7120838.

Domingo, E. and Holland, J.J. (1988). High error rates, population equilibrium and evolution of RNA replication systems p. 3-36. In: E. Domingo, J.J. Holland and P. Ahlquist (ed.), RNA genetics, vol. III. Variability of RNA Genomes. CRC Press, Boca Raton, Fla.

Donaldson, A.I. and Alexandersen, S. (2001). Relative resistance of pigs to infection by natural aerosols of FMD virus. Vet Rec, 148(19): 600-602. doi:10.1136/vr.148.19.600.

Donaldson, A.I., Gibson, C.F., Oliver, R., Hamblin, C. and Kitching, R.P. (1987). Infection of cattle by airborne foot-and-mouth disease virus: Minimal doses with O1 and SAT 2 strains. Res Vet Sci, 43: 339-346. PMID: 2832913.

Donaldson, A.I., Gloster, J., Harvey, L.D.J. and Deans, D.H. (1982). Use of prediction models to forecast and analyse airborne spread during the foot-and-mouth disease outbreaks in Brittany, Jersey and the Isle of Wight in 1981. Vet Rec, 110(3): 53-57. doi: 10.1136/vr.110.3.53.

Dorea, F.C., Noremark, M., Widgren, S., Frossling, J., Boklund, A., Halasa T. and Stahl, K. (2017). Evaluation of strategies to control a potential outbreak of foot-and-mouth disease in Sweden. Front Vet Sci, 4: 118. doi: 10.3389/fvets.2017.00118.

Dukes, J.P., King, D.P. and Alexandersen, S. (2006). Novel reverse transcription loop-mediated isothermal amplification for rapid detection of foot-and-mouth disease virus. Arch Virol, 151(6): 1093-1106. doi: 10.1007/s00705-005-0708-5.

Eigen, M. (1992). Oxford University Press; Oxford: Steps towards Life.

Farooq, U., Latif, A., Irshad, H., Ullah, A., Zahur, A.B., Naeem, K., Khan, S.U., Ahmed, Z., Rodriguez, L.L. and Smoliga, G. (2015). Loop-mediated isothermal amplification (RT-LAMP): a new approach for the detection of foot-and-mouth disease virus and its serotypes in Pakistan. Iran J Vet Res, 16(4): 331-334. PMID: 27175198.

Ferris, N.P., Condy, J.B., Barnett, I.T. and Armstrong, R.M. (1989). Experimental infection of eland (*Taurotrages oryx*), sable antelope (*Ozanna grandicomis*) and buffalo (*Syncerus caffer*) with foot-and-mouth disease virus. J Comp Pathol, 101(3): 307-316. doi: 10.1016/0021-9975(89)90040-6.

Furuta, Y., Gowen, B.B., Takahashi, K., Shiraki, K., Smee, D.F. and Barnard, D.L. (2013). Favipiravir (T705), a novel viral RNA polymerase inhibitor. Antiviral Res, 100(2): 446454. doi: 10.1016/j.antiviral.2013.09.015.

Gajendragad, M.R., Kamath, K.N.Y., Anil, P.Y., Prabhudas, K. and Natarajan, C. (2001). Development and standardization of a piezo electric immunobiosensor for foot and mouth disease virus typing. Vet Microbiol, 78(4): 319-330. doi: 10.1016/s0378-1135(00)00307-2.

Giridharan, P., Hemadri, D., Tosh, C., Sanyal, A. and Bandyopadhyay, S.K. (2005). Development and evaluation of a multiplex PCR for differentiation of foot-and-mouth disease virus strains native to India. J Virol Methods, 126(1-2): 1-11. doi: 10.1016/j.jviromet.2005.01.015.

Gloster, J., Champion, H.J., Sorensen, J.H., Mikkelsen, T., Ryall, D.B., Astrup, P., Alexandersen, S. and Donaldson, A.I. (2003). Airborne transmission of foot-and-mouth disease virus from burnside farm, Heddon-on-the-Wall, Northumberland, during the 2001 epidemic in the United Kingdom. Vet

Rec, 152(17): 525-533. doi: 10.1136/vr.152.17.525. Erratum in: Vet Rec. 2003 May 17;152(20):628. PMID: 12739601.

Grosso, A.M. (1957). Foot and mouth disease in the Buenos Aires Zoo. Gac Vet, 19: 54-55.

Grubman, M.J. and Baxt, B. (2004). Foot-and-Mouth Disease. Clin Microbiol Rev, 17(2): 465-493. doi: 10.1128/CMR.17.2.465–493.2004.

Hardham, J.M., Krug, P., Pacheco, J.M., Thompson, J., Dominowski, P., Moulin, V., Gay, C.G., Rodriguez, L.L. and Rieder, E. (2020). Novel foot-and-mouth disease vaccine platform: formulations for safe and DIVA-compatible FMD vaccines with improved potency. Front Vet Sci, 7: 554305. doi: 10.3389/fvets.2020.554305.

Hedger, R.S. (1981). Foot-and-mouth disease. In: Infectious Diseases of Wild Mammals, 2nd Ed., Davis, J.W., Karstad, L.H., Trainer, D.O., editors. Iowa State University Press, Ames, Iowa, pp. 87-96.

Hedger, R.S., Condy, J.B. and Golding, S.M. (1972). Infection of some species of African wild life with foot-and-mouth disease virus. J Comp Pathol, 82(4): 455-461. doi: 10.1016/0021-9975(72)90045-x.

Höfner, M.C., Carpenter, W.C., Donaldson, A.I. (1993). Detection of foot-and-mouth disease virus RNA in clinical samples and cell culture isolates by amplification of the capsid coding region. J Virol Methods, 42: 53-61. doi: 10.1016/0166-0934(93)90176-R.

House, C. and House, J.A. (1989). Evaluation of techniques to demonstrate foot-and-mouth disease virus in bovine tongue epithelium: Comparison of the sensitivity of cattle, mice, primary cell cultures, cryopreserved cell cultures and established cell lines. Vet Microbiol, 20: 99-109.

Hugh-Jones, M.E. and Wright, P.B. (1970). Studies on the 1967-8 foot-and-mouth disease epidemic. The relation of weather to the spread of disease. J Hyg (Lond), 68(2): 253-271. doi: 10.1017/s0022172400028722.

Jamal, S.M. and Belsham, G.J. (2018). Molecular epidemiology, evolution and phylogeny of foot-and-mouth disease virus. Infect Genet Evol, 59: 84-98. doi: 10.1016/j.meegid.2018.01.020.

Kitching, R.P. (2002). Clinical variation in foot and mouth disease: cattle. In: Foot and mouth disease: facing the new dilemmas (G.R. Thomson, ed Rev. Sci. Tech. Off. Int. Epiz, 21(3): 499-504.

Kitching, R.P. and Alexandersen, S. (2002). Clinical variation in foot and mouth disease: pigs. Rev. Sci. Tech. Off. Int. Epiz, 21(3): 513-518.

Kitching, R.P. and Hughes, G.J. (2002). Clinical variation in foot and mouth disease: sheep and goats. In: Foot and mouth disease: facing the new dilemmas (G.R. Thomson, ed.). Rev. Sci. Tech. Off. Int. Epiz, 21(3): 505-512.

Knight-Jones, T.J., Robinson, L., Charleston, B., Rodriguez, L.L., Gay, C.G., Sumption, K.J. and Vosloo, W. (2016). Global Foot-and-Mouth Disease Research Update and Gap Analysis: 4 - Diagnostics. Transbound Emerg Dis, 63(1): 42-48. doi: 10.1111/tbed.12523.

Li, J., Chen, Q., Xiong, W. and Fang, X.E. (2009). Establishment of RT-LAMP for rapid detection of foot-and-mouth disease virus. Bing Du Xue Bao, 25: 137-142.

Loeffler, F. and Frosch, P. (1898). Report of the commission for research on foot-and-mouth disease.

Lung, O., Fisher, M., Beeston, A., Hughes, K.B., Clavijo, A., Goolia, M., Pasick, J., Mauro, W. and Deregt, D. (2011). Multiplex RT-PCR detection and microarray typing of vesicular disease viruses. J Virol Methods, 175(2): 236-245. doi: 10.1016/j.jviromet.2011.05.023.

Mahajan, S., Mohapatra, J.K., Pandey, L.K., Sharma G.K. and Pattnaik, B. (2013). Truncated recombinant non-structural protein 2C-based indirect ELISA for FMD sero-surveillance. J Virol Methods, 193(2): 405-414. doi: 10.1016/j.jviromet.2013.07.003.

Mahajan, S., Mohapatra, J.K., Pandey, L.K., Sharma, G.K. and Pattnaik, B. (2015). Indirect ELISA using recombinant nonstructural protein 3D to detect foot and mouth disease virus infection associated antibodies. Biologicals, 43(1): 47-54. doi: 10.1016/j.biologicals.2014.10.002.

Mahmoud, M.A.E., Ghazy, A.A. and Shaapan, R.M. (2019). Diagnosis and control of foot and mouth disease (FMD) in dairy small ruminants; sheep and goats. Int J Dairy Sci, 14: 45-870 52. doi: 10.3923/ijds.2019.45.52

Mahmuda, A., Bande, F., Abdulhaleem, N., Zihiry, A.l. K J K Majid, *et al.* (2017) Monoclonal antibodies in immunodiagnostic assays: a review of recent applications. Sokoto J Vet Sci, 15(1): 1-10.

Meyer, R.F., Brown, C.C., House, C., House, J.A. and Molitor, TW. (1991). Rapid and sensitive detection of foot-and-mouth disease virus in tissues by enzymatic RNA amplification of the polymerase gene. J Virol Methods, 34: 161-172. doi: 10.1016/0166-0934(91)90096-I.

Mezencio, J.M., Babcock, G.D., Kramer, E. and Brown, F. (1999). Evidence for the persistence of foot-and-mouth disease virus in pigs. Vet J, 157(3): 213-217. doi: 10.1053/tvjl.1999.0357.

Mohamed, F., Swafford, S., Petrowski, H., Bracht, A., Schmit, B., Fabian, A., Pacheco, J.M., Hartwig, E., Berninger, M., Carrillo, C., Mayr, G., Moran K., Kavanaugh, D, Leibrecht, H., White, W. and Metwally, S. (2011). Foot-and-mouth disease in feral swine: susceptibility and transmission. Transbound Emerg Dis, 58(4): 358-371. doi: 10.1111/j.1865-1682.2011.01213.x.

Mohapatra, A.K., Mohapatra, J.K., Pandey, L.K., Sanyal, A. and Pattnaik, B. (2014). Diagnostic potential of recombinant nonstructural protein 3B to detect antibodies induced by foot-and-mouth disease virus infection in bovines. Arch Virol, 159(9): 2359-2369. doi: 10.1007/s00705-014-2089-0.

Mohapatra, J.K., Pandey, L.K., Sanyal, A. and Pattnaik, B. (2011a). Recombinant non-structural polyprotein 3AB-based serodiagnostic strategy for FMD surveillance in bovines irrespective of vaccination. J Virol Methods, 177(2): 184-192. doi: 10.1016/j.jviromet.2011.08.006.

Mohapatra, J.K., Pandey, L.K., Sharma, G.K., Barik, S.K., Pawar, S.S., Palsamy, R. and Pattnaik, B. (2011). Multiplex PCR for rapid detection of serotype A foot-and-mouth disease virus variants with amino acid deletion at position 59 of the capsid protein VP3. J Virol Methods, 171(1): 287-291. doi: 10.1016/j.jviromet.2010.10.016.

Mohapatra, J.K., Sanyal, A., Hemadri, D., Tosh, C., Rasool, T.J. and Bandyopadhyay, S.K. (2006). A novel genetic lineage differentiating RT-PCR as a useful tool in molecular epidemiology of foot-and-mouth disease in India. Arch Virol, 151(4): 803-809. doi: 10.1007/s00705-005-0673-z.

Mohapatra, J.K., Subramaniam, S., Tosh, C., Hemadri, D., Sanyal, A., Periyasamy, T.R. and Rasool, T.J. (2007). Genotype differentiating RT-PCR and sandwich ELISA: handy tools in epidemiological investigation of foot and mouth disease. J Virol Methods, 143(1): 117-121. doi: 10.1016/j.jviromet.2007.02.008.

Nair, S.P. (1987). Studies on the susceptibility and growth pattern of foot-and-mouth disease virus vaccine strains in two pig kidney cell lines. Indian J Comp Microbiol Immunol Infect Dis, 8(2): 76–81.

Niedbalski, W., Kesy, A., Erkiert-Polguj, A. and Polguj, M. (2006). Foot-and-mouth disease as a zoonosis. Med Weter, 62: 374-376.

Notomi, T., Okayama, H., Masubuchi, H., Yonekawa, T., Watanabe, K., Amino, N. and Hase, T. (2000). Loop-mediated isothermal amplification of DNA. Nucleic Acids Res, 28(12): E63. doi: 10.1093/nar/28.12.e63.

Nowak, M.A. (2006). Evolutionary Dynamics: Exploring the Equations of Life. Harvard University Press. https://doi.org/10.2307/j.ctvjghw98.

Oem, J.K., Ferris, N.P., Lee, K.N., Joo, Y.S., Hyun, B.H., Park, J.H. (2009). Simple and rapid lateral-flow assay for the detection of foot-and-mouth disease virus. Clin Vac Immunol, 216(11): 1660-1664. doi: 10.1128/CVI.00213-09.

Officer, K., Lan, N.T., Wicker, L., Hoa, N.T., Weegenaar, A., Robinson, J., Ryoji, Y., Loukopoulos, P. (2014). Foot-and-mouth disease in Asiatic black bears (*Ursus thibetanus*). J Vet Diagn Invest, 26(5): 705-713. doi: 10.1177/1040638714547256. Erratum in: J Vet Diagn Invest. 2020 Jul 32(4): 627.

OIE. (2012). Foot and Mouth Disease. In: OIE Manual of Diagnostic Tests and Vaccines for Terrestrial Animals (Mammals, Birds and Bees), OIE (Eds.). 7th Edn., Vol. 1, Chapter 2.1.5, World Organization for Animal Health, Paris, France, ISBN: 978-92-9044-879-2, pp. 145-173.

Oleksiewicz, M.B., Donaldson, A.I., Alexandersen, S. (2001). Development of a novel real-time RT-PCR assay for quantitation of foot-and-mouth disease virus in diverse porcine tissues. J Virol Methods, 92(1): 23-35. doi: 10.1016/s0166-0934(00)00265-2.

Paton, D.J., de Clercq, K., Greiner, M., Dekker, A., Brocchi, E., Bergmann, I., Sammin, D.J., Gubbins, S., Parida, S. (2006). Application of non-structural protein antibody tests in substantiating freedom from foot-and-mouth disease virus infection after emergency vaccination of cattle. Vaccine, 24(42-43): 6503-6512. doi: 10.1016/j.vaccine.2006.06.032.

Racaniello, V.R. (2011). Picornaviridae: the viruses and their replication". In: Fields Virology, Lippincott Williams and Wilkins, Philadelphia, Pa, USA, 3rd edition, pp. 685-722.

Radostits, O.M., Blood, D.C. and Gay, C.C. (2007). Veterinary Medicine, A Text Book of the Disease of Cattle, Sheep, Goats, Pigs and Horses. 8th ed. London: Balliere Tindall, pp. 1223-1227.

Reid, S.M., Ferris, N.P., Brüning, A., Hutchings, G.H., Kowalska, Z. and Akerblom, L. (2001). Development of a rapid chromatographic strip test for the pen-side detection of foot-and-mouth disease virus antigen. J Virol Methods, 96(2): 189-202. doi: 10.1016/s0166-0934(01)00334-2.

Reid, S.M., Grierson, S.S., Ferris, N.P., Hutchings, G.H. and Alexandersen, S. (2003). Evaluation of automated RT-PCR to accelerate the laboratory diagnosis of foot-and-mouth disease virus. J Virol Methods, 107(2): 129-139. doi: 10.1016/s0166-0934(02)00210-0.

Rueckert, R.R. (1996). Picornaviridae: the viruses and their replication, In B.N. Fields, D.M. Knipe, and P.H. Howley (ed.), Fields virology, 3rd ed. Lippincott-Raven, Philadelphia, Pa, p.609-654.

Rweyemamu, M., Roeder, P., Mackay, D., Sumption, K., Brownlie, J., Leforban, Y., Valarcher, J.F., Knowles, N.J. and Saraiva, V. (2008). Epidemiological patterns of foot-and-mouth disease worldwide. Transbound Emerg Dis, 55(1): 57-72. doi: 10.1111/j.1865-1682.2007.01013.x.

Salt, J.S. (1993). The carrier state in foot and mouth disease - An immunological review. Br Vet J, 149(3): 207-223. doi: 10.1016/S0007-1935(05)80168-X.

Schrenzel, J., Kostic, T., Bodrossy, L. and Francois, P. (2009). Introduction to Microarray-Based Detection Methods. In Detection of Highly Dangerous Pathogens: Microarray Methods for BSL 3 and BSL 4 Agents, Ed. Kostic, T., Butaye, P., Schrenzel, J. https://doi.org/10.1002/9783527626687.ch1.

Sharma, G.K., Mohapatra, J.K., Pandey, L.K., Mahajan, S., Mathapati, B.S., Sanyal, A. and Pattnaik, B. (2012). Immunodiagnosis of foot-and-mouth disease using mutated recombinant 3ABC polyprotein in a competitive ELISA. J Virol Methods, 185(1): 52-60. doi: 10.1016/j.jviromet.2012.05.029.

Snowdon, W.A. (1966). Growth of foot-and mouth disease virus in monolayer cultures of calf thyroid cells. Nature, 210(5040): 1079-1080. doi: 10.1038/2101079a0.

Stenfeldt, C., Diaz-San Segundo, F., de los Santos, T., Rodriguez, L.L. and Arzt, J. (2016a). The pathogenesis of foot-and-mouth disease in pigs. Front Vet Sci, 3: 41. doi: 10.3389/fvets.2016.00041.

Stenfeldt, C., Pacheco, J.M., Rodriguez, L.L. and Arzt, J. (2014). Early events in the pathogenesis of foot-and-mouth disease in pigs; identification of oropharyngeal tonsils as sites of primary and sustained viral replication. PLoS One, 9(9): e106859. doi: 10.1371/journal.pone.0106859.

Stenfeldt, C., Pacheco, J.M., Smoliga, G.R., Bishop, E., Pauszek, S.J., Hartwig, E.J., Rodriguez, L.L. and Arzt, J. (2016b). Detection of foot-and-mouth disease virus RNA and capsid protein in lymphoid tissues of convalescent pigs does not indicate existence of a carrier state. Transbound Emerg Dis, 63(2): 152-164. doi: 10.1111/tbed.12235.

Subramaniam, S., Gahtori, R., Biswal, J.K., Ranjan, R. and Pattnaik, B. (2018). Development and evaluation of reverse transcription-polymerase chain reaction (RT-PCR) assay for differentiation of foot and mouth disease virus serotype O lineages circulating in India. Indian J Vet Pathol, 42(4): 287-291.

Subramaniam, S., Pattnaik, B., Sanyal, A., Mohapatra, J.K., Pawar, S.S., Sharma, G.K., Das, B. and Dash, B.B. (2013). Status of foot-and-mouth disease in India. Transbound Emerg Dis, 60(3): 197-203. doi: 10.1111/j.1865-1682.2012.01332.x.

Sutmoller, P. and McVicar, J.W. (1976). Pathogenesis of foot-and-mouth disease: the lung as an additional portal of entry of the virus. J Hyg (Lond), 77(2): 235-243. doi: 10.1017/s0022172400024669.

Sutmoller, P., McVicar, J.W. and Cottral, G.E. (1968). The epizootiological importance of foot-and-mouth disease carriers. I. Experimentally produced foot-and-mouth disease carriers in susceptible and immune cattle. Arch Gesamte Virusforsch, 23(3): 227-235. doi: 10.1007/BF01241895.

Thomson, G.R., Vosloo, W. and A.D. Bastos. (2003). Foot and mouth disease in wildlife. Virus Res, 91(1): 145-161. doi: 10.1016/s0168-1702(02)00263-0.

Vallée, H. and Carré, H. (1922). Sur la pluralité du virus aphteux. Comptes rendus de l'Académie des sciences Paris, 174: 1498-1500.

Van Bekkum, J.G., Frenkel, H.S., Frederiks, H.H.J. and Frenkel, S. (1959). Observations on the carrier state of cattle exposed to foot-and-mouth disease virus. Tijdschr Diergeneeskd, 84: 1159-1164.

VanderWaal, K. and Deen, J. (2018). Global trends in infectious diseases of swine. Proc Natl Acad Sci U S A, 115(45): 11495-11500. doi: 10.1073/pnas.1806068115.

Waldmann, D., Kobe, K. and Pyl, G. (1937). Die aktiveImmunisierung des Rindesgegen Maul- und KlauenseuchemittelsFormolimpfstoff. Zentbl Bakteriol Parasitenkd Infektkhrankh, 138: 401-412.

Waldmann, O. and Trautwein, K. (1926). Experimentalleuntersuchungenueber die pluralitet des maul-und klauenseuche virus. Berlin Tierarztl Wochenschr, 42: 569-571.

Waters, R.A., Wadsworth, J., Mioulet, V., Shaw, A.E., Knowles, N.J., Abdollahi, D., Hassanzadeh, R., Sumption, K. and King, D.P. (2021). Foot-and-mouth disease virus infection in the domestic dog (*Canis lupus familiaris*), Iran. BMC Vet Res, 17(1): 63. doi: 10.1186/s12917-021-02769-1.

Weaver, G.V., Domenech, J., Thiermann, A.R. and Karesh, W.B. (2013). Foot and mouth disease: a look from the wild side. J Wildl Dis, 49(4): 759-785. doi: 10.7589/2012-11-276.

Woodbury, E.L. (1995). A review of the possible mechanisms for the persistence of foot-and-mouth disease virus. Epidemiol Infect, 114(1): 1-13. doi: 10.1017/s0950268800051864.

Yamada, M., Fukai, K., Morioka, K., Nishi, T., Yamazoe, R., Kitano, R., Shimada, N., Yoshida, K., Kanno, T., Sakamoto, K. and Yamakawa, M. (2018). Pathogenesis of the attenuated foot-and-mouth disease virus O/JPN/2000 in experimentally infected pigs. J Vet Med Sci, 80(11):1669-1675. doi: 10.1292/jvms.18-0377.

Zhang, Z. and Alexandersen, S. (2003). Detection of carrier cattle and sheep persistently infected with foot-and-mouth disease virus by a rapid real-time RT-PCR assay. J Virol Methods, 111(2): 95-100.doi: 10.1016/s0166-0934(03)00165-4.

Chapter 2

Swine Influenza

K. Nagarajan

Department of Veterinary Pathology, Madras Veterinary College, Vepery, Chennai – 600007, Tamil Nadu, India
e-mail: nagavet@gmail.com

ABSTRACT

Swine influenza (SI) is a highly infectious and contagious viral disease in pigs, primarily caused by the swine influenza virus (SIV). First recognized during the 1918 human Spanish flu pandemic, the virus was isolated in 1930. SI causes significant economic losses due to its impact on the swine industry, particularly through decreased body weight gain in pigs. Clinically, SI presents with symptoms such as coughing, sneezing, nasal discharge, elevated temperatures, lethargy, breathing difficulties, reduced appetite, and sometimes abortion in pregnant sows. There are two forms of SI: an epidemic form with rapid spread and quick recovery and an endemic form with milder and less obvious symptoms. The causative agent, SIV, belongs to the Orthomyxovirus genus within the Orthomyxoviridae family. It has a negative-sense single-stranded RNA genome with eight segments and an enveloped virion featuring glycoproteins haemagglutinin (HA) and neuraminidase (NA). These proteins facilitate virus attachment to host cells and the release of progeny viruses. Influenza A virus subtypes are categorized based on their HA and NA proteins, such as H1N1, H1N2, and H3N2. SIV infects various species, including humans, birds, pigs, horses, and aquatic mammals. Transmission occurs through aerosols from coughing or sneezing, nasal discharge, and contaminated surfaces (fomites). Pigs are particularly notable for their receptor distribution, which allows the virus to cross species barriers. Viral replication involves HA cleavage by respiratory or bacterial enzymes, endocytosis, and the release of genetic material for replication. NA aids in the release of new virions from host cells. Influenza viruses exhibit high genetic variability through antigenic drift (point mutations) and antigenic shift (gene reassortment during co-infection). Pigs can act as mixing vessels for reassorted viruses, potentially leading to new strains. SIV primarily targets respiratory tract epithelial cells, with rapid replication and peak infection occurring at 48–72 hours post-infection. The high contagion rate and ability to cross species barriers make vigilant monitoring, biosecurity measures and vaccination essential for prevention and control. Understanding the virus molecular biology, transmission dynamics, and evolutionary mechanisms is crucial for mitigating its impact on the swine industry. Hence, this chapter dealt with swine influenza, the aetiology of the disease, viral replication, host range and interspecies transmission, variation and emergence of new subtypes, pathogenesis, gross pathology, histopathology, clinical symptoms, immune responses, disease diagnosis, and differential diagnosis.

Keywords: *Swine influenza, Swine influenza virus, Orthomyxovirus, Pathogenesis, Clinical symptom, Pathology.*

Introduction

Swine influenza is a highly infectious and contagious viral disease of pigs caused by the swine influenza virus (SIV) associated with a marked decrease in body weight gain of pigs, resulting in substantial economic loss to the industry. Influenza was first recognized as a respiratory disease entity in swine during the 1918 human Spanish flu pandemic, and the SIV was isolated in 1930. Shope demonstrated that the swine influenza virus could be transmitted between pigs using ultrafiltered material, and three years later, a virus shown to be related to that from the pig was eventually isolated from a human patient. The virus was not isolated until 1931, when nasal discharge from pigs was inoculated into ferrets and subsequently into embryonated chicken eggs. Generally, SIV causes interstitial pneumonia in pigs. Clinically, SI is characterized by coughing, sneezing, nasal discharge, elevated rectal temperatures, lethargy, difficulty breathing, depressed appetite and occasionally abortion in pregnant sows. SI in pigs is manifested in two forms: epidemic and endemic form. The former is associated with rapid spread followed by quick recovery in the absence of secondary complications, whereas in the endemic form, clinical signs may be less obvious, and all the pigs may not show the clinical signs.

Etiology

The influenza virus is a member of the genus Orthomyxovirus in the family Orthomyxoviridae and has a negative sense, single-stranded RNA molecule that contains eight segments in an enveloped virion. Influenza A viruses are small (80 to 120 nm in diameter) pleomorphic particles. These particles consist of a host-derived lipid bilayer envelope in which the virus-encoded glycoproteins haemagglutinin (HA), neuraminidase (NA) and matrix protein (M2) are embedded; an inner shell of matrix protein; and at the centre, the nucleocapsids of the viral genome. Multiple NP molecules loosely encapsulate the RNA. Complexes containing the three viral polymerase proteins (PB1, PB2 and P.A.) are at the nucleocapsids' ends. A single virus particle must contain each of the eight unique RNA segments to be infectious. The genome encodes 11 proteins, including haemagglutinin (HA), neuraminidase (NA), matrix (M1, M2), nucleoprotein (NP), Nonstructural proteins (NS1 and NS2) and a polymerase complex (PA, PB1, PB2 and PB1-F2). The major proteins are HA and NA, seen as spikes on the viral surface. HA is responsible for the attachment of the virus to the host cell, while NA plays a significant role in releasing progeny viruses from the infected cells. HA produces antibodies that can neutralise the virus, but there is no cross-protective immunity among different subtypes.

Based on antigenic differences between their nucleoproteins (NP) and matrix (M) proteins, influenza viruses are divided into four types: A, B, C, and D. Influenza A viruses infect a wide range of avian and mammalian species, with the latter group including man, pigs, horses and aquatic mammals. Influenza A and B viruses cause seasonal epidemics of disease in people (known as flu season) in winter. Influenza A viruses are the only viruses known to cause flu pandemics (*i.e.*, global epidemics of flu disease). A pandemic can occur when a new and different influenza A virus emerges that infects people, can spread efficiently among people, and against which people have little or no immunity. Influenza C virus infections generally cause mild illness and are not thought to cause human epidemics. Influenza D viruses primarily affect cattle with spillover to other animals but are not known to infect people to cause illness.

Influenza A viruses are divided into subtypes based on two proteins on the virus's surface: hemagglutinin (H) and neuraminidase (N). There are 18 different hemagglutinin subtypes and 11 neuraminidase subtypes (H1 through H18 and N1 through N11, respectively). While more than 130 influenza A subtype combinations have been identified in nature, primarily from wild birds, there are potentially many more influenza A subtype combinations given the propensity for virus "reassortment". Subtypes of influenza A virus affecting swine are H1N1, H1N2, H3N2, H3N1, H1N7, H4N6 and H9N2. H1N1, H1N2 and H3N2 viruses found in Europe are antigenically and genetically different from those found in the USA. Further, it has been suggested that the swine H1N2 viruses originated following multiple genetic reassortments, initially involving human H1N1 and 'human-like' swine H3N2 viruses followed by reassortment with 'avian-like' swine H1N1 virus, indicating that multiple reassortments of influenza viruses could have occurred in pigs before their detection as clinical entities. The emergence of the H3N2 virus in swine happened by reassortment with the human-like H3 gene of European porcine H3N2 influenza viruses and the N2 gene of the European porcine H1N2 viruses.

The current system of nomenclature of influenza viruses was introduced in 1980 and designates the type, host, place, strain number (if any), year of isolation and antigenic subtype of a virus. For example, a swine influenza virus isolated in India in 2009 would be designated A/Swine/UP-India/2009(H1N1) (Nagarajan *et al.*, 2010). Three main subtypes currently circulating in the pig population worldwide are classical swine influenza virus (SIV) and reassortant viruses of H1N1, H3N2 and H1N2 (Dee, 2005). However, pigs can also be infected with other subtypes of influenza A viruses. The different variants of influenza A virus affecting pigs have been reported worldwide, including Asian countries like China, Hong Kong, Japan, Russia, Taiwan and Thailand. In India, human influenza A virus infections in pigs were first documented serologically in 1978. Clinical disease (SI) among Indian pigs due to the A/Swine/UP-India/2009(H1N1) pandemic virus was reported in 2009 (Nagarajan *et al.*, 2010).

Viral Replication

The influenza virus's replication cycle starts with HA cleavage into HA1 and HA2 by enzymes in the respiratory tract. The host produces the enzymes but may also be derived from bacteria, which can promote influenza infection. Viruses grown in cells that lack a cleavage enzyme can be activated by treatment with trypsin. After HA cleavage, the receptor-binding site of HA1 can attach to a terminal sialic acid residue of a cell surface receptor; once attached to the host cell, the virus is endocytosed (receptor-mediated endocytosis)-NA functions as a receptor-destroying enzyme by cleaving terminal sialic acid residues from the receptor. Thus, NA releases progeny virions from the host cell in which they arose and facilitates virus spread. The progeny virions can infect other cells or can be transmitted to another individual.

Host Range and Interspecies Transmission

Influenza A viruses are highly contagious pathogens isolated from various animals, including humans, birds, pigs, horses, minks, seals, whales, and most recently, cats and dogs. The transmission generally occurs through contact with aerosols produced during coughing or sneezing, nasal discharge and fomites. The Classical SIV can produce flu in humans, birds and other animals. During the acute febrile stage, shedding of the virus can occur within 24 hours of infection, and the virus may be carried and spread by birds or humans. The incubation period is about 72 hours, and the infection is more common during the fall and winter months due to the psychrophilic nature of the virus.

Generally, there is a functional barrier to infection of people with avian influenza. The functional barrier is partly because avian influenza viruses preferentially use α2,3 linked sialic acid receptors expressed in bird cells, and human viruses use α2,6 linked sialic acid receptors expressed in the human respiratory tract. Although early studies identified both α2,3 and α2,6 receptors on tracheal epithelial cells, setting the stage for establishing the pig's reputation as "a mixing vessel," the more recent studies indicate that very few swine tracheal cells express α2,3 receptors. Receptor distribution in human and swine respiratory tracts is similar, suggesting that swine are not a necessary intermediate host for avian-to-human infections. In the pig, α2,6 receptors are abundant throughout the porcine respiratory tract. However, α2,3 receptors are limited to the small bronchioles and alveoli and are of similar or lesser abundance than α2,6 receptors in this portion of the tract.

Interestingly, many cells other than the airway epithelial cells also appear to express sialic acid receptors, including peribronchial mucosal gland epithelial cells, subepithelial connective tissue cells, and endothelial cells. Thus, a preponderance of receptors for a virus in the smaller bronchioles and alveoli will predispose a host to developing severe disease (pneumonia) rather than bronchitis. However, transmission of the virus between individuals may be limited. Conversely, if most receptors for a particular virus are in the upper respiratory tract and larger airways, virus shedding and, thus, transmission between individuals may be facilitated, but the disease induced will be relatively milder (bronchitis).

Gene mutation and reassortment of genes are the key factors responsible for influenza A virus virulence and host tropism change. The emergence of a new influenza A subtype can even result in modifying receptor binding specificities, which could cause a human pandemic. In humans, α2-6 linked sialic acid receptors are located deep in the lung, making catching the flu infection from other species difficult. Swine can serve as hosts in which avian influenza viruses adapt for replication in mammals. On the other hand, the SIVs are known for their ability to jump from one species to another. Once established in the swine population, the virus evolved through re-assortment with both human and swine viruses.

Ducks and other waterfowl are the principal natural hosts of influenza A viruses, with all 18 HA and 11 NA virus subtypes circulating among them. Unlike in other species, influenza viruses target the gastrointestinal tract of waterfowl rather than the respiratory tract, and the infections are almost without exception, completely subclinical, with some ducks shedding virus for as long as 30 days. This, together with the migratory behaviour of waterfowl and the ability of influenza viruses to persist in cold lake water, contributes to the fact that waterfowl form an immense reservoir for influenza viruses in nature. Viruses from this natural reservoir are sometimes transmitted to other host species, which they continue spreading less often and sporadically cause high mortality. There are reports of direct virus transmission from waterfowl to pigs, horses, mink, domestic poultry and aquatic mammals, with associated infections of varying severity.

Variation and Emergence of New Subtypes

Aquatic birds are known to be the natural reservoir of these viruses. The gene pool of influenza A viruses in aquatic birds provides all the genetic diversity required for the emergence of pandemic influenza viruses for humans, lower animals and birds. The variability results from an accumulation of molecular changes in the eight RNA segments that can occur by several different mechanisms, including (i) point mutations (antigenic drift), (ii) gene reassortment (antigenic shift), (iii) defective interfering particles and (iv) RNA recombination. Each of these mechanisms may contribute to the evolution of influenza viruses. Mutations, including substitutions, deletions and insertions, are among the most important mechanisms for producing variation in influenza viruses. Influenza A viruses in humans, pigs and horses show antigenic drift and antigenic shift. The lack of proofreading among RNA polymerases contributes to replication errors in 1 in 10^4 bases in contrast with the much higher replication fidelity found among DNA polymerases, *i.e.*, errors on the order of 1 in 10^9 bases per replication cycle. Each round of RNA virus replication results in a mixed population with many variants, most of which are not viable but some of which have potentially advantageous mutations that can become dominant under the right selective conditions.

Influenza viruses have segmented genomes; hence, reassortment is an important mechanism for producing diversity rapidly. It occurs among influenza A viruses in nature and is important in the appearance of pandemics in human populations. Genetic reassortment or recombination can occur during mixed infections with other influenza viruses, resulting in the generation of novel reassortant influenza A viruses (genetic or antigenic shift). The exchange of whole segments occurs during the simultaneous infection of a cell by two or more different influenza A viruses, creating viruses with entirely new combinations of genes. This results in various viruses possessing selective advantage compared with their parent viruses. Moreover, antigenic drift may also lead to step-by-step modification of virus proteins by continuously accumulating point mutations over a very long period. Antigenic drift favours the emergence of more pathogenic biovars and allows the viruses to escape the immunity acquired through infection or vaccination and to cause further outbreaks and epidemics. Like all influenza viruses, SI viruses change constantly, and pigs may act as mixing vessels for the reassortment of avian, swine and human influenza viruses, thus playing an essential role in the emergence of novel influenza viruses capable of causing a human pandemic. The S-OIVs probably resulted from the reassortment of recent North American H3N2 and H1N2 swine viruses (avian/human/swine 'triple' reassortant viruses) with Eurasian avian-like swine viruses. As a result, these viruses possess PB2 and PA genes of North American avian virus origin, a PB1 gene of human H3N2 virus origin, HA (H1), NP, and NS genes of classical swine virus origin, and NA (N1) and M genes of Eurasian avian-like swine virus origin (hence their original description as 'quadruple' reassortants). However, the human-like PB1 gene and the avian-like PB2 and PA genes have been circulating in pigs since 1997–1998 (when triple reassortant swine viruses were first isolated) and have probably undergone adaptation to pigs. These viruses do not possess markers associated with high pathogenicity. HA and NA are the most important antigens for inducing protective immunity in the host and show the greatest variation. The emergence of new subtypes of influenza virus and the ensuing pandemics are unmistakable, but epidemics that occur between pandemics may still be severe. They result from the gradual change in antigenicity of the circulating virus after successive point mutations in the HA molecule until the virus is sufficiently different from earlier strains so that a large proportion of the population is susceptible and cases reach epidemic levels. This accumulative variation is termed antigenic drift, and the size and severity of the epidemic will depend on the degree to which the virus is different from those already experienced by the population.

Pathogenesis

The clinical disease and lesions induced by the classic H1N1swine influenza virus, avian-like H1N1 viruses adapted to swine, and various H1N1, H1N2, H3N2, and H2N3 reassortants (including the recent 2009 human pandemic H1N1and subsequent reassortants) are essentially the same. SIVs are spread between pigs through direct contact via nasal secretions and inhalation of aerosolised virus in droplets generated by coughing and exhalation. The attachment and replication of swine influenza virus (SIV) is generally restricted to the respiratory tract of pigs. The cellular targets of infection are the epithelial cells lining the nasal passages, trachea, bronchi, bronchioles, and alveoli. Infection and multiplication in host cells progresses very rapidly with influenza viruses. The first evidence of virus infection is in the nucleus of bronchial epithelial cells, as early as 2 h PI. By 4 h PI, virus antigen was abundant in infected cells' nucleus and cytoplasm. In an ultrastructural study, the virus was observed budding from the surface of Type II pneumocytes as early as 5 h PI. Only low numbers of randomly scattered cells are observed in the nasal turbinates and trachea during the first 24–72 h PI. The most extensive infection occurs in epithelial cells lining the bronchi and bronchioles, with peak infection occurring at 48–72 h PI. Although some virus reaches the alveolar level early, especially with nebulization or high dosages with other methods, more extensive spread of the virus to alveolar epithelial cells tends to occur later in the course of infection, at 72-96 h PI.

The HA protein is an integral membrane protein and is the major surface antigen of the influenza virus virion. It is responsible for the attachment of the virus to host cells, internalisation of the virus, and intracellular fusion of viral capsid with endosomal membranes. Newly synthesized HA undergoes post-translational processing: proteolytic cleavage, glycosylation and fatty acid acylation. The final processing step is the cleavage of the HA into two subunits, HA1 and HA2, the so-called fusion peptide (uncleaved H.A. is HA0), connected by disulfide linkages. This cleavage is accomplished by host-produced trypsin-like proteases or, in some cases, bacterial proteases. It is required for infectivity because the free amino terminus of HA2 mediates virus-cell fusion. An influenza virus particle with cleavage-activated HA binds to cells via interaction between the receptor-binding sites of HA and the terminal sialic acid of the cell surface receptor glycoprotein or glycolipid.

Following binding, the attached virion is endocytosed by the cell. The low pH of the endocytotic vesicle triggers a conformational change in cleaved

HA, which is believed to facilitate the insertion of the hydrophobic free amino terminus of HA2 into the vesicular membrane, initiating fusion of the viral and vesicular membranes. Fusion releases the contents of the virion into the cytoplasm of the cell. The amino acid sequence at the cleavage site determines HA cleavability. For example, Low pathogenic avian viruses and non-avian influenza viruses (except H7N7 equine influenza viruses) possess a single Arg residue at the cleavage site that is cleaved by proteases in the respiratory and intestinal organs and hence restricts viral replication locally. In contrast, highly pathogenic H5 and H7 viruses possess several basic amino acids at the HA cleavage site. This motif is recognised by ubiquitous proteases, such as furin and PC6 (also known as PCSK5), and leads to systemic infections. For several outbreaks in poultry, increased pathogenicity of avian influenza viruses has been linked to the acquisition of multibasic HA cleavage sites. This finding underscores the significance of the HA cleavage motif for virulence. Influenza viruses that naturally infect mammalian hosts do not have a multibasic cleavage site, and systemic infection is not a normal characteristic of influenza virus infection in mammals. The HA from the 1918 human pandemic H1N1 virus does not have a multibasic cleavage site but still appears highly cleavable. Thus, no consistent association has been established between a sequence of multiple basic amino acids in the HA and the virulence of viruses possessing such a sequence for mammalian species.

Neuraminidase (NA) is an external protein that prevents aggregation of viral progeny upon release from infected cells. The balanced function of HA and NA results in the most efficient virus replication, and alterations in the NA protein can interfere with this balance, resulting in decreased virulence. Some NA proteins inherently appear to contribute to greater virulence. NA from the 1918 human pandemic H1N1 virus contributed to high virus replication rates in mouse lungs and human airway cells. Some studies suggest that the NA protein also contributes to apoptosis. After virus entry into the cell and release from the endosome, the viral ribonucleoproteins and template RNA are transported to the nucleus, where viral mRNA is transcribed by the polymerases (RNA-dependent RNA polymerase). PB1 is the catalyst, and PB2 removes 50 caps from host pre-mRNA molecules (cap-snatching), assisted by the endonuclease PA, for use as primers for viral mRNA production. These viral mRNAs are moved to the cytoplasm to translate viral proteins. The polymerases influence virulence by controlling protein synthesis processes, favouring virus production over cell maintenance functions. However, sequence variations between polymerases will likely preclude the identification of definitive virulence markers for all viruses.

The primary transcripts are used to translate viral proteins, which in the early stage of infection are predominantly NP and NS1. Translation of host mRNAs is blocked. Newly synthesised NP and NS1 migrated to the nucleus. It is believed that the increased concentration of free NP triggers the shift from mRNA synthesis to cRNA and vRNA synthesis by the infecting viral genome.

Newly synthesised vRNAs are encapsidated in NP within the nucleus and function as templates for the secondary transcription of viral mRNAs. Later in infection, the principal translation products are M1, HA and NA proteins. HA and NA proteins are post-translationally processed and transported to the cell surface, where they integrate into the cell membrane.

The PB1-F2 protein is a nonstructural protein expressed during infection as an alternative open reading frame of the PB1 gene. The protein associates most extensively with mitochondria within host cells but has also been detected in both cytoplasm and nucleus. This protein contributes to virulence through several functions: (1) apoptosis induction, primarily in cells of the innate immune system; (2) suppression of early interferon response by infected cells; (3) increased viral replication rates or delayed viral clearance; and (4) increased inflammation in host tissues. Nearly all of these functions require cooperative interaction of the PB1-F2 with other cellular or virus-induced proteins. Some effects of the protein involve dampening the host's response to infection.

Localization of the protein to mitochondria in macrophages results in mitochondrial membrane depolarization and permeabilization, initiating the death of these cells via the intrinsic apoptosis pathway. Co-localization of the protein with PB1 in the nucleus leads to increased polymerase activity, and likely consequent increased viral replication. Although the PB1-F2 protein appears to contribute to virulence in many studies, its length (full vs truncated) and its association with virulence are inconsistent for all influenza viruses. The protein is expressed at full length (87–90 aa) by most avian influenza viruses, including H5N1 HPAI viruses, and by the 1918 (H1N1), 1957 (H2N2), and 1968 (H3N2) human pandemic viruses, but in truncated form in many viruses from mammalian hosts, including the pandemic 2009 H1N1 virus.

The NS1 protein is not a structural part of the virion. It contributes to virulence by interference with the host antiviral response, including IFN-a/b production and expression of other cytokines such as TNF-a and IL-6. Both proapoptotic and antiapoptotic activities have been documented for NS1. Inhibiting apoptosis of cells that produce viral progeny and promoting apoptosis of cells of the innate immune system that attack infected cells would benefit the virus. Identifying specific virulence determinants has proven difficult because of polygenic contributions to pathogenicity. The polymerase proteins affect alterations in virulence through misdirection of host cell protein synthetic processes to viral protein production. Viral proteins NS1 and PB1-F2 contribute to virulence through inhibition in innate immune responses of the host. The roles of HA and NA proteins as virulence attributes in the infection of mammalian species are less clear. However, a balanced function of these 2 proteins is necessary for efficient virus multiplication. Thus, not all HA-NA pairings that emerge from reassortment will likely result in viruses of equivalent virulence.

Gross Pathology

The main features of SIV infection in pigs are congestion of the mucous membranes of the respiratory tract, a lobular type of pneumonia, congestion and necrosis of the bronchial mucosa. The virus enters the lung through the airways, and hence, the most consistent macroscopic manifestation of influenza virus infection is cranioventral bronchopneumonia, which affects a variable percentage of the lung. In milder infections, small clusters of medium to dark red lobules are located in the cranial and middle lung lobes and occasionally in the caudal lobes' cranioventral portions and accessory lobes. In some pigs, these foci of consolidation are somewhat linear and confluent and situated in the hilar region at the junction of the lung lobes.

Most cranial and middle lung lobes and cranioventral portions of the caudal lobes may be consolidated in more severe or extensive infections. Marked massive atelectasis are prominent, usually limited to portions of the cephalic, cardiac, and azygos lobes. Right-side lung affection unilaterally is more common. Occasionally, bilateral and irregularly symmetrical types of lesions occur. The adjoining non-atelectatic portions of the lungs are emphysematous, elevated and extremely pale. Occasionally, extreme petechiation of the pleurae overlying the areas of the compensatory emphysematous portion may be seen. The lungs are deep purplish red, noticeably depressed compared to the uninvolved lung tissue, and the line of demarcation between normal and pathological lung substance is usually apparent and definite. Palpation of the involved lung reveals a "leathery" consistency without much consolidation and crepitation.

Less frequently, the severe acute infection presents as a diffusely congested lung with prominent interlobular oedema and extensive foam in the bronchi and trachea. Such cases likely represent an exaggerated cytokine response, and in such lungs, the characteristic cranioventral lobular consolidation is obscured by the more diffuse interstitial pneumonia, fibrinous pleuritis and oedema. Airways may be filled with blood-tinged fibrinous exudates, and lymph nodes (bronchial, cervical, mediastinal and mesenteric) are enlarged and oedematous. Mucosa of the pharynx and larynx are hyperemic and covered by white, glassy, tenacious mucus. The same tenacious mucus appears in moderate to copious amounts on opening the trachea. Secretions are frothy and rarely blood-tinged. On the cut section, the bronchioles protrude from the surface of the lung substance and usually have a purplish red, "beefy," pasty appearance throughout. Frothy exudates are present in bronchi bronchioles and fill the lumen. The exudate in the bronchioles is of firmer consistency than that in the respiratory tract. The pleural sacs are free of excess fluid or fibrin.

Research has shown widespread interstitial pneumonia can exist up to 21 days after infection, and haemorrhagic lymph nodes can be observed. The cervical, mediastinal, and mesenteric lymph nodes are highly enlarged, edematous and rarely congested. Those at the hilum of the lung are sometimes so large as to resemble grapes. On section, they are found to be soft and ooze fluid in relatively large amounts. A slight excess of clear or straw-coloured fluid, sometimes slightly cloudy, containing flecks and fibrin strands, was not infrequently encountered in the peritoneum. The stomach sometimes contained bile-tinged watery mucus, and the gastric mucosa was invariably hyperemic, especially along the greater curvature and at the cardiac end. The mucosa of the colon frequently exhibits hyperemic patches of various sizes. These areas were mildly edematous and overlaid by a scant catarrhal exudate. A new antigenic variant of H1N1 SIV A is associated with porcine proliferative and necrotising pneumonia.

Histopathology

The hallmark microscopic lesion is necrotising bronchitis and bronchiolitis. The characteristic histopathological lung lesions include detachment of large bronchial/bronchiolar epithelium areas and accumulation of necrotic epithelial cell debris and neutrophils in the air spaces. Later, the airways are filled with exudates containing monocytes and lymphocytes. The degree of epithelial cell damage is variable and, depending on the time after infection may consist of necrosis and sloughing of epithelial cells, attenuation of the epithelial layer when it begins to repair or proliferation of the epithelial layer in the chronic condition.

During the first 24 hours of PI, neutrophils accumulate in the vasculature adjacent to bronchioles, and light loose lymphocytic cuffs appear around the airways, sometimes before epithelial damage is evident. Necrosis and sloughing of airway epithelial cells and neutrophil transmigration into lumens of the airway occur at 24 to 48 hours PI. By 48 hours PI, affected airways are lined by flattened epithelial layers, and neutrophils in the luminal debris have been replaced or are accompanied by macrophages. Lymphocyte accumulation around airways and adjacent vessels becomes more prominent. Continuing in repair, the epithelial cells lining the airways proliferate and mature into tightly packed tall columnar cells, resulting in an irregular hyperplastic appearance, often by 72-hour PI. Alveoli usually collapse and frequently contain desquamated epithelial cells, mononuclear wandering cells and occasionally some coagulated plasma. Leucocytes, when present, were most abundant in the alveoli, opening directly into the terminal bronchioles. Marked perivascular and peribronchiolar cuffing of lymphocytes are commonly seen. In severely damaged airways, the proliferation of fibroblasts from exposed subepithelial lamina propria may result in endobronchial polyps or incomplete splitting of repairing airways (bronchiolitis obliterans). Such damage is most likely to occur in the smaller airways.

In both field cases and experimental challenge studies, the sizes of airways that are affected may vary. The largest noncartilaginous airways (primary bronchioles) are most consistently and most severely affected, likely due to the dynamics of airflow that affect droplet suspension and deposition or possibly due to receptor distribution. There is some variation between cases, and the larger lobar or segmental bronchi may be spared or only minimally affected, or conversely, only these larger airways may be damaged. Likewise, the smallest terminal or respiratory bronchioles are frequently spared but may exhibit the characteristic epithelial necrosis in severe, extensive infections. Damage tends to be most severe in medium-sized airways–moderate passive congestion of the atelectatic areas as evidenced by dilated and injected pulmonary capillaries. The lymphatic channels, especially in the interlobular septa, sometimes dilate and fill with lymph and cells. Minor tissue damage occurs in the trachea.

Endothelial cells lining the capillaries are swollen, and neutrophils accumulate in capillary lumens and marginate against vessel walls. Type II pneumocytes become swollen. Neutrophils, macrophages, and lymphocytes lightly infiltrate the alveolar interstitium and migrate into alveolar lumens. As in the airways, neutrophils predominate early in infection, but subsequently (by 48 hours PI), the alveolar luminal exudate becomes predominantly mononuclear. These cells appear to be a mixture of sloughed swollen pneumocytes and macrophages.

By close examination, necrosis of constituent cells may be identified in alveolar walls, but the septa usually remain intact. In severe infections, necrotic cellular debris accumulates in alveolar lumens, or rarely, the necrotic debris and exudative proteins form hyaline membranes. In such cases, alveolar and interlobular oedema is usually evident. Swollen alveolar epithelial cells lining the lumen add to the thickness of the alveolar septa. In contrast to the alveolar lesions described above, alveoli may be affected only minimally in some infections. Consistent with the multifocal macroscopic lesions in the lung, lobules within the same histopathologic section of the lung may differ in the degree of involvement, and unaffected lobules often sit adjacent to severely affected lobules. This likely reflects the irregular distribution of the virus through the branches of the respiratory tract via inhalation of aerosolised virus.

Subsequently, the epithelial hyperplasia resolves and all that remain are peribronchiolar lymphocytic cuffing and partially atelectatic alveoli with variable leukocyte populations. Although somewhat dependent on the extent of damage, lungs return to normal by two weeks PI. However, in pigs that survive severe infection, airways disrupted by bronchiolitis obliterans may provide evidence of earlier infection.

Much of the damage to the respiratory tract by influenza can be attributed to the direct viral destruction of infected cells. Influenza virus infection induces cytolysis of infected cells by suppressing host cell gene expression and protein synthesis. Viral polymerases bind to host cell polymerase transcripts and cleave host messenger RNAs (mRNAs) (50 cap removal), which then degrade. Viral NS1 protein inhibits the export of host mRNAs from the nucleus, significantly reducing translation in the cytoplasm. Host cell protein synthesis blockage, in concert with initiation of viral protein synthesis and virus particle budding from cytoplasmic membranes, results in cytolytic death within 20 to 40 hours of infection. Necrosis is characterized by cell membrane disruption, cytoplasm and mitochondria swelling, and organelles disintegration and total cell lysis. Cell death by necrosis stimulates a strong inflammatory response through cytokine induction.

Cell death induced by influenza virus infection also occurs through apoptosis by both extrinsic and intrinsic pathways. Apoptosis has been most thoroughly studied in established cell lines of both respiratory and non-respiratory origin. However, it has also been detected in respiratory tract cells collected following in vivo infection. Viral components most strongly implicated in the induction and regulation of apoptosis includes nonstructural proteins NS1 and PB1-F2 and the ion channel structural protein M2. Depending on intracellular conditions, NS1 can exert either proapoptotic or antiapoptotic influences. Some studies suggest that viral neuraminidase and M1 proteins and single- or double-stranded RNA intermediates influence apoptosis. Inconsistent results between different studies about the induction of necrosis vs apoptosis appear to be related to the cell types studied and the viruses used. Although apoptosis has been proposed as a host defence mechanism that reduces virus spread through the rapid death of virus-infected cells and dampens inflammation through phagocytosis of apoptotic cell debris, recent studies suggest that viruses can use the process to their advantage. Viruses may enhance their infectivity by inhibiting caspase activity until sufficient replication of viral proteins has been completed. Contrarily, in a case of Highly Pathogenic Avian Influenza (HPAI) zoonosis, an H5N1 virus isolated from a human fatality

Clinical Symptoms

SIV infection is one of the major ailments that cause acute respiratory disease in pigs. The illness is short, varying from 2 to 6 days, and in uncomplicated cases, the recovery is almost as sudden as the onset. The mortality is stated to range from 1 to 4 percent. The virus is inhaled and deposited on the surface of the lower respiratory tract, and animals show symptoms promptly and usually within 24 to 48 hours after infection. Infected animals exhibit the first observable evidence of illness as a sharp rise in temperature up to 40°C or higher. Blood picture usually reveals mild leucopenia. Involvement of the respiratory system is usually manifested by increased respiratory rate and a peculiar type of diaphragmatic breathing (Jerky movements), described popularly as "thumping," and frequently, a paroxysmal type of cough occurs when the animal raises. The animals cry out when handled, interpreted as evidence of muscular tenderness. By 3rd to 6th day, death occurs and, in general, is preceded by an exaggeration of all the respiratory symptoms, such as prostration, incoordination, delirium, *etc.* Most of the animals recover rapidly in uncomplicated cases after 6th day, and recovery is rapid. Reproductive performance is impaired in both male and female swine. Boars will have decreased semen quality and output, and pregnant females may abort or have delayed return to service. Lactating sows may have reduced milk production (Spronk, 2001).

Two forms of the disease are defined as swine epidemic and endemic form. Epidemic SIV is readily apparent in all age groups. The onset of the disease is dramatic. In sow herds that are naïve to SIV, abortions will be widespread, with up to 5 percent to 10 per cent. One point of differentiation between SIV and PRRS is that the abortion storm of SIV will pass in less than 2 weeks, while PRRS abortion storms typically last longer. A typical SIV abortion storm will sweep through a sow herd and be dramatic. It will be characterized by high fevers, off-feed, abortions, coughing and death. The epidemic version of the disease has a seasonal cadence, peaking in the late fall and winter months, most likely due to the colder temperatures and lower ventilation rates. Most commonly, the chilling of the pigs is the triggering event of the disease explosion.

In the endemic form, the clinical signs may be less obvious, and not all pigs may demonstrate traditional clinical signs of infections. Morbidity rates can reach 100 per cent, while mortality rates are generally low. Sow herds with endemic SIV will have sporadic abortions (due to high fever) and low conception rates that are erroneously recorded as "return to heat" events that may be first and second-trimester abortions. Aborting sows are usually off feed for 1 to 3 days and have a fever of up to 105°F (41°C). Secondary bacterial infections can exacerbate the clinical signs. Virus excretions can occur, for example, by coughing, sneezing, and nasal discharges. SIV is a key respiratory pathogen contributing to morbidity, mortality and economic loss associated with the porcine respiratory disease complex.

At some level, there is a universal immunity against SIV against some variants in all herds. There is also the number of subtypes, antigenic clusters within subtypes and reassortants that circulate within swine populations. However, the expression varies widely. Still, Outbreaks occur, but infections are more frequently endemic, with no or mild clinical expression and melded with other concurrent respiratory infections. Pigs of all ages, from nursing pigs to sows, may be infected, with blips of clinical disease appearing at different phases of production that vary with the situation. Veterinary practitioners' and diagnosticians' tasks are to determine whether the increased clinical disease is due to the resurgence of an already circulating virus within the operation or the introduction of a different virus against partially immune animals.

Expression of disease generally falls into one of the following presentations:

1. Acute, "fulminating" SIV: There is rapid spread, high fevers, anorexia, expiratory dyspnea with effort, mortality, and a brief course in the herd/group. Mortality occurs in pigs whose lungs fill rapidly with fluid; a typical 'foam cast' forms in the bronchial tree and is expelled through the nose. There is a rapid spread within and between sites and through pigs of all ages. Even nursing pigs are severely affected, but

essentially all survive. Recovery is rapid and often complete, with very few secondarily infected or chronically affected pigs, and very high levels of specific immunity follow infection. This is rarely confounded with other agents, even though they may be present.

2. Age-associated influenza in growing pigs: Infection can be relatively predictable in specific systems depending on how they are structured. Maternal passively acquired antibody protects weaned pigs against the endemic virus, but that immunity wanes with maternal antibody decay. Usually occurring when sow herd immunity is uniform, and pigs have roughly equivalent levels of passive protection, a clinically evident infection is delayed until the pigs are exposed to a homologous or variant virus that can 'breakthrough' the collective passive immunity. The classic pattern is robust pigs with no problems through the nursery phase until they are 10–12 weeks old and in finishers, where they are exposed to other older growing pigs that are shedding the virus. Approximately 2 weeks into the finishing phase, pigs develop cough and depression that does not explode but works through the group over a protracted 2–3-week timeline as the individual passive antibody decay curves meet up with the various loads of virus that overcome it. The actual SIV-induced clinical disease expressed is usually not dramatic, of variable severity and commonly complicated by concurrent infections endemic in the group.
3. Piglet influenza: A consistent/persistent clinical expression of SIV can be found in nursing and newly weaned pigs. Again, clinically associated with passive protection variance within groups and increasing frequency of SIV variants, the infection is first appreciated as a cough in the 2–3-week-old pig about to be weaned. However, the cough is in scattered pigs in a given room (maybe 10 percent of litters have a pig that sporadically coughs) that are hard to find and almost too subtle to raise any alarm. When these pigs are moved to the nursery, the stress and activity exacerbate clinical expression. During the move and on first entry into the nursery, there is a standard description "...newly weaned pigs came off the truck coughing..." even though the farrowing management would say "...no, we did not notice any cough; maybe a pig or two". Clinically, in this situation, a subset of pigs suffers fever, anorexia, and cough due to SIV infection after a couple of days. Again, clinical disease severity greatly depends on the level and specificity of passive protection. Affected pigs are anorexic during the critical transition phase to solid food, *i.e.*, they are sick for 36 to 48 h after the move and become very hard to start on feed. When just a few pigs in a large group are affected, the infection probably goes unrecognized. However, on many farms, a consistent 5–10 percent or greater are affected, and the problem is economically substantial.

These situations are the purpose of sow immunization, not to protect the sow but to extend the passive shield until pigs are well-started on feed and can handle the infection.

Immune Responses

The absence of clinical signs in most SIV infections suggests that manyinfluenza infections remain subclinical, resulting in immunity for the rest of the pig's life span or might reflect protection by maternal antibodies during early infection. Both innate and adaptive immune responses of pigs to influenza virus contribute to better control of these infections. Alveolar macrophages have been found indispensable in controlling the influenza viruses in pigs. Regarding adaptive immunity, humoral and cellular immune responses are crucial in pigs' inflammatory process against infection. Cross-protection against some SIV subtypes has been observed. Also, it has been noticed that post-infection immunity to H1N1 or H3N2 viruses provides cross-protection against H1N2 infection. However, vaccinating sows to increase piglet protection against SIV infection has yet to be widely accepted. Influenza type A viruses can escape from pre-existing immunity because of the frequent genetic changes in the HA domain.

Diagnosis

In swine, the influenza virus can be detected by different methods. The methods used include typical clinical signs, characteristic pathological lesions, virus isolation, membrane antigen detection, nucleic acid detection {Reverse transcription-polymerase chain reaction (RT-PCR)}, antigen detection in tissues by immunohistochemistry or frozen section fluorescent antibody staining, in-situ hybridization and serology. The most common problem of diagnosing influenza virus in swine remains the timing of sample selection. Sample collection must be based on the differentiation of acutely infected pigs producing large amounts of virus from subacute pigs producing little or no virus. In most experimental studies, virus clearance was highly rapid. SIV could not be isolated from lungs or other respiratory tract tissues on or after day 7. Influenza-specific antibodies have been detected in the serum and nasal swabs as early as 3- and 4 days post-inoculation (DPI), respectively.

Influenza virus titer peaks at 48 hpi, and little virus is shed 6 to 8 days after infection. Identifying acutely febrile pigs (12 to 72 hpi) is the best predictor of virus shedding. Usually, the peak fever is at 24 hpi. Fever from secondary infections may run concurrently or start up at 4 days after the influenza fever has subsided, emphasizing the importance of identifying pigs in the first few days after the onset of infection. Coughing can persist for 1 to 2 weeks, and the absence of fever would be a poor indicator of virus shedding. The best source of virus for detection or culture is bronchial swabs from post-mortem tissues. Nasal swabs are generally less sensitive and would be better evaluated by PCR or virus isolation. When collecting tissues for examination from an affected swine herd, a wide range of samples from several affected pigs in the group should be collected because agents and lesions might be limited to a few animals.

Influenza virus can also be detected in post-mortem lung samples by immunohistochemistry and frozen section fluorescent antibody testing. Bronchial epithelial necrosis characteristic of influenza virus infection will begin 12 to 24 h post-infection, with epithelial regeneration beginning 5 days later. Airways are not simultaneously infected, so that different lung sections may have concurrent necrosis, regeneration or no lesions. Complete resolution of lung injury with normal particular clearance may take 28 days. Influenza virus infection by itself causes little permanent lung damage. The sensitivity of influenza virus detection in frozen or fixed lung samples is strongly influenced by sample size, the number of lung sections examined, and the timing of tissue collection. Different diagnostic methods have been employed for detecting SIV antigens in formalin-fixed paraffin-embedded tissues by monoclonal and polyclonal antibody-based immunohistochemical procedures.

Virus Isolation

The virus can be isolated by inoculating fertilized chicken eggs or cell cultures with nasal or pharyngeal mucus samples collected by swabbing the nasal passage or throat. In veterinary diagnostic laboratories, the detection of type A SIV infection has been routinely carried out by virus isolation in embryonated chicken eggs or Madin-Darby canine kidney (MDCK) cells with subsequent subtype determination by HI and neuraminidase inhibition tests using monospecific antiserum to each subtype. SIV grows readily when injected into embryonated chicken eggs (9-12) days. It may be inoculated by either the intrallantoic or intramniotic routes. The temperature of incubation ranges from 33 to 37°C. The embryos usually do not die, and the virus is demonstrated after 72-96 hours of incubation by haemagglutination tests on the allantoic and amniotic fluids. The cytopathic effect of the virus can be observed in cell culture, and allantoic fluid or cell lysates can be tested for their ability to agglutinate chicken red blood cells, presumptive evidence for the presence of an influenza virus.

Various cell culture systems have been used to grow and assay SIV. These include calf kidney cells, fetal pig lung cells, canine kidney cells, pig kidney cells, chicken embryo fibroblasts, human diploid cells and Chang conjunctival cells. Other cell culture systems include a pig oviduct cell line and a swine testicle cell line. Fetal pig trachea, lung, and nasal epithelial organ cultures and tracheal organ cultures from chickens, horses and ferrets also support the growth of SIV. The primary epithelial cells of human adenoid and primary rhesus monkey kidney tissue support the replication of influenza viruses. Cell lines such as Vero, MRC-5, Madin-Darby canine kidney (MDCK), and baby hamster kidney (BHK) also support the growth of influenza viruses. However,

MDCK is the best for isolating and propagating influenza viruses in these cell lines. A porcine lung epithelial cell line called St. Jude porcine lung (SJPL) cells has been established, which is equally good as MDCK. The haemagglutinin and neuraminidase subtypes are determined by the hemagglutination-inhibition (HI) and neuraminidase-inhibition (NI) assays.

Commercially available antigen capture ELISAs can also detect type A influenza viruses. However, to increase the accuracy of serodiagnosis, the virus isolates representing currently circulating field strains should be incorporated. ELISA performed with recombinant HA protein is also helpful in this respect. Recently, a rapid microchip-based electrophoretic immunoassay has been developed to detect swine influenza virus.

Haemagglutination Inhibition Test

The HI assay is the most common serologic assay for diagnosing swine influenza. One of its advantages is that it can discriminate between different subtypes and antigenic variants within a subtype. It also has several disadvantages; however, it is relatively insensitive and may not detect a response if the infecting strain is antigenically different from the virus used in the assay. Thus, infections with new subtypes or antigenic variants within the subtype will not be detected. For this reason, the virus strains used in the assay must be updated regularly. Furthermore, it should be conducted on paired sera collected 10 – 21 days apart. A fourfold or more significant increase in titer between the first and the second sample suggests a recent SIV infection. In addition, the assay is very laborious, needs many standardized reagents, and is consequently prone to variation.

Immunohistochemistry

The cell culture method is more sensitive but requires a viable virus and several days to observe cytopathic effects. The isolation of the virus can be difficult as pigs typically clear the virus from the respiratory tract within 7 days of infection. Other methods used to detect a virus or viral components include immunofluorescence of lung tissue, nasal epithelial cells or bronchioalveolar lavage contents, immunohistochemistry of fixed tissue samples, enzyme-linked immunosorbent assay (ELISA), polymerase chain reaction (PCR), cell culture and immunoperoxidase staining for determination of virus type and subtype, and a rapid enzyme-immunoassay membrane test. Diagnostic tests for detecting nucleoprotein antigens, such as immunofluorescent staining and enzyme-linked immunosorbent assays, have also been used. A FAT was the standard method for detecting SIV infections for many years. IHC test works well on formalin-fixed lung and has become the first choice in many veterinary diagnostic laboratories in which it is available. It should be performed within approximately one week of infection because the virus clears. A monoclonal antibody against the type A nucleoprotein will detect H1 and H3 viruses.

PCR/RT PCR Reverse-Transcription-Polymerase Chain Reaction

The polymerase chain reaction (PCR) molecular tool efficiently detects SIV nucleic acid in clinical samples. Employing a set of primers specific to the HA subtype, subtyping of viruses can also be achieved by reverse transcription-PCR (RT-PCR). Real-time RT-PCR assays have outplayed most other detection techniques as they are fast, reliable and helpful in screening numerous nasal swab specimens. Multiplex RT-PCR (mRT-PCR) assays can also be employed for differential diagnosis and detecting and identifying subtypes commonly isolated from avian, swine, and human influenza. However, due to the zoonotic nature of SIVs, the potentially infectious specimens must be handled carefully to avoid transmission of viruses to humans. The RT-PCR is more rapid and sensitive than traditional virus isolation methods, embryonating chicken eggs and haemagglutinin assays to detect swine influenza virus. The universal primer has been developed to systematically detect and identify 18HA and 11NA subtypes of influenza A viruses by RT-PCR method. Using PCR to detect SIV in nasal swab specimens was described. This method detects lower amounts of extracted virus from experimental specimens than those detectable by infectivity assays in embryonated chicken eggs. Sensitivity has been increased several times by reamplification with internal primers of primarily amplified PCR products. The use of two multiplex RT-PCR assays that can identify and subtype H1 and H3 or N1 and N2 directly from clinical specimens has been reported. RT-PCR reactions could also determine the HA subtyping of influenza viruses, each using a set of primers specific to one HA subtype.

Real-Time-Polymerase Chain Reaction

Real-time polymerase chain reaction (PCR) is more rapid and sensitive than traditional techniques, including virus isolation by cell culture. In real-time PCR, the concentration of PCR product is measured as the reaction proceeds, enabling the determination of the viral load within a sample. A Real-time RT-PCR assay based on the influenza virus matrix gene was developed to detect type A influenza virus.

In-situ Hybridization

In-situ hybridization (ISH) is a valuable tool in addition to standard RNA extraction techniques for evaluating gene expression in tissue and cells. Its major advantage is determining which tissues or cells in a mixed population express the RNA of interest. The use of the ISH technique has been successful in detecting SIV nucleic acids with DNA probes complementary to viral RNA. Jung *et al.* (2002) demonstrated and localized the SIV in naturally infected formalin fixed paraffin embedded tissue specimens of pigs by ISH using a non-radioactive digoxigenin labelled cDNA probe and by IHC using an influenza virus H1N1 specific monoclonal antibody. Diagnosis of the disease by virological and serological methods is complicated in piglets with maternal antibodies. Virus recovery and the severity of signs of disease are inversely related to the level of maternal antibody, and, depending on this level, it may be tough to isolate the virus. Maternal antibodies also inhibit active antibody production by the piglets, so the serological response may not be detected.

Differential Diagnosis

Other agents that may cause respiratory disease in pigs include porcine reproductive and respiratory syndrome virus, Aujeszky's disease virus, *Actinobacillus pleuropneumoniae* and Mycoplasma pneumonia. Most of these agents show other signs that do not resemble swine influenza. The acute form of *Actinobacillus pleuropneumoniae* infection can show clinical signs similar to SI, such as dyspnoea, tachypnoea, abdominal breathing, coughing, fever, depression and anorexia. SIV is usually associated with *Mycoplasma hyopneumoniae, Salmonella cholera suis*, PRRSV, porcine circovirus II (PCV-II) and several secondary bacteria. The greatest mortality has been reported on co-infection with the above respiratory and multi-systemic pathogens.

REFERENCES

Chrun, T., Maze, E.A., Vatzia, E., Martini, V., Paudyal, B., Edmans, M.D., McNee, A., Manjegowda, T., Salguero, F.J., Wanasen, N., Koonpaew, S., Graham, S.P. and Tchilian, E. (2021). Simultaneous infection with porcine reproductive and respiratory syndrome and influenza viruses abrogates clinical protection induced by live attenuated porcine reproductive and respiratory syndrome vaccination. Front Immunol, 12: 758368. doi: 10.3389/fimmu.2021.758368.

Du, N., Yang, X.X., Wang, M., Lan, Y., Yang, L., Cheng, Y.H., Liu, L.Q., Chen, Y.K., Guo, Y.J., Li, D.X. and Shu, Y.L. (2009). Review on the etiological property of the swine influenza virus. Chinese J Virol. 25 Suppl: 39-47. PMID: 20361600.

Janke, B.H. (2014). Influenza A virus infections in swine: pathogenesis and diagnosis. Vet Pathol, 51(2): 410-426. doi: 10.1177/0300985813513043.

Janke, B.H. (2013). Clinicopathological features of Swine influenza. Curr Top Microbiol Immunol, 370: 69-83. doi: 10.1007/82_2013_308.

Nagarajan, K., Saikumar, G., Arya, R.S., Gupta, A., Somvanshi, R. and Pattnaik, B. (2010). Influenza A H1N1 virus in Indian pigs and its genetic relatedness with pandemic human influenza A 2009 H1N1. Indian J Med Res, 132: 160-167. PMID: 20716816.

Yazawa, S., Okada, M., Ono, M., Fujii, S., Okuda, Y., Shibata, I. and Kida, H. (2004). Experimental dual infection of pigs with an H1N1 swine influenza virus (A/Sw/Hok/2/81) and *Mycoplasma hyopneumoniae*. Vet Microbiol, 98(3-4): 221-228. doi: 10.1016/j.vetmic.2003.11.005.

Chapter 3

Lumpy Skin Disease

G.B. Manjunatha Reddy*, P. Sai Mounica, H.B. Chetan Kumar and N. Shivasharanappa

ICAR-National Institute of Veterinary Epidemiology and Disease Informatics (NIVEDI), Bengaluru, Karnataka, India

****e-mail: gbmpatho@gmail.com***

ABSTRACT

In recent times, a pox viral bovine disease known as Lumpy Skin Disease has caused an unprecedented amount of economic loss and animal deaths in South Asian nations, including India. Though the disease was restricted to African countries initially, it has rapidly shifted its epidemiological trends and raised concerns globally. The disease is caused by the Lumpy Skin Disease virus of the genus Capripoxvirus and is mainly transmitted through mechanical vectors such as biting flies, mosquitoes, ticks, and transmission via movement of the infected animals. The disease is characterised by the development of peculiar skin lesions all over the body, biphasic fever, and enlargement of lymph nodes. LSD causes huge economic loss by a drop in milk production, decreased body condition, reduced hide value, infertility etc. posing a serious threat to the stakeholders and the livestock industry. Prevention and control strategies are the most important factors to limit the impact caused by lumpy skin disease. Hence, this chapter provides an overall review of Lumpy skin disease, its etiology, epidemiology, pathogenesis, clinical signs, diagnostics, prevention and control measures.

***Keywords**: Lumpy skin disease, India, Epidemiology, Diagnosis, Prevention and control.*

Introduction

Lumpy skin disease (LSD) is a vector-borne pox viral disease that is characterized by the appearance of skin nodules and the natural hosts for this virus are considered to be cattle and Asian water buffaloes (Elhaig *et al.*, 2017). LSD has been classified as an important notifiable disease considering the potential for international spread and economic impacts by the World Organization for Animal Health (OIE, 2020). Due to decreased milk production, reduced draught power, skin damage, trade restrictions, loss of body condition, abortions, infertility, and the cost of veterinary care, LSD causes a significant negative economic impact on the livestock industry. The disease was discovered for the first time in Zambia in 1929 and since then it has rapidly spread from its origin in central Africa to the Middle East, Europe, and Asia. In India, the disease was first reported in the year 2019 in Odisha state. During the same period, the surrounding countries of India *viz.*,Bangladesh, China, Nepal, Bhutan, Myanmar, Sri Lanka, and Pakistan have also reported LSD outbreaks. The whole population of cattle (192.49 million) and buffaloes (109.85 million) in India are at risk of getting infected with this dreadful disease.

Etiology

LSD is caused by the Lumpy skin disease virus (LSDV), which belongs to the genus *Capripoxvirus* under the family *Poxviridae*. It is a brick-shaped, enveloped virus consisting of double-stranded DNA (151 kb) with complex symmetry. The average size of the virus is 294±20 nm in length and 262 ±22 nm in width (Kitching and Smale, 1986). There is only one serotype of LSDV and it is closely related to sheep pox virus (SPPV) and goat pox virus (GTPV). However, they are phylogenetically distinct from each other. All capripoxviruses grow slowly on cell cultures and may require several passages. They can be propagated on a variety of cells of bovine and ovine origin, causing easily recognizable cytopathic effects (CPE). The replication of LSDV occurs in the cytoplasm of the host cell causing intracytoplasmic eosinophilic inclusion bodies. In *Capripoxvirus*es, the antibodies produced against one member are serologically cross-reactive with the other. Hence, the differentiation among the members of the genus Capripoxvirus is possible through molecular techniques like PCR and sequencing. The virus is susceptible to sunlight and detergents containing lipid solvents, but in dark environmental conditions, such as contaminated animal sheds, it can persist for many months (OIE, 2014). The virus can survive for a long period under cold conditions and can be recovered from the skin nodules kept at -80°C for 10 years (Ahmed *et al.*, 2020). As per the current information available, there is no evidence regarding the difference in virulence of different LSDV strains.

Epidemiology

The initial description of the LSD clinical indications was made from Zambia, in 1929 (Morris, 1931). In South African countries, there were persistent reports of disease up until 1949 which resulted in significant economic losses (Von Backstrom, 1945; Diesel, 1949). The disease was first appeared in East Africa in 1957 and West Africa in 1974 (Davies, 1991). In a succession of epizootics that were documented, the disease has continued to spread across the majority of the African continent. LSD is now present throughout the Africa except Libya, Algeria, Morocco, and Tunisia (Tuppurainen and Oura, 2012). Outside of the sub-Saharan African region, Egypt (1988), Kuwait (1991), Yemen (1995), United Arab Emirates (2000), Bahrain (2003), and Oman (2010) had all reported outbreaks of LSD (Tageldin *et al.*, 2014). In India, the first outbreak of lumpy skin was reported in the Mayurbhanj and Bhadrak Districts of Odisha in the month of August 2019 (Sudhakar *et al.*, 2020).

Typically, in endemic countries, LSD outbreaks occur in epidemics several years apart. It is unknown how and where the virus survives between epidemics, or if it has any particular reservoirs. LSD outbreaks are caused by an increase in the number of naive animals, the abundance of active blood-feeding vectors, and uncontrolled animal movement. The first outbreak typically results from the movement of the host (cattle/buffaloes) either legally or illegally in an unaffected area. Whereas, generally, local spread is caused by vectors, while long distance jumps are caused by host movement. The majorities of the outbreaks are seasonal and coincide with high vector populations during wet, humid conditions. However, because no season is necessarily vector-free in many affected regions, the disease could occur at any time. The morbidity (10-20 per cent) is high and mortality is very low (1-5 per cent). Infected animal eventually clears the infection and there is no scientific evidence for the establishment of carrier status in the recovered animals. Even though LSD is reported mainly in cattle, the disease has been reported in Asian water buffaloes and other species (Manjunatha Reddy *et al.*, 2023; Reddy *et al.*, 2024). *Bos taurus* is more susceptible to clinical disease than *Bos indicus* (OIE, 2018). However, even within the herd maintaining the same breed, the clinical manifestation varies from subclinical to fatal disease (Carn and Kitching, 1995). Although sheep and goats co-exist with cattle and buffalo, the role of small ruminants as reservoir of LSDV has not yet been confirmed and not implicated as reservoir of LSDV. The role of wildlife in the epidemiology of LSDV requires further investigation. LSDV is not known to infect humans (OIE, 2018).

Risk Factors

The occurrence of the LSD is influenced by various perilous factors which are host, environment and pathogen associated (Table 3.1). Importantly, the risk factors such as agro-climatic conditions, communal grazing or water bodies and unrestricted movement of the animals across different borders, facilitate the spread of outbreaks in various localities. The high incidence of the disease was recorded during wet seasons especially when biting fly populations are abundant (Gari *et al.*, 2010). Necrotic skin and air-dried hides are the potential source of virus since the virus persists for months at ambient temperature.

Imports of any animal product like meat, milk, or hides originating from LSD affected countries have potential to spread in LSD free countries (Kiplagat *et al.*, 2020). During the infective phase, virus secretes in nasal, lachrymal, pharyngeal secretions, semen, milk and blood may act as source of infection to others animals (Gumbe, 2018).

Table 3.1: Various Risk Factors and their Effect on of the Lumpy Skin Disease

Risk Variables	*Effect*	
Species	Cattle are more susceptible than buffaloes	HOST
Sex	Both sexes are equally susceptible	
Age	Very young and old animal are more susceptible than adults	
Breed	The cross breeds are more susceptible than native Indian breeds however the native breeds are more reactive	
Serotype	Only one serotype and not much variation in pathogenicity between strains	PATHOGEN
Stability/viability of virus	Stable and viable in both dry and cold conditions.In biological samples the virus viability varies (Blood: 5-16 days, Skin nodule: 39 days, Scabs: several years and semen: 22-42 days)	
Temperature	LSDV can survive both cold and warm conditions. But, inactivated by heat easily	ENVIRONMENT
pH	Sensitive to both acidic and alkaline pH	
Season	Incidence is more in wet and humid seasons due to abundance of vector populationIncidence is less during the dry and cold seasons and more during migratory season and livestock fairs or sales or ritual/festivals	

Transmission

The transmission of the LSDV occurs through vector and movement of the infected animals. Transmission through direct contact and fomites is rare (Ali *et al.*, 2012). The following are the possible routes of transmission of LSDV:

- ☆ Vector-borne: Blood sucking/biting of vectors - Stable fly (*Stomoxys calcitrans*), *Culicoides* midges, mosquitoes (*Culex and Aedes*) and ticks (*Rhipicephalus, Amblyomma*).
- ☆ Contaminated feed/fodder and water with saliva, nasal and ocular discharges.
- ☆ Direct contact with infected animals.
- ☆ Vertical transmission: Contaminated semen
- ☆ Intrauterine and Trans-mammary

☆ The long-distance jump or transmission of virus is mainly due to movement host and rarely due to transport vehicle is carrying the vectors and short distance jump or transmission due to vectors and other modes of transmission.

Indian Scenario of LSD

The disease was first reported in the eastern part of the country in 2019 during the monsoon season. Since its first report, due to uncontrolled movement of the animals, common grazing and water bodies, poor animal health and preventive measures, the disease soon spread throughout India and adversely affected the livelihood of the marginal and small-scale farmers (Kumar and Tripati, 2022). The major events of the disease outbreaks in India occurred in three waves. During the first wave (2019-2020), the first outbreak was reported in the Mayurbhanj district, Odisha in 2019, where the morbidity rate was higher and mortality was very low. The disease was mild and mostly dermal form was reported. Soon, it spread to other neighbouring states and in the second wave (2020-2021), most of the southern states were affected along with other central states with more deaths reported from Maharashtra compared to other affected states including Madhya Pradesh and Uttar Pradesh. In contrast, the disease became more virulent in the third wave (2021–2023), especially the respiratory form was more prevalent along with the dermal form of the disease. Particularly in the northern states of India, where more of stray or owned cattle populations are present have reported the maximum mortality rate. The mortality was also attributable to co-morbidities in these stray and unowned or even in animal shelters which are mainly due to managemental practices.

Economic Impact

In India, livestock farming is the mainstay of the landless, small and marginal farmers and it contributes approximately 4 percent to the national GDP. India is also the home for largest bovine population in the world with 192.49 million cattle and 109.85 million buffaloes (DAHD, 2019). LSD impacts cattle and buffalo production by causing substantial economic losses. Different stakeholders being affected are animal breeders, dairy farmers, agriculturists (drought power), milk and meat processing and selling industries, wholesale and retail dairy and meat sellers, organic farming community, cattle transport and feed manufacturers *etc.* Among the stakeholders, the highest impact will be on landless or marginal poor, small scale and backyard farmers whose entire earning depends solely on animal farming alone. LSD causes huge economic losses by direct and indirect causes. The economic loses are attributed to decreased milk production, decreased body condition, damage to hide/skin, drought power loss in bullocks, fertility problems and abortions, emaciation, veterinary treatment, secondary complication like maggot wound, mastitis, temporary or permanent blindness. Moreover, the disease also causes mortality upto 5-15 percent of affected animals. Other costs include, national cost of surveillance and control measures like awareness campaigns, monitoring and surveillance program, compensation program for culled or slaughtered animals, cleaning and disinfection of affected farms, sampling and laboratory testing, purchase of vaccines and implantation of a vaccination program, vector control, restriction on movement of live animals and products and carcass disposal. The traders will be affected due to ban on export of live animals and animal products in and out of country.

Pathogenesis

The LSDV which is an epitheliotropic virus enters the host body via skin or the mucosa of the gastro intestinal tract (Sanz-Bernardo *et al.*, 2020). The virus when it enters the local lymph nodes, the primary multiplication of virus causes lymphadenitis and viraemia after the initial febrile reaction which persists for two weeks. Then virus spreads through monocytes and localizes in the skin and develops into inflammatory nodules as a result of its rapid multiplication in the cells. LSDV replicates inside the host cells such as macrophages, fibroblasts, pericytes and endothelial cell in the lymphatics and walls of blood vessels leading to vasculitis and lymphangitis, while thrombosis and infarction may develop in severe cases. Virus multiplication in cells produces hyperplasia and ballooning degeneration of keratinocytes, formation of epidermal micro vesicles, inflammatory cell infiltration into dermis (Manjunathareddy *et al.*, 2024). Four to seven days after infection coalescence of epidermal micro-vesicles occur into large vesicle and ulceration of tissue develop (Namazi and Tafti, 2021).

Clinical Signs

The incubation period may vary between 2-5 weeks in natural condition, whereas, 4-14 days in experimental condition. The clinical sings include nasal and ocular discharge, biphasic fever, brisket oedema, upper limb oedema, drastic reduction in milk production and characteristic nodular skin lesions of 1-5cm in diameter. Initially, subscapular and pre-femoral lymphnodes become enlarged and are easily palpable. High fever which persists for a week usually precedes the appearance of skin lesions and may be accompanied by a sharp drop in milk yield. In Mild clinical cases, the animal may show a few nodules whereas, in severe cases, the entire body may be covered with skin lesions (Figure 3.1a). They are most commonly found at head, neck, udder, genitalia (Figure 3.1b), perineum and lateral parts of ribcages. These nodules turn necrotic and form a recognizable "Sit-fast" which is the sequestrum of necrotic material over the course of the illness, which takes about two to three weeks. Pneumonia and mastitis due to secondary bacterial infection are the common complications. Sometimes ulcerative lesions may also appear in the cornea of the one or both eyes, leading to blindness in some cases. Silent or subclinical infections can occur up to one third of the infected cattle which do not show any clinical sings although all might have viraemia.

Figure 3.1a: Mid-Stage of LSD Affected Animal Showing Various Sized Nodules all over the Body.

Figure 3.1b: Late Stage of LSD Affected Animals Showing LSD Lesions.

The morbidity and mortality depend on the control measures adopted after detection of first case. During the typical course of disease, cattle die several weeks after infection. Thus, the outbreak is likely to have been going on for a minimum of 2-3weeks and probably longer. Severe cases of LSD are highly characteristic and easy to diagnose. However early stages of infection and mild cases may be difficult to distinguish even for the experienced veterinarians.

Malnourished animals, lactating cows, and young calves appear to develop more severe disease, which may be related to a compromised humoral immunity (Namazi and Tafti, 2021). In most animals that recover from clinical illness, a lifelong cell-mediated immunity is developed. Calves that are born from infected cows acquire maternal antibodies that could give them a six-month clinical disease protection.

Diagnosis

The early diagnosis of disease is very important for initiating treatment, control and preventive measures. The early and mid-stage of LSD is difficult to diagnose by clinical sings. However, the late stage of disease is comparatively easy for clinician to diagnose at field level as characteristic clinical signs are exhibited by the affected animal. Hence, quick and early laboratory diagnosis is very important for initiating early treatment and to differentiate from other similar diseases. The details of suitable samples for laboratory diagnosis are provided in table depending on stage of disease (Table 3.2). There are different testes used for virus/antigen and antibody detection in various laboratories depending on the availability of facilities.

Table 3.2: Details of Various Clinical Samples to be Sent for Laboratory Diagnosis

<table>
<tr><th rowspan="2">Sample Type</th><th colspan="3">Clinical Stage of Disease</th><th rowspan="2"></th></tr>
<tr><th>Early</th><th>Mid</th><th>Late</th></tr>
<tr><td>Swabs (Nasal/lacrimal)</td><td></td><td></td><td></td><td rowspan="8">All the samples should be shifted to lab at the quickest time possible at 2-8° C temperature</td></tr>
<tr><td>Blood</td><td></td><td></td><td>-</td></tr>
<tr><td>Skin biopsy</td><td></td><td></td><td></td></tr>
<tr><td>Scab</td><td>-</td><td></td><td></td></tr>
<tr><td>Serum</td><td>-</td><td></td><td></td></tr>
<tr><td>Milk</td><td>-</td><td></td><td></td></tr>
<tr><td>Rectal swabs/faecal</td><td>-</td><td></td><td></td></tr>
<tr><td>Vectors</td><td></td><td></td><td></td></tr>
</table>

Gross and Histopathological Findings

The postmortem examination of the dead animals reveal the nodular skin lesions grossly all over the body depending on the severity of the disease. Internally, the respiratory system will show necrotic pock lesions in the upper respiratory tract and lungs shows typical "gunshot wounds" as seen in case of sheep and goat pox. Liver may show necrotic foci and in rare cases the digestive system may show pock lesions. Histopathology of skin shows varied degree of mononuclear cell infiltration especially dendritic cells, lymphocytes and macrophages in dermis and hypodermis. Even though there are reports of presence of intra-cytoplasmic inclusion bodies in LSD affected organs, we did not observe any inclusions till date which is very peculiar finding especially with respect to LSDV Indian strains.

Virus isolation and characterization: Virus isolation from clinical samples is critical and gold standard for diagnosing LSD. The LSDV can be grown on a number of primary cells or cell lines of bovine, ovine, or caprine origin. Chorioallantoic membrane of embryonated chicken eggs and African green monkey kidney (Vero) cells are the most commonly used for the virus isolation. The virus grows slowly in cell cultures, and produce specific cytopathic effect (CPE) and intracytoplasmic inclusion bodies five to seven days after inoculation (OIE, 2021).

Electron microscopy: Transmission electron microscopy~~ie~~ (TEM) can be used for detection of virus directly from clinical sample. Diagnosis can be confirmed within a few hours of receipt of specimens. Mature capripox virions have an average size 320 x 260 nm and show more of oval profile and larger lateral bodies than orthopox virions (OIE, 2021).

Molecular detection methods: The molecular testing is critical for monitoring the spread of these viruses and develop the strategies to control disease outbreaks. With the advancement of molecular techniques like PCR, the diagnosis of LSD has become easy, less time consuming and specific. The confirmation of LSDV may be done by using Capripoxvirus specific PCR approach on conventional (Zheng *et al.*, 2007) and real-time PCR technologies (Bowden *et al.*, 2008). For differentiating virulent LSDV from the vaccine strain, Restriction Fragment Length Polymorphism (RFLP) has also been used (Menasherow *et al.*, 2014). Molecular assays employing loop-mediated isothermal amplification to identify capripoxvirus genomes have been shown to have sensitivity and specificity comparable to real-time PCR, with a simpler approach and a cheaper cost.

Serological tests: The Virus neutralization test (VNT)is the only approved serological test for monitoring the LSDV antibodies (Milovanoviæ *et al.*, 2019). Besides, indirect fluorescent antibody testing (IFAT), enzyme-linked immunosorbent assays (ELISA), and Western blotting are used in different research laboratories (Namazi and Tafti, 2021).Neutralizing antibodies develop 3-4 days after the onset of clinical symptoms and reach maximal titer levels in 2-3 weeks. The agar gel immuno diffusion test (AGID) and IFAT are less specific than VNT due to cross-reactivity with antibodies to other poxviruses. Western blotting is sensitive and specific, but it is difficult and expensive to perform.

Differential Diagnosis

Lumpy skin disease in severe conditions, the clinical signs are highly characteristic, whereas in mild condition and early stages, it is quite difficult to differentiate from other similar diseases. The differential diagnosis of LSD includes pseudo-LSD, pseudo-cowpox, foot and mouth disease, cowpox, bovine popular stomatitis, infectious bovine rhinotracheitis, dermatophilosis, dermatophytosis, bovine farcy, photosensitisation, actinomycosis, actinobacillosis, urticaria, insect bites, besnoitiosis, nocardiasis, demodicosis, onchocerciasis, bovine viral diarrhoea, and malignant catarrhal fever (OIE, 2018). In case, if homologous vaccination is practiced the vaccine induced mild disease should be considered for differential diagnosis.

Treatment

There is no specific anti-viral treatment is available to treat Lumpy Skin Disease. However, the disease can be managed by providing the supportive care. The following protocols may be followed in case of LSD outbreak.

1. Separation and isolation of suspected/affected animals.
2. Symptomatic treatment of affected animals. The following table may be taken for preparing the treatment schedule based on the experience and availability of medicine in the locality.
3. Feeding management: Providing easily digestible preferably liquid feed/food and succulent green fodder.

4. Proven or indigenous technology knowledge based herbal preparations may be used for wound and dermatitis treatment, as immunomodulators and antioxidants.
5. Application of fly repellents.

Immunization

As the LSDV is quite stable and can remain for months, a long-term vaccination regimen should be developed and made mandatory to control LSD covering the entire bovine population. A live attenuated vaccine has been developed and used in endemic country to control the disease. It is known that the members of the capripoxvirus provide cross-protection. Hence, homologous (Neethling LSDV strain) and heterologous (sheeppox or goatpox virus) live attenuated vaccines are recommended to immunize and protect the cattle against LSD (OIE, 2013). Most of the vaccine manufacturer in African countries use Neethling strain virus for production of LSD vaccine. This LSD vaccine is attenuated by culturing sixty times in lamb kidney cells followed by sub-culturing twenty times in chorio-allantoic membrane of embryonated chicken eggs. Dose of the vaccine varies between manufacturers, generally from $10^{2.5}$ to 10^4 $TCID_{50}$/dose and provides immunity up to three years. Attenuated LSDV vaccinations may produce minor adverse reaction known as the "Neethling response". Besides, use of other capripox virus vaccines in bovines is also recommended due to cross protective nature of vaccine. SPPV vaccines at a higher dose (three, five and ten-fold) have been used in cattle against LSDV in those regions where LSD and SPP are both present. Kenyan or Romanian sheep pox virus strains have been used to confer LSD immunity in bovines. These live vaccines could potentially serve as a source of infection for the susceptible sheep and goat population, hence they are not recommended in nations free of sheep and goat pox. In India, as per the guidelines of Department of Animal Husbandry and Dairying, Ministry of Fisheries, immunization with Uttarkashi strain of goat pox virus is used at $10^{3.0}$ $TCID_{50}$ per animal subcutaneously. Recently, India has also developed homologous LSD vaccine from local strain named Lumpi-ProVac^{Ind}. The vaccination plan should be chosen depending on the needs of the country and the resources that are available. Generally, animals that have been newly introduced to a farm should be immunized and at the age of three or four months, the calves which are born from the cowsthat received a vaccine or have recovered from a natural infection need to be immunized. Vaccines can be administered annually to all susceptible cattle and buffaloes especially in high risk groups and areas.

Prevention and Control

Like several other viral diseases of livestock, the LSDV can be effectively controlled by implanting the following measures.

1. Animal movement restriction: The animal movement in or out from infective zone should be banned. If possible, the live animal markets, fairs, gatherings, migration *etc.*, should be stopped.
2. Preventive vaccination: The ring vaccination (5km radius) should be carried in all (cattle and buffaloes) healthy animals which are aged above 4 months as per the GoI guidelines with goat pox vaccine ($10^{3.0}$ $TCID_{50}$).
3. Bio-security measures: The effective containment of the disease measures such as quarantine, isolation, disinfection of premises, deworming, disposal of farm waste and carcasses, people movement, insemination, health records *etc.*, should be regularly practiced and monitored.
4. Vector Control: The measures should be adopted for control of biting or blood feeding vectors through application of insecticide or repellents and also through reducing the breeding grounds for these vectors in farm premises.
5. Cleaning and disinfection: The affected areas including the farm equipment and fomites should be cleaned and disinfected with appropriate chemicals/disinfectants like ether (20 per cent), chloroform, formalin (1 per cent), phenol (2 percent for 15 minutes), sodium hypochlorite (2-3 per cent), iodine compounds (1:33 dilution) and quaternary ammonium compounds (0.5 per cent).
6. Creating mass awareness among different stake holders through radio, TV, press releases, posters, newspapers/print media, digital media, social media posts, webinars, workshops *etc.*, for better execution of control measures.
7. Preparation of standard operating procedures (SOPs) at state/regional level.
8. Prepare and train the rapid response teams.
9. Effective networking with line departments for better and quick response.
10. Monitoring and surveillance especially in unaffected areas.

Conclusion

India being the home for the world's largest bovine population is greatly affected by the ongoing havoc of lumpy skin disease outbreaks affecting the livelihood of the small and marginal farmers. The disease is an important vector-borne transboundary poxviral disease affecting cattle and buffalo populations. The rapid transboundary spread of the virus and increase in the virulence of the virus leading to rise in mortality rate implies the need to better understand the epidemiology, diagnostics and preventive measures to control the losses incurred by the lumpy skin disease.

REFERENCES

Ahmed, N., Doley, S., Barlaskar, S. A., Nath, A. J., and Yadav, S. N. (2020). Lumpy skin disease: an emerging bovine viral infection in India. Indian J Anim Hlth, 59(2): 137-142.

Bowden, T. R., Babiuk, S. L., Parkyn, G. R., Copps, J. S., and Boyle, D. B. (2008). Capripoxvirus tissue tropism and shedding: A quantitative study in experimentally infected sheep and goats. Virology, 371: 380-393.

Das, A., Babiuk, S. and McIntosh, M.T. (2012). Development of a loop-mediated isothermal amplification assay for rapid detection of capripoxviruses. J Clin Microbiol, 50(5): 1613-1620.

Davies, F.G. (1991). Lumpy skin disease, an African capripox virus disease of cattle. British Vet J, 147(6): 489-503.

Diesel, A.M. (1949). The epizootology of "lumpy skin disease" in South Africa.

Elhaig, M.M., Selim, A. and Mahmoud, M, (2017). Lumpy skin disease in cattle: Frequency of occurrence in a dairy farm and a preliminary assessment of its possible impact on Egyptian buffaloes. Onderstepoort J Vet Res, 84(1): 1-6.

Gari, G., Waret-Szkuta, A., Grosbois, V., Jacquiet, P. and Roger, F. (2010). Risk factors associated with observed clinical lumpy skin disease in Ethiopia. Epidemiol Infect, 138(11): 1657-1666.

Gumbe, A.A.F. (2018). Review on lumpy skin disease and its economic impacts in Ethiopia. J Dairy Vet Anim Res, 7(2): 39-46.

Kiplagat, S.K., Kitala, P.M., Onono, J.O., Beard, P.M. and Lyons, N.A. (2020). Risk factors for outbreaks of lumpy skin disease and the economic impact in cattle farms of Nakuru County, Kenya. Front Vet Sci, 7: 259.

Kitching, R.P. and Smale, C. (1986). Comparison of the external dimensions of capripoxvirus isolates. Res Vet Sci, 41(3): 425-427.

Kumar, N. and Tripathi, B.N. (2022). A serious skin virus epidemic sweeping through the Indian subcontinent is a threat to the livelihood of farmers. Virulence, 13(1): 1943-1944.

Manjunathareddy, G.B., Saminathan, M., Sanjeevakumar, L., Rao, S., Dinesh, M., Dhama, K., Singh, K.P., and Tripathi, B.N. (2024). Pathological, immunological and molecular epidemiological analysis of lumpy skin disease virus in Indian cattle during a high-mortality epidemic. Veterinary Quarterly, 44(1):1-22.

Manjunatha Reddy, G.B., Pabbineedi, S.M., Nagaraj, S., Bijalwan, S., Tadakod, S., Bhutia, Z., Palmu, D., Rai, S., Bhutia, P.D., Bhutia, P.T. and Shenga, E. (2023). Lumpy skin disease (LSD) in yak (Bos grunniens): an evidence of species spillover from cattle in India. Microorganisms, 11(12): 2823.

Menasherow, S., Rubinstein-Giuni, M., Kovtunenko, A., Eyngor, Y., Fridgut, O., Rotenberg, D., Khinich, Y., and Stram, Y. (2014). Development of an assay to differentiate between virulent and vaccine strains of lumpy skin disease virus (LSDV). J Virol Methods, 199: 95-101.

Milovanoviæ, M., Dietze, K., Miliæeviæ, V., Radojièiæ, S., Valèiæ, M., Moritz, T. and Hoffmann, B. (2019). Humoral immune response to repeated lumpy skin disease virus vaccination and performance of serological tests. BMC Vet Res, 15(1): 1-9.

Morris, J.P.A. (1930). Pseudo-urticaria. Northern Rhodesia. Dept Anim Health Ann Rpt, 12.

Namazi, F. and KhodakaramTafti, A. (2021). Lumpy skin disease, an emerging transboundary viral disease: A review. Vet Med Sci, 7(3): 888-896.

OIE, Office International des Epizooties. (2017). Manual of diagnostic tests and vaccines for terrestrial animals 2017. Chapter 2.4.13 Lumpy skin disease. Retrieved from http://www.oie.int/fileadmin/Home/eng/Health standards/tahm/2.04.13_LSD.pdf

OIE, Office International des Epizooties. (2017). Manual of diagnostic tests and vaccines for terrestrial animals 2017. Chapter 2.4.13 Lumpy skin disease. Retrieved from http://www.oie.int/fileadmin/Home/eng/Health_standards/tahm/2.04.13_LSD.pdf.

OIE. (2013). World Organization for Animal Health. Lumpy Skin Disease. Technical Disease Card.

Reddy, G.B.M., Mounica, P.S., Sudeep, N., Vikram, R., Garam, G.B., Lalzampuia, H., Ragulraj, S., Pal, S., Khate, K., Bijalwan, S., and Girish, P.S., (2024). First evidence of lumpy skin disease in mithun (Bos frontalis) in India. Archives of Virology, 169(3): 65.

Sanz-Bernardo, B., Haga, I.R., Wijesiriwardana, N., Hawes, P.C., Simpson, J., Morrison, L.R., MacIntyre, N., Brocchi, E., Atkinson, J., Haegeman, A. and De Clercq, K. (2020). Lumpy skin disease is characterized by severe multifocal dermatitis with necrotizing fibrinoid vasculitis following experimental infection. Vet Pathol, 57(3): 388-396.

Sudhakar, S.B., Mishra, N., Kalaiyarasu, S., Jhade, S.K. and Singh, V.P. (2022). Genetic and phylogenetic analysis of lumpy skin disease viruses (LSDV) isolated from the first and subsequent field outbreaks in India during 2019 reveals close proximity with unique signatures of historical Kenyan NI2490/Kenya/KSGPlike field strains. Transbound Emerg Dis, 69(4): e451-e462.

Sudhakar, S.B., Mishra, N., Kalaiyarasu, S., Jhade, S.K., Hemadri, D., Sood, R., Bal, G.C., Nayak, M.K., Pradhan, S.K. and Singh, V.P. (2020). Lumpy skin disease (LSD) outbreaks in cattle in Odisha state, India in August 2019: Epidemiological features and molecular studies. Transbound Emerg Dis, 67(6): 2408-2422.

Tageldin, M.H., Wallace, D.B., Gerdes, G.H., Putterill, J.F., Greyling, R.R., Phosiwa, M.N., Al Busaidy, R.M. and Al Ismaaily, S.I. (2014). Lumpy skin disease of cattle: an emerging problem in the Sultanate of Oman. Trop Animal Hlth Prod, 46(1): 241-246.

Tuppurainen, E., Alexandrov, T. and Beltrán-Alcrudo, D. (2017). Lumpy skin disease field manual – A manual for veterinarians. FAO Animal Production and Health Manual No. 20. Rome. Food and Agriculture Organization of the United Nations (FAO). 60 pages.

Tuppurainen, E.S.M. and Oura, C.A.L. (2012). lumpy skin disease: an emerging threat to Europe, the Middle East and Asia. Transbound Emerg Dis, 59(1): 40-48.

Von Backstrom, U. (1945). Ngamiland cattle disease: preliminary report on a new disease, the etiological agent being probably of an infectious nature. J South African Vet Assoc, 16(1): 29-35.

World Organisation for Animal Health OIE, (2020). OIE-Listed diseases, infections and infestations in force in 2020. Available from https://www.oie.int/animal-health-in-the-world/oie-listed-diseases-2020/. Accessed on 17 November 2020.

Zheng, M., Liu, Q., Jin, N. Y., Guo, J. G., Huang, X., Li, H. M., Zhu, W. and Xiong, Y. (2007). A duplex PCR assay for simultaneous detection and differentiation of Capripoxvirus and Orf virus. Mol. Cell. Probes, 21: 276-281.

Chapter 4

Contagious Ecthyma

A. Anand Kumar and N. Syaamasundar*

College of Veterinary Science, Sri Venkateswara Veterinary University, Tirupati – 517 502, Andhra Pradesh, India
****e-mail: 7aakumar@gmail.com***

ABSTRACT

Contagious ecthyma is highly contagious, zoonotic, viral skin disease caused by orf virus (ORFV). This disease is geographically widespread where ever sheep and goat raised and has economic significance. Virus can be transmitted through direct contact or indirectly (fomites). As the virus is epitheliotropic, it enters through abraded skin and replicates in regenerating epidermal keratocytes and seats its infection and progress to different stages viz; vesicles, pustules and scabs. This virus also induce apoptosis of APC's through CD95 pathway by inhibiting Bcl – 2 (B cell lymphoma 2) and inhibit apoptosis of virally infected cells. Clincal signs like anorexia, starvation due to presence of painful lesions on skin and in the oral cavity, udder and teats will be observed. Lesions on foot lead to transient lameness and sometimes complicated with secondary bacterial infections like footrot. Microscopically; intra epidermal micro-abscesses with vacuolation and swelling of keratocytes in stratum spinosum and intracytoplasmic eosinophilic inclusion bodies will be observed. Diagnosis of the disease is based on the clinical signs, gross and histopathological lesions. Other methods like electron microscopy, tissue culture, serological and molecular diagnosis can be carried out for confirmation. The limitations in diagnosis can sometimes result as the virus being identified initially as orf virus, but later reassigned to another species of parapox virus. Treatment for contagious ecthyma includes primarily segregation/ isolation of the effected animals from the healthy herd. Symptomatic supportive therapy and diet should be provided. Preventive measures like vaccinating the herd with live vaccine in the locations of past history out breaks or incidence and freshly vaccinated animals should be isolated from unvaccinated animals until the vaccination scabs have fallen off.

***Keywords**: Apoptosis, Keratocytes, ORFV, Stratum spinosum, Vescicles.*

Synonyms: Orf, Sore mouth, Scabby mouth.

Orf virus infection in animals is commonly referred to as sore mouth, scabby mouth, or contagious ecthyma (Jubb *et al.*, 2015; Vegad, 2019). Animals infected with orf virus typically develop scabby lesions around lips, muzzle, and in mouth. Lesions usually involve the lips but can involve other cutaneous junctions such as the coronet, and can extend onto the face and ears as well. Its highly contagious zoonotic pox viral disease (Jubb *et al.*, 2015) and human beings are infected and typically develop ulcerative lesions or nodules on their hands.

Contagious ecthyma is a disease of sheep, goats (Vegad, 2019; Venkatesan *et al.*, 2022) and some other domesticated and wild animals (Jubb *et al.*, 2015). The incidental infection reported in humans, camelids, cattle, many wild ruminants and rarely dogs, cats. This disease has also been observed in other ungulates including alpacas, reindeer, Japanese serorws, musk oxen, bighorn sheep, deer, elk and it is suspected to occur in somewild chamois (Quinn *et al.*, 2011; Schoell, 2015; Dalton *et al.*, 2021; Al-Bayati, 2022). Rabbits and mice could be infected experimentally by some groups, but not others; their susceptibility might be influenced by factors such as the viral strain or dose (Quinn *et al.*, 2011). In humans, however, it usually presents on the dorsal hands or fingers at areas of contact with the infected animals (Santiago *et al.*, 2019). *Orf* virus is notorious for jumping hosts to human populations after close contact with a zoonotic source with active lesions or via fomites (Al-Bayati, 2022).

The synonyms of this disease are; Orf, Infectious Labial Dermatitis, Contagious Pustular Dermatitis, Contagious Pustular Stomatitis, Sore Mouth, Scabby Mouth (Jubb *et al.*, 2015; Vegad, 2019). It is highly contagious, zoonotic, viral skin disease (Jubb *et al.*, 2015) caused by the orf virus (ORFV). The parapox virus that causes contagious ecthyma is related to the viruses that cause pseudo cowpox and bovine papular stomatitis. The virus is transmitted by contact. Enveloped virus family Pox viridae; a member of the genus *Para pox virus* in the subfamily Chordopox virinaea (Upton *et al.*, 2003; Delhon *et al.*, 2004) and virions are brick shaped with surface composed of globular protein and spinal filament having complex symmetry which replicate in cytoplasm, stable in environment and resistant to glycerol, ether (Quinn *et al.*, 2011).

This disease is geographically widespread where ever sheep and goat raised and has economic significance where loos of condition, animal neither graze or suckle with morbidity of 90 percent (susceptible population) and mortality of <1 percent or rarely high in immunosuppressed or stressed or secondary bacterial infection intervenes like *Fusobacterium necrophorum* and *Deratophilus congolensis* or Screwworm fly (*Cochliomyia hominivorax*) (Jubb *et al.*, 2015).

Contagious ecthyma (Orf) virus can be transmitted through direct contact or indirectly (fomites). Transmission through saliva of effected animals having oral lesions is not confirmed. Clinically normal sheep will carry the virus on skin and scabs that might have entered through cuts and abrasions. The virus is also viable on wool and hide of sheep for 1 month approximately even after development/healed up lesions. Virus has been recovered from dried crust after survival for 12 years in laboratory. Humans get affected by contact with infected animals or while handling the live vaccines containing virus. Direct contact with active lesions from affected animals or contact with fomites can infect humans (Santiago *et al.*, 2019). Dogs may acquire infection by eating infected lamb carcasses/cadavers.

The incubation period of this virus is very short and is around 2 to 3 days. The viral virulence gene; Vascular Endothelial Growth Factor (VEGF) plays important role in the pathogenesis of Orf (Jubb *et al.*, 2015). As the virus is epitheliotropic, it enters through abraded skin and replicates in regenerating epidermal keratocytes. There the virus seats its infection followed by edema and granulomatous inflammation of dermal cells which may lead to papules formation and further progress to different stages viz; vesicles, pustules and scabs. Healing will be completed by 4 weeks. This virus also induce apoptosis of APC's through CD95 pathway by inhibiting Bcl – 2 (B cell lymphoma 2) and inhibit apoptosis of virally infected cells (Jubb *et al.*, 2015). The effected animals exhibit the symptoms depending on the virulence of virus/severity of the infection. There will be hyperaemia, small pustules around mouth, muzzle with extensive proliferative and exudative lesions. The scabs that involve mucosa as well as skin with initial papule –vesicle – pustule stages may be observed on muzzle, sometimes eyelids, ears, feet, perineal region (Vegad, 2019).

Clincal signs like anorexia, starvation due to presence of painful lesions on skin and in the oral cavity will be observed. Nursing lambs transmit virus to their dam and lesions will be produced on udder and teats. Lesions on foot leads to transient lameness and sometimes complicated with secondary bacterial infections like foot rot (*Dermatophilus congolenses*) and maggot infestation and predispose to granulation, inflammation, ulceration of skin, bacterial mastitis and gangrenous mastitis. Lesions on the udder may result in the abandonment of offspring and, in rare cases; the lesions may extend into the internal organs. Uncomplicated cases may recover in one or two months. The mortality rate is generally low, deaths can occur fromsequelae such as secondary infections or failure to nurse (Jubb *et al.*, 2015; Vegad, 2019).

Gross lesions will be seen at commissures of lips which spread around lip margins (Figures 4.1a-c) to muzzle and sometimes on face and around eye. In severe cases, lesions on gingiva, dental pad, palate and tongue and rarely lesions may extend to oesophagus, rumen, omasum, lungs and heart. Lesions on limbs are less common that involve coronet, interdigital cleft and bulb of heels. Lesions may also be present in the inner thigh, axilla and edge of wound in earmarked lambs or tail dock sites where skin with less wool is present. Multifocal severe proliferative dermatitis with peripheral lymphadenopathy of the draining lymph nodes in the areas of affected skin (Jubb *et al.*, 2015).

suppurative arthritis, chronic fibrinous pneumonia and premature thymic involution may also be noted (Vegad, 2019).

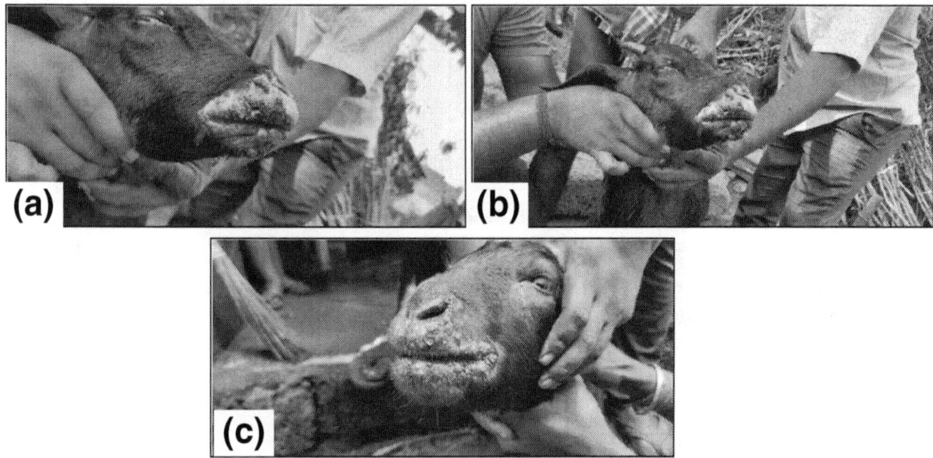

Figure 4.1a-c: Scabs Around the Upper and Lower Lips and Commissure of Goats of different Localities Affected with Contagious Ecthyma.

Microscopically; in the epidermis- marked proliferation of cells, intra epidermal micro-abscesses with vacuolation and swelling of keratocytes in stratum spinosum will be seen. Reticular degeneration, accumulation of scale crust and intracytoplasmic eosinophilic inclusion bodies will be observed (Jubb *et al.,* 2015). Spongiform degeneration in follicular structures and polymorphic infiltrate can be present in the lower dermis (Mirande *et al.,* 1975).

Diagnosis of the disease is based on the clinical signs of papules, vesicles, pustules on muzzle and mouth, gross and histopathological lesions. A punch/ shave biopsy is diagnostic and the gold standard (Groves *et al.,* 1991). Although not always readily available, a PCR assay for orf viral infection is highly sensitive in humans. This assay can be the noninvasive investigation of choice if the diagnosis is in doubt (Thompson *et al.,* 2022). Other diagnostic tools used include cell culture, electron microscopy, virus neutralization, immunohistochemistry, immunofluorescence, enzyme-linked immunosorbent assay, loop-mediated isothermal amplification, and recombinase polymerase amplification (Gill *et al.,* 1990). The limitations in diagnosis can sometimes result as the virus being identified initially as orf virus, but later reassigned to another species of parapox virus.

Cotagious ecthyma has to be differentiated from; Ulcerative dermatosis where matured animals more affected as there is more tissue destruction and crateriform ulcers will be observed, in Sheep pox there will not be downward growth of basal layer of epidermis into dermis, in Blue tongue- transmission is vector borne, in Staphylococcal dermatitis/folliculitis the skin of muzzle and surrounding eyes are affected (Delhon *et al.,* 2004).

Treatment for contagious ecthyma includes primarily segregation/isolation of the effected animals from the healthy herd and providing liquid diets with the help of tubes orally to overcome the deaths due to inanition and may be supported with antibiotics to control. In secondary infections; the antiseptic lotions fly repellents and larvicides can be used to clean the wounds and arrest of invasion of the wounds by flies and subsequently deposition of larvae can be prevented.

In mild cases, treatment may not be required. Softening/mild emollient ointments may be helpful in more severe cases. It is important to make sure that affected animals are eating and drinking. Ointments containing zinc oxide, boric acid or sulphur may be used.

A course of antimicrobials is indicated in very severe cases which are rare as the disease is mild and self-limitingone. Gruel feeding/molasses mixed soft diet may be indicated. In very severely affected animals where food intake is affected badly parenteral fluids such as dextrose, dextrose saline is indicated.

Preventive measures for control of contagious ecthyma includes: restriction of entry of ailing animals in to uninfected herd and quarantine measures to be adopted for new animals as some carriers may not have clinical signs but may introduce the virus. Precautions like disinfection of moving areas, equipment and other fomites should be taken up to prevent virus introduction and soft liquid diets and fodder to be provided to reduce the risk of cuts in the oral cavity both externally and internally. Vaccinating the herd with live vaccine in the locations of past history out breaks or incidence and freshly vaccinated animals should be isolatedfrom unvaccinated animals until the vaccination scabs have fallen off. Small amount of live vaccine brushed over light scarification of skin, usually on inside of thigh or below elbow or caudal fold. Lamb of one month old vaccinated with a booster at one to two months later should be practised in endemic areas (Inoshima *et al.*, 2000).

The economic impact of contagious ecthyma is significant as severe generalized infections have also been described occasionally. Carrier animals may be a factor in flock or herd outbreaks. Contagious ecthyma is a zoonotic disease, and human-to-human transmission can also occur. Lesions in humans are extremely painful and may last as long as 6 weeks (Schoell, 2015). Most infections in humans are localized and heal spontaneously; however, large, poorly healing lesions can occur in people who are immune-suppressed. Mild transitory pustular lesions on forearm, when they come in contact with infected sheep. In slaughter house workers may result due to handling of wool and skin. Typical lesions occur at the site of infection like abrasion infected while handling diseased sheep or accidental means when vaccinating. Lesions are very itchy and respond poorly to treatment. This has to be differentiated from Milker's nodule as they arise from cattle and resolve within 6 weeks, the abattoir worker mainly affected with cutaneous form of anthrax and has to be differentiated (Diven, 2001). The 2 most important differential diagnoses to rule out include those that share zoonotic-exposure history with farm animals and exhibit cutaneous lesions.

Paravaccinia (Milker nodules) is a similar parapoxvirus infection from cows rather than goats or sheep, as in orf. Clinically, the lesions of orf and Milker nodules are indistinguishable and typically resolve spontaneously within 6 weeks (Pang *et al.*, 2023). Cutaneous anthrax is a bacterial infection from *Bacillus anthracis*. An eventual "black eschar" evolution is pathognomonic for this lesion. The more infamous pulmonary symptoms can develop in the systemic spread and can be life-threatening. Spores are transmittable from infected animals, wool, and even meat. Gram stain, culture, and even PCR are all available to aid in diagnosis (Ayhan *et al.*, 2017). Other differential diagnoses include deep fungal infection, squamous cell carcinoma (especially keratoacanthoma-type), pyogenic granuloma, atypical mycobacterial infection, and neutrophilic dermatoses (Diven, 2001).

REFERENCES

Al-Bayati, H.A.M. (2022). Assessment of contagious ecthyma virus in camels of Wasit Province, Iraq. Arch Razi Inst, 77(2): 565-571. doi: 10.22092/ARI.2021.356862.1933.

Ayhan, E. and Aktas, H. (2017). Dermoscopic features and types of orf and milker's nodule. Postepy Dermatol Alergol, 34(4): 357-362. doi: 10.5114/ada.2017.69317.

Dalton, C.S., Workentine, M.L., Leclerc, L.M., Kutz, S. and van der Meer, F. (2023). Next-generation sequencing approach to investigate genome variability of Parapoxvirus in Canadian muskoxen (*Ovibos moschatus*). Infect Genet Evol, 109: 105414. doi: 10.1016/j.meegid.2023.105414.

Delhon, G., Tulman, E.R., Afonso, C.L., Lu, Z., de la Concha-Bermejillo, A., Lehmkuhl, H.D., Piccone, M.E., Kutish, G.F. and Rock, D.L. (2004). Genomes of the parapoxviruses ORF virus and bovine papular stomatitis virus. J Virol, 78(1): 168-177. doi: 10.1128/jvi.78.1.168-177.2004.

Diven, D.G. (2001). An overview of poxviruses. J Am Acad Dermatol, 44(1): 1-16. doi: 10.1067/mjd.2001.109302.

Gill, M.J., Arlette, J., Buchan, K.A. and Barber, K. (1990). Human orf. A diagnostic consideration? Arch Dermatol, 126(3): 356-358.

Groves, R.W., Wilson-Jones, E. and MacDonald, D.M. (1991). Human orf and milkers' nodule: a clinicopathologic study. J Am Acad Dermatol, 25(4): 706-711. doi: 10.1016/0190-9622(91)70257-3.

Inoshima, Y., Morooka, A. and Sentsui, H. (2000). Detection and diagnosis of parapoxvirus by the polymerase chain reaction. J Virol Methods, 84(2): 201-208.

Jubb, K.V.F., Kennedy, P.C. and Palmer, N. (2015). Pathology of Domestic Animals, 6th Edn. 2015.

Mirande, L.M, Valera, J.A., Vestfrid and M.A. (1975). ORF. Epidemiological, histopathological and clinical considerations, apropos of a case. Med Cutan Ibero Lat Am, 3(3):233-239.

Pang, F. and Long, Q. (2023). Recent advances in diagnostic approaches for orf virus. Appl Microbiol Biotechnol, 107(5-6): 1515-1523. doi: 10.1007/s00253-023-12412-8.

Quinn, P.J., Leonard, F.C. Hartigan, P., Markey, B.K. Fanning, S. and Fitzpatrick, E.S. (2011). Book on Veterinary Microbiology and Microbial Disease. 2nd Edn.

Santiago, L., Oliveira, D., Cardoso, J.C. and Figueired, A. (2019). Human Orf: An Under-recognized Entity. Acta Dermatovenerol Croat, 27(4):280-281.

Schoell, A. (2015). Laboratory Animal Medicine (3rd Edition).

Thompson, H.J., Harview, C.L., Swick, B. and Powers, J.G. (2022). Orf Virus in Humans: case series and clinical review. Cutis, 110(1): 48-52.

Upton, C., Slack, S., Hunter, A.L., Ehlers, A. and Roper, R.L. (2003). Poxvirus orthologous clusters: toward defining the minimum essential poxvirus genome. J Virol, 77(13): 7590-7600. doi: 10.1128/jvi.77.13.7590-7600.2003.

Vegad, J.L. (2019). Textbook of Veterinary Special Pathology, Infectious Diseases of Livestock and Poultry, 2nd Edn, 2019.

Venkatesan, G., Kushwaha, A., Kumar, A., Bora, D.P. and Sasikumar, P. (2022). An improved visual closed tube loop mediated isothermal amplification (LAMP) assay for rapid identification of orf virus in sheep and goats. Vet Ital, 58(2). doi: 10.12834/VetIt.2426.15340.2. PMID: 36586114.

Chapter 5

Global Scenario of Canine Parvovirus Infections in Canine: Epidemiology, Diagnostics and Immunoprophylactic Agents

S. Nandi, G.K. Sharma, S.K. Biswas, Ragini Mishra, Sushmita Nautiyal, K.G. Sai Balaji and K.P. Singh*

Centre for Animal Disease Research and Diagnosis, Indian Veterinary Research Institute, Izatnagar, U.P.

****e-mail: sukdebnandi@yahoo.in***

ABSTRACT

Canine parvovirus 2 (CPV-2) is one of the most important enteropathogen of dogs emerged in 1978 and manifested by two forms namely gastroenteritis and myocarditis. It is characterized by depression, loss of appetite, vomiting and leucopenia. CPV-2 is probably evolved from a very closely related virus in cats, feline panleukopaenia virus (FPLV) or a closely related carnivore parvovirus. It caused high morbidity of 100 percent and low mortality of 10 percent in adult dogs and 91 percent in pups. Over the years a number of variants namely CPV-2a, CPV-2b, CPV-2c, New CPV-2a, New CPV-2b and Asp 300 (2a/2b) have been reported from different countries in the world with varying degree of pathogenic potential. Although CPV-2 differs from FPV by 6 amino acids in the VP2 protein, subsequent variants differ from CPV-2 only in one or two places. Further, CPV-2 affects only dogs, new variants expanded their host range to cat as well. There are a number of different serological and molecular tests (PCR, nested PCR, SYBR Green based real time PCR, Taqman based real time PCR, Minor grove binding assay based PCR) available for prompt, specific and accurate diagnosis of the disease. Some molecular tests not only detect the CPV-2 but also identify the variant of CPV-2 involved in disease outbreak. Further, both live attenuated and inactivated vaccines are available to control the disease in animals. Besides, new generation vaccines namely recombinant vaccine, peptide vaccine and DNA vaccine have been developed for control of the disease in canines effectively and efficiently. However, new generation vaccines have not been issued license to be used in the field condition. Again, the presence of maternal antibodies often interferes with the active immunization with live attenuated vaccine and there always exists a window of susceptibility in spite of following proper immunization regimen. Lastly, judicious use of the vaccines in pet dogs, stray dogs and wild canids keeping in mind the new variants of the CPV-2 along with the proper sanitation and disinfection practices must be implemented for the successful control the disease.

Keywords: *Canine parvovirus-2 (CPV-2), CPV-2a, CPV-2b, CPV-2c, New CPV-2a, New CPV-2b, Asp300 (2a/2b), Hemorrhagic enteritis, Myocarditis, Canine, Vaccination, Feline panleukemia virus (FPV), Raccoon parvovirus (RPV), Mine enteritis virus (MEV), Blue fox parvovirus (BFPV).*

Introduction

CPV-2 is the most important enteric virus affecting domestic and wild carnivores around the world (McCaw and Hoskins, 2006). In the late 1970s, feline panleukopenia virus or related carnivore parvovirus was found to be ancestor of the CPV-2 (Decaro and Buonavoglia, 2012). There are two forms of the disease: enteritis in dogs of all ages and myocarditis in pups of 6 weeks to 6 months old with 100 percent morbidity and mortality of 10 percent in adult dogs and 91 percent in pups. The virus is very sturdy and usually transmitted by faecal oral route. The virus has the propensity to replicate in the rapidly dividing cells such as lymphoid tissues, intestinal epithelium and bone marrow as well as myocardium in neonatal pups. Vomiting, haemorrhagic diarrhoea, depression, loss of appetite, fever, dehydration are the prominent signs of the disease but myocarditis is seen mainly in neonatal pups. Due to infection of bone marrow and lymphoid tissue there is sharp drop of WBC count. Although a galaxy of immunoprophylactic agents are available to prevent the CPV-2 infections in dogs, lot of outbreaks are still reported in many countries around the world due to presence of unvaccinated dogs, interference of active immunization by maternally derived antibodies and emergence of a number of different mutants over the years. Because of sudden and widespread outbreaks of severe contagious enteritis in dogs, canine parvovirus (CPV) has become an issue of great concern to veterinarians.

Canine parvovirus2 (CPV2) is a single stranded, small, DNA virus and imposes serious health problem in the canine population worldwide (Shackelton *et al.*, 2005). It was emerged in late 70s from feline panleukopniavirus (FPLV) as a host range variant due to point mutation and had established itself as a highly contagious and infectious pathogen in canines (Burtonboy *et al.*, 1979; Appel *et al.*, 1985). The disease is characterized with acute heamorrhagic gastroenteritis, depression, diarrhoea, loss of appetite, frequent vomition, leukopenia and myocarditis (Kelly, 1979; Appel *et al.*, 1979; Decaro and Buonavoglia, 2012). The virus has wide host range and effects all age group of canids but young ones are more susceptible. Morbidity is about 100 percent but mortality is quite higher (91 per cent) in puppies compared to adults (10 per cent) (Nandi and Kumar, 2010a).

Etiology

Canine parvovirus-2 belongs to genus *Protoparvovirus* under *Parvovirinae* subfamily of the family *Parvoviridae* (Kings *et al.*, 2012). Other members of genus are feline panleukopenia virus (FPV), mink enteritis virus (MEV) and raccoon parvovirus (RPV) (Tattersall *et al.*, 2005; Decaro and Buonavoglia, 2012). All these are host variants of the carnivore parvovirus subgroup (Parrish *et al.*, 1991).

Virion Morphology

Parvoviruses are small, non-enveloped viruses of 25 nm in diameter and infect vertebrates and insects. X-ray crystallography reveals that the parvovirus capsid is made of 60 protein subunits (5-6 copies of VP-1 and 54-55 VP-2 copies) and exhibits T= 1 icosahedral symmetry (Tsao *et al.*, 1991). The surface of virus capsid is characterized with a large spike-like protrusion (22-A°) at or surrounding the icosahedral three fold, a 15-A° canyon-like depression at the fivefold and a dimple-like depression at the two fold axes. The protein residues at three fold axes are most immune-dominant in nature. (Tsao *et al.*, 1991; Agbandje *et al.*, 1993).

Genome Organization

Canine parvovirus-2 contains a linear, autonomously replicating, single stranded negative sense DNA molecule of a ~5200-nucleotides with two promoters (Murphy *et al.*, 1999). Both termini of the linear genome have palindromic sequences (about 150 bases) that can fold into self priming duplex "hairpin" like structure and used as origin of DNA replication as well as facilitate in encapsidation (packaging) of viral DNA within nascent virion (Parrish, 1999). Genome is comprised of two main open reading frame (ORF); longer ORF encodes two non-structural proteins (NS1 and NS2) while smaller one encodes two structural (VP1 and VP2) proteins through alternative splicing of the same mRNAs (Cotmore *et al.*, 1995).

Physico-chemical Properties

Parvoviral virion comprises of 70-80 percent of proteins and 20-30 percent nucleic acid. The mature virion is compact and dense structure with molecular mass ranging 5.5-6.2 X 10^6daltons, sedimentation coefficients of 110S-122S, and buoyant densities of 1.39-1.43 g/cm^3 in Cesium Chloride. All parvoviruses in concentrated suspension and in tissues are highly stable, as they are extremely resistant to pH (3-9), prolonged incubation temperature at 56°C and treatment with lipid solvents, trypsin and most disinfectants. But, in diluted solution they are metastable, undergoing an inactivating conformational transition in response to heat or denaturants. However, under natural conditions, infectious virions are exceptionally durable, surviving for weeks or months at room temperature or for several years at 4 °C. Virions can be inactivated by formalin, sodium hypochlorite, β-propiolactone, hydroxylamine, oxidizing agents and ultraviolet irradiation. Several parvoviruses are diagnosed on the basis of their agglutinating activity of RBCs of different mammals and avian species (Muzyczka and Berns, 2001). Canine parvovirus-2 and its variants have the property to agglutinate the RBCs of cat, monkey and pig.

Emergence of Canine Parvovirus and its Variants

In early 1978, world faced a pandemic in the canine population characterized by hemorrhagic gastroenteritis and high mortality, caused by a new, unknown virus designated as canine parvovirus-2 to distinguish it from the distantly related canine minute virus (also known as canine parvovirus -1). Phylogenetic studies and analysis of retrospective samples revealed that CPV emerged in early to mid 1970s, just few years before recognition. Detection of CPV-2 antibodies in the dogs sera showed the first report of positive titre in dogs of Europe or Eurasia in between 1974 and 1976. Since then, the virus has been ubiquitously distributed in different countries (Parrish, 1999). It is believed that CPV-2 possibly emerged as a host variant from feline panleukopenia virus (FPV), likely through adaptation of an FPV-like parvovirus of wild carnivores (Truyen *et al.*, 2006). Sequence analysis of FPV and CPV-2 revealed that both have 98 percent sequence homology and differ by 5 to 6 amino acids 80 (Lys to Arg), 93 (Lys to Asn), 103 (Val to Ala), 323 (Asp to Asn), 564 (Asn to Ser) and 568 (Ala to Gly) change in VP-2 domain (at threefold spike) interacting with the host-cell transferrin receptor (TfR). These non-synonymous substitutions, 93 (Lys to Asn), 103 (Val to Ala) and 323 (Asp to Asn) were sufficient to successful cross species adaptation of CPV-2 in canine host, while 80 (Lys to Arg) 564 (Asn to Ser) and 568 (Ala to Gly) substitutions deprived the CPV-2 to replicate in feline host (Chang *et al.*, 1992; Truyen and Parrish, 1992). CPV-2 and FPV also differ some other properties such as antigenicity when analyzed with monoclonal antibodies, TfR binding affinity, haemagglutination properties (Truyen *et al.*, 1995). Another important difference between these two carnivore parvoviruses is that CPV evolves much more rapidly than FPV (Battilani *et al.*, 2006) showing genomic substitution rates similar to those of RNA viruses, with values of about 10^{-4} substitutions per site per year (Shackelton *et al.*, 2005). Due to prolong active virus circulation and vaccination, host immunity pressure might have also contributed further emergence of CPV-2 variants.

Host Range, Susceptibility and Age

CPV-2 has a wide host range: *Mustelidae* (weasels, ferrets, minks, and badgers), *Canidae* (dogs, foxes, and wolves), *Procyonidae* (raccoons), and *Felidae* (cats, lions, tigers, and cheetahs). The high prevalence of anti-parvovirus antibodies in wildlife sanctuaries indicate that cross-infections can occur and may be responsible for interspecies transmission of canine and feline parvoviruses. All breeds of dog are susceptible. The crossbred dogs are less susceptible in comparison to pure bred like Rottweiler, Doberman Pinchers, German Shepherd, Labrador and English Springer Spaniels. The exception to this is Toy Poodles and Cocker Spaniels (Houston *et al.*, 1996). The CPV infection is more severe in young puppies especially those younger than 3 months of age (Appel *et al.*, 1979, Jacob *et al.*, 1980).

Epidemiology of CPV-2

CPV-2 was reported in May, 1978 after an epizootic of gastroenteritis and myocarditis occurring worldwide in pups (Kelly, 1979; Appel *et al.*, 1979; Burtonboy*et al.*, 1979), and in 1979, CPV-2a which subsequently becomes predominant subtype of CPV-2 worldwide was identified. In 1984 another variant, CPV-2b was emerged in canine population (Parrish *et al.*, 1985). Later, in 2000 an antigenic type CPV-2c was reported in Italy (Buonavoglia *et al.*, 2001). Presently original antigenic type CPV-2 has been completely replaced by its antigenic variants. Molecular epidemiology of CPV revealed that different geographical locations of the world are facing canine parvovirus enteritis outbreaks caused by CPV-2 variants in recent years. In Europe all strains with wide spread presence of CPV-2c, have been reported from Germany (Decaro *et al.*, 2012), Poland (Majer-Dziedzic *et al.*, 2011), and Portugal (Miranda and Parrish, 2016). Unlike other European country, in Greece, Hungary, and Bulgaria, CPV-2a remains most common variant with sporadic presence of CPV-2b and CPV-2c (Filipov *et al.*, 2016). South American countries like Ecuador (Aldaz *et al.*, 2013), Brazil (Pinto *et al.*, 2010), and Argentina (Calderón *et al.*, 2015) have reported presence of all antigenic variants but CPV-2c is most dominant strain. North American continent like in USA and Canada all antigenic variants are circulating but CPV-2b is dominant one while the current dominant viral variant in Mexico is CPV-2c. In Africa presence of both CPV-2a and 2b in South Africa, only CPV-2a in Nigeria and in Tunisia all type of antigenic variants were reported. New Zealand had reported only CPV-2a where as in Australia CPV-2a was the most prevalent strain with sporadic presence of CPV-2b. In Asia, China and Taiwan have reported all type of CPV-2 variants, while South Korea, Japan, Thailand, Iraq and Turkey reported both CPV-2a and CPV-2b strains.

In India canine parvovirus-2 was first reported by Ramadass *et al.* (1982). Different canine parvovirus variants (CPV-2a, CPV-2b, new CPV-2a and new CPV-2b) have been isolated from dogs of almost all the states of India (Chinchkar *et al.*, 2006; Kumar and Nandi, 2010). Only single report was documented about the presence of CPV-2c strain (Nandi *et al.*, 2010). Presently new CPV -2a has surpassed the earlier circulating strain CPV-2b or new CPV-2b and established as a predominant strain of CPV-2 in the canine population of India (Srinivas *et al.*, 2013; Mittal *et al.*, 2014).

Clinical Signs

Canine parvoviral enteritis is most dangerous and contagious viral disease that affects unprotected dogs (Nandi and Kumar, 2010a). It has 3-7 days of incubation period and followed by clinical outcomes. The clinical outcomes can be classified in two forms (i) acute gastroenteritis form and (ii) cardiac form. The acute gastroenteritis form affects all age groups of dog but it is more severe in pups. Early symptoms are depression, loss of appetite, vomition, high fever, severe diarrhoea and pain. There is slight rise of temperature in the initial stage of the disease but gradually turn to subnormal level with advancement of vomition and diarrhoea (Kramer *et al.*, 1980). There is no consistent character of the stool, it may be watery, yellowish or tinged with frank blood in severe cases. Due to excessive loss of fluid through vomition and diarrhoea, dogs become dehydrated leading to death within 3-4 days after onset of clinical symptoms. Puppies may die suddenly due to shock as early as 2 days of illness. Leukepenia is the constant finding with white blood cell (WBC) counts dropping below 2000–3000 cells/ml of blood. Subclinical and in-apparent infections are frequently detected, mainly in pups with intermediate MDA titers and in adult dogs (Decaro *et al.*, 2005a). The mortality rates can be high (up to 91 per cent) in pups, but are usually less than 10 percent in adult dogs (Nandi and Kumar, 2010a).

Pathology

CPV-2 mainly infects and replicates in rapidly diving cells such as hematopoietic and intestinal tissues of young and adult animals but virus can spread all tissue including brain (Elia *et al.*, 2007; Decaro *et al.*, 2009a). The main route of canine parvovirus infection is via fecal/oral route and after penetration virus replicates in oropharyngeal lymphoid tissue followed by development of viremia. Virus has more affinity toward rapidly dividing cells of the gastrointestinal tract, thymus, lymph nodes, bone marrow and leading to extensive cellular damage. Affected gastrointestinal tissues include the epithelium of the tongue, oral cavity, esophagus, intestinal tract and especially the germinal epithelial cells of the intestinal crypts. Malabsorption and increased intestinal permeability lead to diarrhoea. Infection of leukocytes, mainly circulating and tissue-associated lymphocytes, induce acute lymphopenia (often associated with neutropenia) (Pollock, 1982). Infection of the dam by CPV-2 variants early in gestation can lead to infertility and resorption, or abortion of puppies. Puppies that are infected *in-utero* or in the age of 2–3-weeks, develop a fatal myocarditis and congestive heart failure. Damage to the developing myocardium usually occurs up to the first 2 weeks of life, but clinical signs of myocardial damage may be delayed until up to 2 months of age. However, this form is no longer observed as almost all young pups are protected by maternally derived antibodies (MDA). Generalized infection has been reported in neonatal puppies, with hemorrhage and necrosis within the brain, liver, lungs, kidneys, lymphoid tissues, and gastrointestinal tract. Neurologic signs in puppies with parvoviral enteritis may result from hypoxia secondary to myocarditis, hypoglycemia, or intracranial thrombosis or hemorrhage. The possibility of co-infection with CDV should also be considered. Hemorrhagic enteritis of the small intestine and enlargement of mesenteric lymph nodes and Peyer's patches are the main gross lesions observed in dogs that die as a consequence of CPV infection. Histopathologically, the small intestine is affected by multifocal cryptic necrosis and intranuclear inclusion bodies, whereas extensive depletion of lymphocytes is seen in Peyer's patches, lymph nodes, spleen and thymus.

Diagnosis CPV-2 Infection

The rapid and accurate diagnosis of CPV-2 infection in dogs is of paramount importance to isolate them from shelter or kennel, curtail the secondary infection and stop chances of further spread of the disease to other susceptible dogs. Merely, clinical signs are not decisive to diagnose CPV-2 infection as other viral infections such as coronaviruses, adenoviruses, morbilliviruses, rotaviruses, reoviruses, noroviruses show same types of clinical signs (Hirasawa *et al.*, 1994). Therefore, a suspected case of CPV-2 infection must be confirmed through laboratory tests. There are several types of laboratory tests which mainly rely on the faecal samples collected from suspected dog. In late stage of infection, whole blood have been proven more suitable for detection of CPV-2 as there is long lasting viremia in affected dog (Decaro *et al.*, 2005a,b). There are broadly two types of diagnostics approaches for CPV-2 and its variants infection; namely (1) conventional and serological and (2) molecular tests.

Conventional and Serological Methods

Conventional methods detect viral antigens or whole virus in clinical samples. These are electron microscopy (Teramoto *et al.*, 1984), virus isolation (Mochizuki *et al.*, 1993), latex agglutination (Veijalainen *et al.*, 1986), hemagglutination (Teramoto *et al.*, 1984; Mochizuki *et al.*, 1993), and enzyme-linked immunosorbent assay (Teramoto *et al.*, 1984; Drane *et al.*, 1994) and immunochromatographic test. Virus isolation (VI) requires cell culture facility with expertise and knowledge in cell culture. In addition to this, VI is cumbersome and requires long incubation time approximately 5-10 days. Moreover, cell lines show negligible CPE following CPV infection and other techniques such as immunoflourescence or HA test are needed to confirm the presence of virus.

Electron Microscopy

During severe illness, parvoviral virions or parvovirus-like morphology can be readily indentified in faeces by negative staining and use of electron microscopy (Burtonboy *et al.*, 1979; McCandlish *et al.*, 1981).

Virus Isolation (VI)

A number of primary cell cultures and cell lines like MDCK (Madin-Darby

Canine Kidney) or CRFK (Crandall Resse Feline Kidney) were reported that support the replication of canine parvovirus-2 and used for virus isolation from suspected clinical samples. A new cell line named A-72 established from canine fibroma, can be used for isolation of virus particularly from field samples. Generally for virus isolation from clinical samples required 1-3 blind passages each of 5-10 days. Except A-72 cell line, CPE in other susceptible cell lines are not appreciable, so for the confirmation of CPV-2 in cell culture supernatant, HA or IF techniques is used.

Antigen Detection

Viral antigens can be also detected by means of immunohistochemistry carried out on intestinal, brain or tongue sections. The test adopted by field veterinarians as well as dog owners for the diagnosis of CPV infection is immunochromatographic test (IC) (Esfandiari and Klingeborn, 2000). Although, it is simple, rapid and easy to use but require large amount of virus antigen to produce appreciable band. ELISA techniques utilizes specific monoclonal antibodies coated on the plastic, nitrocellulose membranes for the detection of viral antigen (Waner *et al.*, 1996). The SNAP Parvo test from IDEXX Laboratories is a faecal CPV antigen ELISA, which is popular and used worldwide as a rapid in house test. Although it is highly specific, but very variable data on its sensitivity 81.8 percent (Marcovich *et al.*, 2012), 56.2 percent (Desario *et al.*, 2005) and 18.4 percent (Schmitz *et al.*, 2009) has been reported.

Slide Agglutination Test (SAT) and Slide Inhibition Test (SIT)

SAT and SIT are the test used for detection of CPV antigen in feces and antibody in clinical serum respectively (Marulappa *et al.*, 2009). In SAT presence of antigen is demonstrated with hemagglutination activity of RBCs, while in SIT presence of antibody in serum is detected by hemagglutination inhibition of hyper-immune sera of CPV-2b. Results revealed that SAT is comparatively less sensitive than standard HA test while in SIT, titers of hyper–immune serum is much higher for homologous CPV-2b isolate than for heterologous CPV-2c isolates.

Molecular Diagnostic Test

Polymerase Chain Reaction (PCR)

PCR has been considered as the most reliable and widely applied diagnostic technique because of high degree of sensitivity and specificity in detecting CPV from faecal samples (Schunck *et al.*, 1995; Murphy *et al.*, 1999; Decaro*et al.*, 2005). PCR has been found to detect fewer number of virion particles than that of HA, EM and ELISA from faecal samples. The PCR assay combined with the gel filtration was able to detect the virus from faeces containing as little as 10^3 PFU/g fresh faeces, whereas the virus was detected by either the ELISA test or the culture method only from faeces containing more than 10^6 PFU/g fresh faeces (Uwatoko *et al.*, 1995). To increase the sensitivity and specificity of the reaction, the nested PCR has been employed (Hirasawa *et al.*, 1994). The conventional PCR is 100 times less sensitive compared to nested PCR.

Real Time PCR

Real-time quantitative PCR is based on continuous optical monitoring of a fluorogenic PCR reaction (Holland *et al.*, 1991; Heid *et al.*, 1996). It is more sensitive, rapid and reproducible compared to conventional PCR assays and allowing a precise DNA quantitation even when a high sample throughput is screened (Aberham *et al.*, 2001; Gruber *et al.*, 2001). Real time PCR is run in 96 well format and most of the steps are automated. Due to use of inexpensive procedure for preparation of DNA by simple boiling of faecal homogenates, the total time required for analysis of 20-30 samples was about 6 hrs and 3 hrs for conventional and real-time PCR, respectively (Decaro *et al.*, 2005d). Decaro *et al.* (2005) designed a simple TaqMan probe based real time PCR assay for the detection and quantification of canine parvovirus on the basis of VP-2 DNA from dog faeces. It was found that the TaqMan CPV-2 real time assay is more sensitive than both HA and conventional PCR and being able to detect as few as 10^2 copies of CPV-2 DNA.

Nucleic Acid Sequencing

The PCR product as it is or cloned in the suitable cloning vector can be sequenced using the suitable primer with the help of automated DNA sequencer for typing of CPV strains. The sequence is analyzed using the appropriate software. This is an important technique to know with certainty the particular variant of the CPV present in the field sample. Both the nucleotide and amino acid sequence data can also be used to know the percent homology and for phylogenetic analysis of CPV-2 isolates from different geographical regions (Nandi *et al.*, 2009).

Isothermal Amplication of DNA

Insulated Isothermal PCR Method

An insulated isothermal PCR (iiPCR) method was developed for on-site detection of all circulating CPV-2 strains and FPV (Wilkes *et al.*, 2015). Originally the iiPCR technology was developed by Tsai *et al.* (2012). It utilizes a hydrolysis probe (similar to Taq-Man based qPCR) and runs in a commercially available device, POCKITTM Nucleic Acid Analyzer (POCKITTM; GeneReach USA, Lexington, MA, USA). Limit of detection, reaction sensitivity and performance of CPV-2 iiPCR test was determined and compared with in house real time PCR by testing of serial dilution of CPV-2b strain and 100 canine fecal samples. The 95 percent limit of detection of the iiPCR method was 13 copies of standard DNA and detection limits for CPV-2b DNA were equivalent for iiPCR and real-time PCR. The iiPCR have 98.41 percent sensitivity and 100 percent specificity to detect CPV-2 in feces compare to real time PCR.

Loop Mediated Isothermal Amplification Assay (LAMP)

LAMP assay is a novel, highly sensitive, rapid and reliable technique for the detection of genomic material of different microorganisms. The assay was reported by Notomi *et al.* (2000). The whole procedure generally require two sets of primers, simple boiling on water bath at 60-65°C and approximately 1 hour of incubation. The end-product can be visualized as a white precipitate of magnesium pyrophosphate, colour change of different dye (SYBR green-1, Calcein) under UV-rays or visible light and specific ladder pattern in gel electrophoresis. The established LAMP assay targeting VP-2 gene of CPV-2 has 100 percent relative sensitivity and 76.9 percent relative specificity than conventional PCR (Cho *et al.*, 2006). The detection limit of the LAMP method was 10^{-1} median tissue culture infective doses (TCID50)/ml, compared with 10 $TCID_{50}$/ml for PCR analysis (Cho *et al.*, 2006).

Characterization and Typing of Canine Parvovirus

Since the first emergence of canine parvovirus-2 in 1978, different antigenic variants of canine parvovirus-2 have been reported and they have replaced completely the original canine parvovirus-2 (old type) (Decaro *et al.*, 2005). These antigenic variants namely (CPV-2a, CPV-2b, CPV-2c, new CPV 2a and new CPV2b) differ for the presence of one or two single nucleotide polymorphisms (SNPs) in the sequence of the VP-2 gene. SNPs at positions 4062 and 4064 of the viral genome determines the presence of a specific amino acid; Asn, Asp, and Glu at residue 426 of the capsid protein in types 2a, 2b and 2c, respectively (Parrish *et al.*, 1991; Buonavoglia *et al.*, 2001). PCR-RE analysis based characterization of different strains employed amplification of VP1/VP2 gene (2.2 kb) and its RE digestion with *Hpa*I and *Rsa*I, can differentiate between original CPV-2 and CPV-2a/2b. Kumar *et al.* (2004) used *Alu* I to differentiate CPV-2a and CPV-2b. The typing of field samples using PCR followed by *Rsa*I based PCR-RE mapping showed that the vaccine strain used in India are CPV-2 type while field isolates are either of CPV-2a/2b (Savi *et al.*, 2009). *Mbo*II restriction enzyme was used efficiently to discrimination between CPV2b and CPV2c (Buonavoglia *et al.*, 2001). However, the nucleotide sequence analysis of VP2 gene would provide the accurate typing of different mutants of CPV.

Vaccines and Immunity

Vaccines are most effective and cost effective means to control any viral diseases. Effective vaccines are available for the prevention of CPV-2 infections. Both modified live (ML) and inactivated parvovirus vaccines are available to protect dogs against CPV infections Attenuated strains of CPV have been developed by repeated passage of the viruses in cell culture. The vaccine viruses replicate mildly in the intestine and virus is shed at much lower titres in the faeces. Experimentally live virus vaccines have been shown to protect dogs for at least 2 years or longer (Carmichael *et al.*, 1993) and in exceptional cases for 9 years and more (Schultz, 2006).

A single dose of ML virus when administered at 16 weeks or older could provide long-term immunity in large number of animals thus increasing the herd immunity. Inactivated vaccines however, provide only several month protection in vaccinated dogs (Carmichael *et al.*, 1993; Schultz, 2006). During the first few weeks pups are protected from CPV infection by maternally derived antibodies. The duration of immunity depends on amounts of colostrum consumed by pups during first 2-3 days of life. The maternally derived antibodies have average half life of 9-11 days (Decaro *et al.*, 2005b). The decline of maternal antibody level starts from first week to 13 weeks in the pups. Vaccines particularly MLV are not affected if administered in the presence of maternally derived antibodies. There is a strong correlation between HI or serum neutralizing antibody titers and protection against CPV infection. The HI test has been useful to measure antibodies which correlated with immunity. Dogs vaccinated with killed vaccine developed a serum antibody titre of less than 1:80 in HI test and shed virulent CPV when challenged orally. It is indicated that dogs with low antibody titre support viral replication in the intestine and are a source of infection for susceptible contacts. Dogs that recover from the infection have standard HI titres ranging from 1: 2,560 to 1: 20,480 which persist for at least one year and are solidly immune (Appel *et al.*, 1979; Carmichael *et al.*, 1993). The HI titre of 80 or more is considered protective although some authors set the threshold for protection at 1: 100 or above (Larson, 1996). The highest rate of infection is reported in pups older than 6 weeks of age (Pratelli *et al.*, 2000). Pups are fully susceptible to challenge CPV infection when the HI titre falls below 80 (Carmichael *et al.*, 1993) and such low titres commonly interfere with vaccination. This leaves a period of several weeks called 'immunity gap' when the young pups are susceptible to infection but refractory to vaccination. After a single vaccination either at 6 weeks of age with a CPV monovalent vaccine or at 8 weeks of age with a multivalent vaccine around 60 percent pups are seroconverted. At 12 weeks of age when the pups had received 2-3 inoculation about 90 percent of the pups respond to vaccination (Pratelli *et al.*, 2000, Decaro *et al.*, 2005b; Schultz, 2006). The principal reason for the non-responders was the persistence of interfering levels of maternal antibodies. None of the vaccines tested were capable of breaking through a maternal antibody titer of 1:160 or higher, regardless whether the vaccines were high titered or not (Decaro *et al.*, 2005b). Vaccination of dogs is generally performed using multivalent vaccines, which contain canine parvovirus along with canine distemper virus (CDV), canine adenovirus (CAV), rabies virus and leptospira bacterin. Monovalent CPV-2 vaccines are also available, some of them containing very high titer (10^7 $TCID_{50}$) virus and widely recommended for initial vaccination of pups. The American Animal Hospital Association (AAHA) recommends vaccinating dogs at 6 to 8 weeks of age with repeat vaccination every 3 to 4 weeks until the age of 16 weeks and possibly 24 weeks in high risk breeds. All dogs should receive a booster one year after completion of the initial series followed by booster every 3 years (Paul *et al.*, 2006). However, it is wise to determine the antibody titer of one or two pups in the litter at 5 or 6 weeks of age. The time of vaccination of the litter may be calculated on the basis of antibody titer taking half life of 9.5 days. Vaccination is likely to be successful when the maternal antibody titer has declined to less than 1:10.

Conclusion

In summary, parvovirus is a very common problem of canines that is a huge killer of puppies. Due to its ability to be transmitted through hands, clothes, and most likely rodents and insects, it is virtually impossible to have a kennel that will not eventually be exposed to the disease. Modified live vaccines are safe and effective, but despite the best vaccination protocol, all puppies will have a window of susceptibility of at least several days where they will be at risk. In addition, the newer CPV-2c strain presents new challenges as the current vaccines may not be as effective in providing protection against it. Again, commercially available FPV or CPV-2 based vaccines might also protect animals from the new virus infection. However, if the new virus gains wider host ranges, deadly outbreaks could be observed as when CPV-2 emerged in dogs. In that case, recent isolates need to be investigated to anticipate and assess the risk caused by newly emerging viruses. Further the homologous vaccine should be based on current or newer variant to tackle the disease in a very efficient and effective manner. Also, zoo sanitary measures should be employed to prevent the disease in wild animals.

Although potent and efficacious live attenuated and inactivated vaccines are available in India but large numbers of cases are diagnosed by HA, HI, ELISA or PCR, mostly from the unvaccinated dogs as the stray dogs are not usually vaccinated against the disease and they remain carrier of the virus and source of infection to other susceptible dogs. Extensive studies must be undertaken to know the molecular epidemiology of the canine parvovirus infections in different canine species and the variants of the CPV involved in the outbreak of the disease. The necessary preventive measures must be undertaken to immunize the susceptible dogs including the stray dogs with the potent and efficacious vaccines against the disease to check the spread of the disease. Prompt symptomatic treatment, restoration of fluid and antibiotics to prevent bacterial infection by a veterinarian will increase survivability in infected puppies but vaccination program should be considered the best way to control the disease in dog.

REFERENCES

Aberham, C., Pendl, C., Gross, P., Zerlauth, G. and Gessner, M. (2001). A quantitative, internally controlled real-time PCR assay for the detection of parvovirus B19 DNA. J Virol Methods, 92: 183-191.

Aldaz, J., García-Díaz, J., Calleros, L., Sosa, K., Iraola, G. and Marandino, A. (2013). High local genetic diversity of canine parvovirus from Ecuador. Vet Microbiol, 166: 214-219.

Appel, M.J.G., Scott, F.W. and Carmichael, L.E.(1979). Isolation and immunization studies of a canine parvo-like virus from dogs with hemorrhagic enteritis. Vet Res, 105: 156-159.

Appel, M., Scott, F.W. and Carmichael, L.R. (1985). Isolation and immunization studies of a canine parvovirus from dogs with haemorrhagic enteritis. Vet Rec, 105: 156-159.

Battilani, M., Scagliarini, A., Ciulli, S., Morganti, L. and Prosperi, S. (2006). High genetic diversity of the VP2 gene of a canine parvovirus strain detected in a domestic cat. Virology, 352: 22-26.

Buonavoglia, C., Martella, V., Pratelli, A., Tempesta, M., Cavalli, A., Buonavoglia, D., Bozzo, G., Elia, G., Decaro, N. and Carmichael, L. (2001). Evidence for evolution of canine parvovirus type 2 in Italy. J Gen Virol, 82: 3021-3025.

Burtonboy, G., Goignoul, F., Delferriere, N. and Pastoret, P.P. (1979). Canine hemorrhagic enteritis: detection of viral particles by electron microscopy. Arch Virol, 61: 1-11.

Calderón, M.G., Romanutti C., Wilda M., D' Antuono, A., Keller, L. and Giacomodonato, M.N. (2015). Resurgence of canine parvovirus 2a strain in the domestic dog population from Argentina. J Virol Methods, 222: 145-149.

Carmichael, L.E., Joubert, J.C. and Pollock, R.V. (1993). A modified live canine parvovirus vaccine and Immune response. Cornell Vet, 73: 13-29.

Chang, S.F., Sgro, J.Y. and Parrish, C.R. (1992). Multiple amino acids in the capsid structure of canine parvovirus coordinately determine the canine host range and specific antigenic and haemagglutination properties. J Virol, 66: 6858-6867.

Chinchkar, S.R., Mohan, S.B., Hanumantha, R.N., Rangarajan, P.N., Thiagarajan, D. and Srinivasan V.A. (2006). Analysis of VP2 gene sequences of canine parvovirus isolates in India. Arch Virol, 151: 1881-1887.

Cho, H.S., Kang, J. and Park. N.Y. (2006). Detection of canine parvovirus in fecal samples using loop-mediated isothermal amplification. J Vet Diagn Invest, 18: 81-84.

Cotmore, S.F., Christensen, J, Nüesch, J.P. and Tattersall, P. (1995). The NS1 polypeptide of the murine parvovirus minute virus of mice binds to DNA sequences containing the motif [ACCA] 2-3. J Virol, 69(3): 1652-1660.

Decaro, N. and Buonavoglia, C. (2012). Canine parvovirus - A review of epidemiological and diagnostic aspects, with emphasis on type 2c. Vet Microbiol, 155(1): 1-12.

Decaro, N., Campolo, M., Desario, C., Elia, G., Martella, V., Lorusso, E. and Buonavoglia, C. (2005a). Maternally-derived antibodies in pups and protection from canine parvovirus infection. Biologicals, 33: 259-265.

Decaro, N., Cirone, F., Desario, C., Elia, G., Lorusso, E., Colaianni, M.L., Martella, V. and Buonavoglia, C. (2009a). Severe parvovirus in a 12-yearold dog that had been repeatedly vaccinated. Vet Rec, 164: 593-595.

Decaro, N., Desario, C., Campolo, M., Elia, G., Martella, V., Ricci, D., Lorusso, E. and Buonavoglia, C. (2005b). Clinical and virological findings in pups naturally infected by canine parvovirus type 2 Glu-426 mutant. J Vet Diagn Invest, 17: 133-138.

Decaro, N., Desario, C., Parisi, A., Martella, V., Lorusso, A., Miccolupo, A., Mari, V., Colaianni, M.L., Cavalli, A., Di Trani, L. and Buonavoglia, C. (2009b). Genetic analysis of canine parvovirus type 2c. Virology, 385: 5-10.

Decaro, N., Elia, G., Campolo, M., Desario, C., Lucente, M.S., Bellacicco, A.L. and Buonavoglia, C. (2005). New approaches for the molecular characterization of canine parvovirus type 2 strains. J Vet Med B Infect Dis Vet Public Health, 52: 316-319.

Decaro, N., Elia, G., Martella, V., Desario, C., Campolo, M., Di Trani, L., Tarsitano, E., Tempesta, M. and Buonavoglia, C. (2005d). A real-time PCR assay for rapid detection and quantitation of canine parvovirus type 2 DNA in the feces of dogs. Vet Microbiol, 105: 19-28.

Desario, C., Decaro, N., Campolo, M., Cavalli, A., Cirone, F., Elia, G., Martella, V., Lorusso, E., Camero, M. and Buonavoglia, C. (2005). Canine parvovirus infection: which diagnostic test for virus? J Virol Methods, 121: 179-185.

Drane, D.P., Hamilton, R.C. and Cox, J.C. (1994). Evaluation of a novel diagnostic test for canine parvovirus. Vet Microbiol, 41(3): 293-302.

Elia, G., Cavalli, A., Desario, C., Lorusso, E., Lucente, M.S., Decaro, N., Martella, V. and Buonavoglia, C. (2007). Detection of infectious canine parvovirus type 2 by mRNA real-time RT-PCR. J Virol Methods, 146: 202-208.

Esfandiari, J. and Klingeborn, B. (2000). A comparative study of a new rapid and one-step test for the detection of parvovirus in faeces from dogs, cats and mink. J Vet Med B Infect Dis Vet Public Health, 47(2): 145-153.

Filipov, C., Desario, C., Patouchas, O., Eftimov, P., Gruichev, G. and Manov, V. (2016). A ten-year molecular survey on parvoviruses infecting carnivores in Bulgaria. Transbound Emerg Dis, 63: 460-464.

Gruber, F., Falkner, F.G., Dorner, F. and Hammerle, T. (2001). Quantitation of viral DNA by real-time PCR applying duplex amplification, internal standardization, and two-color fluorescence detection. Appl Env Microbiol, 67: 2837-2839.

Heid, C.A., Stevens, J., Livak, K.J. and Williams, P.M. (1996). Real time quantitative PCR. Genome Res, 6: 986-994.

Hirasawa, T., Kaneshige, T. and Mikazuki, K. (1994). Sensitive detection ofcanine parvovirus DNA by the nested polymerase chain reaction. Vet Microbiol, 41: 135-145.

Holland, P.M., Abramson, R.D., Watson, R. and Gelfand, D.H. (1991). Detection of specific polymerase chain reaction product by utilizing the 50-30 exonuclease activity of *Thermus aquaticus* DNA polymerase. Proc Nat Acad Science (USA), 88: 7276-7280.

Houston, D.M., Ribble, C.S. and Head, L.L. (1996). Risk factors associated with parvovirus enteritis in dogs. J Am Vet Med Assoc, 208: 542-548.

Hueffer, K., Parker, J.S., Weichert, W.S., Geisel, R.E., Sgro, J.Y. and Parrish, C.R. (2003). The natural host range shift and subsequent evolution of canine parvovirus resulted from virus-specific binding to the canine transferrin receptor. J Virol, 77: 1718-1726.

Jacob, R.M., Weiser, M.G., Hall, R.L. and Kowalski. J.J. (1980). Clinicopathogenic features of canine parvoviral enteritis. J Am Anim Hosp Assoc, 16: 809-813.

Kang, B.K., Song, D.S., Lee, C.S., Jung, K.I., Park, S.J. and Kim, E.M. (2008). Prevalence and genetic characterization of canine parvoviruses in Korea. Virus Genes, 36: 127-133.

Kapil., S., Cooper, E., Lamm, C., Murray, B., Rezabek, G. and Johnston, L. (2007). Canine parvovirus types 2c and 2b circulating in North American dogs in 2006 and 2007. J Clin Microbiol, 45: 4044-4047.

Kelly, W.R. and Atwell, R.B. (1979). Diffuse subacute myocarditis of possible viral etiology: a cause of sudden death in pups. Aust Vet J, 55: 36-37.

King, A.M.Q., Adams, M.J., Carstens, E.B. and Lefkowitz, E.J. (2012). Virus taxonomy: classification and nomenclature of virus: Ninth reports of the international committee on Taxonomy of viruses. Ed: San Diego: Elsevier Academics Press.

Kramer, J.M., Meunter, P.C. and Pollock, R.V.H. (1980). Canine parvovirus: update. Vet. Med. Small Anim. Clin, 175: 1541-1545.

Kumar, M. and Nandi, S. (2010). Development of a SYBR Green based real-time PCR assay for detection and quantitation of canine parvovirus in faecal samples. J Virol Methods, 169: 198-201.

Larson, L.J. and Schultz, R.D. (2008). Do two current canine parvovirus type 2 and 2b vaccines provide protection against the new type 2c variant. Vet Ther, 9: 94-101.

Majer-Dziedzic, B., Jakubczak, A. and Zietek, J. (2011). Phylogenetic analysis of canine parvovirus CPV-2 strains and its variants isolated in Poland. Polish J Vet Sci, 14: 379-384.

Markovich, J.E., Stucker, K.M., Carr, A.H., Harbison, C.E., Scarlett, J.M. and Parrish C.R. (2012). Effects of canine parvovirus strain variations on diagnostic test results and clinical management of enteritis in dogs. J Am Vet Med Assoc, 241: 66-72.

Martella, V., Decaro, N. and Buonavoglia, C. (2006). Evolution of CPV-2 and implication for antigenic/genetic characterization. *Virus Genes,* 33: 11-13.

Marulappa, S.Y. and Kapil, S. (2009). Simple tests for rapid detection of canine parvovirus antigen and canine parvovirus specific antibodies. Clin Vac Immunol, 16: 127-131.

McCandlish, I.A.P., Thompson. H., Fisher, E.W., Cornwell, H.J.C., Macartney, J. and Walton, I.A. (1981). Canine parvovirus infection. In Pract, 3: 5-14.

McCaw, D.L. and Hoskins J.D. Canine viral enteritis. In: Greene, C.E. (Ed.), Infectious diseases of the dog and cat, third ed. WB Saunders, Philadelphia, pp. 63-73. 2006.

Mittal, M., Chakravarti, S., Mohapatra, J.K., Chug, P.K., Dubey, R., Upmanuyu, V., Narwal, P.S., Kumar, A., Churamani, C.P. and Kanwar, N.S. (2014). Molecular typing of canine parvovirus strains circulating from 2008 to 2012 in an organized kennel in India reveals the possibility of vaccination failure. Infect Genet Evolut, 23: 1-6.

Mochizuki, M., SanGabriel, M.C., Nakatani, H. and Yoshida, M. (1993). Comparison of polymerase chain reaction with virus isolation and haemagglutination assays for the detection of canine parvoviruses in faecal specimens. Res Vet Sci, 55: 60-63.

Mukhopadhyay, H.K., Amsaveni, S., Matta, S.L., Antony, P.X., Thanislass, J., Pillai, R.M. (2012). Development and evaluation of loop-mediated isothermal amplification assay for rapid and sensitive detection of canine parvovirus DNA directly in faecal specimens. Lett Appl Microbiol, 55(3): 202-209.

Murphy, F.A., Gibbs, E.P.J., Horzinek. M.C. and Studdert, M.J. (1999). Parvoviridae. In: Vet Virol, 3rd ed. 343-356. San Diego, CA.

Muzyczka, N. and Berns, K.I. (2001). Parvoviridae: The viruses and their replication. In: Knipe, D.M., Howley, P.M. (Eds.), Fields Virology.4th ed. Lippincott Williams and Wilkins, Philadephia, PA. 2327-2359.

Nandi, S. and Kumar, M. (2010a). Canine Parvovirus: Current Perspectives. Indian J Virol, 21(1): 31-44.

Nandi, S., Anbazhagan, R., Kumar, M. and Chauhan, R.S. (2009). Molecular characterization of canine parvovirus strains in vaccines by polymerase chain reaction and restriction endonuclease analysis. Indian J Virol, 20(1): 12-15.

Nandi, S., Chidri, S., Kumar, M. and Chauhan, R.S. (2010). Occurrence of canine parvovirus type 2c in the dogs with haemorrhagic enteritis in India. Res Vet Sci, 88: 169-171.

Nandi, S., Pandey, A.B., Sharma, K. and Chauhan, R.S. (2008). Polymerase chain reaction for the detection of canine parvoviral DNA in vaccines. Indian J Virol, 19(1): 9-11.

Notomi, T., Okayama, H., Masubuchi, H., Yonekawa, T., Watanabe, K., Amino, N. and Hase, T. (2000). Loop-mediated isothermal amplification of DNA. Nucleic Acids Res, 28: E63.

Parrish, C.R. (1991). Mapping specific functions in the capsid structure of canine parvovirus and feline panleucopenia virus using infectious plasmid clones. J Virol, 183: 195-205.

Parrish, C.R. (1999). Host range relationships and the evolution of canine parvovirus. Vet Microbiol, 69: 29-40.

Parrish, C.R., Aquadro, C.F. and Carmichael, L.E. (1988). Canine host range and a specific epitope map along with variant sequences in the capsid protein gene of canine parvovirus and related feline, mink and racoon parvoviruses. Virology, 166: 293-307.

Pereira, C.A.D., Leal, E.S. and Durigon, L.E. (2007). Selection regimen shift and demographic growth increase associated with the emergence of high-fitness variants of the canine parvovirus. Infect Genet Evolut, 7: 399-409.

Phromnoi, S., Sirinarumitr, K. and Sirinarumitr, T. (2010). Sequence analysis of VP2 gene of canine parvovirus isolates in Thailand. Virus Genes, 41: 23-29.

Pinto, L.D., Streck, A.F., Gonçalves, K.R., Souza, C.K., Corbellini, Â.O., Corbellini, L.G. *et al.* (2010). Typing of canine parvovirus strains circulating in Brazil between 2008 and 2010. Virus Res, 165: 29-33.

Pollock, R.V. (1982). Experimental canine parvovirus infection in dogs. Cornell Vet, 72: 103-119.

Pratelli, A.A., Cavalli, G., Normanno, M.G., Palma, G.D., Pastorelli, Buonavoglia, C. and Martella, V. (2000). Immunization of pups with maternally derived antibodies to canine parvovirus (CPV) using a modified-live variant (CPV-2b). J Vet Med, 47: 273-276.

Ramadass, P. and Khadher, T.G.A. (1982). Diagnosis of canine parvovirus infection by agar gel precipitation test and fluorescent antibody techniques. Cherion, 11: 323-326.

Savi, J., Minakshi, P. and Prasad, G. (2009). Genotyping of field strain of canine parvovirus in Haryana using PCR and RFPL. Indian J Anim Sci, 79(10): 971-973.

Schmitz, S., Coenen, C., Konig, M., Thiel, H.J. and Neiger, R. (2009). Comparison of three rapid commercial canine parvovirus antigen detection tests with electron microscopy and polymerase chain reaction. J Vet Diagn Invest, 21: 344-345.

Schultz, R. (2006). Duration of immunity for canine and feline vaccines: a review. Vet Microbiol, 117(1): 75-79.

Schunck, B., Kraft, W. and Truyen, U. (1995). A simple touch-down polymerase chain reaction for the detection of canine parvovirus and feline panleukopenia virus in faeces. J Virol Methods, 55: 427-433.

Shackelton, L.A., Parrish, C.R., Truyen, U. and Holmes, E.C. (2005). High rate of viral evolution associated with the emergence of carnivore parvovirus. Proc Nat Acad Sci (USA), 102: 379-384.

Tattersall, P., Bergoin, M., Bloom, M.E., Brown, K.E., Linden, R.M., Muzyczka, N., Parrish, C.R. and Tijssen, P. (2005). Family Parvoviridae. In: Fauquet, C.M., Mayo, M.A., Maniloff, J., Desselberger, U., Ball, L.A. (Eds.), Virus Taxonomy: VIII Report of the International Committee on Taxonomy of Viruses. Elsevier Academic Press.

Teramoto, Y,A., Mildbrand, M.M., Carlson, J., Collins, J.K. and Winston, S. (1984). Comparison of ELISA, DNA hybridization, hemagglutination, and electron microscopy for detection of canine parvovirus infections. J Clin Microbiol, 20(3): 373-378.

Truyen, U., Platzer, G. and Parrish, C.R. (1996). Antigenic type distribution among canine parvoviruses in dogs and cats in Germany. Vet Rec, 138: 365-366.

Truyen, U. (2006). Evolution of canine parvovirus-A need for new vaccines? Vet Microbiol, 117: 9-13.

Truyen, U., Gruneberg, S.F., Chang, B., Obermaier, P., Veijalainen, S. and Parrish C.R. (1995). Evolution of the feline-subgroup of parvoviruses and the control of canine host range *in vivo*. J Virol, 69: 4702-4710.

Truyen, U.A. (2006). Evolution of canine parvovirus - a need for new vaccines. Vet Microbiol, 117: 9-13.

Tsai, Y.L., Wang, H.T., Chang, H.F., Tsai, C.F., Lin, C.K., Teng, P.H., Su, C., Jeng, C.C. and Lee, P.Y. (2012). Development of TaqMan probe-based insulated isothermal PCR (iiPCR) for sensitive and specific on-site pathogen detection. PLoSONE, 7: e45278.

Uwatoko, K., Sunairi, M., Nakajima, M. and Yamaura, K. (1995). Rapid method utilizing the polymerase chain reaction for detection of canine parvovirus in faeces of diarrheic dogs. Vet Microbiol, 43: 315-323.

Veijalainen, P.M., Neuvonen, E., Niskanen, A. and Juokslahti, T. (1986). Latex agglutination test for detecting feline panleukopenia virus, canine parvovirus, and parvoviruses of fur animals. J Clin Microbiol, 23(3): 556-559.

Waner, T., Mazar, S., Nachmias, E., Keren-Kornblatt, E. and Harrus, S. (2003). Evaluation of a dot ELISA kit for measuring immunoglobulin M antibodies to canine parvovirus and distemper virus. Vet Rec, 152: 588-591.

Wange, J., Lin, P., Zhao, H., Cheng, Y., Jiang, Z., Zhu, H., Wu, Hua. and Cheng, S. (2016). Continuing evolution of canine Parvovirsus in China: isolation of novel variants with an Ala5Gly mutation in the VP-2 protein. Infect Genet Evolut, 38: 73-78.

Wilkes, R.P., Lee, P.Y., Tsai, Y.L., Tsai. C.F., Chang, H.H., Chang, H.F. and Wang, H.T. (2015). An insulated isothermal PCR method on a field-deployable device for rapid and sensitive detection of canine parvovirus type 2 at points of need. J Virol Methods, 220: 35-38.

Xu, G., Hu, L., Zhong, H., Wang, H., Yusa., S., Weiss, T.C., Paul., J.R., Pickerill, Sam. and You, Q. (2012). Cross Priming Amplification (CPA): Mechanism and optimization for isothermal DNA amplification. Scientific Rep, 2: 246-253.

Chapter 6

Japanese Encephalitis in Swine

Satyendu Nandy and Sachin Kumar*

Department of Biosciences and Bioengineering, Indian Institute of Technology Guwahati, Guwahati, Assam, India
****e-mail: sachinku@iitg.ac.in***

ABSTRACT

The Japanese encephalitis virus (JEV), a flavivirus transmitted by mosquitoes, is the primary cause of pediatric encephalitis in Southeast Asia. JEV spreads in the wild through two types of hosts, namely swine and avian species. Pigs are a suitable indicator for forecasting the occurrence of Japanese encephalitis virus (JEV) outbreaks in humans. Pigs not only develop viremia that supports the virus's transmission but also display neurological and reproductive symptoms. New information on JEV's epidemiology indicates that pigs with viremia serve as a reservoir of infectious viruses for hematophagous mosquito species, notably Culex tritaeniorhynchus, where the disease is prevalent.

***Keywords**: Japanese encephalitis virus, Flavivirus, Genotype, Encephalitis, Zoonoses.*

Introduction

Japanese encephalitis (JE) is a prominent cause of viral encephalitis in Southeast Asia, posing a grave public health concern in India. Notably, new cases are surfacing in previously unaffected regions of the country. The Government of India introduced the Japanese encephalitis vaccination in 2006 for children aged 1 to 15 years and later implemented through the Universal Immunization Program (UIP) in 2011 (Murhekar *et al.*, 2017). Some Indian states, such as Assam, Uttar Pradesh, and West Bengal, have witnessed a rising trend in JE incidence. New cases with significant death rates due to acute encephalitis syndrome (AES) were recorded in Odisha in 2016 (Sahu *et al.*, 2018). Even with the implementation of mass JE vaccination programs in prioritized endemic areas in 2011, there has been a shift in the age group affected by JE virus (JEV) infection, impacting the adult population in West Bengal (Rajaiah and Kumar, 2022). The recent identification of genotype I (GI) circulation in Gorakhpur, Uttar Pradesh, and the coexistence of GI and genotype III (GIII) in West Bengal should serve as a warning sign for public health authorities, urging them to fortify surveillance systems in all endemic hotspots across the country.

JEV is a member of the *Flavivirus* genus within the *Flaviviridae* family, characterized as an enveloped, positive-sense RNA virus (Solomon *et al.*, 2003). The majority of virulent flaviviruses, like JEV, are arthropod-transmitted viruses sustained through biological transmission between vertebrate hosts with viremia and blood-feeding arthropods like mosquitoes (Kemenesi and Bányai, 2018). In simpler terms, arboviruses need susceptible vertebrate hosts capable of generating enough viremia. This viremia is essential for infecting competent arthropods when they bite these hosts, allowing the virus to be transmitted to new vertebrate hosts within the same cycle. In the case of JEV, pigs and birds are crucial in fulfilling this role as amplifying hosts, which aids in maintaining the ongoing transmission cycle of the virus in regions where it is endemic (Le Flohic *et al.*, 2013). JEV infection can affect various vertebrate host species, such as cows, goats, horses, dogs, bats, reptiles, chickens, and humans (Gould *et al.*, 1964; Cleton *et al.*, 2014). Nevertheless, most are typically viewed as incidental or terminal hosts because they cannot generate the necessary level of viremia for transmission. It has been demonstrated that JEV can be spread without the need for both viremic hosts and capable vectors to be present simultaneously. Instances of these alternate transmission routes that are not reliant on vectors include the oral transmission of tick-borne encephalitis virus through the consumption of raw milk or cheese derived from infected animals (Holzmann *et al.*, 2009; Hudopisk *et al.*, 2013), the transmission of Zika virus through sexual contact (Duggal *et al.*, 2017; Duggal *et al.*, 2018) and the direct transmission of West Nile viruses (WNV) among birds, possibly through oral and cloacal shedding (Banet-Noach *et al.*, 2003). The increasing body of evidence that has been steadily mounting indicates that the transmission of Japanese encephalitis can also occur through oronasal shedding between pigs (Cappelle *et al.*, 2016; Diallo *et al.*, 2018), underscoring the potential complexity of the JEV enzootic transmission cycle, which may be more intricate than initially thought.

The sudden appearance of JEV genotype V (GV) in China and Korea in 2009, following its initial detection in Malaya in 1952, has prompted endemic countries to bolster their surveillance efforts. GV is suspected of efficiently spreading to other parts of Asia. Moreover, the reduced effectiveness of the JEV GIII-based vaccine against JEV genotype V underscores the need to thoroughly evaluate ongoing vaccination strategies in endemic countries, anticipating the possible introduction of GV and its implications for future control measures.

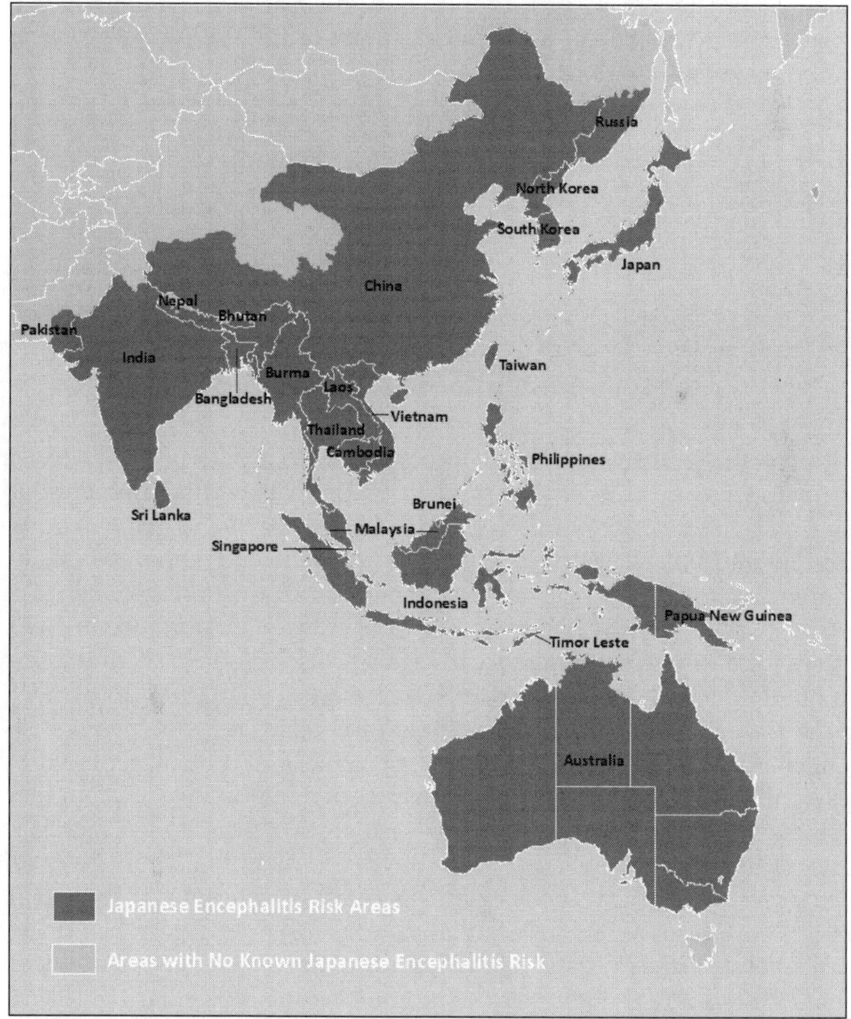

Figure 6.1: Global Picture of Japanese Encephalitis-Affected Area

[*Source*: Centers for Disease Control and Prevention, National Center for Emerging and Zoonotic Infectious Diseases (NCEZID), Division of Vector-Borne Diseases (DVBD)].

Genome Structure of JEV

JEV is a virus with a lipid envelope, a single-stranded positive-sense RNA genome, and a size ranging from 40 to 50 nm in diameter. It falls within the JEV sero-complex of the *Flaviviridae* family. The viral genome, approximately

12,000 nucleotides long, codes for three structural proteins (C, M, and E) and seven non-structural proteins (NS1, NS2a, NS2b, NS3, NS4a, NS4b, and NS5) (Kalia *et al.*, 2013). This genome contains a single open reading frame (ORF) flanked by two non-coding regions (NCRs) at 5′ and 3′ ends. At the 5′ end of the viral genome, there is a type I cap(**m7GpppAmp**), while at the 3′ end, there is a characteristic dinucleotide CU, as mosquito-borne Flaviviruses lack a poly (A) tail (Westaway, 1987).

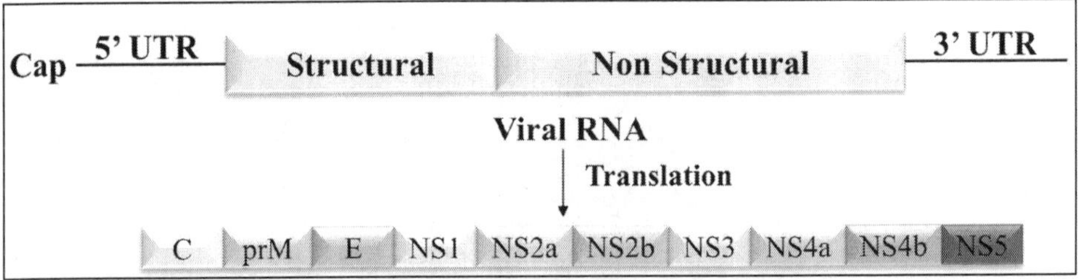

Figure 6.2: JEV Genome Representation.

Among the viral proteins, the 'E' protein (envelope protein) is vital for virus-cell interactions and triggers the production of virus-neutralizing antibodies. Serologically, there is only one recognized serotype of JEV. However, when examining the 'E' gene region, JEV is categorized into five genotypes: GI to GV (Solomon *et al.*, 2003). There are notable molecular distinctions between GV and the others (GI-GIV), with GV having a distinctive insertion mutation in its 3′UTR and showing a 3.4 percent difference in amino acids, including 117 unique amino acids not found in the other known JEV genotypes (Gao *et al.*, 2015). GIII has been the most commonly detected genotype among these genotypes worldwide (Erlanger *et al.*, 2009). However, the recent trend of GIII being replaced by GI raises significant global concerns, emphasizing the importance of strengthening surveillance systems (Wang *et al.*, 2007). Reports of the limited neutralizing efficacy of the GIII vaccine against the newly dominant GI strain have raised alarms regarding ongoing JE vaccination strategies in various countries (Fan *et al.*, 2012). Furthermore, the substantial differences observed in the nucleotide sequence between the circulating JEV strain and the currently used vaccine strain in Korea underscore the necessity of monitoring the molecular epidemiology of JEV in India to implement effective control strategies in the future. In Figure 3, the classification of flavivirus is given.

Evolution of JEV

Being an RNA virus, JEV is susceptible to mutations because it lacks a proofreading mechanism during replication. Researchers have estimated the average evolutionary rate of JEV to be about 4.35×10^{-4} (with a range of 3.4906×10^{-4} to 5.303×10^{-4}) nucleotide substitutions per site per year. Interestingly, among its various genotypes, the evolutionary rate of the complete JEV genome

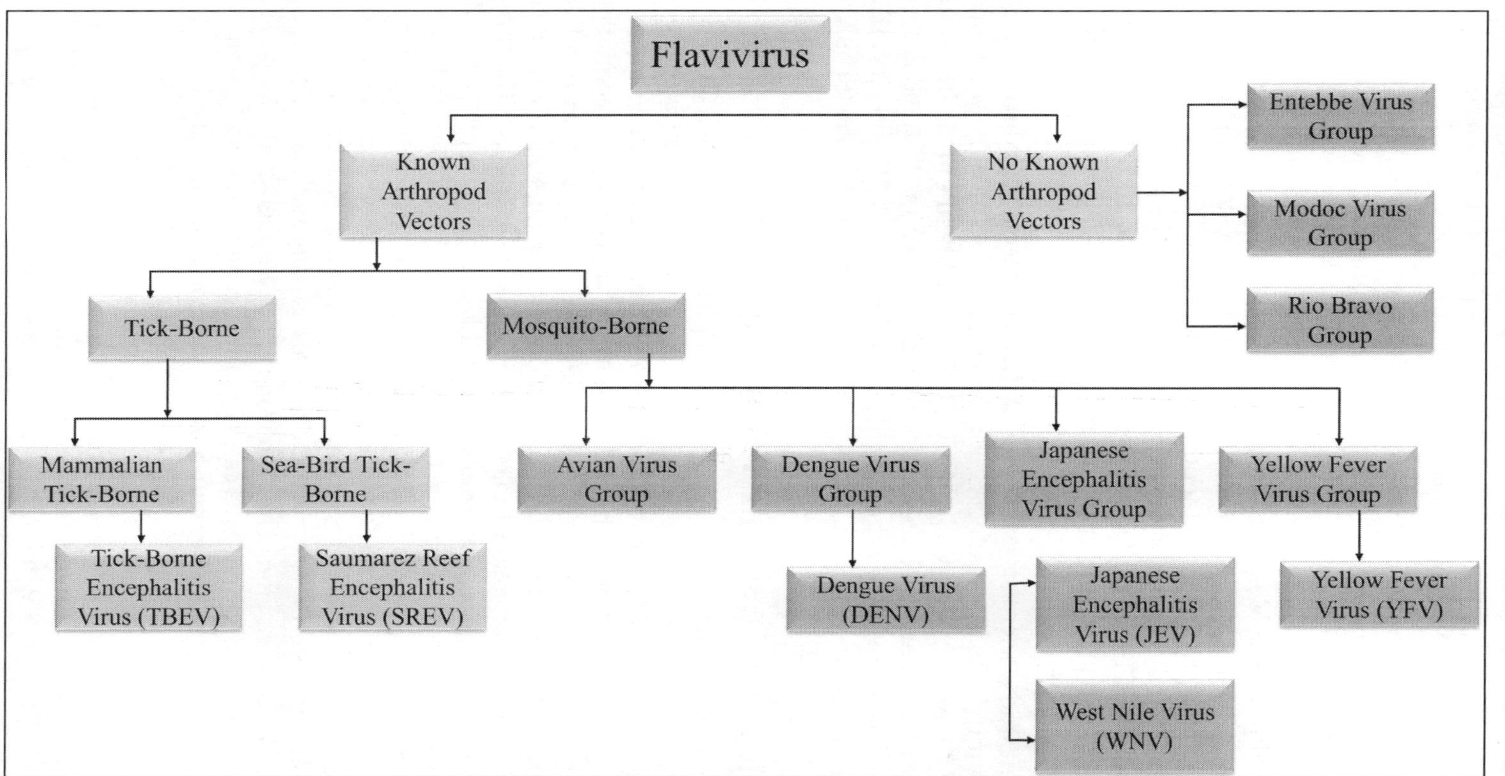

Figure 6.3: Categorization of Flaviviruses.

Flaviviruses are viruses transmitted by arthropods. They are divided into two groups: those with identified vectors and those with unidentified vectors. Ticks and mosquitoes play a crucial role in transmitting these viruses. The flaviviruses transmitted by mosquitoes are further categorized into different sero-complexes. (Picture Adapted from Unni *et al.* (2011)

is lower for GIII (6.029×10^{-4}) compared to GI and GII. This has been confirmed in scientific references (Mohammed *et al.*, 2011).

Experiments have shown that when multiple genotypes of JEV co-circulate, they can potentially undergo recombination, leading to genetic diversity (Chuang and Chen, 2009; Carney *et al.*, 2012). Phylogenetic analysis has traced the origin of JEV GV as the oldest lineage, with its ancestor believed to have evolved in the Indonesia–Malaya region, likely around the 1550s. Subsequently, it diverged into GIV around 1736. A hypothesis also suggests that JEV and the Murray Valley encephalitis virus (another member of the Flaviviridae family, found exclusively in Australia) share a common ancestry (Erlanger *et al.*, 2009). From genotype IV, GII and GIII diverged around 1821 and 1851, respectively, while GI is predicted to have emerged from GII approximately during the 1900s. These findings are based on reference (Mohammed *et al.*, 2011).

In a recent study, scientists estimated that JEV emerged approximately 3255 years ago, giving rise to five distinct genotypes (Gao *et al.*, 2015). The suggested order of JEV divergence is GV-GIV-GIII-GII-GI. Notably, the JEV GII strain, also known as the 'Bennett isolate,' was isolated in Korea in 1951 and shares a close nucleotide sequence similarity with GI, followed by GIII and GIV. JEV GIV is limited to Indonesia, with only six isolates identified to date (Schuh *et al.*, 2010; Schuh *et al.*, 2013a).

Epidemiology and Transmission of JEV

The Japanese encephalitis virus, a type of encephalitic flavivirus, can lead to deadly infections in people with no prior immunity, affecting the central nervous system. It is presently found primarily in the Asian-Pacific area. Since no targeted treatments exist, safeguarding against Japanese encephalitis through vaccination is paramount. Currently, multiple approved vaccines, either inactivated or live attenuated, are administered via intramuscular or subcutaneous injection, which offers adequate protection against infection and subsequent illness. The three categories of licensed vaccines currently accessible are inactivated Vero cell-based vaccines (derived from Beijing-1 or SA14-14-2 strains), a live attenuated vaccine (JEV SA14-14-2), and a live chimeric vaccine (ChimeriVax-JE) (Chen *et al.*, 2015; Hegde and Gore, 2017; Hills *et al.*, 2019). Even though vaccination is the most powerful method for managing the Japanese encephalitis virus (JEV), endemic nations still report around 68,000 cases annually. Among these cases, roughly 75 percent affect children under 15 (Campbell *et al.*, 2011). The clinical signs in infected individuals can vary from general flu-like symptoms like headaches, high fever, and fatigue to severe outcomes like paralysis, motor and memory issues, and seizures (Monath, 2002). Researchers have identified five separate genotypes of the Japanese encephalitis virus (JEV) by analyzing the sequences of the virus's envelope protein. Most cases of human disease are attributed to JEV strains categorized within clade b of genotype I (GI-b) and genotype III (GIII). Notably, there have been reports of the emergence of JEV

GI-b in various regions, resulting in the displacement of previously prevalent GIII strains (Fan *et al.*, 2013; Schuh *et al.*, 2013b, 2014).

The transmission of JEV through capable mosquito species meets the requirements for the biological transmission of arboviruses, involving the replication of JEV in both mosquitoes and amplifying hosts. So far, over ten mosquito species have been confirmed as effective vectors for transmitting JEV (Pearce *et al.*, 2018; Auerswald *et al.*, 2021). Among these mosquito species, *Culex* mosquitoes stand out as the primary competent carriers of JEV, primarily because they prefer feeding on animals (Tsai, 1993). Competent vector species, by definition, must be able to naturally acquire the virus, transmit it by feeding on susceptible vertebrate hosts, and be abundant enough to play a significant role. The primary vectors meeting these criteria include *Cx. tritaeniorhynchus, Cx. gelidus, Cx. vishnui, and Cx. annulirostris*. Among these, *Cx. tritaeniorhynchus* is identified as the key vector for JEV due to its high susceptibility, transmission efficiency, and widespread distribution (Gresser *et al.*, 1958; Rosen, 1986; Clark *et al.*, 2012). The importance of these primary mosquito vectors may shift over time due to the influence of climate change, such as global warming and increased flooding, which can affect vector populations and their geographic range (Longbottom *et al.*, 2017; Sakamoto *et al.*, 2019). While Culex species mosquitoes play a significant role as primary carriers, the main reservoirs of infectious viruses are viremic swine and avian hosts, notably the water-wading birds belonging to the Ardeidae family (Boyle *et al.*, 1983). Japanese encephalitis is primarily recognized as a rural agricultural ailment, with the overflow of JEV from animals to humans primarily driven by the close interaction between humans and amplifying hosts. The intensive backyard rearing of domestic pigs is responsible for the heightened risk of human JEV infection in endemic areas (Tsai, 1993). The constant production of susceptible pig populations due to their rapid reproduction and turnover on these farms serves to amplify JEV and sustain its presence in the region. Additionally, the preference of *Cx. tritaeniorhynchus* mosquitoes for feeding on pigs further underscores the significance of pigs as crucial amplifying hosts (Van den Hurk *et al.*, 2009; Misra and Kalita, 2010).

JEV Infection in Pigs

Much of our current knowledge regarding the way JEV affects domestic pigs is based on the detection of the virus or its genetic material in the blood and brain of naturally infected animals (Morimoto *et al.*, 1972; Nerome *et al.*, 2007; Desingu *et al.*, 2016). JEV infection and resulting disease in pigs are typically mild and age-dependent, with neurotropic disease being more common in young pigs. However, natural piglet infections are not commonly reported in endemic areas, possibly due to maternal antibodies that can last up to six months of age (Scherer *et al.*, 1959). In the early stages of JEV infection, pigs may display non-specific clinical signs such as fever, reduced appetite, and depression, while neurological symptoms like hind limb tremors or ataxia can develop after about five days post-infection (Park *et al.*, 2018; Park *et al.*, 2021). Some infected pigs may also develop wasting-like syndrome (Yamada *et al.*, 2004).

JEV infection in sexually mature adult pigs can lead to reproductive issues such as abortions and temporary infertility, which has significant implications for the pork and swine industry (Fujisaki, 1971). Abnormal farrowing, mummified fetuses, and weak piglets are commonly observed if a pregnant sow becomes infected before reaching 60 to 70 days of gestation (Shimizu *et al.*, 1954). Reports indicate that during epidemic seasons in Japan between 1947 and 1969, approximately 40 percent to 53 percent of unvaccinated pregnant sows experienced stillbirths and abortions (Imoto *et al.*, 2010). Boars can also be affected, developing conditions like edematous or congested orchitis with abnormal spermatozoa, but they typically recover completely (Ogasa *et al.*, 1977).

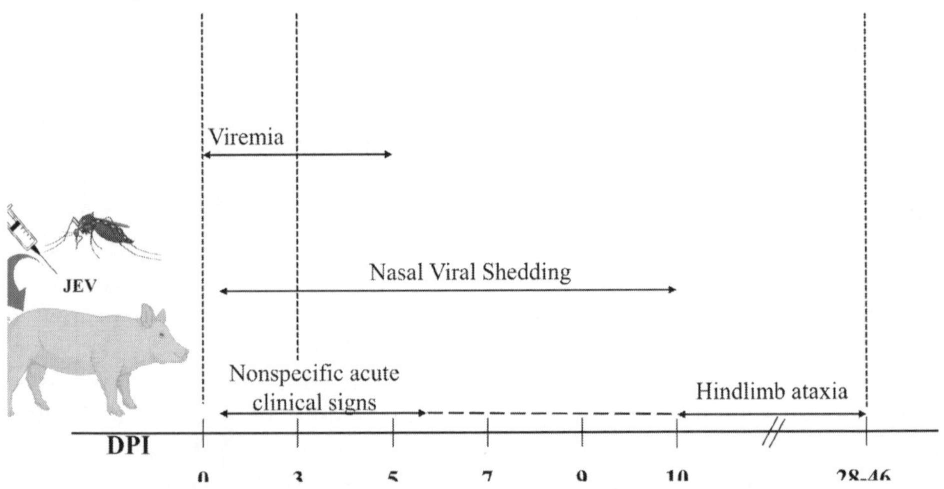

Figure 6.4: A Visual Representation Illustrates the Infection Consequences of JEV in Piglets.

When young piglets are infected through various methods of entry into the body, they may experience comparable pathological outcomes, such as the presence of the virus in the bloodstream, the shedding of the virus through the nasal passages, the development of clinical symptoms, and the continued presence of the virus (DPI represents the days after infection) (Picture adapted from (Park *et al.*, 2022).

Despite the potential for reproductive issues and the role of JEV-infected pigs in spreading the virus, there are currently no approved JEV vaccines for pigs. Only regionally approved vaccines, including live attenuated at 222, ML17, and anyang 300 vaccines, are available locally in Japan, China, and Korea (Lee *et al.*, 2012; Nah *et al.*, 2015). While immunizing pigs could potentially reduce JEV disease and viremia in swine, it is important to note that relying solely on pig vaccination is insufficient to prevent human JEV infection to the same extent as direct human vaccination. This is due to the rapid turnover of the pig population, associated costs, and logistical challenges of implementing a swine vaccination program. Additionally, other hosts like wild feral pigs or birds can still amplify the virus and maintain the transmission cycle in an area, diminishing the impact of swine immunization and reducing the transmission risk to humans (Erlanger *et al.*, 2009).

Research on JEV pathogenesis in pigs has primarily focused on understanding the kinetics of viremia, which directly relates to JEV transmission, and characterizing neurotropic disease, which resembles human Japanese encephalitis. With the advancement of molecular detection techniques, studies have detected JEV in various sample types, including oronasal shedding during the acute phase of infection and in lymphoid and nervous tissues during recovery. While viremia in pigs is crucial to the enzootic transmission of JEV as a mosquito-borne flavivirus, other pathological outcomes related to JEV transmission may have yet to be thoroughly explored.JEV infection in pigs involves viremia, systemic disease, neuro-invasion, and persistent infection. Viremia can be detected as early as one-day post-infection (dpi) and persists for 4 to 5 days. Interestingly, there are no significant differences in the level and duration of viremia between different infection routes (Redant *et al.*, 2020). Viremic titers typically exceed 10^5 50 percent tissue culture infectious dose ($TCID_{50}$)/mL or 10^6 plaque-forming units (PFU)/mL between 1 and 5 dpi, sufficient for infecting the mosquito vector *Cx. tritaeniorhynchus*. Studies have demonstrated that over 50 percent of *Cx. tritaeniorhynchus* can become infected through the per os route using blood meals containing infective doses of approximately 10^5 and 10^6 PFU/ML (Turell *et al.*, 2006). These findings confirm the role of domestic pigs in supporting the biological transmission of JEV.

Neuro-Invasive Phenotype of JEV in Pigs

The emergence of viremia leads to the rapid and widespread distribution of JEV throughout the pig's body. The virus's dispersion and entry into the nervous system co-occur with the peak of viremia, resulting in the identification of infectious viruses and viral genetic material in various tissues (Young *et al.*, 2020). Unlike in other mammalian species, the neuro-invasive nature of JEV does not lead to fatal illnesses in pigs. While significant pro-inflammatory cytokine responses are observed in the brains of humans, primates, and mice following JEV infection, the virus's replication in pig brains is efficiently suppressed, mainly due to the activation of OAS1 (20-50-oligoadenylate synthetase 1) expression and increased interferon-gamma activity, independent of type I interferon (Redant *et al.*, 2020). For instance, the Nakayama strain, which has a 50 percent lethal dose of 0.5 PFU when injected intraperitoneally into ICR mice, does not cause fatal disease in pigs when administered via injection or the oronasal route (Ricklin *et al.*, 2016).

JEV can be detected in various tissues of experimentally challenged pigs even weeks after they recover from the acute infection and develop neutralizing antibodies. Although JEV infections are typically characterized as critical, the presence of JEV RNA can still be found in the tonsils and brains of infected pigs nearly a month after the initial infection, suggesting that the virus may somehow evade the host's immune response. Research has shown that infected pigs can shed live infectious viruses from their nasal passages, which can facilitate the direct transmission of JEV to immunologically naive pigs, especially in a laboratory setting(Fan *et al.*, 2019). Studies by (Ricklin *et al.*, 2016) have demonstrated that pigs are highly susceptible to oronasal infection, even with viral titers as low as 10 $TCID_{50}$. Exposure to the virus through the oronasal route, either through intranasal inoculation or direct nose-to-nose contact with infected pigs, results in viremia, systemic infection, and antibody production similar to pigs infected through needle-based challenges (Redant *et al.*, 2020). There is also theoretical potential for mucosal transmission in humans, as recent evidence has shown the oral shedding of viral RNA in JE patients through throat swab sampling (Bharucha *et al.*, 2018). Nevertheless, the exact mechanism behind the oronasal shedding of JEV remains unclear. The source of shedding likely involves a combination of the virus being released directly from the nasal or olfactory epithelium (García-Nicolás *et al.*, 2018) and indirectly from blood as oral mucosal transudate (Thompson and Benjamin, 2019). This oronasal shedding of JEV has significant implications for veterinary diagnosis and virological surveillance, as collecting oral fluid and nasal secretions from infected pigs can be achieved using simple techniques. Based on the detection of viral genetic material and virus isolation, oronasal shedding of JEV can persist for almost two weeks, significantly longer than the duration of the viremic phase (Lyons *et al.*, 2018).

Conclusion

Infected pigs release enough virus in their mouth and nasal secretions that could infect other pigs directly and can also develop persistent infections in their lymphoid and nervous tissues (Park *et al.*, 2018; Park *et al.*, 2021). Most of the JEV isolates obtained from pig samples are derived from blood, brain, or fetal materials, while there is a limited representation of viruses found in oronasal secretions. The rope method for collecting oronasal fluids has been proven effective in field studies (Chiou *et al.*, 2021). Therefore, more research can be carried out using this approach to provide evidence of natural direct transmission of JEV and assess its significance in the overall transmission and maintenance of the virus. Furthermore, the inconspicuous and swift evolution of JEV could lead to the possible emergence of virus variations in regions where it is already prevalent and in new areas, particularly in states like Uttar Pradesh, West Bengal, Assam, Kerala, Karnataka, and Tamil Nadu. Various JEV genotypes/strains, potentially introduced by migratory birds from multiple global locations within and between countries, underscore the need for immediate attention to investigate its significance in shaping Japanese encephalitis control strategies in India.

REFERENCES

Auerswald, H., Maquart, P.O., Chevalier, V. and Boyer, S. (2021). Mosquito vector competence for Japanese encephalitis virus. Viruses, 13: 1154.

Banet-Noach, C., Simanov, L. and Malkinson, M. (2003). Direct (non-vector) transmission of West Nile virus in geese. Avian Pathol, 32: 489-494.

Bharucha, T., Sengvilaipaseuth, O., Seephonelee, M., Vongsouvath, M., Vongsouvath, M., Rattanavong, S., Piorkowski, G., Lecuit, M., Gorman, C., Pommier, J.D., Newton, P.N., de Lamballerie, X. and Dubot-Pérès, A. (2018). Detection of Japanese Encephalitis Virus RNA in Human Throat Samples in Laos - A Pilot study. Sci Rep, 8(1): 8018.

Boyle, D.B., Dickerman, R.W. and Marshall, I.D. (1983). Primary viraemia responses of herons to experimental infection with Murray Valley encephalitis, Kunjin and Japanese encephalitis viruses. Aust. J. Exp. Biol. Med. Sci, 61: 655-664.

Campbell, G.L., Hills, S.L., Fischer, M., Jacobson, J.A., Hoke, C.H., Hombach, J.M., Marfin, A.A., Solomon, T., Tsai, T.F. and Tsu, V.D. (2011). Estimated global incidence of Japanese encephalitis: a systematic review. Bull World Health Organ, 89: 766-774.

Cappelle, J., Duong, V., Pring, L., Kong, L., Yakovleff, M., Prasetyo, D.B., Peng, B., Choeung, R., Duboz, R., Ong, S.J.P.n.t.d., 2016. Intensive circulation of Japanese encephalitis virus in peri-urban sentinel pigs near Phnom Penh, Cambodia. PLoS Negl Trop Dis, 10, e0005149.

Carney, J., Daly, J.M., Nisalak, A. and Solomon, T. (2012). Recombination and positive selection identified in complete genome sequences of Japanese encephalitis virus. Arch Virol, 157: 75-83.

Chen, H.L., Chang, J.K. and Tang, R.B. (2015). Current recommendations for the Japanese encephalitis vaccine. J Chin Med Assoc, 78: 271-275.

Chiou, S.S., Chen, J.M., Chen, Y.Y., Chia, M.Y. and Fan, Y.C. (2021). The feasibility of field collected pig oronasal secretions as specimens for the virologic surveillance of Japanese encephalitis virus. PLoS Negl Trop Dis, 15: e0009977.

Chuang, C.K. and Chen, W.J. (2009). Experimental evidence that RNA recombination occurs in the Japanese encephalitis virus. Virol, 394: 286-297.

Clark, D.C., Brault, A.C. and Hunsperger, E. (2012). The contribution of rodent models to the pathological assessment of flaviviral infections of the central nervous system. Arch Virol, 157: 1423-1440.

Cleton, N.B., Bosco-Lauth, A., Page, M.J. and Bowen, R.A. (2014). Age-related susceptibility to Japanese encephalitis virus in domestic ducklings and chicks. Am J Trop Med Hyg, 90: 242.

Desingu, P., Ray, P.K., Patel, B., Singh, R., Singh, R. and Saikumar, G. (2016). Pathogenic and genotypic characterization of a Japanese encephalitis virus isolate associated with reproductive failure in an Indian pig herd. PloS one, 11: e0147611.

Diallo, A.O.I., Chevalier, V., Cappelle, J., Duong, V., Fontenille, D. and Duboz, R. (2018). How much does direct transmission between pigs contribute to Japanese Encephalitis virus circulation? A modelling approach in Cambodia. PloS one, 13: e0201209.

Duggal, N.K., McDonald, E.M., Ritter, J.M. and Brault, A.C. (2018). Sexual transmission of Zika virus enhances in utero transmission in a mouse model. Sci Rep, 8: 4510.

Duggal, N.K., Ritter, J.M., Pestorius, S.E., Zaki, S.R., Davis, B.S., Chang, G.J.J., Bowen, R.A. and Brault, A.C. (2017). Frequent Zika virus sexual transmission and prolonged viral RNA shedding in an immunodeficient mouse model. Cell Rep, 18: 1751-1760.

Erlanger, T.E., Weiss, S., Keiser, J., Utzinger, J. and Wiedenmayer, K. (2009). Past, present, and future of Japanese encephalitis. Emerg Infect Dis, 15: 1.

Fan, Y.C., Chen, J.M., Chen, Y.Y., Lin, J.W. and Chiou, S.S. (2013). Reduced neutralizing antibody titer against genotype I virus in swine immunized with a live-attenuated genotype III Japanese encephalitis virus vaccine. Vet Microbiol, 163: 248-256.

Fan, Y.C., Chen, J.M., Chiu, H.C., Chen, Y.Y., Lin, J.W., Shih, C.C., Chen, C.M., Chang, C.C., Chang, G.J.J. and Chiou, S.S. (2012). Partially neutralizing potency against emerging genotype I virus among children received formalin-inactivated Japanese encephalitis virus vaccine. PLoS Negi Trop Dis. 6(9):e1834.

Fan, Y.C., Liang, J.J., Chen, J.M., Lin, J.W., Chen, Y.Y., Su, K.H., Lin, C.C., Tu, W.C., Chiou, M.T. and Ou, S.C. (2019). NS2B/NS3 mutations enhance the infectivity of genotype I Japanese encephalitis virus in amplifying hosts. PLoS Pathog, 15: e1007992.

Fujisaki, Y. (1971). Japanese encephalitis in swine. Natl Inst Anim Health Q, 62: 16-24.

Gao, X., Liu, H., Li, M., Fu, S. and Liang, G. (2015). Insights into the evolutionary history of Japanese encephalitis virus (JEV) based on whole-genome sequences comprising the five genotypes. Virol, 12: 1-7.

García-Nicolás, O., Braun, R.O., Milona, P., Lewandowska, M., Dijkman, R., Alves, M.P. and Summerfield, A. (2018). Targeting of the nasal mucosa by Japanese encephalitis virus for non-vector-borne transmission. J Virol, 92(24): e01091-18.

Gould, D.J., Byrne, R.J. and Hayes, D.E. (1964). Experimental infection of horses with Japanese encephalitis virus by mosquito bite. Am J Trop Med Hyg, 13: 742-746.

Gresser, I., Hardy, J., Hu, S. and Scherer, W. (1958). Factors influencing transmission of Japanese B encephalitis virus by a colonized strain of Culex tritaeniorhynchus Giles, from infected pigs and chicks to susceptible pigs and birds. Am J Trop Med Hyg, 7: 365-373.

Hegde, N.R. and Gore, M.M. (2017). Japanese encephalitis vaccines: Immunogenicity, protective efficacy, effectiveness, and impact on the burden of disease. Hum Vaccin Immunother, 13: 1320-1337.

Hills, S.L., Walter, E.B., Atmar, R.L., Fischer, M., Walter Jr, E., Barnett, E., Barrett, A., Bocchini, J.A., Chen, L. and Deussing, E. (2019). Japanese encephalitis vaccine: recommendations of the advisory committee on immunization practices. MMWR Recomm Rep, 68: 1.

Holzmann, H., Aberle, S.W., Stiasny, K., Werner, P., Mischak, A., Zainer, B., Netzer, M., Koppi, S., Bechter, E. and Heinz, F.X. (2009). Tick-borne encephalitis from eating goat cheese in a mountain region of Austria. Emerg Infect Dis, 15: 1671.

Hudopisk, N., Korva, M., Janet, E., Simetinger, M., Grgiè-Vitek, M., Gubenšek, J., Natek, V., Kraigher, A., Strle, F. and Avšiè-Županc, T. (2013). Tick-borne encephalitis associated with consumption of raw goat milk, Slovenia, 2012. Emerg Infect Dis, 19: 806.

Imoto, J.I., Ishikawa, T., Yamanaka, A., Konishi, M., Murakami, K., Shibahara, T., Kubo, M., Lim, C.-K., Hamano, M. and Takasaki, T. (2010). Needle-free jet injection of small doses of Japanese encephalitis DNA and inactivated vaccine mixture induces neutralizing antibodies in miniature pigs and protects against fetal death and mummification in pregnant sows. Vaccine, 28: 7373-7380.

Kalia, M., Khasa, R., Sharma, M., Nain, M. and Vrati, S. (2013). Japanese encephalitis virus infects neuronal cells through a clathrin-independent endocytic mechanism. J Virol, 87: 148-162.

Kemenesi, G. and Bányai, K. (2018). Tick-borne flaviviruses, with a focus on Powassan virus. Clin Microbiol Rev, 32: 10.1128/cmr. 00106-00117.

Le Flohic, G., Porphyre, V., Barbazan, P. and Gonzalez, J.P. (2013). Review of climate, landscape, and viral genetics as drivers of the Japanese encephalitis virus ecology. PLOS Negl Trop Dis, 7: e2208.

Lee, J.A., Yang, D.K., Kim, H.H., Kim, S.Y., Nah, J.J., Cho, S.D. and Song, J.Y. (2012). Evaluation of Japanese encephalitis virus vaccine strains currently used in pigs by molecular characterization. 35: 169-174.

Longbottom, J., Browne, A.J., Pigott, D.M., Sinka, M.E., Golding, N., Hay, S.I., Moyes, C.L. and Shearer, F.M. (2017). Mapping the spatial distribution of the Japanese encephalitis vector, Culex tritaeniorhynchus Giles, 1901 (Diptera: Culicidae) within areas of Japanese encephalitis risk. Parasit Vectors, 10: 1-12.

Lyons, A.C., Huang, Y.J.S., Park, S.L., Ayers, V.B., Hettenbach, S.M., Higgs, S., McVey, D.S., Noronha, L., Hsu, W.W. and Vanlandingham, D.L. (2018). Shedding of Japanese encephalitis virus in oral fluid of infected swine. Vector Borne Zoonotic Dis, 18(9): 469-474.

Misra, U.K. and Kalita, J. (2010). Overview: japanese encephalitis. Prog Neurobiol, 91: 108-120.

Mohammed, M.A., Galbraith, S.E., Radford, A.D., Dove, W., Takasaki, T., Kurane, I. and Solomon, T. (2011). Molecular phylogenetic and evolutionary analyses of Muar strain of Japanese encephalitis virus reveal it is the missing fifth genotype. Infect Genet Evolut, 11: 855-862.

Monath TP. (2002). Japanese encephalitis vaccines: current vaccines and future prospects. Curr Top Microbiol Immunol, 267: 105-138.

Morimoto, T., Kurogi, H., Miura, Y., Sugimori, T. and Fujisaki, Y.J.N.I.o.A.H.Q. (1972). Isolation of Japanese encephalitis virus and a hemagglutinating DNA virus from the brain of stillborn piglets. Natl Inst Anim Health Q, 12: 127-136.

Murhekar, M.V., Oak, C., Ranjan, P., Kanagasabai, K., Shinde, S., Pandey, A.K., Mittal, M., Gore, M. and Mehendale, S.M. (2017). Coverage and missed opportunity for Japanese encephalitis vaccine, Gorakhpur division, Uttar Pradesh, India, 2015: implications for Japanese encephalitis control. Indian J Med Res, 145: 63.

Nah, J.J., Yang D.K., Kim, H.H. and Song, J.Y. (2015). The present and future of veterinary vaccines for Japanese encephalitis in Korea. Clin Exp Vaccine Res, 4: 130-136.

Nerome, R., Tajima, S., Takasaki, T., Yoshida, T., Kotaki, A., Lim, C.K., Ito, M., Sugiyama, A., Yamauchi, A. and Yano, T. (2007). Molecular epidemiological analyses of Japanese encephalitis virus isolates from swine in Japan from 2002 to 2004. J Gen Virol, 88: 2762-2768.

Ogasa, A., Yokoki, Y., Fujisaki, Y. and Habu, A. (1977). Reproductive disorders in boars infected experimentally with Japanese encephalitis virus. Japanese J Anim Reprod, 23: 171-175.

Park, S.L., Huang, Y.J.S., Lyons, A.C., Ayers, V.B., Hettenbach, S.M., McVey, D.S., Burton, K.R., Higgs, S. and Vanlandingham, D.L. (2018). North American domestic pigs are susceptible to experimental infection with Japanese encephalitis virus. Sci Rep, 8: 7951.

Park, S.L., Huang, Y.J.S., Lyons, A.C., Ayers, V.B., Hettenbach, S.M., McVey, D.S., Noronha, L.E., Burton, K.R., Hsu, W.W. and Higgs, S. (2021). Mosquito saliva modulates Japanese encephalitis virus infection in domestic pigs. Front Virol, 1: 724016.

Park, S.L., Huang, Y.J.S. and Vanlandingham, D.L. (2022). Re-examining the importance of pigs in the transmission of Japanese encephalitis virus. Pathog, 11: 575.

Pearce, J.C., Learoyd, T.P., Langendorf, B.J. and Logan, J.G. (2018). Japanese encephalitis: the vectors, ecology and potential for expansion. J Travel Med, 25: S16-S26.

Rajaiah, P. and Kumar, A. (2022). Japanese encephalitis virus in India: An update on virus genotypes. Indian J Med Res, 156: 588.

Redant, V., Favoreel, H.W., Dallmeier, K., Van Campe, W. and De Regge, N. (2020). Efficient control of Japanese encephalitis virus in the central nervous system of infected pigs occurs in the absence of a pronounced inflammatory immune response. J Neuroinflammation, 17: 1-17.

Ricklin, M.E., Garcia-Nicolas, O., Brechbuhl, D., Python, S., Zumkehr, B., Nougairede, A., Charrel, R.N., Posthaus, H., Oevermann, A. and Summerfield, A. (2016). Vector-free transmission and persistence of Japanese encephalitis virus in pigs. Nature, 7: 10832.

Rosen, L. (1986). The natural history of Japanese encephalitis virus. Ann Rev Microbiol, 40: 395-414.

Sahu, S.S., Dash, S., Sonia, T., Muthukumaravel, S., Sankari, T., Gunasekaran, K. and Jambulingam, P. (2018). Entomological investigation of Japanese encephalitis outbreak in Malkangiri district of Odisha state, India. Mem Inst Oswaldo Cruz, 113(6): e170499.

Sakamoto, R., Tanimoto, T., Takahashi, K., Hamaki, T., Kusumi, E. and Crump, A. (2019). Flourishing Japanese encephalitis, associated with global warming and urbanisation in Asia, demands widespread integrated vaccination programmes. Ann Glob Health, 85(1):111.

Scherer, W.F., Moyer, J.T. and Izumi, T. (1959). Immunologic studies of Japanese Encephalitis Virus in Japan: V. Maternal antibodies, antibody responses and viremia following infection of swine. J Immunol, 83: 620-626.

Schuh, A.J., Guzman, H., Tesh, R.B., Barrett, A.D.J.V.B. and Diseases, Z. (2013a). Genetic diversity of Japanese encephalitis virus isolates obtained from the Indonesian archipelago between 1974 and 1987. Vector Borne Zoonotic Dis, 13: 479-488.

Schuh, A.J., Li, L., Tesh, R.B., Innis, B.L. and Barrett, A.D. (2010). Genetic characterization of early isolates of Japanese encephalitis virus: genotype II has been circulating since at least 1951. J Gen Virol, 91: 95.

Schuh, A.J., Ward, M.J., Leigh Brown, A.J. and Barrett, A.D. (2013b). Phylogeography of Japanese encephalitis virus: genotype is associated with climate. PLoS Negl Trop Dis, 7: e2411.

Schuh, A.J., Ward, M.J., Leigh Brown, A.J. and Barrett, A.D. (2014). Dynamics of the emergence and establishment of a newly dominant genotype of Japanese encephalitis virus throughout Asia. J Virol, 88: 4522-4532.

Shimizu, T., Kawakami, Y., Fukuhara, S. and Matumoto, M. (1954). Experimental stillbirth in pregnant swine infected with Japanese encephalitis virus. Jpn J Exp Med, 24: 363-375.

Solomon, T., Ni, H., Beasley, D.W., Ekkelenkamp, M., Cardosa, M.J. and Barrett, A.D. (2003). Origin and evolution of Japanese encephalitis virus in southeast Asia. J Virol, 77: 3091-3098.

Thompson, D. and Benjamin, M. (2019). Sampling oral and processing fluids to identify disease in pigs. Michigan State University Extension.

Tsai, T.F. (1993). Inactivated Japanese encephalitis virus vaccine; recommendations of the Advisory Committee on Immunization Practices (ACIP). CDC. gov.

Turell, M.J., Mores, C.N., Dohm, D.J., Lee, W.J., Kim, H.C. and Klein, T.A. (2006). Laboratory transmission of Japanese encephalitis, West Nile, and Getah viruses by mosquitoes (Diptera: Culicidae) collected near Camp Greaves, Gyeonggi Province, Republic of Korea, 2003. J Med Entomol, 43: 1076-1081.

Unni, S.K., Rùžek, D., Chhatbar, C., Mishra, R., Johri, M.K. and Singh, S.K. (2011). Japanese encephalitis virus: from genome to infectome. Microbes Infect, 13: 312-321.

Van den Hurk, A.F., Ritchie, S.A. and Mackenzie, J.S. (2009). Ecology and geographical expansion of Japanese encephalitis virus. Ann Rev Entomol, 54: 17-35.

Wang, H.Y., Takasaki, T., Fu, S.H., Sun, X.H., Zhang, H.L., Wang, Z.X., Hao, Z.Y., Zhang, J.K., Tang, Q. and Kotaki, A. (2007). Molecular epidemiological analysis of Japanese encephalitis virus in China. J Gen Virol, 88: 885-894.

Westaway, E. (1987). Flavivirus replication strategy. Adv Virus Res, 33: 45-90.

Yamada, M., Nakamura, K., Yoshii, M. and Kaku, Y. (2004). Nonsuppurative encephalitis in piglets after experimental inoculation of Japanese encephalitis flavivirus isolated from pigs. Vet Pathol, 41: 62-67.

Young, C.L., Lyons, A.C., Hsu, W.W., Vanlandingham, D.L., Park, S.L., Bilyeu, A.N., Ayers, V.B., Hettenbach, S.M., Zelenka, A.M. and Cool, K.R. (2020). Protection of swine by potent neutralizing anti-Japanese encephalitis virus monoclonal antibodies derived from vaccination. Antivir Res, 174: 104675.

Chapter 7

Sheeppox and Goatpox

***Amit Kumar*[1]**, *Aswin P. Kumar*[1], *Abhishek M. Todkari*[1], *Ankit Prasad Kelwan*[1] *and Gnanavel Venkatesan*[2]**

[1]Pox Virus Laboratory, Division of Virology, ICAR-Indian Veterinary Research Institute, Mukteswar – 263138, Uttarakhand, India
[2]ICAR-Indian Veterinary Research Institute, Bengaluru, Karnataka, India
****e-mail: amitvet87@gmail.com***

ABSTRACT

Capripoxvirus (CaPV) infections of small ruminants, namely, sheeppox and goatpox (SGP) are transboundary and OIE notifiable diseases that are enzootic in parts of Africa, the Middle East, and Asia including the Indian sub-continent with occasional outbreaks in Europe. These infections are associated with significant production losses affecting small and marginal farmers of the developing world due to high morbidity, mortality, and animal trade restrictions. The diseases are characterized by fever, papules, nodular lesions on the skin and sometimes internal organs, and lymphadenopathy. The etiological agents, goatpox virus (GTPV) and sheeppox virus (SPPV) are serologically indistinguishable but can be differentiated by molecular means. In affected countries, vaccination along with effective bio-security measures and quick diagnosis are the main control measures to contain the disease. The development of rapid pen-side diagnostics, recombinant protein-based sero-diagnostic assays, and DIVA-enabled vaccines is necessary for improved disease control. Here, we discuss the updates on epidemiology, diagnostics, and control strategies including vaccines to control SGP infections.

Keywords: *Capripox infections, Sheeppox virus (SPPV), Goatpox virus (GTPV), Epidemiology, Diagnosis, Vaccines.*

Introduction

Sheeppox and goatpox (SGP) infections are economically important viral diseases of small ruminants caused by sheeppox virus (SPPV) and goatpox virus (GTPV), respectively. SGP infections pose significant production losses due to high morbidity, mortality, hide damage along with costs associated with disease treatment and control and international trade restrictions (Garner *et al.*, 1995; Madhavan *et al.*, 2016). The young animals are more susceptible to SGP infections. The economic loss due to SGP in Maharashtra state (India) alone was estimated at over INR 107.5 million with average morbidity and mortality of 63.5 percent and 49.5 per cent, respectively, and it took nearly six years for a flock to recover from the outbreak (Garner *et al.*, 2000). By extrapolation of this data, the total estimated annual loss at the national level amounts to INR 1250 million (Bhanuprakash *et al.*, 2011). SGP infections are enzootic in Central and Northern Africa, the Middle East, and Asia including the Indian sub-continent along with sporadic outbreaks in Europe (Bhanuprakash *et al.*, 2006a; Babiuk *et al.*, 2008). In India, regular outbreaks of SGP are reported in almost all states and cause high annual losses due to their enzootic nature (Bhanuprakash *et al.*, 2006a; Mirzaie *et al.*, 2015). Transmission of sheeppox and goatpox is associated with either direct or indirect contact with infected animals (Kitching and Taylor, 1985). SPPV and GTPV cause more severe disease in the homologous host which is characterized by pyrexia, oculo-nasal discharge and pock lesions on the skin and internal mucosa (Babiuk *et al.*, 2008). Control and eradication of SGP infections depend on correct diagnosis and suitable prophylaxis. In the future, there is a need to develop DIVA vaccines with companion serological assay along with molecular epidemiology tools such as differentiation of SPPV and GTPV along with the development of robust high-throughput diagnostics.

History and Geographic Distribution

The history of sheeppox dates back to the 2nd century A.D., originating in central Asia (Hutyra *et al.*, 1946). The etiological agent, SPPV was first isolated by Borrel in 1902 followed by subsequent pathogenicity studies of SPPV in the 19th century. Later, the disease spread to many African, Asian and European countries in the mid-20th century (Bhanuprakash *et al.*, 2006a). Since then, the disease has been eradicated from many countries but is still present in central and Northern Africa, Western Asia, Middle-east and has spread to Turkey and Greece (Mangana-Vougiouka *et al.*, 2008). In India, sheeppox outbreaks have been reported since the pre-independence era in Tamil Nadu, Bombay and United Provinces (Sathe, 1931), and the disease is now enzootic in the country affecting almost all states (Joshi *et al.*, 1992; Madhavan *et al.*, 2016).

Goatpox was first reported in Norway in 1879 by Hansen. The distribution of goatpox is similar to sheeppox *viz.* Asia, Africa, the Middle East, and parts of Europe. In India, goatpox was first reported in 1936 from Indian Veterinary Research Institute (then Imperial Institute of Veterinary Research annual report 1936-37). Later, the disease outbreaks were reported from various states including the North-Eastern part (Rao and Bandyopadhyay, 2000; Madhavan *et al.*, 2016; Bora *et al.*, 2018).

SGP infections are enzootic throughout Southwest and central Asia, the Indian subcontinent, and Northern and central Africa. Occasional outbreaks have been reported in regions of Europe like Turkey, Greece, and Bulgaria (Bhanuprakash *et al.*, 2011). The introduction of affected animals through trade is the major mode of spread in naive areas (Babiuk *et al.*, 2008).

Etiological Agents - SPPV and GTPV

SPPV, GTPV along with lumpy skin disease virus (LSDV) belong to the genus *Capripoxvirus*, subfamily *Chordopoxvirinae*, and family *Poxviridae*. All three capripoxviruses (CaPVs) cannot be differentiated based on morphology and serology (Kitching and Smale, 1986). Although CaPVs were originally classified based on host origin, presently it has been replaced by molecular methods of differentiation (Hosamani *et al.*, 2004a; Venkatesan *et al.*, 2012, 2014a, 2014b). The majority of SPPV and GTPV strains show a host preference for severe disease in homologous hosts as compared to milder disease in heterologous hosts (Madhavan *et al.*, 2016).

The SPPV/GTPV virions are brick-shaped with a complex symmetry and size of 212-260 nm x 250-280nm and consist of an envelope, a surface membrane, a biconcave core, and lateral bodies (Tantawi and Al-Falluji, 1979; Kitching and Smale, 1986). Poxviruses are generally stable in the environment and resistant to drying and remain infective in dried scabs/wool for months. The heat sensitivity of SPPV/GTPV differs between strains/isolates (Pandey and Singh, 1970; Tantawi and Al-Falluji, 1979). These are inactivated within 2 h at 56°C or within 30 min at 65°C. They can survive at pH 6.6-8.6 but are susceptible to highly acidic or alkaline pH (Kitching, 2004). The infectivity of the virus is sensitive to ether, trypsin, chloroform, and formalin as well as repeated freeze-thawing (Pandey and Singh, 1970).

Genome and Molecular Characterization

SPPV/GTPV contain linear double-stranded DNA genomes of ~150 kbp size with covalently closed ends encoding nearly 147-148 ORFs (Tulman *et al.*, 2002). Like other poxviruses, the genome contains a central coding region flanked by two identical inverted terminal repeats (Gershon and Black, 1988; Tulman *et al.*, 2002). The genome possesses the highest A-T content (73-75 per cent) among poxviruses. The central core region contains genes involved in replication and morphogenesis, whereas terminal genomic regions are associated with virulence and host range and differ more in amino acid identity and ORF size between SPPV and GTPV (Tulman *et al.*, 2002; Zeng *et al.*, 2014). Both SPPV and GTPV share 96 percent nucleotide identity among them over the entire length of the genome (Tulman *et al.*, 2002). Although some strains show high sequence variation, intra-species genomic identity is more than 99 percent (Zeng *et al.*, 2014). All genes present in SPPV and GTPV are also present in LSDV, whereas 8-9 genes, which are intact in LSDV with probable virulence and host range functions, are disrupted in SPPV and GTPV genomes (Tulman *et al.*, 2001, 2002). Phylogenetic analysis of P32, GPCR, and RPO30 genes revealed that SPPV and GTPV are phylogenetically distinct entities and likely derived from an LSDV-like ancestor (Tulman *et al.*, 2002; Zhou *et al.*, 2012). GTPV is more closely related to LSDV than SPPV (Tulman *et al.*, 2002; Zeng *et al.*, 2014). The full genome sequences for several SPPV and GTPV strains are now available (Biswas *et al.*, 2020; Kumar *et al.*, 2022). Few genetic changes between virulent and attenuated the different CaPVs have been identified (Tulman *et al.*, 2002). SPPV-019 (kelch-like gene) deleted SPPV showed marked attenuation in sheep *in vivo* implying that it may be a virulence determinant in sheep (Balinsky *et al.*, 2007). It has been shown that K3L host range gene of CaPVs inhibits double-stranded RNA-dependent protein kinase in a species-specific manner (Zhao *et al.*, 2018). Gene duplications, losses, gains by horizontal gene transfer, and recombination occur frequently between different CaPV species especially at terminal regions (Bratke and McLysaght, 2008). CaPVs display a relatively low rate of accumulation of point mutations. SPPV and GTPV antigens are very closely related to each other due to partial sharing of antigens.

Epidemiology and Transmission

The majority of SPPV and GTPV strains are considered host-specific (Kitching, 1983). Apart from sheep and goats, GTPV infection has been reported from wild serows in Mizoram state, India (Dutta *et al.*, 2019). SGP affect goats and sheep of all ages, both sexes, and all breeds, but clinical symptoms are more common and severe in young and old animals. Indigenous sheep and goats exhibit some natural immunity, whereas Merino and European breeds are more susceptible to infection with SPPV (Karim, 1983; Heine *et al.*, 1999). Outbreaks are recorded throughout the year but are considered to be more prevalent during the rainy season (Bhanuprakash *et al.*, 2005; Yeruham *et al.*, 2007).

SPPV/GTPV virions are present in saliva, nasal discharge, conjunctival secretions, urine, and feces of infected animals, including milk, and in scabs from skin lesions. The short-distance transmission by direct contact through aerosols and droplets from nasal secretions and saliva is the major mode of spread (Kitching and Taylor, 1985; Bowden *et al.*, 2008), but indirect transmission also occurs by contamination of the environment or fomites like pens or yards, lorries and shearing clippers, *etc.* The extensive involvement of skin and the high amount of associated viral load in skin lesions may also contribute to mechanical transmission of sheeppox and goatpox via insect vectors like *Stomoxys calcitrans* under experimental conditions (Kitching and Mellor, 1986; Mellor *et al.*, 1987; Bowden *et al.*, 2008). Poxviruses are resistant viruses with the capability of "long-interval" transmission cycles and remain viable for long periods without depending upon continuing direct animal-to-animal transmission. The presence of a resistant virus on the infected/recovered animal hosts and in the shed environment ensures virus persistence (Rao and Bandyopadhyay, 2000). There is no evidence of animals in a carrier state that are persistently infected with goat and sheep poxvirus. Goatpox has also been reported as a mixed infection with PPR or orf simultaneously or by co-infection in goats (Chu *et al.*, 2011; Malik *et al.*, 2011).

Pathology and Symptomatology

The severity of the clinical disease is variable depending on the host species, breed, age, nutritional and immune status of the susceptible host, virulence of virus strain, the nature of the secondary infection, and organs involved (Davies, 1981; Bhanuprakash *et al.*, 2006). In nature, infection is transmitted mostly through the respiratory route by aerosols. However, most experimental pathogenesis studies have utilized intradermal inoculation of the virus (Kitching and Taylor, 1985; Bowden *et al.*, 2008; Embury-Hyatt *et al.*, 2012).

The SPPV and GTPV have tropism for skin and lung as well as discrete sites within the mucosal surface of oro-nasal tissues and gastrointestinal tract. Following experimental intradermal inoculation, the virus replicates in cells at the base of hair follicles and in mesenchymal cells of the dermis. SPPV antigen can be detected as early as 4 days post-infection (dpi) in macrophages at the intradermal inoculation site, often adopting the appearance of sheep pox cells, which are large round or spindle-shaped cells with vacuolated cytoplasm and nucleus, marginated chromatin, and often multiple cytoplasmic viral inclusions (Bowden *et al.*, 2008). Virus-infected cells disseminate from the primary lesion to the draining lymph node and then to the blood via the draining lymph node. The nature of viremia is cell-associated, mainly monocytes and macrophages (Kitching and Taylor, 1985; Gulbahar *et al.*, 2000). Cell-associated viremia starts at 4-6 dpi and disappears at 14-21 dpi in most surviving sheep until the time of seroconversion, when neutralizing antibodies develop in the host (Kitching and Taylor, 1985). Following primary viremia, the virus is found in all organs including the lung, skin, and spleen at 8 dpi and rumen and liver at 13 dpi due to hematogenous spread of the virus to mucosal surfaces, virus multiplication and development of lesions at these sites. High virus titers are detected in skin nodules, lungs, tongue lesions, and, to a lesser extent, in lymph nodes, liver, and spleen (Bowden *et al.*, 2008). The virus is consistently shed in nasal, oral, and conjunctival secretions at 6-8 dpi with peak shedding between 10-14 dpi lasting up to 3-6 weeks followed by a decrease in viral shedding. Low level of viral shedding also occurs in urine and feces (Bowden *et al.*, 2008; Embury-Hyatt *et al.*, 2012). Animals are most contagious soon after the appearance of papules, during 10 dpi before the development of significant levels of neutralizing antibodies. The skin lesions are probably the most significant source of the virus for horizontal infection, as lesions may persist for 3-4 weeks before healing. The immunity in SGP infections is generally cell-mediated, although humoral immunity plays a role in limiting the spread of the virus in the host.

The disease develops in 4 clinical phases: incubation, invasive (the general symptoms phase), eruptive, and convalescent (recovery). In natural cases, the incubation period of SGP is 8-13 days but in experimentally infected animals, it may be as short as 4 days (Rao and Bandyopadhyay, 2000; Bhanuprakash *et al.*, 2006a). The invasive or prodromal phase lasts 2- 4 days and is characterized by signs like pyrexia (40°C) which is followed by depression, arched back, inappetence, edema of the eyelids, trembling, accelerated breathing, rhinitis (sero-mucosal discharge, then muco-purulent) and conjunctivitis. Superficial lymph nodes become enlarged, especially pre-scapular lymph nodes. After this phase, the eruptive phase starts with the appearance of circular hyperemic spots (macula). Lesions progress to the papules having erythema and edema in the central part. Lesions are mainly found on the perineum, inguinal area, scrotum, udder, muzzle, eyelids, axillae, and sometimes the neck, thorax, abdomen, and the internal part of hind legs. Then the center of the lesion becomes depressed and gray (necrotic) and surrounded by an area of hyperemia. Late in the course of the disease, the lesions become dry, and form scabs. Papules don't progress to vesicles and pustules, instead become necrotic to form scabs. Scabs persist for 3-4 weeks and after healing, cicatrix may remain. Secondary bacterial infection may complicate the healing process. Scabs that have been shed by the animals remain infective for several months. Generalization of the disease is often noticed in kids and lambs aged 4-5 months, whereas in older animals, mild infections are seen. Other clinical signs such as bronchial pneumonia, keratitis, and abortion may also occur. The mortality rate in goatpox can be as high as 50-70 percent especially in young animals (Hailat *et al.*, 1994; Garner *et al.*, 2000). In sheeppox, morbidity is 75-100 percent and mortality is 10-58 percent depending on the virulence of the isolate (Bhanuprakash *et al.*, 2006a). Generally, an epidemic in a susceptible flock can affect over 75 percent of goats, with mortality as high as 50 per cent, and case fatality rates in young stock may approach 100 per cent.

Diagnosis

The clinical signs of SGP may vary from mild to severe depending on breed, age, immunity, *etc.* The disease needs to be differentiated from orf, bluetongue, dermatophilosis, streptothricosis, mange, photosensitization/urticaria, PPR, multiple insect bites, caseous lymphadenitis, *etc.* The mixed infections like SPPV with PPR and BT may complicate the diagnosis (Ozmen *et al.*, 2009). Skin papules, blood, lung lesions, or lymph nodes are used as specimens for virus isolation and antigen detection. These samples are to be collected within the first week of the occurrence of clinical signs or before the development of neutralizing antibodies (OIE, 2008; Mirzaie *et al.*, 2015). However, samples for nucleic acid detection can be collected in the presence of neutralizing antibodies. For histopathological studies, tissue from the surrounding area is collected in 10 percent formalin.

Identification of Agent/Antigen

1. ***Electron microscopy*** - EM can be used for rapid identification of SPPV/GTPV virions in vesicular fluid, and skin lesions but these are not easily distinguishable morphologically from orthopox viruses and need to be distinguished by immunostaining. However, CaPVs are distinguishable from ORFV (Kitching and Smale, 1986). EM requires expensive equipment and trained laboratory staff and lacks sensitivity. However, results are obtained within a day and it can be considered mainly as a confirmative method.
2. ***Virus isolation*** - SPPV/GTPV can be grown using a variety of sheep, goat, and cattle cells, but failed to grow on chorioallantoic membrane (CAM), infant mice, rabbits, and guinea-pigs (Adlakha *et al.*, 1971). SPPV/GTPV has been adapted in the primary embryonic/young lamb/kid lung, testis, kidney cells as well as MDBK and Vero cell lines (Plowright and Ferris, 1958; Adlakaha *et al.*, 1971; Madhavan *et al.*, 2016). Currently, primary lamb kidney or primary lamb testis cells are the most commonly used cells for primary isolation due to their sensitive nature (Adlakha *et al.*, 1971; Kalra and Sharma, 1981). Commercially available lamb testis cell line (OA3.Ts) has been evaluated for the propagation of CaPV isolates (Babiuk *et al.*, 2007). CaPVs generally produce a more localized cytopathic effect (CPE) with maximum CPE observed by the 5th-6th day post-infection (dpi). Both SPPV and GTPV will produce similar cytopathic effects like ballooning, increased refractivity, rounding, clumping, tract formation, detachment from monolayer, and sometimes foci formation. Cytopathic changes are associated with intra-cytoplasmic eosinophilic inclusion bodies. The typical cell inclusions become obvious 3-5 days after inoculation and isolation may be confirmed by immunofluorescent or immunoperoxidase test using anti-CaPV serum (Gulbahar *et al.*, 2000; Babiuk *et al.*, 2007) or PCR (Mangana-Voigiouka *et al.*, 2000). Several blind passages may be required for virus isolation.
3. ***Serological assays*** - Conventionally agar gel immunodiffusion (AGID), counter immunoelectrophoresis (CIE), single radial immunodiffusion, reverse phase passive hemagglutination (RPHA) test, spot agglutination test, coagglutination test, latex agglutination test, complement fixation test, and dot ELISA assays have been demonstrated for detection of antibodies or antigens of SPPV/GTPV (Rao and Bandyopadhyay, 2000). The fluorescent antibody test (FAT) is reliable and simple for the detection of CaPV antigen in infected edema fluid, skin, and cell monolayers (Soman, 1986). Immunocapture ELISA can be regarded as a relatively simple assay for detecting CaPV antigen in scab samples. Various capture ELISA assays employing polyclonal sera against whole virus (Rao *et al.*, 1997) or monospecific sera against

P32 protein (Carn, 1995; Nagichabe *et al.*, 1999) have been described. It cannot detect virus particles bound to neutralizing antibodies and is relatively less sensitive with a detection limit of 10^3 $TCID_{50}$ per 50 µl of virus (Carn, 1995; Rao *et al.*, 1997).

4. ***Nucleic acid-based assays***

   i) ***Polymerase chain reaction* (*PCR*)** - PCR test provides the most rapid and accurate laboratory confirmation of SGP diagnosis. Several conventional PCR assays targeting DNA polymerase gene (Balamurugan *et al.*, 2009), fusion and attachment protein genes (Ireland and Binepal, 1998), inverted terminal repeats (Mangana-Vougiouka *et al.*, 2000), P32 (Heine *et al.*, 1999), ORF 026 (Orlova *et al.*, 2006) have been described. Multiplex PCRs for simultaneous detection and differentiation of SPPV and GTPV (Fulzele *et al.*, 2006; Lamien *et al.*, 2011a) and SPPV/GTPV and ORFV have been described (Zheng *et al.*, 2007; Venkatesan *et al.*, 2014a). A PCR to differentiate SPPV field isolates from Romanian and Yugoslavian RM/65 SPPV strains vaccine strains has been developed (Chibssa *et al.*, 2018). The conventional PCR method is cheaper to perform than real-time PCR and less vulnerable to technical problems but is not quantitative. However, sensitivity and specificity are still superior to other methods available.

   ii) ***Quantitative PCR*** - Several realtime PCR assays have been developed for the detection of CaPVs targeting the DNA pol gene (Balamurugan *et al.*, 2009) and poly (A) polymerase (Balinsky *et al.*, 2008). Real-time PCR methods for simultaneous detection, quantitation, and differentiation of CaPVs have also been reported (Lamein *et al.*, 2011b; Venkatesan *et al.*, 2014b). Advantages of real-time PCR include speed, sensitivity, specificity, and real-time monitoring to determine exact concentrations (Bowden *et al.*, 2008; Balamurugan *et al.*, 2009). A real-time PCR assay using snapback primers and DNA intercalating Eva Green dye is useful for genotyping of CaPVs (Gelaye *et al.*, 2013).A triplex real-time PCR assay for the differentiation of SPPV, GTPV, and LSDV has also been reported (Haegeman *et al.*, 2013).

   iii) ***PCR-restriction fragment length polymorphism* (*PCR-RFLP*)** - Although genetic relatedness is observed among CaPVs, RE digestion profiles, especially *Hind* III digest can differentiate between CaPV species (Black *et al.*, 1986). PCR-RFLP assays targeting attachment gene digested with *EcoR* I (Fulzele *et al.*, 2006), full-length P32 gene digested with *Hinf* I (Hosamani *et al.*, 2004a) and partial P32 gene digested with *Dra* I and *EcoR* I Venkatesan *et al.*, 2012) allow differentiation of GTPV and SPPV.

*iv) **High-resolution melting PCR assay*** - High-resolution melting PCR assay for the simultaneous detection and differentiation of CaPVs has been developed (Chan *et al.*, 2014; Gelaye *et al.*, 2017; Pestova *et al.*, 2018).

*v) **Loop-mediated isothermal amplification*** **(*LAMP*)** ***assays*** - LAMP assays are simple, efficient, cost-effective, and convenient, making them a useful diagnostic tool for clinical samples (Zhao *et al.*, 2014). LAMP-based on P32 (Murray *et al.*, 2013), Poly (A) polymerase (Das *et al.*, 2012), and DNA polymerase (Venkatesan *et al.*, 2016) genes of CaPVs have been optimized and evaluated for rapid, specific and sensitive detection of members of CaPVs. Also, a LAMP assay has been developed for the differentiation of GTPV and SPPV (Zhao *et al.*, 2014). LAMP assays are highly specific, with no apparent cross-reactivity to other related viruses. LAMP assay result is based on color changes and therefore the interpretation of the results may be a source of misdiagnosis. The suitable pen-side extraction method required for the LAMP assay still needs to be developed.

*vi) **Recombinase polymerase amplification assay*** - A recombinase polymerase amplification assay combined with a real-time fluorescent detection and lateral flow dipstick has been developed (Yang *et al.*, 2017).

*vii) **Sequencing of host-range/virulence genes***- Sequencing of P32, RPO30, and GPCR genes (Le Goff *et al.*, 2009; Lamien *et al.*, 2011a) can be used to detect the virus strain. However, sequencing is expensive and requires expensive equipment as well as competent and highly trained staff.

Serological Detection

Different members of the genus *Capripoxvirus* (SPPV, GTPV, and LSDV) cannot be distinguished serologically. Conventionally, single radial hemolysis test, RPHA test, and indirect fluorescent antibody test have been employed successfully to diagnose SGP.

1. ***Serum neutralization test*** **(*SNT*)** - The current gold standard for determining anti-CaPV antibodies is SNT. Although, SNT is very specific, but is slow, labor-intensive and not sufficiently sensitive to detect low levels of antibodies present in animals with mild clinical disease or vaccination. Also, it requires handling of live CaPV which may not be suitable during the eradication phase and in countries, where the disease is exotic. Standardization of the assay is difficult and the sensitivity of the assay may vary in different laboratories. Other drawbacks include partial neutralization, low serum neutralization indices, and virus breakthrough due to variable susceptibility of cell cultures. SNT using lamb testis or lamb kidney tissue-culture system

shows incomplete neutralization with low titers (Plowright and Ferris, 1958). The use of Vero cells in neutralization has been reported to give more consistent results than LT or LK cells.

2. ***Western blot*** - Western blotting of test sera using P32 protein against CaPVs provides a sensitive and specific system for the detection of antibodies to CaPVs but it is expensive and difficult to carry out (Chand *et al.*, 1994).

3. ***Indirect ELISAs*** - ELISAs for the detection of antibodies against CaPVs have been described using inactivated whole virus antigen (Babiuk *et al.*, 2009) and recombinant P32 protein (Heine *et al.*, 1999; Bhanot *et al.*, 2010; Venkatesan *et al.*, 2018) and core antigens (Bowden *et al.*, 2009). However, currently, there is no validated ELISA for the detection of antibodies to CaPVs.

    i) ***Whole virus-based ELISA*** - The inactivated whole virus-based ELISA used to detect CaPV-specific antibodies has been found more sensitive than SNT and CIE tests (Sharma *et al.*, 1988; Bhanuprakash *et al.*, 2006b; Babiuk *et al.*, 2009). Whole virus ELISA is suited to screen sera from all three host species, but the viral antigen is difficult and expensive to produce in large quantities. A whole-virus antigen-based ID Screen® Capripox Double Antigen Multi-species ELISA assay for the detection of antibodies against capripoxviruses is commercially available by IdVet.

    ii) ***Recombinant antigen-based ELISA*** - Convenient and effective tests suitable for high through put CaPV sero-surveillance are not readily available. It would be a significant advance if the immunodominant CaPV antigens were identified and utilized in an ELISA format. Several attempts have been made since long to express various CaPV proteins in different expression systems *viz.* bacterial, mammalian, and yeast cells, and to assess the potential use of recombinant proteins in various diagnostic assays (Chervyakova *et al.*, 2016). Development of a recombinant protein-based ELISA is impeded by difficulties in identifying a single immuno-dominant CaPV antigen which is stable and easy to purify (Babiuk *et al.*, 2009). Development of such techniques is needed for high throughput screening of a large number of infected and vaccinated serum samples. Recombinant P32 protein expressed in prokaryotic systems (Heine *et al.*, 1999; Qiang *et al.*, 2007), yeast (Bhanot *et al.*, 2009) and mammalian (Chen *et al.*, 2008) systems has been used for detection of antibodies to CaPVs, but difficulties with expression and stability of the recombinant antigen have compromised these tests (Heine *et al.*, 1999). Peptide-based antigenic region of P32 were utilized to develop an indirect ELISA method for detecting SPPV and GTPV antibodies in animals (Tian *et al.,* 2010). Among 42 ORFs, ORF 95 and ORF 103 have been selected as antigens for ELISA based on their reactivity (Bowden *et al.,* 2009). ORF117 (A27L) has also been evaluated for diagnostic potential (Dasprakash *et al.,* 2019).

Prevention and Control

Vaccination is the only effective method to prevent and reduce the incidence of SGP infections in endemic countries. A variety of attenuated live and inactivated capripox vaccines have been shown to protect sheep and goats against SGP infections (Bhanuprakash *et al.,* 2012). SPPV and GTPV are phylogenetically distinct viruses displaying a host preference. Cross-protection among sheep and goats due to sheeppox and goatpox vaccine or vice versa is partial (Hosamani *et al.,* 2004b). For these reasons, the use of indigenous homologous vaccines based on indigenous or locally prevalent strains is highly desirable (Bhanuprakash *et al.,* 2011) as genetic differences between field isolates and vaccine strains may lead to vaccine failure (Gelaye *et al.,* 2015). Historically, ovination/caprination *i.e.* inoculation of infected scab material suspended in glycerol saline has been used in many countries including India to prevent SPPV and GTPV infection, respectively (Bhanuprakash *et al.,* 2004).

Inactivated vaccines: Phenol-inactivated Mysore strain (Dubey and Sahney, 1978), formalin-inactivated Ranchi (Goswami and Soman, 1988), Sambalpur and Mukteswar strains (Yadav *et al.,* 1986) have been evaluated for goatpox. For sheeppox, aluminum hydroxide adsorbed inactivated vaccines (Boumart *et al.,* 2016) have been used. Inactivated vaccines confer a low level of immunity and give variable results (Awad *et al.,* 2003). Short duration of immunity, shorter shelf-life, cost, and large antigenic mass are the inherent drawbacks of inactivated vaccines.

Live attenuated vaccines: Several live virus vaccines have been used for protection against sheeppox and goatpox (Kitching, 1983; Kitching, 1986; Bhanuprakash *et al.,* 2004). For goatpox, live attenuated Gorgan and Mysore strains have been tried (Dubey and Sahney, 1978; Kitching, 2003). An indigenous, live, attenuated Vero cell-based goatpox vaccine has been developed in India by the Indian Veterinary Research Institute (IVRI) Mukteswar using Uttarkashi strain, P-60) (Hosamani *et al.,* 2004b). The vaccine is found to be safe, and potent and provides complete protection against GTPV for 52 months (Bhanuprakash *et al.,* 2022). The vaccine has undergone extensive field validation in different parts of the country after successful in-house trials (Hosamani *et al.,* 2004b). For sheeppox, a number of strains like KSGPV O-240 (Kitching *et al.,* 1983), RM/65 (Sharma *et al.,* 1987), Karnal (Singh *et al.,* 1987), Ranipet (Bhanuprakash *et al.,* 2004) and Roumanian-Fanar (RF) (Ramyar and Hessami, 1967; Bhanuprakash *et al.,* 2003) strains have been used with variable success. A Vero cell-adapted SPPV vaccine (SPPV-Srinagar, 38/00, P40) developed at IVRI, Mukteswar is safe, potent, and efficacious (Yogisharadhya *et al.,* 2011). Problems with live vaccines include variable protection, characterization of vaccine virus, and safety issues (post-vaccinal reactions). There is a great possibility of generating vaccines with reduced virulence by genetic sequencing and by knocking out the gene/s involved in virulence such as kelch-like protein and ankyrin proteins in CaPVs. In this line, GTPV-AV41 deleted with thymidine kinase and ORFs 8-18 has shown satisfactory immunogenicity along with good safety (Zhu *et al.*, 2018). IL-10 deleted LSDV is safe for sheep and goats against virulent CaPV challenge (Boshra *et al.*, 2013).

Combined vaccines: Combined bivalent vaccines consisting of GTPV/SPPV and PPRV have been developed successfully (Hosamani *et al.*, 2006).

Vectored vaccines: CaPV genome as a vector for cloning the genes of other pathogens, like F and H genes of peste des petis ruminants (PPR) (Caufour *et al.*, 2014; Diallo *et al.*, 2002; Chen *et al.*, 2010; Caufour et al., 2014; Fakri *et al.*, 2018), VP7 of bluetongue virus (Wade-Evans *et al.*, 1996) have been evaluated. ORF 135 of GTPV encoding an immunomodulatory protein may serve as an optimized insertion site to generate capripoxvirus-vectored live dual vaccines (Zhang *et al.*, 2018).

Currently, no vaccines with a differentiation of infected from vaccinated animals (DIVA) component are commercially available against CaPVs. Identification of a single, suitable, immunodominant marker antigen may prove to be a more significant challenge in developing a DIVA-enabled vaccine (Tuppurainen *et al.*, 2017).

Conclusions and Future Perspectives

SGP infections pose a socio-economic threat to small ruminants in enzootic countries like India as well as the possible threat of expansion of their geographical range into naive places due to animal trade. Mixed or co-infections of these diseases are common and cannot be ruled out during the control program. In developing countries like India, the occurrence of these mixed infections in sheep and goats is usually either undiagnosed or diagnosed as sheeppox or goatpox based on the host species involved or as orf infections by its characteristic lesions. In these situations, early and rapid diagnosis using PCR/LAMP assays as a multiplex format for simultaneous detection and differentiation of CaPV as well orf will be useful. In addition to prompt and unequivocal diagnosis, a well-organized vaccination and effective biosecurity measures are the main control elements to contain the disease. The development of high-throughput ELISA with enhanced sensitivity and specificity as well as a DIVA vaccine along with a companion test is need of the hour.

REFERENCES

Adlakha, S.C., Bansal, M.P. and Malik, B.S. (1971). Studies on sheep and goatpox viruses. I. Adaptation of the viruses in laboratory animals and tissue culture. Indian J Anim Sci, 41(3): 171-175.

Awad, M., Michael, A., Soliman, S.M., Samir, S.S. and Daoud, A.M. 2003. Trials for preparation of inactivated sheep pox vaccine using binary ethyleneimine. Egypt J Immunol, 10(2): 67-72.

Babiuk, S., Bowden, T.R., Boyle, D.B., Wallace, D.B. and Kitching, R.P. (2008). Capripoxviruses: an emerging worldwide threat to sheep, goats and cattle. Transbound Emerg Dis, 55(7): 263-272.

Babiuk, S., Parkyn, G., Copps, J., Larence, J.C., Sabara, M.I., Bowden, T.R., Boyle, D. B. and Kitching, R. P. 2007. Evaluation of an ovine testes cell line (OA3.Ts) for use in the propagation and detection of capripox virus and development of immunostaining technique for viral plaque visualization. J. Vet. Diagn. Invest. 5: 486-491.

Babiuk, S., Wallace, D.B., Smith, S.J., Bowden, T.R., Dalman, B., Parkyn, G., Copps, J. and Boyle, D.B. (2009). Detection of antibodies against capripoxviruses using an inactivated sheeppox virus ELISA. Transbound Emerg Dis, 56(4): 132-141.

Balamurugan, V., Jayappa, K.D., Hosamani, M., Bhanuprakash, V., Venkatesan, G and Singh, R.K. (2009). Comparative efficacy of conventional and Taqman polymerase chain reaction assays in the detection of capripoxviruses from clinical samples. J Vet Diagn Invest, 21(2): 225-231.

Balinsky, C.A., Delhon, G., Afonso, C.L., Risatti, G.R., Borca, M.V., French, R.A., Tulman, E. R., Geary, S.J. and Rock D.L. (2007). Sheeppox virus kelch-like gene SPPV-019 affects virus virulence. J Virol, 81(20): 11392-11401.

Balinsky, C.A., Delhon, G., Smoliga, G., Prarat, M., French, R.A., Geary, S.J., Rock, D.L. and Rodriguez, L.L. (2008). Rapid preclinical detection of sheeppox virus by a real-time PCR assay. J Clin Microbiol, 46(2): 438-442.

Bhanot, V., Balamurugan, V., Bhanuprakash, V., Venkatesan, G., Sen, A., Yadav, V., Yogisharadhya, R. and Singh, R.K. (2009). Expression of P32 protein of goatpox virus in Pichia pastoris and its potential use as a diagnostic antigen in ELISA. J Virol Methods, 162(1-2): 251-257.

Bhanuprakash, V., Moorthy, A.R.S., Krishnappa, G., Srinivasa Gowda, R.N. and Indrani, B.K. (2003). A live attenuated vaccine for sheep pox. Indian J Anim Sci, 73(10): 1093-1098.

Bhanuprakash, V., Indrani, B.K., Hegde, R., Kumar, M.M. and Moorthy, A.R.S. (2004). A classical live attenuated vaccine for sheep pox. Trop Ani Health Prod, 36(4): 307-320.

Bhanuprakash, V., Moorthy, A.R.S., Krishnappa, G., Srinivasa Gowda, R.N. and Indrani, B K. 2005. An epidemiological study of sheeppox infection in Karnataka state, India. Rev Sci Tech Off Int Epiz 24(3): 909-920.

Bhanuprakash, V., Indrani, B.K., Hosamani, M. and Singh, R.K. (2006a). The current status of sheep pox disease. Comp Immunol Microbiol Infect Dis, 29: 27-60.

Bhanuprakash, V., Hosamani, M., Juneja, S., Kumar, N. and Singh, R.K. (2006b). Detection of goat pox antibodies: comparative efficacy of indirect ELISA and counter immunoelectrophoresis. J Appl Anim Res, 30(2): 177-180.

Bhanuprakash, V., Hosamani, M. and Singh, R.K. (2011). Prospects of control and eradication of capripox from the Indian subcontinent: a perspective. Antiviral Res, 91(3): 225-232.

Bhanuprakash, V., Hosamani, M., Venkatesan, G., Balamurugan, V., Yogisharadhya, R. and Singh, R.K. (2012). Animal poxvirus vaccines: a comprehensive review. Exp Rev Vaccines, 11(11): 1355-1374.

Bhanuprakash, V., Hosamani, M., Venkatesan, G. and Singh, R.K. (2022). Long-term protective immunity to goatpox in goats after a single immunization with a live attenuated goatpox vaccine. Arch Virol, 167(10): 2035-2040.

Biswas, S., Noyce, R.S., Babiuk, L.A., Lung, O., Bulach, D.M., Bowden, T.R., Boyle, D.B., Babiuk, S. and Evans DH. (2020). Extended sequencing of vaccine and wild-type capripoxvirus isolates provides insights into genes modulating virulence and host range. Transbound Emerg Dis, 67(1): 80-97.

Black, D.N., Hammond, J.M. and Kitching, R.P. (1986). Genomic relationship between capripoxviruses. Virus Res, 5(2-3): 277-292.

Bora, D.P., Venkatesan, G., Neher, S., Mech, P., Ralte, E., Sarma, D. and Das, S.K. (2018). Goatpox outbreak at a high altitude goat farm of Mizoram: possibility of wild life spill over to domestic goat population. VirusDis, 29(4): 560-564.

Boshra, H., Truong, T., Nfon, C., Gerdts, V., Tikoo, S., Babiuk, L.A., Kara, P., Mather, A., Wallace, D.B. and Babiuk, S. (2013). Capripoxvirus-vectored vaccines against livestock diseases in Africa. Antiviral Res, 98(2): 217-227.

Boumart, Z., Daouam, S., Belkourati, I., Rafi, L., Tuppurainen, E., Tadlaoui, K.O. and El Harrak, M. (2016). Comparative innocuity and efficacy of live and inactivated sheeppox vaccines. BMC Vet Res, 12(1): 133.

Bowden, T.R., Babiuk, S.L., Parkyn, G.R., Copps, J.S. and Boyle, D. (2008). Capripoxvirus tissue tropism and shedding: a quantitative study in experimentally infected sheep and goats. Virol, 371(2): 380-393.

Bowden, T.R., Coupar, B.E., Babiuk, S.L., White, J.R., Boyd, V., Duch, C.J., Shiell, B.J., Ueda, N., Parkyn, G.R., Copps, J.S. and Boyle, D.B. (2009). Detection of antibodies specific for sheeppox and goatpox viruses using recombinant capripoxvirus antigens in an indirect enzyme-linked immunosorbent assay. J Virol Methods, 161(1): 19-29.

Bratke, K. A. and McLysaght, A. 2008. Identification of multiple independent horizontal gene transfers into poxviruses using a comparative genomics approach. BMC Evol Biol, 8: 67.

Carn, V.M. (1995). An antigen trapping ELISA for the detection of capripoxvirus in tissue culture supernatants and biopsy samples. J Virol Methods, 51(1): 95-102.

Caufour, P., Rufael, T., Lamien, C.E., Lancelot, R., Kidane, M., Awel, D., Sertse, T., Kwiatek, O., Libeau, G., Sahle, M. and Diallo, A. (2014). Protective efficacy of a single immunization with capripoxvirus-vectored recombinant peste des petits ruminants vaccines in presence of pre-existing immunity. Vaccine, 32(30): 3772-3779.

Chan, K.W., Lee, M.L., Yang, W.C., Wong, M.L., Hsu, W.L., Ho, C.F., Hsieh, Y.C. and Wang, C.Y. (2014). Differential diagnosis of goatpox virus in Taiwan by multiplex polymerase chain reaction assay and high-resolution melt analysis. J Vet Diagn Invest, 26(2): 195-202.

Chand, P., Kitching, R.P and Black, D.N. (1994). Evaluation of the Western blot analysis of virus-specific antibody responses for capripox and contagious pustular dermatitis viral infections in sheep. Epidemiol Infect, 113(2): 377-385.

Chen, W., Hu, S., Qu, L., Hu, Q., Zhang, Q., Zhi, H., Huang, K. and Bu, Z. (2010). A goat poxvirus-vectored peste-des-petits-ruminants vaccine induces long-lasting neutralization antibody to high levels in goats and sheep. Vaccine, 28(30): 4742-4750.

Chen, Y.X., Cai, X.P., Jing, Z.Z., Ding, J.T., Wang, Y., Meng, X.L., Zhang, Y., Jia, W.Z., Qiao, J., Yan, H.B., Fang, Y.X., Chen, G.H. and Luo, X.N. (2008). Construction, expression and immunogenicity of eukaryotic vectors based on goat pox virus P32 gene. Bing Du Xue Bao, 24(3): 133-137.

Chervyakova, O.V., Zaitsev, V.L., Iskakov, B.K., Tailakova, E.T., Strochkov, V.M., Sultankulova, K.T., Sandybayev, N.T., Stanbekova, G.E., Beisenov, D.K., Abduraimov, Y.O., Mambetaliyev M., Sansyzbay A.R., Kovalskaya, N.Y., Nemchinov, L.G., Hammond and R.W. (2016). Recombinant sheep pox virus proteins elicit neutralizing antibodies. Viruses, 8(6): 159.

Chibssa, T.R., Grabherr, R., Loitsch, A., Settypalli, T.B.K., Tuppurainen, E., Nwankpa, N., Tounkara, K., Madani, H., Omani, A., Diop, M., Cattoli, G., Diallo, A. and Lamien, C.E. (2018). A gel-based PCR method to differentiate sheeppox virus field isolates from vaccine strains. Virol J, 15(1): 59.

Chu, Y., Yan, X., Gao, P., Zhao, P., He, Y., Liu, J. and Lu, Z. (2011). Molecular detection of a mixed infection of goatpox virus, orf virus, and *Mycoplasma capricolum* subsp. *capripneumoniae* in goats. J Vet Diag. Invest, 23: 786-789.

Das, A., Babiuk, S. and McIntosh, M.T. (2012). Development of a loop-Mediated isothermal amplification assay for rapid detection of Capripoxviruses. J Clin Microbiol, 50(5): 1613-1620.

Dashprakash, M., Venkatesan, G., Kumar, A., Sankar, M., Arya, S., Ramakrishnan, M.A., Pandey, A.B. and Mondal, B. (2019). Prokaryotic expression, purification and evaluation of goatpox virus ORF117 protein as a diagnostic antigen in indirect ELISA to detect goatpox. Arch Virol, 164(4): 1049-1058.

Davies, F.G. (1981). Sheep and goatpox and Lumpy skin disease. In: Gibbs, E. P. J. ed. Virus diseases of food animals. Vol 2. Academic Press, London, pp. 733-748.

Diallo, A., Minet, C., Berhe, G., Le Goff, C., Black, D.N., Fleming, M., Barrett, T., Grillet, C. and Libeau, G. (2002). Goat immune response to capripox vaccine expressing the hemagglutinin protein of peste des petits ruminants. Ann N Y Acad Sci, 969: 88-91.

Dubey, S C. and Sawhney, A.N. (1978). Live and inactivated tissue culture vaccines against goat-pox. Indian Vet J, 55: 926-927.

Dutta, T. K., Roychoudhury, P., Kawlni, L., Lalmuanpuia, J., Dey, A., Muthuchelvan, D., Mandakini, R., Sarkar, A., Ramakrishnan, M. A. and Subudhi, P. K. 2019. An outbreak of Goatpox virus infection in Wild Red Serow (*Capricornis rubidus*) in Mizoram, India.Transbound Emerg Dis, 66(1): 181-185.

Embury-Hyatt, C., Babiuk, S., Manning, L., Ganske, S., Bowden, T.R., Boyle, D.B. and Copps, J. (2012). Pathology and viral antigen distribution following experimental infection of sheep and goats with capripoxvirus. J Comp Pathol, 146(2-3): 106-115.

Fakri, F., Bamouh, Z., Ghzal, F., Baha, W., Tadlaoui, K., Fihri, O.F., Chen, W., Bu, Z. and Elharrak, M. (2018). Comparative evaluation of three capripoxvirus-vectored peste des petits ruminants vaccines. Virol, 514: 211-215.

Fulzele, S., Singh, R.K., Hosamani, M., Mondal, B. and Yadav, M.P. (2006). Evaluation of duplex PCR and PCR-RFLP for diagnosis of sheeppox and goatpox. Int J Trop Med, 1(2): 66-70.

Garner, M.G and Lack, M.B. (1995). Modeling the potential impact of exotic diseases on regional Australia. Aust Vet J, 72(3): 81-87.

Garner, M.G., Sawarkar, S.D., Brett, E.K., Edwards, J.R., Kulkarni, V.B., Boyle, D.B. and Singh, S.N. (2000). The extent and impact of sheep pox and goat pox in the state of Maharashtra, India. Trop Anim Health Prod, 32(4): 205-223.

Gelaye, E., Belay, A., Ayelet, G., Jenberie, S., Yami, M., Loitsch, A., Tuppurainen, E., Grabherr, R., Diallo, A. and Lamien, C.E. (2015). Capripox disease in Ethiopia: Genetic differences between field isolates and vaccine strain, and implications for vaccination failure. Antiviral Res, 119: 28-35.

Gelaye, E., Lamien, C.E., Silber, R., Tuppurainen, E.S., Grabherr, R. and Diallo, A. (2013). Development of a cost-effective method for capripoxvirus genotyping using snapback primer and dsDNA intercalating dye. PloS one, 8(10): 75971.

Gelaye, E., Mach, L., Kolodziejek, J., Grabherr, R., Loitsch, A., Achenbach, J.E., Nowotny, N., Diallo, A. and Lamien, C.E. (2017). A novel HRM assay for the simultaneous detection and differentiation of eight poxviruses of medical and veterinary importance. Sci Rep, 7: 42892.

Gershon, P. D. and Black, D.N. (1988). A comparison of the genomes of capripoxvirus isolates of sheep, goats, and cattle. Virol, 164(2): 341-349.

Goswami, T.K. and Soman, J.P. (1988). Development and evaluation of an inactivated goat-pox vaccine. Indian J Ani Sci 58(2): 200-203.

Gulbahar, M.Y., Cabalar, M., Gul, Y. and Icen, H. (2000). Immunohistochemical detection of antigen in lamb tissues naturally infected with sheep pox virus. J Vet Med Series B, 47 (3): 173-181.

Haddow, J.R. and Idnani, J.A. (1946). Goat dermatitis: A new virus disease of goats in India. Indian J Vet Sci, 16: 181-184.

Haegeman, A., Zro, K., Vandenbussche, F., Demeestere, L., Van Campe, W., Ennaji, M.M. and De Clercq, K. (2013). Development and validation of three Capripoxvirus real-time PCRs for parallel testing. J Virol Methods, 193(2): 446-451.

Hailat, N., Al-Rawashdeh, O., Lafi, S. and Al-Bateineh, Z. (1994). An outbreak of sheep pox associated with unusual winter conditions in Jordan. Trop Anim Health Prod, 26(2): 79-80.

Heine, H.G., Stevens, M.P., Foord, A.J. and Boyle, D.B. (1999). A capripox virus detection PCR and antibody ELISA based on the major antigen P32 the homolog of the vaccinia virus H3L gene. J Immunol Methods, 227(1-2): 187-196.

Hosamani, M., Mondal, B., Tembhurne, P.A., Bandyopadhyay, S.K., Singh, R.K. and Rasool, T.J. (2004a). Differentiation of sheep pox and goat poxviruses by sequence analysis and PCR-RFLP of P32 gene. Virus Genes, 29(1): 73-80.

Hosamani, M., Nandi, S., Mondal, B., Singh, R. K., Rasool, T.J. and Bandyopadhyay, S.K. (2004b). A Vero cell-attenuated Goatpox virus provides protection against virulent virus challenge. Acta Virol, 48(1): 15-21.

Hosamani, M., Singh, S.K., Mondal, B., Sen, A., Bhanuprakash, V., Bandyopadhyay, S.K., Yadav, M.P. and Singh, R K. (2006). A bivalent vaccine against goat pox and Peste des Petits ruminants induces protective immune response in goats. Vaccine, 24(35-36): 6058-6064.

Hutyra, F., Marek, J. and Manninger, R. (1946). Special pathology and therapeutics of the diseases of domestic animals. 5th Ed. Bailliere, Tindall and Cox, London. pp. 353-366.

Imperial Institute of Veterinary Research (1936-37). Annual report. Imperial Institute of Veterinary Research, Mukteswar (UP), India. pp. 18-19.

Ireland, D.C. and Binepal, Y S. (1998). Improved detection of capripoxvirus in biopsy samples by PCR. J Virol Methods, 74(1): 1-7.

Joshi, R.K., Chandra, R., Rao, V.D.P. and Garg, S.K. (1992). Goatpox: a review. J Appl Anim Res, 2: 57-72.

Kalra, S.K. and Sharma V.K. (1981). Adaptation of Jaipur strain of sheeppox virus in primary lamb testicular cell culture. Indian J Exp Biol, 19(2): 165-169.

Karim, M.A. (1983). Pox among Sannen goats in Iraq. Trop Anim Health Prod, 15(1): 62.

Kitching, P. (1983). Progress towards sheep and goat pox vaccines. Vaccine, 1(1): 4-9.

Kitching, R.P and Smale, C. (1986). Comparison of the external dimensions of capripoxvirus isolates. Res Vet Sci, 41(3): 425-427.

Kitching, R.P. (2004). Sheep pox and goat pox. In: Coetzer, J. A.W., Tustin, R. C. eds. Infectious Disease of Livestock, Vol. 2. Oxford University Press, Southern Africa. pp. 1277-1281.

Kitching, R.P. and Mellor, P S. (1986). Insect transmission of capripoxvirus. Res Vet Sci, 40(2): 255-258.

Kitching, R.P. and Taylor, W.P. (1985). Transmission of capripoxvirus. Res Vet Sci, 39(2): 196-199.

Kumar, A., Venkatesan, G., Hosamani, M., Bhanuprakash, V., Balamurugan, V., Ramakrishnan, M.A. and Singh, R.K. (2022). The complete genome sequence of Indian sheeppox vaccine virus and comparative analysis with other capripoxviruses. Gene, 810: 146085.

Lamien, C.E., Le Goff, C., Silber, R., Wallace, D.B., Gulyaz, V., Tuppurainen, E., Madani, H., Caufour, P., Adam, T., El Harrak, M. and Luckins, A.G. (2011a). Use of the Capripoxvirus homologue of Vaccinia virus 30 kDa RNA polymerase subunit (RPO30) gene as a novel diagnostic and genotyping target: Development of a classical PCR method to differentiate Goat poxvirus from Sheep poxvirus. Vet Microbiol, 149(1): 30-39.

Lamien, C.E., Lelenta, M., Goger, W., Silber, R., Tuppurainen, E., Matijevic, M., Luckins, A.G. and Diallo, A. (2011b). Real time PCR method for simultaneous detection, quantitation and differentiation of capripoxviruses. J Virol Methods, 171(1): 134-140.

Le Goff, C., Lamien, C.E., Fakhfakh, E., Chadeyras, A., Aba-Adulugba, E., Libeau, G., Tuppurainen, E., Wallace, D.B., Adam, T., Silber, R., Gulyaz, V., Madani, H., Caufour, P., Hammami, S., Diallo, A. and Albina, E. (2009). Capripoxvirus G-protein-coupled chemokine receptor: a host-range gene suitable for virus animal origin discrimination. J Gen Virol, 90(8): 1967-1977.

Madhavan, A., Venkatesan, G. and Kumar, A. (2016). Capripoxviruses of small ruminants: current updates and future perspectives. Asian J Anim Vet Adv, 11: 757-770.

Malik, Y.S., Singh, D., Chandrashekar, K.M., Shukla, S., Sharma, K., Vaid, N. and Chakravarti, S. (2011). Occurrence of dual infection of peste-des-petits-ruminants and goatpox in indigenous goats of central India. Transbound Emerg Dis, 58(3): 268-273.

Mangana-Vougiouka, O., Kottaridi, C. and Nomikou, K. 2008. The epidemiology of sheep pox in Greece from 1987 to 2007. Rev Sci Tech OIE, 27(3): 899-905.

Mangana-Vougiouka, O., Markoulatos, P., Koptopoulos, G., Nomikou, K., Bakandritsos, N. and Papadopoulos, P. 2000. Sheep poxvirus identification from clinical specimens by PCR, cell culture, immunofluorescence and agar gel immunoprecipitation assay. Mol Cell Probes, 14(5): 305-310.

Mellor, P.S., Kitching, R.P. and Wilkinson, P J. (1987). Mechanical transmission of capripox virus and African swine fever virus by *Stomoxys calcitrans*. Res Vet Sci, 43(1): 109-112.

Mirzaie, K., Barani, S.M. and Bokaie, S. (2015). A review of sheep pox and goat pox: perspective of their control and eradication in Iran. J Adv Vet Anim Res, 2(4): 373-381.

Murray, L., Edwards, L., Tuppurainen, E.S., Bachanek-Bankowska, K., Oura, C.A., Mioulet, V. and King, D.P. (2013). Detection of capripoxvirus DNA using a novel loop-mediated isothermal amplification assay. BMC Vet Res, 9: 90.

Ngichabe, C.K., Binepal, Y.S., Njiru, J.W. and Carn, V.M. (1999). Evaluation of an immunocapture enzyme-linked immunosorbent assay for detection of capripoxviruses. Vet Rec, 145: 231-232.

OIE. (2008). Sheeppox and goat pox. In: Manual of diagnostic tests and vaccines for terrestrial animals (mammals, birds and bees). 6th ed. Paris, France. pp. 1058-1068.

Orlova, E.S., Shcherbakova, A.V., Diev, V.I. and Zakharov, V.M. (2006). Differentiation of capripoxvirus species and strains by polymerase chain reaction. Mol Biol (Mosk), 40(1): 139.

Ozmen, O., Kale, M., Haligur, M. and Yavru, S. (2009). Pathological, serological, and virological findings in sheep infected simultaneously with bluetongue, peste-des-petits-ruminants, and sheeppox viruses. Trop Anim Health Prod, 41(6): 951-958.

Pandey, R. and Singh, I.P. (1970). A note on heat, chloroform and ether sensitivity of sheep pox and goat pox viruses. Acta Virol, 14(4): 318-319.

Pestova, Y., Byadovskaya, O., Kononov, A. and Sprygin, A. (2018). A real time high-resolution melting PCR assay for detection and differentiation among sheep poxvirus, goat pox virus, field and vaccine strains of lumpy skin disease virus. Mol Cell Probes, 41: 57-60.

Plowright, W. and Ferris, R.D. (1958). The growth and cytopathogenicity of sheep-pox virus in tissue cultures. Br J Exp Path, 39(4): 424-435.

Qiang, Z., Wei-min, M., Guo-hua, W., Xin-min, Y., Jian, L., Hai-xia, Z., Cai-zhu, Z., Run, W., Xiang-tao, L., Xue-peng, C. (2007). Prokaryotic expression of structural protein P32 gene of goat pox virus. J Gan Agric Univ, 42(5): 8.

Ramyar, H. and Hessami, M. (1967). Development of an attenuated live virus vaccine against sheeppox. Zentbl Vet Med, 14B: 516-519.

Rao, T.V.S. and Bandyopadhyay, S.K. (2000). A comprehensive review of goat pox and sheep pox and their diagnosis. Anim Health Res Rev, 1(2): 127-136.

Rao, T.V.S., Malik P., Nandi, S. and Negi, B.S. (1997). Evaluation of immunocapture ELISA for diagnosis of goat pox. Acta Virol, 41(6): 345-348.

Sathe, R.G. (1931). Pox in sheep. Indian Vet J, 8: 118-121.

Sharma, B., Negi, B.S., Yadav, M.P., Shankar, H. and Pandey, A.B. (1988). Application of ELISA in the detection of goat pox antigen and antibody. Acta Virol, 32: 65-69.

Sharma, P.C., Prasad, S. and Kalra, S.K. (1987). Immunological response of RM/65 sheeppox vaccine in lambs. Indian J Anim Sci, 57(9): 923-928.

Singh, I.P., Chandra, R., Rao, V.D.P., Garg, S.K. and Thapliyal, D.C. (1987). Immunogenicity of attenuated 'Karnal' strain of sheeppox virus. Indian J Anim Sci, 57(2): 119-121.

Soman, J.P. (1986). Antigenic detection of sheep poxvirus in the sheep kidney and testes cell cultures by immunoflourescent test. Indian Vet J, 63(10): 793-795.

Tantawi, H.H. and Al-Falluji, M.M. (1979). Laboratory characteristics of four strains of goatpox virus. Acta Virol, 23(6): 455-460.

Tian, H., Chen, Y., Wu, J., Shang, Y. and Liu, X. (2010). Serodiagnosis of sheeppox and goatpox using an indirect ELISA based on synthetic peptide targeting for the major antigen P32.Virol J, 7: 245.

Tulman, E.R., Afonso, C.L., Lu, Z., Zsak, L., Kutish, G.F. and Rock, D.L. (2001). Genome of lumpy skin disease virus. J Virol, 75(15): 7122-7130.

Tulman, E.R., Afonso, C.L., Lu, Z., Zsak, L., Sur, J.H., Sandybaev, N.T., Kerembekova, U.Z., Zaitsev, V.L., Kutish, G.F. and Rock, D.L. (2002). The genomes of sheeppox and goatpox viruses. J Virol, 76(12): 6054-6061.

Tuppurainen, E.S.M., Venter, E.H., Shisler, J.L., Gari, G., Mekonnen, G.A., Juleff, N., Lyons, N. A., De Clercq, K., Upton, C., Bowden, T.R., Babiuk, S.

and Babiuk, L. (2017). Review: Capripoxvirus diseases: current status and opportunities for control. Transbound Emerg Dis, 64(3): 729-745.

Venkatesan, G., Balamurugan, V. and Bhanuprakash, V. (2014a). Multiplex PCR for simultaneous detection and differentiation of sheeppox, goatpox and orf viruses from clinical samples of sheep and goats. J Virol Methods, 195: 1-8.

Venkatesan, G., Balamurugan, V. and Bhanuprakash, V. (2014b). TaqMan based real-time duplex PCR for simultaneous detection and quantitation of capripox and orf virus genomes in clinical samples. J Virol Methods, 201: 44-50.

Venkatesan, G., Balamurugan, V., Bhanuprakash, V. Singh, R.K. and Pandey, A.B. (2016). Loop-mediated isothermal amplification assay for rapid and sensitive detection of sheep pox and goat pox viruses in clinical samples. Mol Cell Probes, 30(3): 174-177.

Venkatesan, G., Teli, M.K., Sankar, M., Kumar, A., Dashprakash, M., Arya, S., Madhavan, A., Ramakrisnan, M.A. and Pandey, A.B. (2018). Expression and evaluation of recombinant P32 protein based ELISA for sero-diagnostic potential of capripox in sheep and goats. Mol Cell Probes, 37: 48-54.

Venkatesan, G., Balamurugan, V., Yogisharadhya, R., Kumar A. and Bhanuprakash, V. (2012). Differentiation of sheeppox and goatpox viruses by polymerase chain reaction-restriction fragment length polymorphism. Virol Sin, 27(6): 352-358.

Wade-Evans, A.M., Romero, C.H., Mellor, P., Takamatsu, H., Anderson, J., Thevasagayam, J., Fleming, M.J., Mertens, P.P., and Black, D.N. (1996). Expression of the major core structural protein (VP7) of bluetongue virus, by a recombinant capripox virus, provides partial protection of sheep against a virulent heterotypic bluetongue virus challenge. Virol, 220(1): 227-231.

Yadav, M.P., Pandey, A.B., Negi, B.S., Sharma, B. and Shanker, H. (1986). Studies on inactivated goatpox vaccine. Indian J Virol, 2(2): 207-221.

Yang, Y., Qin, X., Zhang, X., Zhao, Z., Zhang, W., Zhu, X., Cong, G., Li, Y. and Zhang, Z. (2017). Development of real-time and lateral flow dipstick recombinase polymerase amplification assays for rapid detection of goatpox virus and sheeppox virus. Virol J, 14(1): 131.

Yeruham, I., Yadin, H., Van Ham, M., Bumbarov, V., Soham, A. and Perl, S. (2007). Economic and epidemiological aspects of an outbreak of sheeppox in a dairy sheep flock. Vet Rec, 160: 236-237.

Yogisharadhya, R., Bhanuprakash, V., Hosamani, M., Venkatesan, G., Balamurugan, V., Bora, D.P., Bhanot, V., Prabhu, M. and Singh, R.K. (2011). Comparative efficacy of live replicating sheeppox vaccine strains in Ovines. Biologicals, 39(6): 417-423.

Zeng, X., Chi, X., Li, W., Hao, W., Li, M., Huang, X., Huang, Y., Rock, D.L., Luo, S. and Wang, S. (2014). Complete genome sequence analysis of goatpox virus isolated from China shows high variation. Vet Microbiol, 173(1-2): 38-49.

Zhang, M., Sun, Y., Chen, W. and Bu, Z. (2018). The 135 gene of goatpox virus encodes an inhibitor of nuclear factor-κb and apoptosis and may serve as an improved insertion site to generate vectored live vaccine. J Virol, 92(18): e00190-18.

Zhao, Z., Fan, B., Wu, G., Yan, X., Li, Y., Zhou, X., Yue, H., Dai, X., Zhu, H., Tian, B. and Li, J. (2014). Development of loop-mediated isothermal amplification assay for specific and rapid detection of differential goat pox virus and sheep pox virus. BMC Microbiol, 14: 10.

Zhao, Z., Zhu, X., Wu, N., Qin, X., Huang, C., Wu, G., Zhang, Q., Zhang, Z. (2018). Species-specific inhibition of capripoxvirus replication by host antiviral protein kinase R. Ann NY Acad Sci, 1438(1): 18-29.

Zheng, M., Liu, Q., Jin, N., Guo, J., Huang, X., Li, H., Zhu, W. and Xiong, Y. (2007). A duplex PCR assay for simultaneous detection and differentiation of capripox virus and orf virus. Mol Cell Probes, 21(4): 276-281.

Zhou, T., Jia, H., Chen, G., He, X., Fang, Y., Wang, X., Guan, Q., Zeng, S., Cui, Q. and Jing, Z. (2012). Phylogenetic analysis of Chinese sheeppox and goatpox virus isolates. Virol J, 9: 25.

Zhu, Y., Li, Y., Bai, B., Fang, J., Zhang, K., Yin, X., Li, S., Li, W., Ma, Y., Cui, Y., Wang, J., Liu, X., Li, X., Sun, L. and Jin, N. (2018). Construction of an attenuated goatpox virus AV41 strain by deleting the TK gene and ORF8-18. Antiviral Res, 157: 111-119.

Chapter 8

Epidemiology, Advances in Diagnosis and Control of Bluetongue Disease

K. Chand, S. Gautam, A.P. Madhusoodan, N.S. Kharayat, D. Bisht and B. Mondal*

ICAR-Indian Veterinary Research Institute, Mukteswar Campus, Nainital – 263 138, Uttarakhand, India
**e-mail: virusshield@gmail.com*

ABSTRACT

Bluetongue is World Organization for Animal Health (WOHA) listed notifiable disease of sheep, but it also affects most of the domestic and wild ruminants. The disease is caused by Bluetongue Virus (BTV) of family Reoviridae of genus Orbivirus. In India, bluetongue is endemic and most of serotypes have been reported one or another time. Outbreak in India mostly follow monsoon pattern, as transmission depends upon breeding conditions of vectors. Currently, over all 29 serotypes were defined worldwide based on serological responses to major serotype determinant VP2 protein. Tentative diagnosis of disease rely on postmortem finding, clinical signs and by determining epidemiological scenario which is further confirmed by virus isolation, antigen and antibody detection. Control strategies include prophylactic immunization and vector control. Currently in bluetongue endemic regions, serotype specific live attenuated and inactivated vaccines are in use. In India, pentavalent vaccine containing BTV-1, 2, 10, 16 and 23 serotypes is used in the some part of southern India.

***Keywords**: Bluetongue, Sheep, Serotype, VP2 protein, Diagnosis, Vaccine.*

Introduction

According to the 20th livestock census, India secured 2nd and 3rd position in terms of goat (148.88 million) and sheep (74.26 million) population in world (DAHD, 2019). In India distribution of sheep and goat is 13.87 percent and 27.80 percent respectively and are mostly reared for the purpose of meat and wool along with milk and manure in some area which makes these animals contributing toward economy. However, there are several diseases affecting livestock, leading to the serious disease in animals consequently making both animal and owners suffer because of huge losses of economy affecting their livelihood. Bluetongue is one of arbovirus borne disease of domestic and wild ruminants, which is transmitted by *Culicoides spp.* with symptom being most devastating in sheep.

Bluetongue is first reported in South Africa after an introduction of a sheep from Europe (Spreull, 1905). The word "Bluetongue" has been taken from African term "bloutong", that was mainly used by farmers of South-Africa as they noticed the cyanosis of the tongue in sheep that were diseased. Till, 1940 disease was limited to South Africa, subsequently epidemic recorded beyond Africa in Cyprus and in due course disease was also found in Middle East Asia and Southern Europe countries (Gambles, 1949; MacLachlan, 2004). Because of proficient vector population for distribution of virus, disease outbreak occurs all over the temperate regions and the tropical and subtropical regions of the globe as well. Overall 27 serotypes of BTV are known in the world. Isolation of BTV-28 is done from contaminated sheep pox vaccine and BTV-29 was isolated from alpaca. China and Italy has also reported putative new serotypes. The first outbreak of the bluetongue disease in small ruminants in India was reported from Maharashtra in 1964 (Sapre, 1964). A number of outbreaks in Northern and particularly in South India were reported thereafter.

Transmission and Clinical Signs

Vector responsible for transmission of Bluetongue virus, is a blood sucking midges of the *Culicoides* of family *Ceratopongonidae.* Among which C. *oxystoma* and C. *imicola* are the most common vectors for transmission of disease. Among numerous species of *Culicoides, C. variipennis, C. imicola, C. fulvus, C. actoni, C. wadai, C. nubeculosus,* proven to transmit the virus. However, sometime virus is transmitted vertically or horizontally between animals via ingestion of infected tissues and infected placenta was also reported. In the majority of *Culicoides* species, adult females are haematophagous and they take a blood meal about every 3 or 4 days. They mostly found near ground level, close to animals. The most of *Culicoides* species are active and thus bite in the twilight and at night (Mellor *et al.*, 2000).

Introduction of virus from one region to another region can occur by movements of domestic and wild ruminants, transport of animal product (semen, embryos), infected vector carried by living (plants or animals) or inanimate (ship, aircraft), active flight of infected vector and passive flight of vector by wind. The establishment of virus in new region will be determined by the distribution and number of susceptible hosts, titre of virus in host blood, vector population and the ambient temperature. BTV usually affects ruminants both wild as well as domestic, elephants, llamas and camelids. Among these sheep, pronghorn and also the white tail deer are clinically affected, whereas, camelids, goat and cattle are mostly asymptomatic. Cattle and goats usually exhibit subclinical infections and therefore may serve as key and covert viral reservoirs for sheep. However, BTV-8 caused epidemics in central and Western Europe with development of clinical signs in cattle between 2007 and 2008 (Darpel *et al.*, 2007; DeClercq, 2008). In sheep clinical manifestations ranges from subclinical to acute or even fatal. The morbidity and mortality rates, and the case fatality rate, depend on various factors, such as the breed and the age of the infected animals (older animals are more susceptible) as well as the serotype and strain involved. In sheep, viraemia is usually seen 3-5 days post infection. Facial edema; mucosal erosions and ulcerations of the upper G.I tract, including the oral mucosa, esophagus, and fore stomachs; coronitis associated with a reluctance to move; and fever reaching 42°C are the main clinical signs observed in sheep. Cyanosis of the tongue was rarely reported. Lameness, oedema of limbs, congestion of the coronary bands as well as stiffness of the limbs along with wasting of muscle tissues is commonly encountered. There is sometimes a focal loss of wool, together with dermatitis lesions. Diarrhoea is occasionally observed, essentially when the animals regain their normal feeding behaviour during the recovery period. Gross lesions of bluetongue in sheep are extremely variable. Externally, they include serous nasal discharge becoming, in some cases, mucopurulent with dry crusts on the nostrils, congestive to ulcerative muzzle, froth around the mouth and swelling of the lips, nose, face, intermandibular area, eyelids and sometimes ears. Skin hyperaemia can extend to the whole body, including axillary and inguinal areas. Congestion, oedema, petechiae or ecchymoses may be observed on mucous membranes of the oral cavity. In rumen and reticulum there may be hypermia and rarely erosion on papillae and laminae. In respiratory tract, lesions include odema, congestion, haemorrhages and sometime cyanosis of nasal mucosa, larynx and pharynx. Lesion in lung includes hypermia with alveolar and interstitial odema and sometime fluid in thoracic cavity. A variable size haemorrhage in the tuinca meida of pulmonary artery, near the heart is pathognomic lesion. Subepicardial and subendothelial petechiae and ecchymoses are common, mainly in the left ventricle. Focal grey-white areas of necrosis and haemorrhagic foci may be present in the myocardium, often in papillary muscles (Erasmus *et al.*, 1975; Brewer *et al.*, 1994).

Pathogenesis

Transmission of virus by *Culicoides* vector, depends upon several factors like daily bite rate of vector, extrinsic incubation period, temperature variation, number of female vector per host, host to vector transmission and host choice. The virus persists in the vector during their lifespan. After a blood meal, the BTV passes through the intestinal wall of vector and distributed via hemocoel to different tissues and then to salivary gland, where virus continues to replicate. Then virus subsequently excreted in the saliva of insect. The insect reaches its maximum infective capability 10 day after having blood from viraemic animal. Virus inoculation occurs via biting of infected midges on the skin. The bluetongue virus is routed via skin to a local lymph node, where it begins to replicate. After reaching to circulation primary viremia starts, which lead to the dissemination of virus to secondary lymphoid organs. Then the virus is disseminated in the entire body and mainly starts its multiplication in endothelial and mononuclear phagocytic cells, lymphocytes and other cells. Outer capsid VP2 protein of virus is responsible for cell attachment and transmission of virus by RBCs, as it remains sequestered in invagination of membrane of RBC resulting in prolonged viraemia even in the existence of neutralizing antibodies. Due to production of TNF-α, IL-1, 8, 6, IFN-1, cox2 there is increase in inflammatory response and eventually damage of cell and tissue. The virus causes vascular obstruction and infarction of tissue by causing damage in small blood vessels. This results in buccal ulcers, inflammation of coronary band and extravasations leading to facial and pulmonary oedema, pericardial and pleural effusions. Bluetongue virus can be recovered for several days or years from the blood of an infected animal. The incubation period in sheep is 6 to 8 days and suspected to be the same in cattle. The pathogenesis can vary depend upon the BTV serotype and ruminant species. There is large difference between cattle and sheep in expression of disease. This may be due to different response of endothelial cells to infection. As compare to sheep, cattle usually develop only a subclinical infection except BTV-8 serotype in northern Europe which caused clinical disease (MacLachlan, 2004; Vogtlin *et al.*, 2013).

Bluetongue Virus

Bluetongue virus is non enveloped virus having 10 distinct segmented dsRNA genome having icosahedral symmetry of approximately 80-90nm in diameter. BTV belongs to family *Reoviridae* of genus *Orbivirus* (Mertens *et al.*, 1989). The virus genome size is approximately 19.2kbp consists of 10 segments of dsRNA with 57 percent AU and 43 percent GC content. BTV genome encodes seven structural proteins (VP1-VP7) and five non structural proteins (NS1, NS2, NS3/NS3A, NS4 and NS5). Each genome encodes for single protein with the exception of seg-9 and 10. Out of that Segment-9 encodes for VP6 and NS4 protein, similarly seg-10 codes for NS3, NS3A and NS5 protein. Each genome segment size varies in order of decreasing molecular weight from 3954-822bp. Outer (VP2 and VP5), intermediate (VP3 and VP7), and innermost subcore (VP1, VP4, VP6) capsid layers are present in mature virion. The VP2 protein having 60 trimers facilitate clathrin mediated endocytosis, interspersed with 120 trimer of VP5 which contribute in BTV core entry into the cells. VP2 protein is responsible for hemagglutination and the induction of serotype-specific neutralizing antibodies. The strong affinity of recombinant VP2 for glycophorin A is responsible for its interaction with erythrocytes. The VP5 penetrates the membrane and directs the release of virus particles from endosomal compartment into cytoplasm. Moreover, pH-induced change in VP5 conformation allows fusion of membranes and formation of synctia. During infection VP5 influences the neutralizing antibodies elicited by VP2. The VP7 protein surrounds subcore surface, is a major serogroup specific antigen, VP3 protein binds with RNA molecules and transcriptase complex formed by VP1, VP4, VP6 among which VP1 act as RdRp, VP4 as RNA capping enzyme and VP6 having helicase activity. The VP7 directs the attachment and penetration in insect cells. The VP3/VP7 complex protects the genome of the virus by preventing the type-I interferon production activity through sensors in cytoplasm like cytosolic helicases and interactions with RNA silencing mechanisms.NS1 is an early viral protein expressed at high levels in infected host cells. NS1 plays a significant role in formation of tubules which are a conspicuous intracellular morphological feature of BTV infection. NS2 protein possesses a high affinity for single-stranded RNA and has phosphohydrolase activity, so plays role in BTV replication and assembly. NS3 acts as viroporin and increases membrane permeabilization thus facilitating virus release. NS4 protein has nuclear localization and associates with lipid droplets of cell and is suggested to be a significant defence against interferons. The BTV remains stable in proteins and can stay alive in blood for many years when stored at -20°C. The virus is sensitive to 3 percent NaOH, phenol, organic iodine compound and beta-propiolactone (Roy *et al.*, 1990; Stewart *et al.*, 2015).

Disease Scenario in India

Bluetongue is endemic in India and amidst the main sources of bluetongue virus in Asia due to huge animal population. Outbreak in India mostly follow monsoon pattern, as transmission depends upon breeding conditions of *Culicoides*. In south India, incidence of outbreak is relatively high, followed by northern plains, rarely observed in western India and also in Himalayan Mountain. In 1995, an outbreak was reported in Dehradun (Uttrakhand) which led to the death of goats. A number of outbreaks in South India were reported between 1986 to 1995. More than 2000 outbreaks in sheep occurred in India between 1997 to 2005 due to endemic circulation of various BTV serotypes. The main factor responsible for the outbreaks in different region of India is distribution and density of ruminants' population, rainfall, climatic conditions and *Culicoides* vectors. So far from India 23 serotypes has been reported among which 15 serotypes were reported by virus isolation and 22 (1-20, 23 and 24) by serology. Isolates of BTV in India show reassortments at higher degree between varying serotypes as well as different topotypes of same serotypes, leading to the dissemination of virus with multiple serotypes. Infection of host with serotypes exceeding one facilitates interchange of the genetic material, resulting in evolution of novel strains of BTV (Sushmitha *et al.*, 2012).

In India, seroprevalence of BT was reported in sheep, goat, cattle and buffalo. In 1984, Mehrotra and Shukla recorded the seroprevalence in sheep between 16.4- 61.1 percent in various states *viz.* Maharashtra, Andhra Pradesh, Karnataka, Jammu and Kashmir and Himachal Pradesh. Other workers also reported the seroprevalence of BT from Gujarat, Maharashtra, Madhya Pradesh, West Bengal, Rajasthan and Tamil Nadu. In Punjab, seroprevalence of 6.64 percent was reported. In Kerala, seroprevalence of 5.1 percent was recorded in sheep and goats. A seroprevalence of 47 percent was observed in goats of Sunderbans area in West Bengal. A high prevalence (86 percent) of BTV in Mithun in the northeast region of the country was also reported through a study. The overall prevalence of BT seroconversion of 47.58 percent and 20.61 percent of antigen positivity was found among the goats in Tripura during the year 2014 to 2017. In a study conducted in 2016 prevalence of BTV antibody and antigen in goats of Andaman and Nicobar islands was 78.49 percent and 72.5 percent respectively (Chand *et al.,* 2015; Rao *et al.,* 2016).

Diagnosis

A tentative diagnosis can be determined based on epizootiology, pathological lesion, vector dispersion, and clinical symptoms. Laboratory diagnostic techniques such as viral antigen and antibody detection to diagnose asymptomatic infection are used for confirmatory diagnosis. As BTV was found to be associated with RBCs, blood is commonly used for isolation and tissue sample like lymph node, lungs, liver, brain, bone marrow could also be used. Virus propagation in embryonated chicken egg (ECE) via intravenous route is more sensitive method for isolation of virus. For intravenous inoculation, washed blood cells are inoculated into 5-12 ECE that are 9-12 days old and eggs are incubated in humid chamber at 32-33.5 °C and candled daily. Embryo remaining alive at 7 days post inoculation is killed and infected embryos may have haemorrhagic appearance. For virus isolation BTV susceptible cell lines such as mouse L, *Culicoides sonorensis* (KC) or CuVa cells, *Aedes albopictus* clone C6/36 (AA), baby hamster kidney (BHK-21), African green monkey kidney (vero) and calf pulmonary artery endothelial (CPAE) are used. The efficiency of virus isolation in cell culture is often significantly lower with diagnostic samples as compared to ECE. Maximum recoveries are achieved by primary isolation of virus in ECE followed by passage in insect or mammal cells for further replication of virus. A cytopathic effect (CPE) is not necessarily observed in AA, KC or CuVa cells but appears in mammalian cells. Isolated virus may be detected after each ECE or cell culture passage by antigen detection or nucleic acid detection methods. Isolation of virus in sheep can be performed; this procedure is not routinely used but is useful where laboratory facilities are not available (Clavijo *et al.,* 2000; MacLachlan, 2004).

Bluetongue virus isolates are serogrouped on the basis of their reactivity with specific antisera that detect proteins, such as VP7, that are conserved within each serogroup. Antigen capture enzyme-linked immunosorbent assay (ELISA) is used for the detection of viral antigen present in ECE and culture medium harvests, infected insects, tissue and blood. Serogroup-specific monoclonal antibodies (MAbs) and polyclonal antibody have been used successfully to detect captured virus. For serotyping of BTV isolates, virus neutralization test (VNT) is used. There is a variety of cell lines (BHK, Vero and L929) commonly available to detect the presence of neutralising anti-BTV antibodies. The methods use to seotyping of BTV is plaque reduction, plaque inhibition and microtitre neutralization assay. RT-PCR methods give rapid identification of viral nucleic acid in blood and other tissues of infected animals. Multiple RT-PCR formats are available that can be used to detect 'serogroup' and to identify 'serotype' of BTV. These molecular methods are quicker than traditional virological and immunological assays, which may require up to 3-4 weeks generating information on serogroup and serotype. For international trade as per WOHA, NS1 gene-based RT-PCR test and nested PCR has been recommended. The nested assay may be useful for laboratories without the facility to perform real-time PCR. For first line diagnosis of BTV, TaqMan fluorescence probe based real time RT-PCR is highly sensitive. Advantages of real-time PCR assay over conventional PCR assay include speediness of testing, quantitation of the virus, and the reduced opportunity for contamination to occur as no post-amplification handling such as gel electrophoresis is needed. Antibodies against BTV in infected animals can be detected in a variety of ways that vary in sensitivity and specificity. Both serogroup-specific and serotype-specific antibodies are elicited. Agar gel immunodiffusion (AGID) is one of the standard testing method for international movement of ruminants, but is no longer considered accurate for use in the support of international trade. Major disadvantage of the AGID is lack of specificity in that it does not exclusively differentiate between antibodies to the BT and Epizootic hemorrhagic disease (EHD) serogroups. Therefore, C-ELISA is a preferred quick method, for detecting antibodies in serum samples. The BT competitive or blocking ELISA was developed to measure BTV-specific antibody without detecting cross-reacting antibody to other Orbiviruses. The specificity is the result of using one of a number of BT serogroup-reactive MAbs. A number of commercially produced C-ELISAs based on recombinant VP7 and anti-VP7 MAbare now available. These commercial ELISA kits are routinely used in many laboratories across the world and have been proved to be fit for purpose in ring-trials. During trade, for national and international surveillance C-ELISA is commonly used. The indirect ELISA has been also shown to be reliable and useful for surveillance purposes. Serum neutralization test (SNT) plays a major role which identifies antibodies against varying BTV serotypes.SNT is able to identify serotype-specific neutralising antibodies as well as determine their titre. It is also useful for epidemiological surveillance, transmission studies and for determining useful antibody response to vaccination.The use of SNT is often most useful in concurrence with other virological methods that, in combination, can provide a more definitive basis for resolving serotype distribution. Cell lines commonly used are for Serum neutralization assay are BHK-21 and Vero. BTV serotype-specific antisera raised in guinea-pigs or rabbits have been reported to have less serotype cross-reactivity than those raised in cattle or sheep (Chand *et al.*, 2019; OIE, 2021).

Prevention and Control

As Bluetongue virus leads to massive economic losses and restrict trade/ animal movement, control and prevention have become a matter of concern. Strategies includes, vector control, proper therapeutic measure to reduce mortality and morbidity, DIVA strategies for sustainable surveillance and monitoring. To gain a better understanding of the epidemiological condition and of the risks related with BT, and to create balanced way, it is essential to establish BT monitoring and surveillance programmes adapted to the risks. In disease free areas, animal movement control, vector control particularly during transportation, quarantine and serological survey should be done. Movements of animals that have been confined from vector attacks for 14 or 28 days and give negative PCR or ELISA tests results are considered to be safe. In addition, vaccinated animals or naturally immunized animals are considered to be safe and are therefore allowed to move to a BT-free zone. In infected areas, vector control both during transportation and establishments and animal housing facilities should be carried out. Physical barriers, insecticides, alarm systems and elimination of breeding sites should be applied (Tacken *et al.*, 2015).

Vaccination is the mainly efficient measure that can be applied in an infected region. The foremost point of vaccination is to avoid clinical cases and therefore to restrict the losses incurred by farmers. Vaccination is also used to control the disease, and can be used to facilitate safe trade or even to eradicate the disease. Presently serotype specific live attenuated and inactivated BTV vaccines are available for use in sheep and sheep and cattle, respectively. Recombinant vaccines based on different approaches are under development but none has been licensed. The foremost vaccines developed for prophylactic vaccination of livestock against BTV infection were modified-live virus (MLV) vaccines. MLV vaccines can be formed and administered in a cost-effective approach and have many intrinsic benefits. One dose of MLV is sufficient to induce virus-specific neutralizing antibodies, and low dose of attenuated virus are usually sufficient to stimulate a strong protective immune response. As monovalent live attenuated fails to provide cross protection against many of other serotypes, five different combinations of serotypes of attenuated BTV has been used in vaccine. Although Live attenuated vaccine having disadvantages such as reversion to virulence, genetic reassortment of the wild viruses with vaccine. Live attenuated vaccines should not be used during *Culicoides* vector seasons because these vectors may transmit the vaccine viruses from vaccinated animals to non vaccinated animals which may result in reassortment of gene and give rise to new and potentially more pathogenic strain. Inactivated vaccines for BTV were initially produced as killed whole virus preparations, using heat, ultraviolet (UV) radiation, or chemical methods of inactivation such as using hydroxylamine or binary ethylenimine. The main disadvantages of inactivated vaccines are that they are more costly to produce than MLV vaccines and they require complete inactivation to ensure residual infectious virus. The method of inactivation can reduce immunogenicity that may necessitate multiple vaccinations to produce protective immunity. In India currently inactivated adjuvanted pentavalent vaccines are in use, contains BTV-1, 2, 10, 16 and 23 serotypes. Baculovirus expressed VP2 protein based sub unit vaccine were also developed. Other recombinant vaccines such as Virus like particles (VLP), DNA vaccine, disabled infectious single animal vaccine (DISA), disabled infectious single-cycle vaccine (DISC), giving promising results, but most of them remains up to laboratory levels. Virus-like particle (VLP) vaccines have been shown to be proficient vaccine candidates against BTV. However, there has been no commercialization of VLP-based vaccines to date. Moreover presence of numerous serotypes poses problems for its control and eradication. Neutralization resistant VP2 phenotypes have been reported in a particular serotype which makes escape of virus from neutralization, leading to failure of cross protection. Thus before use of BT vaccines in a concerned area, serotype prevalent in particular area has to be taken into account (Bhanuprakash *et al.*, 2009; Reddy *et al.*, 2010; Van Rijn *et al.*, 2019).

To prevent the spread of disease through international trade in animals and animal products is one of the main objectives of the World Organisation for Animal Health. This is achieved by establishing international standards for a wide range of diseases. The OIE standards relating to bluetongue are published in the Terrestrial Animal Health Code (Terrestrial Code) and the Manual of Diagnostic Tests and Vaccines for Terrestrial Animals (Terrestrial Manual). These standards consist of guidelines and recommendations on declaring a territory, country or zone free from BTV, needs for seasonal freedom from BTV and recommendations for the safe importation of semen, embryos and live animals, into a BTV-free zone or country, along with specific surveillance guidelines for bluetongue.

REFERENCES

Bhanuprakash, V., Indrani, B.K., Hosamani, M., Balamurugan, V. and Singh, R.K. (2009). Bluetongue vaccines: the past, present and future. Exp Rev, 8(2): 191-204.

Brewer, A.W. and MacLachlan, N.J. (1994). The pathogenesis of bluetongue virus infection of bovine blood cells in vitro: Ultrastructural characterization. Arch Virol, 136: 287-298.

Chand, K., Biswas, S.K., Pandey, A.B., Muthuchelvan, D. and Mondal, B. (2015). Bluetongue in India: A review. Adv Anim Vet Sci, 3: 605-612.

Chand, K., Biswas, S.K. and Ramakrishnan, M.A. (2019). A multi-species indirect ELISA for detection group-specific antibodies against VP7 protein of bluetongue virus. Small Rumin Res, 180: 6-8.

Clavijo, A., Heckert, R.A., Dulac, G.C. and Afshar, A. (2000). Isolation and identification of bluetongue virus. J Virol Methods, 87(1-2): 3-23.

Darpel, K.E., Batten, C.A., Veronesi, E., Shaw, A.E., Anthony, S., Bachanek-Bankowska, K., Kgosana, L., Bin-Tarif, A., Carpenter, S., Muller-Doblies, U.U. and Takamatsu, H.H. (2007). Clinical signs and pathology shown by British sheep and cattle infected with bluetongue virus serotype 8 derived from the 2006 outbreak in northern Europen. Vet Rec, 161(8): 253-261.

De Clercq, K., De Leeuw, I., Verheyden, B., Vandemeulebroucke, E., Vanbinst, T., Herr, C., Méroc, E., Bertels, G., Steurbaut, N., Miry, C., De Bleecker, K., Maquet, G., Bughin,J., Saulmont, M., Lebrun, M., Sustronck, B., De Deken, R., Hooyberghs, J., Houdart,P., Raemaekers, M., Mintiens, K., Kerkhofs, P., Goris, N. and Vandenbussche, F. (2008). Transplacental Infection and Apparently Immunotolerance Induced by a Wild-type Bluetongue Virus Serotype 8 Natural Infection. Transbound Emerg Dis, 55: 352-359.

Erasmus, B.J. (1990). Bluetongue virus. Virus infections of ruminants. 3rd ed. Pp. 227-237.

Gambles, R.M. (1949). Bluetonge of sheep in Cyprus. J Comp Pathol, 59: 176-190.

MacLachlan, N.J. (2004). Bluetongue: pathogenesis and duration of viraemia. Vet Ital, 40(4): 462-467.

Mellor, P.S., Boorman, J. and Baylis, M. (2000). Culicoides biting midges: their role as arbovirus vectors. Annu Rev Entomol, 45(1): 307-340.

Mertens, P.P.C., Pedley, S., Cowley, J., Burroughs, J.N., Corteyn, A.H., Jeggo, M.H., Jennings, D.M. and Gorman, B.M. (1989). Analysis of the roles of bluetongue virus outer capsid proteins VP2 and VP5 in determination of virus serotype. Virol, 170(2): 561-565.

OIE Terrestrial Manual OIE Chapter 3.1.3. 2021. Bluetongue. In Manual of diagnostic tests and vaccines for terrestrial animal. 1-19.

Rao, P.P., Hegde, N.R., Reddy, Y.N., Krishnajyothi, Y., Reddy, Y.V., Susmitha, B., Gollapalli, S.R., Putty, K. and Reddy, G.H. (2016). Epidemiology of bluetongue in India. Transbound Emerg Dis, 63(2): 151-164.

Reddy, Y.K.M., Manohar, B.M., Pandey, A.B., Reddy, Y.N., Prasad, G. and Chauhan, R.S. (2010). Development and evaluation of inactivated pentavalent adjuvanted vaccine for Bluetongue. Indian Vet J, 87: 434-436.

Roy, P., Adachi, A., Urakawa, T., Booth, T.F. and Thomas, C.P. (1990). Identification of bluetongue virus VP6 protein as a nucleic acid-binding protein and the localization of VP6 in virus-infected vertebrate cells. J Virol, 64(1): 1-8.

Sapre, S.N. (1964). An outbreak of bluetongue in goats and sheep. Vet Rev, 15: 78-80.

Spreull, J. (1905). Malarial catarrhal fever (bluetongue) of sheep in South Africa. J Comp Pathol Therap, 18: 321-337.

Stewart, M., Hardy, A., Barry, G., Pinto, R.M., Caporale, M., Melzi, E., Hughes, J., Taggart, A., Janowicz, A., Varela, M. and Ratinier, M. (2015). Characterization of a second open reading frame in genome segment 10 of bluetongue virus. J Gen Virol, 96: 3280-3293.

Susmitha, B., Sudheer, D., Rao, P.P., Uma, M., Prasad, G., Minakshi, P., Hegde, N.R. and Reddy, Y.N. (2012). Evidence of bluetongue virus serotype 21 (BTV-21) divergence. Virus Genes, 44(3): 466-469.

Tacken, M.G., Daus, F.J., Feenstra, F., van Gennip, R.G. and van Rijn, P.A. (2015). Development of a competitive ELISA for NS3 antibodies as DIVA test accompanying the novel Disabled Infectious Single Animal (DISA) vaccine for Bluetongue. Vaccine, 33(42): 5539-5545.

Van Rijn, P.A. (2019). Prospects of next-generation vaccines for bluetongue. Front Vet Sci, 6: 407.

Vögtlin, A., Hofmann, M. A., Nenniger, C., Renzullo, S., Steinrigl, A., Loitsch, A. and Thür, B. (2013). Long-term infection of goats with bluetongue virus serotype 25. Vet Microbiol, 166(1-2): 165-173.

Chapter 9

Peste des Petits Ruminants (Small Ruminants Plague)

Balamurugan Vinayagamurthy

WOAH Reference Laboratory for PPR, Indian Council of Agricultural Research National Institute of Veterinary Epidemiology and Disease Informatics (ICAR-NIVEDI), Post Box No. 6450, Yelahanka, Bengaluru – 560 064, Karnataka, India e-mail: b.vinayagamurthy@icar.gov.in

ABSTRACT

Peste des Petits Ruminants (PPR), also known as "Small ruminants plague", is a severe viral disease affecting sheep and goats, which impedes small ruminant productivity. It is characterized by symptoms such as high fever, erosive stomatitis, conjunctivitis, gastroenteritis, and pneumonia, resulting in significant morbidity and mortality. PPR is caused by the PPR virus (PPRV), which belongs to the Morbillivirus capriniae species under the Morbillivirus genus in the Paramyxoviridae family. PPRV has a lipid envelope and a single-stranded RNA genome and encodes six structural proteins: nucleoprotein (N), viral RNA-dependent polymerase (L), RNA-polymerase phosphoprotein co-factor (P), matrix protein (M), fusion protein (F), and hemagglutinin protein (H). Various lineages of PPRV have been identified worldwide, with Lineage IV being the dominant one in India. PPR poses a significant economic threat due to its impact on livestock morbidity, mortality, and productivity, leading to trade restrictions. The virus primarily affects sheep and goats, with occasional infections in other animals, such as camels and wildlife small ruminants. Infected animals excrete the virus in secretions, discharges, and faeces, especially during the initial days of the disease. This can form infectious droplets in the air, leading to transmission through inhalation. PPRV spreads through close contact between infected and susceptible animals, primarily via inhalation of infectious aerosols. Although the virus can contaminate various surfaces, indirect transmission is less significant due to the virus's limited survival outside the host. Animal migration plays a critical role in the spread of PPR, particularly in endemic regions like India, where the active circulation of PPRV contributes to its transmission including trade of small ruminants in markets also increases the risk. The primary epidemiological factors contributing to the risk of PPR in sheep and goats include bringing animals into the flocks from unknown sources, as well as various management practices and rearing methods. While the virus can naturally spread to other species such as cattle and buffaloes, the clinical manifestation of the disease is typically not observed in these animals. Diagnosis of PPR is based on clinical signs, pathological lesions, and specific detection of the antibodies, virus or its genome using serological tests and molecular assays. The most effective strategy for preventing and controlling PPR is through vaccination. Two live attenuated vaccines, Nigeria 75/1 and Sungri/96, developed from cell culture, are available globally. Nevertheless, in India, only the Sungri/96 strain is commercially available. Symptomatic treatments are provided to infected animals, including antibiotics to prevent secondary bacterial infections, anti-diarrheal medicines, and fluids. The Government of India initiated a centrally sponsored PPR control program in 2011, which has now evolved into a PPR eradication program in 2022 aligned with the global strategy of the World Organization of Animal Health/Food and Agriculture Organization of the United Nations for Control and eradication by 2030.

Keywords: *PPR, Pathogen, Pathogenesis, Epidemiology, Diagnosis, Management, Prevention, Control.*

Introduction

Small ruminants are integral to the sustainable livelihoods of rural farmers in developing and underdeveloped countries, often serving as a critical source of income for landless farmers. Peste des Petits Ruminants (PPR), commonly referred to as the 'Small ruminants plague,' poses a significant threat to the national economies of countries where it is endemic. PPR is a highly contagious viral disease that primarily affects domestic and wild small ruminants, and it has more recently been observed in camels. The disease manifests with clinical signs such as fever, necrotic stomatitis, catarrhal inflammation of the ocular and nasal mucosa, enteritis, and bronchopneumonia, leading to either recovery or death of the infected animals. The disease was first described in 1942 in Côte d'Ivoire (Ivory Coast), West Africa (Gargadennec and Lalanne 1942); since then, the disease has spread to different regions in sub-Saharan Africa, the Arabian Peninsula, the Middle East, Southwest Asia, Indian subcontinent and other countries (Balamurugan *et al.*, 2014; Kumar *et al.*, 2014; Muthuchelvan *et al.*, 2015). The causative agent of PPR is the PPR virus (PPRV), a member of the *Morbillivirus caprinae*, which belongs to the genus *Morbillivirus* of the family *Paramyxoviridae*. PPRV is genetically classified into four lineages (I, II, III, and IV) based on fusion (F) and nucleocapsid (N) gene sequence analyses. Historically, lineages I to III were prevalent in Africa, while lineage IV was generally found in Asia. However, the introduction of lineage IV in some African countries has raised concerns about its transboundary nature. The virus's ability to cross borders and infect various lineages of PPRV presents a significant challenge to improving small ruminant productivity in endemic regions (Banyard *et al.*, 2010; Kwiatek *et al.*, 2011; Balamurugan *et al.*, 2014).

To combat the disease's impact, the World Organisation for Animal Health (WOAH) and the Food and Agriculture Organization (FAO) have initiated a global plan for PPR control and eradication by 2030. Currently, approximately >76 countries with a combined population of 2.1 billion sheep and goats are affected by PPR in Africa, the Middle East, and the Indian subcontinent. This underscores the importance of controlling and eradicating PPR, especially in regions where small ruminants are vital for food security and socio-economic growth. Given the critical role that small ruminants play in ensuring food security and driving socio-economic development, particularly in regions like Africa and Asia, the imperative to completely control and eradicate PPR cannot be overstated. Successful elimination of PPR would mark a significant achievement, making it only the second animal disease to be eradicated; following rinderpest (Parida *et al.*, 2015b). In a recent development, the Food and Agriculture Organization (FAO) and the World Organisation for Animal Health (WOAH) organized an International Conference in Abidjan, Côte d'Ivoire. During this event, high-level authorities from countries affected by PPR came to a consensus on a global plan aimed at controlling and eradicating PPR by the year 2030. Global initiatives, including the PPR Global Eradication Programme (PPR-GEP) and the PPR Global Control and Eradication Strategy (GCES) have been launched to combat PPR (WOAH and FAO, 2015). These efforts aim for a PPR-free world by 2030, with strategies and blueprints covering phases up to 2030 (FAO and WOAH, 2022).

This chapter provides a comprehensive overview of PPR, delving into various aspects such as the pathogen, pathogenesis, epidemiology, advancements in disease diagnosis and management, and measures for prevention, while also offering insights into future perspectives.

Peste Des Petits Ruminants Virus

PPR is caused by an enveloped RNA virus recognized as PPR Virus (PPRV), the only member of the *Morbillivirus caprinae* species, which belongs to the genus *Morbillivirus* of the family *Paramyxoviridae* (a subfamily of *Paramyxovirinae*) and classified under the order *Mononegavirales*. Other members within this *morbillivirus* genus consist of viruses that exhibit immunological cross-reactivity, including rinderpest virus (RPV), measles virus (MV), canine distemper virus (CDV), phocine distemper virus (PDV), and dolphin and porpoise morbillivirus (DMV) (Barrett *et al.*, 1993). The virus possesses a variable structure, featuring a lipoprotein envelope encasing a ribonucleoprotein core, housing the RNA genome. The PPRV genome closely resembles that of the Rinderpest Virus (RPV) at the nucleotide (nt) level. It is a single-stranded RNA genome, approximately 16 kilobases (kb) in length, with negative polarity. The genes are organised in the order of 3ʹ N–P/C/V–M–F–H–L 5ʹ, separated by intergenic regions, following the "rule of six". The genome is divided into six transcriptional units that encode eight proteins (Baron *et al.*, 2016). These include two non-structural proteins (V and C) and six structural proteins: surface glycoproteins fusion (F) and haemagglutinin (H), the matrix protein (M), the nucleocapsid protein (N), and the phosphoprotein (P), which forms the polymerase complex along with the large (L) protein. The N protein is the most abundant viral protein, and it plays a significant role in virus transcription and replication. The P gene encodes three proteins: the phosphoprotein (P) and two non-structural proteins (V and C) by overlapping open reading frames. The V protein is involved in the regulation of viral replication and is a strong inhibitor of interferon actions, contributing to immunosuppression during morbillivirus infections. The V protein of PPRV interferes with host innate immunity by blocking interferon signalling. This role is essential for further understanding the virus's mechanism to evade the host's immune response. The C protein is a small, basic, non-phosphorylated protein, and although it is the least conserved protein among morbilliviruses, there is significant conservation at the C-terminus of the protein. The M protein has an intrinsic ability to bind to lipid membranes and plays a vital role in viral assembly and budding. It forms a bridge between the viral glycoproteins (H and F) on the viral surface and the nucleocapsid-ribonucleoprotein core. A defective or mutated M protein can prevent the virus from completing its infectious cycle. The surface glycoproteins, H and F proteins, mediate virus attachment and entry into host cells. The F protein, involved in cell fusion, is highly immunogenic and important for protective immunity. PPRV F protein can induce both fusion and hemifusion. The H protein facilitates binding to host cell receptors, especially the signalling lymphocyte activation molecule (SLAM) or CD150.The L protein, also known as RdRp (RNA-dependent RNA polymerase), is the largest virus protein, well-conserved among morbilliviruses and plays a key role in virus replication. Overall, these proteins are essential for various stages of the virus's lifecycle and interaction with the host immune system (Balamurugan *et al.*, 2020).

In terms of PPRV's ability to survive outside a host, the virus is relatively fragile and cannot persist for extended periods. It can, however, survive for longer durations in chilled and frozen tissues. PPRV's stability varies with temperature, as it has a short half-life at higher temperatures at 56°C but can remain stable for hours at moderate temperature at 37°C. It is inactivated at 50°C for 60 minutes and is sensitive to ether and similar lipid solvent agents. PPRV's hemagglutination properties, its ability to adhere to erythrocytes, and other characteristics play a role in its infection and transmission (WOAH, 2013).

Epidemiology

Geographical Distribution

In 1942, the first scientific report of the disease known as PPR was documented by Gargadennec and Lalanne. This outbreak occurred in Ivory Coast

(Cote d'Ivoire) in West Africa, exhibiting clinical similarities to Rinderpest (RP) but affecting only small ruminants, while in-contact cattle remained healthy. This disease was referred to by various names, including "kata," "pseudorinderpest," "pneumoenteritis complex," and "stomatitis-pneumoenteritis syndrome" in French-speaking West African countries. Historical evidence, dating back to outbreaks in Senegal in 1871 and French Guinea in 1927, suggested that PPR might have existed for a more extended period than initially thought. PPR quickly spread to neighboring African countries, including Nigeria, Senegal, and Ghana. Until the early 1980s, documented outbreaks were reported across various parts of West Africa. In 1984, PPR's presence in Sudan marked the beginning of its spread. Since the late 1980s, PPR extended its reach to Central and Eastern Africa and parts of Asia. In India, the first confirmed outbreak of sheep occurred in Arasur village, Tamil Nadu, in 1987 (Shaila *et al.*, 1989). While initially believed to be limited to southern India, severe epidemics swept through the rest of the country in 1994 and beyond, establishing PPR as an enzootic disease in many northern Indian states (Balamurugan *et al.*, 2014, 2021). PPR infection, based on serological tests and clinical investigations, has been reported in various countries. The disease has since spread to new countries in Southern Africa, Central Asia, Southeast Asia, China, Southern Europe (Bulgaria), and Western Turkey, and recently in 2024 to Southern East Europe countries namely Georgia, Greece, and Romania. The introduction of the Asian lineage in some African nations and the presence of PPR in Europe through Western Turkey has raised significant concerns in global animal health (reviewed by Balamurugan, 2017). Recently, numerous PPR outbreaks have been reported in additional African countries such as Tanzania, the Democratic Republic of Congo, Angola, Egypt's Ismailia Province, Morocco, and Algeria, as well as in Asian regions like China. PPR is now endemic in many countries, with its geographic range expanding in recent years, including areas across China, Kazakhstan, Tunisia, Ismailia Province in Egypt, Tanzania, Mauritania, Eastern Asia and others (Balamurugan, 2017).

Molecular Epidemiology

Reports on molecular evolution, emergence, re-emergence, and the characterization of various PPRV isolates and strains have been documented across regions in Africa, the Middle East, Asia, and Southern Europe, originating from different countries. Sequences and phylogenetic analyses based on the F and N genes of diverse PPRV isolates from various parts of the world have led to the identification of four distinct lineages (I, II, III, and IV) of the virus (Shaila *et al.*, 1996; Dhar *et al.*, 2002; Balamurugan *et al.*, 2010). Lineage I is represented by viruses isolated in West Africa during the 1970s, originating from Nigerian isolates and the Senegalese strain. Later, viruses from Guinea, Senegal, Cote d'Ivoire, and Burkina Faso were also classified under this lineage. Lineage II comprises viruses isolated in West Africa during the 1980s, including Ivory Coast (Cote d'Ivoire), Guinea, and Ghana. Notably, this is the sole African lineage that did not cross the Red Sea to reach Asian countries. Lineage III consists of isolates from Eastern African countries, such as Meilig, Sudan, Ethiopia, Yemen, and Oman. Lineage IV, known as the Asian lineage, encompasses PPRV isolates from the Arabian Peninsula, the Middle East, and Asian countries, including India (Shaila *et al.,* 1996). As of now, there have been no reports of other PPRV lineages circulating in India. Furthermore, the PPRV isolated from an outbreak in the Indo-Bangladesh border of Tripura state in North-East India suggests the transboundary circulation of the virus between India and the Bangladesh border (Muthuchelvan *et al.,* 2014). Similarly, recent reports have indicated the spread of the disease to numerous new countries in Africa and Asia, involving different lineages of PPRV. Recently, lineage IV has also been identified in Turkey and several African countries, including Morocco, Cameroon, and Gabon. There has been evidence of PPR infection reported in several other countries, and the spread of the disease to new regions in Southern Africa, Central Asia, Southeast Asia, China, and recently in Southern East Europe countries namely Georgia, Turkey, Greece, and Romania, involving various lineages of PPRV, has raised concerns in global animal health (WOAH and FAO reports). Recent reports have also highlighted the co-circulation of PPRV Asian lineage IV with lineage II in Nigeria (Woma *et al.,* 2016) and the emergence of lineage IV virus in Ethiopia (Muniraju *et al.,* 2014). Recent introductions of the Asian lineage into some African countries and the presence of PPR in Europe through Western Turkey have further heightened these concerns (Kwiatek *et al.,* 2011; Balamurugan *et al.,* 2014).

Seasonal Occurrence

Seasonal variations in PPR outbreaks have been observed in various regions around the world. The occurrence of the disease can be influenced by factors such as sheep and goat husbandry practices, agroclimatic conditions, and geographical locations. Seasonal patterns of PPR outbreaks have been reviewed by several researchers (Balamurugan, 2017). PPR outbreaks are not confined to a specific time of the year but are most frequently encountered during particular seasons. These peak seasons typically coincide with the wet season or the cold dry season (Taylor, 1984). Several researchers have associated the onset of PPR outbreaks with the introduction of new animals into existing flocks. PPR has been observed to persist throughout the year, with a higher frequency of outbreaks during the lean periods, which include the wet season, rainy season, or cold dry season (Balamurugan *et al.,* 2021). This seasonal pattern in disease occurrence has been correlated with animal movements (migration patterns) and climate-related factors. Climatic conditions conducive to the survival and spread of the virus may also contribute to the seasonal distribution of PPR outbreaks. It's important to note that in many developing countries like India, most farmers have small and marginal land holdings. Small ruminants are often raised on free-range pastureland, shrubs, and forest areas. As available pastureland and forest areas continue to decrease, these animals are compelled to travel long distances during the dry season in search of fodder and water. However, with the onset of the monsoon season, nutritional conditions improve, and the migration of animals is restricted due to the availability of local fodder, resulting in a significant reduction in the frequency of PPR outbreaks.

Host Susceptibility

The primary factors determining the host range and tissue tropism of PPRV are cellular receptors. PPRV exhibits specific receptor interactions that influence its host range and tropism. Understanding these interactions and the virus's stability is essential for developing diagnostic tests, treatment strategies, and control measures for PPR. PPRV is a relatively recent addition to the group of morbilliviruses responsible for various diseases. The interaction between the virus and the host begins with specific receptor binding, primarily mediated by the H protein of PPRV and sialic acid on the host cell membrane (Munir *et al.*, 2013). Currently, the only established host receptor for this virus is SLAM (Signaling Lymphocyte Activation Molecule), which is found on epithelial cells, activated lymphocytes, and macrophages (Tatsuo *et al.*, 2000). Other morbilliviruses also target SLAM-positive cells for infection and destruction. Notably, the B95a marmoset B cell line is sensitive to both virulent and vaccine strains of PPRV, emphasizing the importance of SLAM in PPRV infection (Sreenivasa *et al.*, 2006). siRNA studies have further confirmed SLAM as a putative co-receptor for PPRV (Pawar *et al.*, 2008). Additionally, the ovine Nectin-4 protein has been proposed as another potential receptor. When overexpressed in epithelial cells, Nectin-4 enables efficient replication of PPRV and is predominantly expressed in epithelial tissues. Tumour-associated marker PVRL4 (Nectin-4) has been identified as an epithelial receptor for various morbilliviruses, including PPRV (Birch *et al.*, 2013). Research has also shown that receptor tyrosine kinase signaling regulates PPRV replication. Blocking receptor tyrosine kinase signaling in Vero cells impairs PPRV replication, suggesting its role in the virus's lifecycle (Chaudhary *et al.*, 2015).

PPRV primarily affects small ruminants, including sheep and goats, which are its main natural hosts. Occasionally, other artiodactyls, such as camels and small ruminant wildlife, can also be affected. Notably, the clinical symptoms are often more severe in goats than in sheep. The seroprevalence of PPRV antibodies in small ruminants varies from low to high, depending on the endemicity of the disease. Antibodies to PPRV have been observed in both small and large ruminants under field conditions. Recent findings have also shown serological evidence of camel exposure to PPRV in Nigeria. In general, the prevalence of PPRV antibodies in young and adult sheep and goats indicates subclinical or non-lethal infections, as vaccinations in endemic developing/underdeveloped countries are limited and irregular. However, the presence of anti-PPR antibodies in adult sheep or goats may not necessarily indicate PPR infection, as these animals might have received vaccination during their lifetime. The role of wildlife, particularly wild ruminants, in the epidemiology of PPR remains uncertain, and more research is needed to understand their contribution to disease transmission. The involvement of wildlife in PPR transmission has been suggested, but the extent and specific species involved may vary. Research has indicated that wildlife, especially certain wild ruminants, might play a role in the epidemiology of PPR. Some studies have reported the presence of PPRV antibodies in certain wildlife species, such as Gazelles, Camels, Wild Bharal, and Chowsingha (Balamurgan *et al.*, 2015, 2020). In Asia, outbreaks of PPR have adversely affected wildlife populations, as evidenced by the documented case in Mongolia in 2017, where there was significant mortality among the critically endangered saiga antelope (*Saiga tataricamongolica*) (Pruvot *et al.*, 2020). The potential role of wildlife species as maintenance hosts for the PPR virus (PPRV) in various ecosystems remains uncertain (Fine *et al.*, 2020). Additionally, the factors responsible for the apparent differences in disease manifestation between Asian and African wildlife are not well understood. The spread of PPR to free-ranging wildlife in continental Asia and eastern Europe raises significant concerns, posing a threat to biodiversity, undermining the goal of achieving a PPR-free world by 2030 (Parida *et al.*, 2016) and jeopardizing the realization of some of the UN Sustainable Development Goals (Fine *et al.*, 2020) as well as PPR virus threatens wildlife conservation (Aguilar *et al.*, 2018)

Detection of PPRV in wildlife, however, has been limited compared to its prevalence in domestic small ruminants. While cattle and pigs are susceptible to PPRV, they do not show clinical signs and are not considered as sources of virus transmission. There is no evidence of a carrier state for PPRV in these species. The breed of the animal may also influence the outcome of PPRV infection, with some breeds being more susceptible to the disease than others. However, extensive studies on the impact of breeds on PPRV susceptibility are limited. Age has been associated with the severity of PPR, with younger animals being more susceptible than adults. PPRV strains can vary in their virulence, causing differing levels of morbidity and mortality in affected animal populations (Balamurugan *et al.*, 2015).

Prevalence

Various studies worldwide have reported the prevalence of virus antibodies and the detection of virus antigen/genome in small ruminants, cattle, buffaloes, and camels (Balamurugan *et al.*, 2012, 2020). The presence of PPRV antibodies in both small and large ruminants indicates sub-clinical/inapparent/non-lethal infection, particularly in areas where vaccination is irregularly practised or restricted. Although cattle and swine seroconvert to the virus, they neither exhibit clinical symptoms nor transmit the infection. Recent studies suggest that Suids (swine) are a likely source of infection, while cattle serve as dead-end hosts, playing no epizootiological role in the virus's maintenance and spread. PPRV transmission from small to large ruminants has been observed, providing a mechanism for the virus's survival in the environment outside the primary hosts. The role of wildlife in the epizootiology of PPR has been highlighted, suggesting potential interactions between wildlife and domestic animals that could contribute to the virus's spread. However, the exact role of wildlife in PPR's epizootiology remains uncertain and requires further investigation. Moreover, the presence of PPRV antibodies in bovines and wild ruminants, besides sheep and goats, indicates natural transmission. However, the detection of PPRV in wildlife, apart from Gazelles, Camels, Wild Bharal, saiga antelope and Chowsingha, is limited (Fine *et al.,* 2020). Instances of PPRV detection in the nasal swabs of dogs and the virus genome in lion tissues raise concerns about potential species barriers being crossed. Considering the risk of PPR, vaccination is crucial to prevent virus circulation between wild and domestic small ruminants (Balamurugan *et al.,* 2020).

Risk Factors

The virus primarily impacts domestic sheep and goats, with occasional occurrences in camels and other wildlife with high severity in goats. Infected animals, despite no evidence of a carrier state, excrete PPRV in secretions and excretions for up to 16 weeks post-infection, serving as a significant source of silent infection. Toll-like receptors and cytokines influence species/breed susceptibility to PPRV, and understanding these factors alongside host genetics can offer insights into susceptibility polymorphisms. Breed, age (6 months to 1 year), and sex influence disease outcomes, with younger animals being significant predictors in rural or pastoral settings. Maternal antibodies affect vaccination response, with heritability influencing vaccine response in goats. Response to PPR vaccine varies due to environmental and genetic factors, including MHC class II. Risk factors for PPR include species, age, sex, and herd size. Seasonal incidence correlates with small ruminant movement and climate factors, with more frequent occurrences in enzootic settings during lean periods. Introduction of new animals is a major risk factor, and abortion is associated with PPR in small ruminants. Young animals aged 4 months to one year are severely affected, emphasizing the need for public awareness of biosecurity measures (Balamurugan *et al.,* 2020).

Feeding patterns, including stall feeding and grazing, are not significant risk factors, although stall feeding may contribute to transmission risk through animal purchase. Grazing, though not a substantial risk factor, still poses a likelihood of domestic animal interfaces in pastures and aquatic sources. Space-time clusters and determinants of PPR have been identified, with vaccination recommended before winter and monsoon onset in disease hotspots, particularly targeting high-risk groups (young animals aged 4-24 months). Differences in PPRV seropositivity based on species, sex, age, season, and geographical location have been observed, with some studies noting higher seropositivity in females, possibly linked to physiological differences and prolonged exposure due to productivity potential. PPRV actively circulates in endemic regions, with animal migration being a significant source of disease spread (Balamurugan *et al.,* 2020).

Socioeconomics

Rearing sheep and goats serves as a crucial source of social, financial, and economic stability for landless, marginal, and small-scale farmers in developing and underdeveloped nations. The occurrence of PPR in these animals not only jeopardizes the well-being of farmers already grappling with socio-economic and climate-related challenges but also leads to significant losses in small ruminant populations. As a TAD, PPR can hinder trade, livestock development, and access to new breeds, ultimately reducing human consumption of animal protein. PPR's economic impact is likely underestimated due to other diseases, yet it remains a significant obstacle to small ruminant production. Epidemics affect individual farmers, the agricultural industry, and the national economy, straining government resources and hampering sector growth. PPR outbreaks can result in mortality rates ranging from 50 percent to 80 percent among susceptible small ruminants, causing direct production losses such as live weight reduction, decreased milk production, and occasional instances of abortion within affected flocks. Farmers, in their efforts to control the disease, incur expenses for treatment, labor, and, in some cases, resort to distress sales, especially when confronted with a substantial number of animal deaths. At a macro level, the repercussions extend throughout the entire small ruminant value chain, impacting various interlinked sectors of the economy, both directly (production loss) and indirectly (treatment costs, distress sales, market restrictions, *etc.*). Measuring the impact of PPR involves tangible variables like mortality rates, reduced meat and milk production, and forgone revenue, as well as intangible factors such as diminished functionality, asset building, diversification, and income smoothing. The disease's far-reaching consequences encompass food access, changes in consumption patterns, adjustments in income and expenditure among affected flocks, and effects on social variables like women's participation, employment, asset loss, and family education (Balamurugan *et al.*, 2020).

The economic toll of PPR is substantial at the farm, regional, national, and global levels, amounting to an estimated annual loss of USD 1.45 - 2.1 billion according to WOAH and FAO in 2015. Assessing the disease's impact in different countries, geographic regions, production environments, and management types is vital for formulating effective PPR control plans, and making it essential to prioritize research and control programs. This information is crucial for garnering support and financial aid from governments and international organizations to implement a global PPR eradication campaign. The projected cost for global PPR eradication is estimated at US$ 3080 million, encompassing expenses related to global, regional, and national coordination, institutional development, epidemiology and surveillance, establishing diagnostic laboratories, vaccination implementation, training and research, socio-economic considerations, and contingency and emergency responses. In India alone, recent estimates indicate losses amounting INR 1,800 to 16,110 million at a 5-10 percent annual incidence level (Govindaraj *et al.*, 2016). Studies on the investment benefits of PPR control and eradication highlight substantial gains for various stakeholders. Ultimately, eradicating PPR will bring significant benefits to millions of landless, small, and marginal farmers in vulnerable societies across Africa and Asia. The disease directly threatens the income security of marginalized farmers, impacting their livelihoods, consumption, savings, and overall socio-economic development. Urgent action is required, focusing on PPR prevention through vaccination and biosecurity measures to uplift these communities from economic hardship (Balamurugan *et al.*, 2020).

Transmission

PPRV primarily spreads through close contact between infected and susceptible animals, mainly due to the virus's limited stability outside the host and its low environmental resistance (Balamurugan, 2017). The main mode of transmission is through inhalation of infectious aerosol materials between closely interacting animals. The virus is excreted in secretions and discharges from the eyes, nose, mouth, and feces of infected animals, especially during infection, and these act as a significant source of virus transmission (Balamurugan, 2020). When affected animals cough and sneeze, these discharges can form fine infectious droplets in the air, which are inhaled by animals close, making them susceptible to infection. Although infectious aerosols can also contaminate water, feed troughs, and bedding, this indirect mode of transmission is less significant since PPRV cannot survive for extended periods outside the host and is sensitive to lipid solvents. The trade of small ruminants at markets, where animals from various sources come into close contact, increases the chances of PPRV transmission. While cattle and pigs can seroconvert after contact with infected sheep and goats, clinical disease in these animals is rarely reported. However, this contact provides a means for the virus to persist in the environment through these unnatural hosts. The presence of PPRV antibodies in camels, cattle, and wild ruminants, in addition to sheep and goats, suggests natural transmission of PPRV infection among these animals in field conditions. Nomadic animals often interact with local sheep and goat populations, potentially contracting and transmitting the virus. Therefore, animal movement plays a crucial role in the transmission and maintenance of PPRV in nature. The incursion, persistence, and spread of PPRV in Tanzania, along with its epidemiological patterns and predictions, have also been documented.

Pathogenesis

Experimental studies and field observations in sheep, goats, and cattle provide clear evidence that natural PPRV infection typically occurs when the virus enters the upper respiratory tract epithelium (Parida *et al.*, 2015a). Pope *et al.* (2013) illustrated that the initial location for virus replication is within the tonsillar tissue and lymph nodes connected to the site of inoculation. Their proposal suggests that immune cells carry the PPRV to lymphoid tissues, where the primary virus replication takes place. Subsequently, the virus enters circulation, leading to primary viraemia, and eventually, clinical signs manifest in accordance with the virus's multiplication in the target cells or tissues. The virus's proliferation and pathogenicity are interrelated with various epizootiological determinants of the host (Munir *et al.*, 2013; Balamurugan *et al.*, 2015). In PPRV-infected small ruminants, characteristic lesions associated with epithelial infection and necrosis are observed. The virus's multiplication and pathogenicity are influenced by various factors, including the host's resistance or innate resistance, immune response, parasitic infections, nutritional status, breed, sex, and age of the animal (Munir *et al.*, 2013). Host genetics and non-genetic factors may also play a significant role in determining disease susceptibility, innate resistance, and the immune response to the virus.The virus spreads from its initial replication sites through the lymph and blood (primary viremia), with clinical signs appearing as PPRV replicates in target tissues. Both RPV and PPRV have an affinity for lymphoid and epithelial tissues, especially in the mucosa of the digestive tract. As the mucosal phase of the disease progresses, dramatic cytolytic changes occur in primary lymphoid organs and epithelial tissues, affecting both thymus-dependent and thymus-independent areas. Lymphocyte destruction is widespread in lymphoid tissues like Peyer's patches and the mesenteric lymph nodes (MLN). PPRV's affinity for lung parenchymal cells is distinct from RPV. PPRV antigen can be found in bronchial and bronchiolar epithelium, pneumocytes, macrophages, and other immune cells in the airway. Additionally, the virus is detected in the ileal epithelial cells, exhibiting nuclear and cytoplasmic viral aggregation of fibrillar strands when observed through an electron microscope. Macrophages in the mucosal lamina propria may also contain viral antigens. Various tissues and organs, including the oral cavity, sebaceous glands, intestinal epithelium, Peyer's patches, MLNs, and lymphoid follicles, can harbor PPRV antigens (Chowdhury *et al.*, 2014).

In PPRV infection, immune activation occurs simultaneously with immune suppression, continuing during the convalescence stage. The activation of blood mononuclear cells and tissue macrophages is thought to occur later in the incubation period at the primary replication sites. Many sensitized lymphocytes and reticular cells may undergo cytolysis, while the surviving populations experience immune activation and immune suppression. Although the exact mechanism of immunosuppression in PPRV infection is not fully understood, it is known that PPRV can induce apoptosis in goat PBMCs *in vitro*. Apoptotic lymphoid cells have been observed in goats naturally infected with PPRV. Immunosuppressed goats have shorter viremia periods and more severe disease progression with a higher mortality rate. Genomic analysis of host and PPR vaccine viral transcriptomes has revealed transcription factors modulating immune regulatory pathways under PPRV infection, with various genes involved in immune system regulation and apoptosis pathways (Manjunath *et al.*, 2015). Several studies have detected PPRV antigen and nucleic acid during different stages of infection, providing insight into the virus's spread and transmission. In general, PPRV infection is a complex interplay between the virus and the host's immune response. The virus's affinity for specific tissues and its ability to induce immunosuppression play a crucial role in the pathogenesis of the disease. Understanding these mechanisms is essential for developing effective control strategies.

Clinical Presentations

Clinically, the disease is characterized by elevated body temperature (pyrexia), discharges from the eyes and nose, ulcerative and erosive mouth lesions, gastroenteritis, diarrhoea, and bronchopneumonia. This clinical presentation of PPR can be summarized as the "3Ds" - discharge, diarrhoea, and death, with an additional fourth component, which is bronchopneumonia. Typically, the incubation period is 4 to 6 days, but it can vary from 3 to 10 days (Tripathi *et al.,* 1996; Kumar *et al.,* 2004). Studies have shown different stages of the disease (Fig. 1. A-E), including a short incubation period of 5 to 7 days a prodromal phase with fever, a mucosal phase characterized by fever, ocular and nasal discharges, redness of the conjunctiva, and erosions in the oral mucosa, a diarrheal stage with diarrhoea, pneumonia, dehydration, and potential death, and finally, in non-fatal cases, a "recovery stage" where sheep and goats develop lifelong immunity after recovering from PPR. Numerous studies have documented clinical signs and subsequent complications associated with spontaneous PPR occurrences in goats and sheep.

PPR can manifest in various forms, such as peracute, acute, and mild, depending on the severity of the disease (WOAH, 2022). The peracute form is observed in young goats with a short incubation period of 2 days, leading to symptoms like high fever, prostration, piloerection, oral and ocular discharges, and constipation, ultimately resulting in profuse diarrhoea and death within 5 to 6 days of infection. The acute form is characterized by fever, oculonasal discharges, partial blockage of the nose due to nasal discharges, and the development of necrotic lesions in the oral cavity. The mortality rate is high, around 70-80 per cent, and death typically occurs within 10-12 days after the onset of the disease. The severity of the disease depends on several factors, including the virulence/lineage of the PPR virus, the species and breed of the animals, their immune status, and age (Couacy-Hymann *et al.,* 2007).

Pathology

PPR pathology is characterized by retrogressive and necrotic changes in lymphoid tissues, as well as the epithelial cells of the gastrointestinal and respiratory systems. Gross pathological changes have been observed in both naturally occurring and experimentally infected goats and sheep, with variations in disease severity depending on species, breed, and host immune status (Chowdhury *et al.,* 2014).Prominent lesions in PPR-infected animals involve lung consolidation, color changes (Figure 9.1F), and sometimes frothy mucus in the lungs. Lung involvement is a common feature in PPR, with pneumonia being a predominant sign of the disease. The trachea may contain mucopurulent or frothy exudates, with multifocal bronchiolitis and bronchitis along with pleuritis and hydrothorax. Lymph nodes associated with the lung (mediastinal) and intestine (mesenteric) are typically enlarged, edematous, and congested (Figure 9.1G). The spleen may show congestion and engorgement, sometimes with petechiae on the capsular surface. Severe congestion and necrotic lesions can be found in the gastrointestinal tract, including the rumen, reticulum, and abomasum. The small intestine (duodenum, jejunum, or ileum) may exhibit black coloration of the mucosal lining with haemorrhages and erosions in the

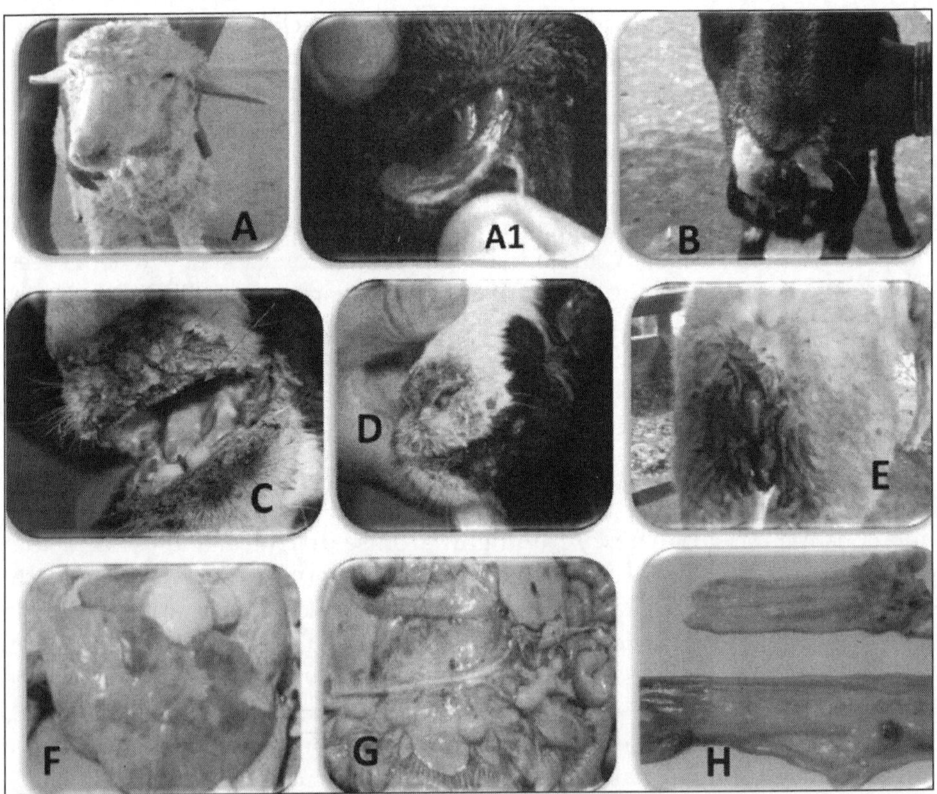

Figure 9.1: PPRV-Infected Animals

Showing Specific Cardinal Clinical Signs (Case definition of PPR) include

A) Pyrexia, or congestion of mucosa in the eye (reddened eye with mild conjunctivitis (A1), B) Oculo-nasal discharge-serus or mucopurulent discharges from nostrils C) oral lesions-ulcers (diphtheritic-plaques)-bran like deposits and necrotic ulcerations on tongue D) Respiratory signs - Nodular lesions around the nostrils and mouth and E) Diarrhoea - severe diarrhoea with solid hind quarters (frank diarrhoea). Post-mortem lesions of PPR affected animal's lungs showing congestion and consolidation of lobes (F), Enlarged mesenteric lymph node (G) and Intestine showing inflammatory erosion, streaks of congestion and hemorrhages on the mucosal folds (H).

terminal ileum (Figure 9.1H).Lesions may also affect the liver and other internal organs. In the posterior part of the colon and rectum, discontinuous streaks of congestion, known as "zebra stripes" or "zebra markings," are observed, which are typical of PPR (Balamurugan *et al.*, 2014; Chowdhury *et al.*, 2014).

In PPR-infected animals, show lesions like oedema and excoriation of the lips, erosions, ulceration and necrosis on various parts of the oral cavity, white necrotic deposits on the palates, tongue, and oral mucosa, erosions on the gums, soft and hard palates, tongue, cheeks, and esophagus. Other signs include emaciation, conjunctivitis, and erosive stomatitis inside the lower lips and adjacent gum near the commissures and the free portion of the tongue. The disease can cause multinucleate giant cells and cytoplasmic and nuclear inclusion bodies in permissive cell cultures and target tissues. Lymphoid tissues, including lymph nodes, show variable degrees of lymphoid cell depletion, while the liver displays swollen hepatocytes, sinusoid dilation, and disrupted hepatic cords. Kidneys may exhibit swollen epithelial lining cells in proximal convoluted tubules, and the intestine shows mild to moderate infiltration with mononuclear cells and vascular congestion, reducing the height of the villi. Peyer's patches can also experience lymphoid depletion, and submucosal oedema is common. The large intestine typically displays congestion of blood vessels in the mucosa and submucosa, along with mononuclear cell infiltration (Chowdhury *et al.*, 2014). PPR is often associated with lymphopenia (leucopenia), with lymphoid necrosis in reticuloendothelial organs. This is concurrent with the onset of pyrexia in the prodromal phase and contributes to lymphopenia. Serum analysis may reveal changes in total protein and enzyme levels, along with immunosuppression characterized by lymphocytic leucopenia and reduced antibody response.

Diagnosis

PPR diagnosis relies on specific clinical signs and distinctive pathological lesions in affected animals, such as lung consolidation, enlarged spleen, and swollen intestinal mesenteric lymph nodes, along with hemorrhagic streaks on the mucosa of the large intestines (Figures 9.1 F-H). Confirmatory diagnosis involves a range of laboratory techniques. While various diagnostic assays exist for detecting viral antigens, antibodies, or genome, conventional tests, including the gold standard virus isolation and virus neutralization test, are typically slow, less sensitive, labour-intensive, and necessitate technical expertise and specialized facilities. These may not be optimal for primary diagnostics. For large-scale sample screening, ELISA has become widely adopted for disease monitoring, surveillance, and diagnosis (Singh *et al.*, 2004a, b; Balamurugan *et al.*, 2014; Gomes *et al.*, 2016).

Advancements in virus biology and molecular techniques have led to the development of rapid, highly specific, and sensitive molecular assays such as RT-PCR, real-time RT-PCR, and loop-mediated isothermal amplification (LAMP), real-time and lateral flow strip reverse transcription recombinase polymerase amplification (RT-RPA) assays (Yang *et al.*, 2017) for viral genome detection. On-the-field diagnostic tools, including lateral flow assays, have also been created for less equipped laboratory settings, though their usage remains below expectations. Recombinant DNA technology has facilitated the production of safe recombinant viral proteins, serving as a better alternative to live antigens for immunoassay-based disease diagnosis. Different researchers have developed various modern diagnostic approaches, detecting PPRV antibodies in serum and PPRV antigen/nucleic acid in secretions, blood, tissues, and even milk. The application of these diagnostics depends on resource availability and the number of samples to be screened at different stages of PPR control and eradication phases (Libeau, 2015; Santhamani *et al.*, 2016; Gomes *et al.*, 2016).

Differential Diagnosis

PPR is often mistaken for other diseases that exhibit fever and similar clinical signs, especially when it is newly introduced. Diseases that may have similar signs for differential diagnosis include bluetongue (BT), contagious ecthyma (Orf), foot and mouth disease (FMD), contagious caprine pleuropneumonia (CCPP), pasteurellosis, among others (Balamurugan, 2017; Gomes *et al.*, 2016). Occasionally, coexisting or mixed infections of PPR with other diseases like goat pox, sheep pox, Orf, BT, border disease virus (BDV), or adenovirus have been reported. Mixed infections of PPRV with other respiratory viruses, such as parainfluenza virus 3 (PIV 3), respiratory syncytial virus (RSV), bovine herpes virus 1 (BoHV-1), bovine viral diarrhoea (BVD), and adenovirus, have also been documented, particularly in dromedary camels in Sudan during an abattoir study (Saeed *et al.*, 2015). Identifying sets of these viruses or mixed infections that are epizootiological and economically significant can contribute to establishing better guidelines for animal trade, preventing further transmission of infection and epizootics (Balamurugan *et al.*, 2020).

Disease Management

Specific treatment for PPR is not available, making the management of the disease crucial for infection control. Nevertheless, managing affected animals involves the administration of antibiotics (such as long-acting oxytetracycline and chlortetracycline) to prevent secondary bacterial infections. Anti-diarrheal medications are also employed, alongside supportive therapy, which includes B complex vitamins and dextrose saline, typically given for duration of 5 to 7 days. This therapy approach can be beneficial in reducing the severity of the disease. Treating and managing clinical cases of PPR during outbreaks in sheep and goats is essential to minimize economic losses for farmers (Balamurugan *et al.*, 2014). Implementing robust vaccination initiatives can effectively manage the disease. Nonetheless, PPR continues to be inadequately reported, especially in developing nations with constrained animal disease reporting and surveillance systems. Effective management of PPR, crucial for poverty reduction in endemic areas, has been achieved through the use of a safe and efficient live attenuated

PPR vaccine. However, complete control and eradication of the disease can only be realized through the successful implementation of a strategic vaccination program.

Prevention and Control

Given the significant impact of PPR on production and trade in endemic countries, prioritizing prevention and control measures is essential. Vaccination is a recommended tool in endemic settings, as isolation and quarantine are challenging to implement at the field level in underdeveloped countries (Singh *et al.*, 2009; Balamurugan *et al.*, 2016). In enzootic countries, where hygienic control measures are difficult to sustain, mass vaccination becomes the primary option. Alongside vaccination, practices like quarantining newly purchased animals, monitoring their health, and thorough cleansing of contaminated areas contribute to disease control. Efficient and timely control and eradication of PPR hinge on diagnosis, surveillance, monitoring, and vaccination in susceptible populations. Widely used vaccines like Sungri 96 in India and Nigeria 75/1 worldwide help in the prevention and control of PPR in sheep and goats, although they don't differentiate infected from vaccinated animals. Despite the absence of a specific treatment for PPR, affected animals receive broad-spectrum antibiotics and fluid therapy, coupled with flock vaccination during the initial outbreak in endemic regions (Abubakar *et al.*, 2017) is important for disease management.

Control strategies for PPR vary between countries depending on disease prevalence, with limited options in underdeveloped and developing nations due to impractical culling policies (Singh, 2011). Vaccination is a recommended preventive measure to avoid immediate economic losses for farmers. The success of rinderpest eradication has instilled confidence in implementing similar programs for PPR, relying on available expertise, scientific manpower, and trained veterinarians. The commitment of professionals involved in mass immunization programs is crucial. Prophylactic services are expanding through public-private partnerships, involving NGOs, cooperatives, and private veterinary practitioners. Establishing a network and database can enhance coordinated control program implementation. Some countries have initiated PPR control measures independently or with international agency support to boost small ruminant production, making the launch of control and eradication programs technically feasible, economically viable, and practically achievable.

PPR mass vaccination strategies aim for 80 percent flock/herd immunity, considering the dynamic small ruminant population, farming practices, and agro-climatic conditions (Singh, and Bandyopadhyay, 2015). Comprehensive vaccination of the entire population within a specified area, followed by targeting 4-month-old naïve animals, is essential to avoid susceptibility windows in newborns. Vaccination should also focus on high-risk groups, such as young animals, and migratory or trade market flocks. Rigorous vaccinations in identified areas can create "PPR-free zones" in enzootic settings. Detailed discussions on prophylaxis, control measures, and vaccination strategies are available in the literature (Balamurugan *et al.*, 2016, 2020; Balamurugan, 2017).

India's Control and Eradication Programme

India has instituted a national program to control PPR, encompassing vaccination, swift diagnosis, surveillance, monitoring, and disease prevention, and Mass vaccination initiatives, resulting in a reduction in reported outbreaks across several states (Balamurugan *et al.*, 2021, 2023). The implementation of a well-planned vaccination program, which specifically targets particular populations and age groups, has been instrumental in decreasing the prevalence of the disease. India initiated its vaccination efforts in 2006, but it wasn't until 2011 that nationwide vaccination campaigns were launched. Over the years, reported PPR outbreaks in India have significantly decreased in most states, particularly Andhra Pradesh, Telangana, Karnataka, and Chhattisgarh. This reduction is primarily attributed to the implementation of a well-planned mass vaccination program since 2011.During the years 1995-2010, the states with the highest reported PPR outbreaks were Andhra Pradesh, West Bengal, and Karnataka. However, in the periods 2011-2015 and 2015-2019, Jharkhand and West Bengal reported more outbreaks (Balamurugan *et al.*, 2021). This shift in disease patterns and reduced outbreak numbers can be attributed to the effectiveness of vaccines, timely vaccination, the circulation of a single lineage virus, and, most importantly, the efficient planning and execution of the vaccination program for sheep and goats.

Recognizing the significance of sheep and goats in food security and socio-economic development, and the availability of safe live attenuated cell culture PPR vaccines and diagnostics, the Department of Animal Husbandry and Dairying (DAHD) in the Government of India introduced the PPR Eradication Programme (PPR-EP) under the LH and DCP scheme. In 2022, a centrally sponsored program was launched in alignment with the PPR Global Control and Eradication Strategy, aiming to control and eradicate PPR by 2030.The National Strategic Plan for PPR control and eradication has been devised to comprehensively tackle the PPR challenge. This plan involves extensive vaccination of sheep and goats until 2025-26, the attainment of herd immunity by 2027-28, and the declaration of freedom from PPRV infection within specified timelines. This strategic plan involves mass vaccination, with a focus on populations older than 4 months, to achieve immunity levels of 80-90 per cent. Furthermore, the plan encompasses seromonitoring, investigations into outbreaks, and surveillance of various host species, including wildlife (Balamurugan *et al.*, 2021, 2023). In addition to these measures, targeted vaccination efforts will concentrate on migratory populations and border regions to ensure thorough coverage and containment of the disease. The establishment of stakeholder networks is also critical to the program's success. These networks should include a laboratory network for surveillance and monitoring, a veterinarian network for vaccination coverage and data collection, expert core teams networks, and a network for vaccine supply to ensure a consistent and quality vaccine supply for field use, *etc.*, Finally, a central, state, and the district-level technical working group should be established to oversee various elements of PPR control and eradication, including diagnostics, surveillance, disease prevention and control, legal frameworks, and stakeholder involvement, as outlined in the national strategic plan for PPR eradication (Balamurugan *et al.*, 2023). This group will ensure effective implementation and overall monitoring of the program, ultimately leading to the successful eradication of PPR. A recent study on surveillance study conducted by ICAR-NIVEDI as part of the PPR Eradication Program of the LH and DCP scheme in 2023 and 2024 unveiled varying levels of seroprevalence in different states, ranging from 49.5 percent to 82 percent in studied states (n=28) with an average of 43.3% of animals showed PPRV antibodies in the pre-vaccination phase on screening of 73,693 serum samples of sheep and goats by PPR c-ELISA kit (unpublished data).

Conclusion and Perspectives

This chapter offers insights into the geographical distribution of PPR, alongside discussing aspects such as disease burden, host specificity, diagnosis, prevention and control measures, and the control program. The purpose is to assist policymakers in making informed decisions for the control and eradication of PPR, a top priority among transboundary animal viral diseases affecting sheep and goats. Effective management of PPR, crucial for poverty reduction in endemic areas, has been achieved through the use of a safe and efficient live attenuated PPR vaccine. However, complete control and eradication of the disease from the world can only be realized through the successful implementation of a strategic vaccination program. To achieve this goal, vaccination efforts need to be intensified and conducted more frequently in line with the strategic plan for PPR control and eradication in India.

PPR, a transboundary viral disease affecting small ruminants, poses a significant challenge in underdeveloped and developing countries. Controlling PPR is crucial for poverty alleviation, as the disease hampers domestic and international trade, impacting farmers, the livestock industry, and national economies. The economic impact of PPR may be underestimated due to its coexistence with other diseases. While effective vaccines have controlled PPR in some developing nations, the dynamic nature of sheep and goat populations requires frequent and intensive vaccination. Addressing the rapid spread of PPR necessitates studies on the impact of agro-climatic changes, disease transmission dynamics, risk factors, and the development of forecasting models. Adapting vaccination strategies to local epidemiological settings can optimize resource allocation and enhance the chances of eradication. Swift eradication is possible with knowledge of the PPRV infection origin, as demonstrated by sponsored vaccination programs. Research efforts should focus on applying conventional virology and epidemiology for efficient control and eradication. In the current scenario, with PPRV capable of infecting large ruminants, the disease gains even higher priority. Collaboration among international agencies such as FAO, OIE, and others supports the notion that PPR could be controlled and eliminated more rapidly than rinderpest.

REFERENCES

Abubakar, M., Mahapatra, M., Muniraju, M., Arshed, M.J., Khan, E.H., Banyard, A.C., Ali, Q. and Parida, S. (2017). Serological Detection of Antibodies to Peste des Petits Ruminants Virus in Large Ruminants. Transbound Emerg Dis, 64(2): 513-519.

Aguilar, X.F., Fine, A.E., Pruvot, M., Njeumi, F., Walzer, C., Kock, R. and Shiilegdamba, E. (2018). PPR virus threatens wildlife conservation. Science (New York, N.Y.), 362(6411): 165-166. https://doi.org/10.1126/science.aav4096

Balamurugan, V. (2017). Peste des Petits Ruminants. In Emerging and re-emerging infectious diseases of livestock. Editor Jagadeesh Bayry, Springer International Publishing AG 2017. Pp.55-98. ISBN978-3-3-19-47424-3

Balamurugan, V., Govindaraj, G. and Roy, P. (2020). Peste des petits ruminants virus, Chapter 14 In Emerging and Transboundary Animal Viruses. Editor Yashpal Singh Malik, R. K. Singh, M.P. Yadav. Springer publishers 2020. Pp. 315-343 ISBN978-981-15-0401-3

Balamurugan, V., Govindaraj, G.N. and Rahman, H. (2016). Planning, implementation of peste des petits ruminants control programme and strategies adopted for disease control in India. Br J Virol, 3(3): 53-62. doi:10.17582/journal.bjv/2016.3.3s.53.62

Balamurugan, V., Hemadri, D., Gajendragad, M.R., Singh, R.K. and Rahman, H. (2014). Diagnosis and control of peste des petits ruminants: A comprehensive review. Virus Dis, 25(1): 39-56. doi:10.1007/s13337-013-0188-2

Balamurugan, V., Krishnamoorthy, P., Veeregowda, B.M., Sen, A., Rajak, K.K., Bhanuprakash, V., Gajendragad, M.R. and Prabhudas, K. (2012). Seroprevalence of Peste des petits ruminants in cattle and buffaloes from Southern Peninsular India. Trop Anim Health Prod, 44(2): 301-306. doi:10.1007/s11250-011-0020-1

Balamurugan, V., Rahman, H. and Munir, M. (2015). Host susceptibility to Peste des Petits ruminants virus In:, M. Munir (ed), Peste Des Petits Ruminants Virus,(Springer: Berlin Heidelberg), 39-50.

Balamurugan, V., Saravanan, P., Sen, A., Rajak, K.K., Venkatesan, G., Krishnamoorthy, P., Bhanuprakash, V. and Singh, R.K. (2012). Prevalence of peste des petits ruminants among sheep and goats in India. J Vet Sci, 13(3): 279-285. doi:10.4142/jvs.2012.13.3.279

Balamurugan, V., Sen, A., Venkatesan, G., Yadav, V., Bhanot, V., Riyesh, T., Bhanuprakash, V. and Singh, R.K. (2010). Sequence and phylogenetic analyses of the structural genes of virulent isolates and vaccine strains of peste des petits ruminants virus from India. Transbound Emerg Dis, 57(5): 352-364. doi:10.1111/j.1865-1682.2010.01156.x

Balamurugan, V., Vinod Kumar, K., Dheeraj, R., Kurli, R., Suresh, K.P., Govindaraj, G., Shome, B.R. and Roy, P. (2021). Temporal and Spatial Epidemiological Analysis of Peste Des Petits Ruminants Outbreaks from the Past 25 Years in Sheep and Goats and Its Control in India. Viruses, 13(3): 480. doi:10.3390/v13030480

Balamurugan,V., Vinod Kumar, K., Govindaraj, G., Suresh, K.P., Shome, B.R. and Gulati, B.R. (2023). Status Paper on Peste des petits ruminants Indian Perspective, ICAR-National Institute of Veterinary Epidemiology and Disease Informatics (NIVEDI), pp: 1-47.

Banyard, A.C., Parida, S., Batten, C., Oura, C., Kwiatek, O. and Libeau, G. (2010). Global distribution of peste des petits ruminants virus and prospects for improved diagnosis and control. J Gen Virol, 91(12): 2885-2897. doi:10.1099/vir.0.025841-0

Baron, M.D., Diallo, A., Lancelot, R. and Libeau, G. (2016). Peste des Petits Ruminants Virus. Adv Virus Res, 95: 1-42. doi:10.1016/bs.aivir.2016.02.001

Barrett, T., Amarel-Doel, C, Kitching, R.P. and Gusev, A. (1993). Use of the polymerase chain reaction in differentiating rinderpest field virus and vaccine virus in the same animals. Rev Sci Tech, (12): 865-872.

Birch, J., Juleff, N., Heaton, M.P., Kalbfleisch, T., Kijas, J. and Bailey, D. (2013). Characterization of ovine Nectin-4, a novel peste des petits ruminants virus receptor. J Virol, (87): 4756-4761.

Chaudhary, K., Chaubey, K.K., Singh, S.V. and Kumar, N. (2015). Receptor tyrosine kinase signaling regulates replication of the peste des petits ruminants virus. Acta Virol, (59): 78-83.

Chowdhury., E.H., Bhuiyan, A.R., Rahman, M.M., Siddique, M.S. and Islam, M.R. (2014) Natural peste des petits ruminants virus infection in Black Bengal goats: virological, pathological and immuno-histochemical investigation. BMC Vet Res, (10): 263.

Couacy-Hymann, E., Bodjo, C., Danho, T., Libeau, G. and Diallo, A. (2007). Evaluation of the virulence of some strains of peste-des-petits-ruminants virus (PPRV) in experimentally infected West African dwarf goats. Vet J, (173): 178-183.

Dhar, P., Sreenivasa, B.P., Barrett, T., Corteyn, M., Singh, R.P. and Bandyopadhyay, S.K. (2002). Recent epidemiology of peste des petits ruminants virus (PPRV). Vet Microbiol, 88(2): 153-159. doi:10.1016/s0378-1135(02)00102-5

Diallo, A. (1990). Morbillivirus group: genome organisation and proteins. Vet Microbiol, 23(1-4): 155-163. doi:10.1016/0378-1135(90)90145-l

FAO and WOAH. (2022). Overview of the plan of action Peste des Petits Ruminants Global Eradication Programme II and III.

Fine, A.E., Pruvot, M., Benfield, C.T.O., Caron, A., Cattoli, G., Chardonnet, P., Dioli, M., Dulu, T., Gilbert, M., Kock, R., Lubroth, J., Mariner, J.C., Ostrowski, S., Parida, S., Fereidouni, S., Shiilegdamba, E., Sleeman, J.M., Schulz, C., Soula, J.J., Van der Stede, Y., Tekola, B.G., Chris Walzer., Steffen Zuther and Njeumi, F. (2020). Eradication of peste des petits ruminants virus and the wildlife-livestock interface. Front Vet Sci, (7): 50. https://doi.org/10.3389/fvets.2020.00050

Gargadennec, L. and Lalanne, A. (1942). La peste des petits ruminants. Bull. Serv. Zootech. Epiz. AOF, 5: 16-21.

Gomes, A.R., Veeregowda, B.M., Byregowda, S.M. and Balamurugan, V. (2016). Comprehensive review on recent developments in the diagnostics and vaccines against Peste des petits ruminants. Br J Virol, 3(3s): 90-104.

Govindaraj, G.N., Balamurugan, V. and Rahman, H. (2016). Estimation of economic Loss of PPR in sheep and goats in India: An annual incidence based analysis. Br J Virol, (3), 77-85. doi:10.17582/journal.bjv/2016.3.3s.77.85

Kumar, N., Maherchandani, S., Kashyap, S.K., Singh, S.V., Sharma, S., Chaubey, K.K. and Ly, H. (2014). Peste des petits ruminants virus infection of small ruminants: a comprehensive review. Viruses, 6(6): 2287-2327. https://doi.org/10.3390/v6062287

Kumar, P., Kumar, R., Sharma, A.K. and Tripathi, B.N. (2002). Pathology of peste des petits ruminats (PPR) in goats and sheep: spontaneous study. Indian J Vet Pathol, 26(1 and 2): 15-18.

Kumar, P., Tripathi, B.N., Sharma, A.K., Kumar, R., Sreenivasa, B.P., Singh, R.P., Dhar, P. and Bandyopadhyay, S.K. (2004). Pathological and immunohistochemical study of experimental peste des petits ruminants virus infection in goats. J Vet Med B Infect Dis Vet Public Health, 51(4): 153-159. https://doi.org/10.1111/j.1439-0450.2004.00747.x

Kwiatek, O., Ali, Y.H., Saeed, I.K., Khalafalla, A.I., Mohamed, O.I., Obeida, A.A., Abdelrahman, M.B., Osman, H.M., Taha, K.M., Abbas, Z., El Harrak, M., Lhor, Y., Diallo, A., Lancelot, R., Albina, E. and Libeau, G. (2011). Asian lineage of peste des petits ruminants virus, Africa. Emerg Infect Dis, 17(7): 1223-1231. doi:10.3201/eid1707.101216

Libeau, G. (2015). Current advances in serological diagnosis of PPRV. In Peste des petitis Ruminants Virus. Editor Muhammad Munir, Spinger-Verlag Berlin Heidelberg. Pp.133-154

Libeau, G., Diallo, A. and Parida, S. (2014). Evolutionary genetics underlying the spread of peste des petits ruminants virus. Animal Front, 4: 14-20.

Manjunath, S., Kumar, G.R., Mishra, B.P., Mishra, B., Sahoo, A.P., Joshi, C.G., Tiwari, A K., Rajak, K.K. and Janga, S.C. (2015). Genomic analysis of host - Peste des petits ruminants vaccine viral transcriptome uncovers transcription factors modulating immune regulatory pathways. Vet Res, (46): 15. https://doi.org/10.1186/s13567-015-0153-8

Munir, M., Zohari, S. and Berg, M. (2013). Molecular Biology and Pathogenesis of Peste des Petits Ruminants Virus. Springer Briefs Anim Sci, 23-32.

Muniraju, M., Munir, M., Parthiban, A.R., Banyard, A.C., Bao, J., Wang, Z., Ayebazibwe, C., Ayelet, G., El Harrak, M., Mahapatra, M., Libeau, G., Batten, C. and Parida, S. (2014). Molecular evolution of peste des petits ruminants virus. Emerg Infect Dis, 20(12): 2023-2033. https://doi.org/10.3201/eid2012.140684

Muthuchelvan, D., De, A., Debnath, B., Choudhary, D., Venkatesan, G., Rajak, K.K., Sudhakar, S.B., Himadri, D., Pandey, A.B. and Parida, S. (2014). Molecular characterization of peste-des-petits ruminants virus (PPRV) isolated from an outbreak in the Indo-Bangladesh border of Tripura state of North-East India. Vet Microbiol, 174(3-4): 591-595. https://doi.org/10.1016/j.vetmic.2014.10.027

Muthuchelvan, D., Rajak, K.K., Ramakrishnan, M.A., Choudhary, D., Bhadouriya, S., Saravanan, P., Pandey, A.B. and Singh, R.K. (2015). Peste-des-petits-ruminants: an Indian perspective. Adv Anim Vet Sci, (3): 422-429.

Parida, S., Couacy-Hymann, E., Pope, R.A., Mahapatra, M., El Harrak, M., Brownlie, J. and Banyard, A.C. (2015a). Pathology of PPR. In: Peste des petitis Ruminants Virus. Editor Muhammad Munir, Spinger-Verlag Berlin Heidelberg. Pp.51-68.

Parida, S., Muniraju, M., Altan, E., Baazizi, R., Raj, G.D. and Mahapatra, M. (2016). Emergence of PPR and its threat to Europe. Small Rumin Res, (142): 16-21. https://doi.org/10.1016/j.smallrumres.2016.02.018

Parida, S., Muniraju, M., Mahapatra, M., Muthuchelvan, D., Buczkowski, H. and Banyard, A.C. (2015b). Peste des petits ruminants. Vet. Microbiol, 181(1): 90-106. doi:https://doi.org/10.1016/j.vetmic.2015.08.009

Pawar, R.M., Raj, G.D., Kumar, T.M., Raja, A. and Balachandran, C. (2008). Effect of siRNA mediated suppression of signaling lymphocyte activation molecule on replication of peste des petits ruminants virus in vitro. Virus Res, 136(1-2): 118-123. doi:10.1016/j.virusres.2008.04.026

Pope, R.A., Parida, S., Bailey, D., Brownlie, J., Barrett, T. and Banyard, A.C. (2013). Early events following experimental infection with peste-des-petits ruminants virus suggest immune cell targeting. PloS one, 8(2): e55830. https://doi.org/10.1371/journal.pone.0055830

Pruvot, M., Fine, A.E., Hollinger, C., Strindberg, S., Damdinjav, B., Buuveibaatar, B., Chimeddorj, B., Bayandonoi, G., Khishgee, B., Sandag, B., Narmandakh, J., Jargalsaikhan, T., Bataa, B., McAloose, D., Shatar, M., Basan, G., Mahapatra, M., Selvaraj, M., Parida, S., Njeumi, F. and Shiilegdamba, E. (2020). Outbreak of peste des petits ruminants among critically endangered Mongolian saiga and other wild ungulates, Mongolia, 2016-2017. Emerg Infect Dis, 26(1): 51-62. https://doi.org/10.3201/eid2601.181998

Saeed, I.K., Ali, Y.H., Abdul Rahman, M.B., Mohammed, Z.A., Osman, H.M., Taha, K.M., Musa, M.Z. and Khalafalla, A.I. (2015). Mixed infection of peste des petits ruminants virus (PPRV) and other respiratory viruses in dromedary camels in Sudan, an abattoir study. Trop Anim Health Prod, 47(5): 995-998. https://doi.org/10.1007/s11250-015-0798-3

Santhamani, R., Singh, R.P. and Njeumi, F. (2016). Peste des petits ruminants diagnosis and diagnostic tools at a glance: perspectives on global control and eradication. Arch Virol, 161(11), 2953-2967. https://doi.org/10.1007/s00705-016-3009-2

Sen, A., Saravanan, P., Balamurugan, V., Rajak, K.K., Sudhakar, S.B., Bhanuprakash, V., Parida, S. and Singh, R.K. (2010). Vaccines against peste des petits ruminants virus. Exp Rev Vac, 9(7): 785-796. doi:10.1586/erv.10.74

Shaila, M.S., Purushothaman, V., Bhavasar, D., Venugopal, K. and Venkatesan, R.A. (1989). Peste des petits ruminants of sheep in India. Vet Rec, 125(24): 602.

Shaila, M.S., Shamaki, D., Forsyth, M.A., Diallo, A., Goatley, L., Kitching, R.P. and Barrett, T. (1996). Geographic distribution and epidemiology of peste des petits ruminants virus. Virus Res, 43(2): 149-153.

Singh, R.K., Balamurugan, V., Bhanuprakash, V., Sen, A., Saravanan, P. and Pal Yadav, M. (2009). Possible control and eradication of peste des petits ruminants from India: technical aspects. Vet Ital, 45(3): 449-462.

Singh, R.P. (2011). Control strategies for peste des petits ruminants in small ruminants of India. Rev Sci Tech, 30(3): 879-887. doi:10.20506/rst.30.3.2079

Singh, R.P. and Bandyopadhyay, S.K. (2015). Peste des petits ruminants vaccine and vaccination in India: sharing experience with disease endemic countries. Virus Dis, 26(4): 215-224. doi:10.1007/s13337-015-0281-9

Singh, R.P., Sreenivasa, B.P., Dhar, P. and Bandyopadhyay, S.K. (2004a). A sandwich-ELISA for the diagnosis of Peste des petits ruminants (PPR) infection in small ruminants using anti-nucleocapsid protein monoclonal antibody. Arch Virol, 149(11): 2155-2170. https://doi.org/10.1007/s00705-004-0366-z

Singh, R.P., Sreenivasa, B.P., Dhar, P., Shah, L.C. and Bandyopadhyay, S.K. (2004b). Development of a monoclonal antibody-based competitive-ELISA for detection and titration of antibodies to peste des petits ruminants (PPR) virus. Vet Microbiol, 98(1): 3-15. https://doi.org/10.1016/j.vetmic.2003.07.007

Sreenivasa, B.P., Singh, R.P., Mondal, B., Dhar, P. and Bandyopadhyay, S.K. (2006). Marmoset B95a cells: a sensitive system for cultivation of Peste des petits ruminants (PPR) virus. Vet Res Commun, 30(1): 103-108. https://doi.org/10.1007/s11259-005-3200-5

Tatsuo, H., Ono, N. and Yanagi, Y. (2001). Morbilliviruses use signaling lymphocyte activation molecules (CD150) as cellular receptors. J Virol, 75(13): 5842-5850. https://doi.org/10.1128/JVI.75.13.5842-5850.2001

Taylor, W.P. (1984). The distribution and epidemiology of peste des petits ruminants. Prev Vet Med, 2(1): 157-166. doi:https://doi.org/10.1016/0167-5877(84)90059-X

Tripathi, B.N., Chhattopadhyay, S.K. and Parihar, N.S. (1996). Clinicopathological studies on an outbreak of peste des petits ruminants in goats and sheep. Indian J Vet Pathol, 20: 99.

WOAH and FAO. (2015). Manual on global strategy for the control and eradication of Pesti des petits ruminants. Retrieved from (http://www.fao.org/3/a-i4460e.pdf)

WOAH - World Organisation for Animal Health (2013). Manual of Diagnostic Tests and Vaccines for Terrestrial Animals. WOAH, Paris 1-14.

WOAH - World Organisation for Animal Health Terrestrial Manual (2022). Peste Des Petits Ruminants (Infection with Peste des Petits Ruminants Virus, Chapter 3.8.9) pp-1-17.

WOAH, FAO (2015). Manual on global strategy for the control and eradication of Pesti des petits ruminants (http://www.fao.org/3/a-i4460e.pdf).

Woma, T.Y., Ekong, P.S., Bwala, D.G., Ibu, J. O., Ta'ama, L., Dyek, D.Y., Saleh, L., Shamaki, D., Kalla, D.J., Bailey, D., Kazeem, H.M. and Quan, M. (2016). Serosurvey of peste des petits ruminants virus in small ruminants from different agro-ecological zones of Nigeria. The Onderstepoort J. Vet Res, 83(1): 1035. https://doi.org/10.4102/ojvr.v83i1.1035

Yang, Y., Qin, X., Song, Y., Zhang, W., Hu, G., Dou, Y., Li, Y. and Zhang, Z. (2017). Development of real-time and lateral flow strip reverse transcription recombinase polymerase Amplification assays for rapid detection of peste des petits ruminants virus. Virol J, 14(1): 24. https://doi.org/10.1186/s12985-017-0688-6

Chapter 10

Porcine Circoviruses

C. Ravishankar*, R. Rajasekhar, S. Shashank, K.V. Jayanth and K. Sumod

Department of Veterinary Microbiology, College of Veterinary and Animal Sciences, Kerala Veterinary and Animal Sciences University, Pookode, Lakkidi P.O. – 673 576, Kerala India

***e-mail: chintu@kvasu.ac.in**

ABSTRACT

Porcine circoviruses (PCVs) are small DNA viruses that cause economically important disease conditions in swine. The most important and well-studied among them is PCV2 which is responsible for important clinical conditions known as porcine circovirus-associated diseases (PCVADs), which include post-weaning multisystemic wasting syndrome (PMWS) characterized by wasting, increased mortality, or respiratory signs. In clinical conditions, PCV2 has been detected along with other porcine viruses like porcine reproductive and respiratory syndrome virus (PRRSV), porcine parvovirus (PPV), and classical swine fever virus (CSFV) with overlapping clinical symptoms thereby making clinical diagnosis difficult. Porcine circovirus 3 (PCV3) is also increasingly associated with clinical illnesses similar to that caused by PCV2. Vaccination is helpful in reducing the economic losses due to PCV2 and this, along with good management and biosecurity practices, seems to be the prudent strategy to control circovirus infection in pig farms.

Keywords: *Porcine circovirus, PCVADs, PMWS, PDNS, Coinfection.*

Introduction

Porcine circoviruses (PCVs) are small DNA viruses and have been classified under the genus *Circovirus* of the family *Ciroviridae* (MacLachlan and Dubovi, 2016). The virions are non-enveloped, have an icosahedral symmetry, and enclose a circular single-stranded ambisense DNA genome. Currently, four PCVs have been recognized (PCV1 to 4) (ICTV, 2022). Porcine circovirus 1 was first detected in the 1970s as a contaminant in PK-15 cell line and experimental infections of pigs with this virus did not produce clinical disease (Allan *et al.*, 1995). However, PCV2 which was first detected in Canada in 1991, is responsible for a myriad of economically important clinical conditions called as porcine circovirus-associated diseases (PCVADs), which includes post-weaning multisystemic wasting syndrome (PMWS) also known as PCV2-systemic disease (PCV2-SD), porcine dermatitis and nephropathy syndrome (PDNS) and reproductive failure also known as PCV2-reproductive disease (PCV2-RD). Also important is the PCV2-subclinical infection (PCV2-SI), which is linked with growth retardation without overt clinical signs. Compared to PCV2, not much research has been carried out on PCV3 and PCV4. Porcine circovirus 3 was initially detected in 2016 in tissues of pigs suffering from respiratory and neurological signs, cardiac and multi-systemic inflammation, reproductive failure, and porcine dermatitis and nephropathy syndrome in the United States of America (Phan *et al.*, 2016; Palinski *et al.*, 2017). Porcine circovirus 4 was discovered in 2019 in domestic pigs in China and retrospective studies have shown that the infection has happened at least from 2008 (Holgado-Martín *et al.*, 2023). In this chapter, focus is on the circoviruses that are associated with clinical disease in pigs.

Porcine circovirus 2 virions are small and have a genome size of approximately 1.7 kb. The genome encodes four ORFs, ORF1, and ORF2 being the two main ones that together constitute most of the viral sequence. The ORF1 encodes two proteins that play a crucial role in the replication cycle called Rep and Rep′. The ORF2 encodes for the structural capsid protein Cap, which is involved in eliciting the host immune response. The Cap gene is a reliable phylogenetic and epidemiological marker for PCV2 strains since it is able to reconstruct the same tree as the whole viral genome. An intergenic region between the 5′ ends of Rep and Cap genes contains the origin of virus replication (Ori) (Meng, 2013). The ORF1 was also found to encode a group of RNAs and minor non-structural proteins (NS) associated RNAs (Rep3a, Rep3b, Rep3c, NS0, NS515, NS672). These RNAs share common 5′ and 3′ sequences and are probably derived from the full Rep RNA by alternative splicing. The ORF3 codes for a protein, which is thought to be associated in apoptosis. The ORF4 encodes a protein that plays a role in upregulating caspase activity and down-regulating the activity of CD4+ and CD8+ T lymphocytes (Baron *et al.*, 2021).

Previously, four genotypes of PCV2 were recognized *viz.* PCV2a (previously known as group II) with five subtypes or clusters (2A, 2B, 2C, 2D, and 2E), PCV2b (previously known as group I) with three clusters (1A, 1B, and 1C), PCV2c and PCV2d (Segales *et al.*, 2008; Guo *et al.*, 2010). Currently, eight genotypes, namely PCV2a, PCV2b, PCV2c, PCV2d, PCV2e, PCV2f, PCV2g, and PCV2h have been reported on the basis of complete ORF2 sequences (Franzo and Segalés, 2018). In India, the genotypes of PCV2 prevalent are 2a, 2b, 2d, 2f, and 2h, and the clusters prevalent are 2, 3, 10, 11, 13, and 18 (Bhattacharjee *et al.*, 2021). Inter and intra-genotypic recombination has been reported in PCV2 and is responsible for the emergence of recombinant strains (Cadar *et al.*, 2012; Anoopraj *et al.*, 2015).

PCV3 is also included under the *Circovirus* genus and has a slightly longer (2 kb) genome when compared to PCV2. Though the virus has a genomic structure similar to PCV2, the capsid and replicase proteins of PCV3 share only 37.0 percent identity with that of PCV2 (Palinski *et al.*, 2017). It has been detected in pigs suffering from respiratory and neurological signs, cardiac and multi-systemic inflammation, reproductive failure, and porcine dermatitis and nephropathy syndrome (PDNS-like condition). Based on ORF1 and ORF2 sequences, two genotypes of PCV3, PCV3a and PCV3b, have been recognized.

Porcine circoviruses are resistant to inactivation when exposed to chloroform, solutions with pH 3, chlorhexidine, iodine, and phenol. However quaternary ammonium compounds and oxidizing agents inactivate the virus. Porcine circovirus 2 is completely inactivated after treatment at temperatures of 80°C and above for 15 min.

Distribution

Clinical disease caused by PCV2 was first detected in farrow-to-finish swine farms in Northeastern Saskatchewan, Western Canada in the early 1990s and was characterized by multi-systemic wasting (Ellis *et al.*, 1998). In the following years, PMWS and other PCVADs were observed in Spain, France, Hungary, the United Kingdom, Brazil, Uruguay, Korea, China, Japan, Malaysia, Thailand, Indonesia, Australia, Cuba, Israel, South Africa, and Uganda. In India, PCV2 was first detected in 2006 in cases of PMWS-associated PCV2 –SD (Kumar *et al.*, 2006). Later the presence of the virus was reported from many of the northeastern and southern states of India (Parthiban *et al.*, 2022a; Somashekara *et al.*, 2023).

As far as PCV3 is concerned, since its first detection of the virus in the USA in 2016, it has been reported in Europe, China, South Korea, and Japan. Porcine circovirus 3 was associated with reproductive failure in sows and piglet mortality and the virus was also detected in pigs in India. In the country, during 2017-2018, PCV3 was identified in tissue samples collected from Maharashtra (Mumbai), Chhattisgarh, Assam (Guwahati) and Mizoram states. The Indian PCV3 strains (designated as PCV3/India/Chhattisgarh/SK1/2017; PCV3/India/Guwahati/KH25/2017; PCV3/India/Mizoram/MZ8/2017 and PCV3/India/Mumbai/AB10/2018) shared 99.8 percent similarity at the genome level. The genomic sequences of the PCV3 strains found in India showed 91.4 to 99.8 percent of their nucleotide identity with other sequences of PCV3 strains circulating worldwide. Phylogenetic analyses indicated that the Indian strains were closely related to the strains circulating in the USA, China, and South Korea

(Bera *et al.*, 2020). The virus has also been detected in three states of southern India *viz.* Tamil Nadu, Kerala, and Telangana (Parthiban *et al.*, 2022b).

Epidemiology

Disease caused by PCV2 has been recorded in domestic pigs and wild boar. Experimental induction of disease in species of livestock other than swine has not been successful. Domestic pigs serve as the natural host for PCV3. However, the virus has been detected in several domestic and wild species, including cattle, dogs, mice, chamois, roe deer, wild boar, ticks, and mosquitoes. Porcine circovirus 2 is shed through colostrum, milk, feces, and nasal and oral secretions of infected pigs and the mode of transmission is direct. Airborne transmission from diseased to healthy pigs has been proven to be an alternate route of transmission. Transplacental transmission of PCV2 has also been demonstrated. However, experiments using pregnant sows inoculated intranasally have yielded variable results. When successful, those studies have shown that PCV2 may cause fetal death, similar to that caused by porcine parvovirus, with the birth of live pigs together with dead piglets and mummies of different sizes. Late abortion has also been described in PCV2 natural infection. Viral DNA has also been detected in semen. However, it is believed that a very high titre of the virus in semen, which is unlikely under natural conditions, is required for reproductive problems to develop in the sow. The transmission of PCV3 is similar to that of PCV2 with direct contact and transplacental routes.

Porcine circovirus 2 may persist in pigs for several months and convalescent pigs can carry the virus for extended periods and are important in the epidemiology of the disease. The decrease in titre of colostral antibody is associated with the onset of PCV2-SD in the late nursery or finishing pigs (Segales, 2021).

Pathogenesis

The incubation period of disease caused by PCV2 is relatively long (around 18–25 days). The virus is shed through respiratory secretions, oral secretions, urine, and feces of infected pigs (Opriessnig *et al.*, 2007). In subclinical infection of PCV2, the interaction of the virus with the host developed a specific adaptive response that clears the acute infection or allows virus persistence in some animals (Segales, 2012).

The virus usually enters through the oronasal route and preferentially infects lymphoid tissue in pigs. The virus has been shown to replicate in different cell types like epithelial cells, and endothelial cells as well as macrophages, dendritic cells and peripheral blood mononuclear cells. Internalization of the PCV2 virus in dendritic cells helps it to elude the host immune system and spread throughout the body due to the dendritic cells' migratory ability. Glycosaminoglycans, including heparan sulfate and chondroitin sulfate B, have been identified as attachment receptors necessary for PCV2 infection (Misinzo *et al.*, 2006). However, the heparan sulfate-binding motif is absent in the capsid protein of PCV3.

Systemic infection with PCV2 leads to damage to the immune system and can lead to immunodeficiency. Replication of PCV2 has also been detected in T-lymphocytes, B-lymphocytes, and monocytes that were isolated from the peripheral blood mononuclear cells (PBMCs) of infected animals. The B-cells from bronchial lymph nodes displayed a higher amount of PCV2 replication as compared to the B-cells from PBMCs. Depletion of lymphocytes in lymphoid tissues, changes in peripheral blood mononuclear cell subpopulations, and altered cytokine expression patterns have been demonstrated in pigs naturally and experimentally affected with the condition. Lymphoid depletion and histiocytic replacement in lymphoid tissues are characteristic lesions of PCV2 infection and PCVAD. The internalization of the virus and its presence inside the macrophages decreased the phagocytic and microbicidal activities of the macrophages and increased the survival and spread of the virus (Chang *et al.*, 2006). In the case of PCV2-SD, the affected animals suffered from a deficiency of neutralizing antibody response and an altered cellular response (Meerts *et al.*, 2006; Fort *et al.*, 2007; Kekarainen *et al.*, 2010).

In PCV3 infection, induction of myocarditis, nephritis, and vasculitis possibly occur via virus-induced or immune-response-mediated mechanisms because the presence or replication of PCV3 has been observed in these tissues (Phan *et al.*, 2016; Palinski *et al.*, 2017; Arruda *et al.*, 2019; Bera *et al.*, 2020). High titres of PCV3 were also found in the thymus and lymph nodes of aborted and weak-born piglets, suggesting that PCV3 has extra tissue tropism in fetuses.

Clinical Signs and Histopathological Findings

Only a small proportion of the pigs infected with PCV2 actually develop clinical PCVAD (Meng, 2013). To definitively diagnose PCVAD, there should be i) clinical manifestation of wasting, increased mortality, or respiratory signs, ii) presence of characteristic histological lesions in lymphoid tissues, and iii) detection of PCV2 antigen or DNA in the tissues with pathological lesions. It is also reported that PCV2 causes overt clinical disease usually in coinfections with other infectious agents affecting pigs.

PCVAD is primarily seen in grower pigs of 7 to 16 weeks of age when the maternal antibodies wane. In infection with PCV2, the clinical signs include weight loss, enlarged lymph nodes, jaundice, and diarrhea. Reproductive failure is another clinical manifestation where the aborted or stillborn fetuses exhibit necrotizing myocarditis. Respiratory distress is also manifested in some infected pigs and interstitial pneumonia and bronchiolitis with mononuclear infiltration are observed in the lungs. In pigs suffering from diarrhea due to PCV2 infection, thickened intestinal mucosa with enlarged mesenteric lymph nodes and granulomatous enteritis has been reported. In PDNS, coalescing red to purple skin lesions, glomerular and interstitial nephritis, and vasculitis are observed. It has been observed that the tissues of pigs diagnosed with PMWS had a higher virus load when compared to those of animals with PDNS.

Anorexia, weight loss, swollen joints, respiratory disease, coughing, panting, tachypnea, dyspnea, diarrhea, and neurologic disease are the clinical manifestations associated with PCV3 infections. Increased sow mortality, decreased conception rates, fetal mummification, poor oestrus, abortions, stillbirths, congenital tremors, PDNS-like skin lesions and decreased conception rates have also been observed in PCV3 infections. The histopathological lesions reported include broncho interstitial pneumonia, cuffing of small and medium-sized airways and small blood vessels with aggregates of lymphocytes and plasma cells, and necrotizing vasculitis in the dermis and subcutis. In the lymph nodes diffuse granulomatous lymphadenitis, characterized by moderate cortical lymphoid depletion, occasionally replaced by histiocytes and numerous multinucleated giant cells have been observed. In the kidneys, diffuse membrane proliferative glomerulonephritis characterized by glomerulosclerosis and thickening of the Bowman's capsules have been reported.

Coinfection

Most PCVAD cases are observed along with infections with other swine pathogens, including PRRSV, *Mycoplasma hyopneumoniae*, bacterial septicemia or pneumonia, and swine influenza virus, and the former is usually the most frequent coinfection agent identified. It has been reported that PRRSV possibly acts synergistically with PCV2 to induce PCVAD, exaggerate immune cell depletion in pigs, and increase the risk of developing the condition. Besides this, it has also been reported that swine pathogens such as pseudorabies virus, and Torque teno virus also may play a role in PCV2 pathogenesis (Pogranichniy *et al.*, 2002; Ramamoorthy and Meng, 2009; Ge *et al.*, 2012; Opriessnig and Halbur, 2012). Coinfection with porcine parvovirus has also been reported (Ellis *et al.*, 2000; Pallares *et al.*, 2002; Wellenberg *et al.*, 2004; Dorr *et al.*, 2007; Ramamoorthy and Meng, 2009; Opriessnig and Halbur, 2012). It has also been demonstrated that *M. hyopneumoniae* potentiates the severity of PCV2-induced lung and lymphoid lesions (Opriessnig *et al.*, 2004). Non-suppurative polio encephalomyelitis in a co-infection with porcine teschovirus and PCV2 has also been recorded.

Diagnosis

The clinical symptoms exhibited in PCV2 infection are mostly non-specific and overlap with those seen in infections with other swine pathogens. Hence, diagnosis of PCV2 infection often requires laboratory testing. Clinical samples that can be collected for detection of the virus include nasal fluids, saliva, joint fluids, tracheobronchial swabs, serum, and tonsillar,fecal, and urinary swabs. The virus can also be detected in the semen. From dead pigs, organ samples such as lungs, lymph nodes, spleen, and tonsils can be collected for diagnostic purposes.

The virus can be isolated in PK-15 cell line and it has been reported that the permissiveness of PCV2 by PK-15-C1 cell clone is high when compared with PK-15 parent cells. Detection of antibodies against the virus has been carried out by using, among other tests, ELISA aimed at detecting anti-Cap antibodies.

The polymerase chain reaction (PCR) and its variants have become the most widely used sensitive and rapid technique for the detection of pathogens. Conventional PCR has been used as a sensitive test by many researchers for detecting PCV2 (Ellis *et al.*, 1999; Calsamiglia *et al.*, 2002). Besides this, SYBR Green and TaqMan-based real-time PCR for detection and quantitation of viral load in the tissues of pigs have been developed (Olvera *et al.*, 2004; Brunborg *et al.*, 2004; Yang *et al.*, 2007; McIntosh *et al.*, 2009; Zhao *et al.*, 2010). Due to the propensity of PCV2 to occur along with other porcine viruses, multiplex PCR (mPCR) to detect two or more swine pathogens in clinical samples has been developed (Xu *et al.*, 2012). Tests to simultaneously detect swine viruses like CSF virus (CSFV), African swine fever virus (ASFV), PCV2, PPV, and PRRSV; PCV2, PPV, CSFV, and PRRSV; PRRSV, CSFV, PCV2, PPV, porcine pseudorabies virus, and Japanese encephalitis virus have been developed. Besides this, nested PCR for detection of PCV2 and a duplex nested reverse transcriptase polymerase chain reaction (RT-PCR) for detection of genomic RNA of type 2 PRRSV and the mRNA of PCV2 in infected tissues has been developed.

Other techniques for the detection of the virus include Loop-Mediated Isothermal Amplification (LAMP), Restriction Fragment Length Polymorphism (RFLP) and PCR-based RFLP; but the widespread use of PCR overshadows their use.

The titre of PCV3 is reported to be higher in lungs and lymph nodes when compared to other organs like the spleen, kidney, and heart. Besides these, the virus can be detected in the liver, tonsils, brain, heart, intestines, muscle, joint fluid, aborted fetus and fecal samples. The virus can be isolated in primary porcine kidney cells or PK-15 cells. The most commonly used test for detection of PCV3 is PCR or real-time PCR targeting either Cap or Rep gene. Multiplex qPCR assays have been developed for differential detection of PCV3 and other swine pathogens including PCV2, CSFV, PRRSV, and pseudorabies virus (Kim *et al.*, 2017; Tian *et al.*, 2020; Zheng *et al.*, 2020). A multiplex PCR assay to simultaneously discriminate PCV1, PCV2, and PCV3 in a single reaction has been developed. LAMP assay for the detection of PCV3 has also been developed.

Prevention and Control

General management strategies to reduce pathogen load in pigs should be stressed to prevent or control circovirus infections. This includes efficient disinfection, "all-in" and "all-out" protocols, minimizing animal contact, restricting batch mixing and cross-fostering, and isolating or euthanizing sick pigs. The maintenance of the environmental factors such as appropriate temperature, air-flow, and space within pens; and in addition, appropriate use of anti-parasitic treatments and vaccination, control of co-infections, monitoring breeding and semen quality, and improvement of herd nutrition are also important in prevention and control of the infection.

Vaccination

Commercial vaccines are available for the control of PCV2 in pigs and have contributed remarkably to reducing the economic losses due to PCVAD. Vaccination reduces the detrimental effects of subclinical infection as well. The available vaccines include inactivated vaccines, inactivated and attenuated chimeric virus vaccines, and recombinant subunit vaccines. Initial vaccines were based on the PCV2a genotype that was prevalent at that time. All commercial vaccines seem to be effective in preventing the most severe outcomes of PCV2 infection. Though the development of homologous vaccines is most advantageous, it is generally accepted that some degree of cross-protection is offered by heterologous vaccines. However, it is also observed that there has been a predominance of some PCV2d viruses after vaccination which may be due to vaccine-induced replacement of a subset of genetic variants.

Conclusions

Porcine circoviruses, especially PCV2, are important viral pathogens of pigs which can bring about economic losses due to reduced weight gain, reproductive dysfunction, and predisposition to other diseases. The virus has been detected along with other important pathogens of pigs leading to severe, often overlapping symptoms which complicate diagnosis and treatment. Research should focus on developing effective vaccines for the conditions and vaccination, together with good managemental practices and biosecurity will help to control the infection in farms.

REFERENCES

Allan, G.M., McNeilly, F., Cassidy, J.P., Reilly, G.A.C., Adair, B.M., Ellis, W.A. and McNulty, M.S. (1995). Pathogenesis of porcine circovirus - experimental infections of colostrum deprived pigs and examination of pig foetal material. Vet Microbiol, 44(1): 49-64. doi: 10.1016/0378-1135(94)00136-k

Anoopraj, R., Rajkhowa, T.K., Cherian, S., Arya, R.S., Tomar, N., Gupta, A., Ray, P.K., Somvanshi, R. and Saikumar, G. (2015). General characterisation and phylogenetic analysis of PCV2 isolates from India: Indications for emergence of natural inter-genotypic recombinants. Infect Genet Evolut, 31: 25-32. doi: 10.1016/j.meegid.2015.01.006.

Arruda, B., Piñeyro, P., Derscheid, R., Hause, B., Byers, E., Dion, K., Long, D., Sievers, C., Tangen, J., Williams, T. and Schwartz, K. (2019). PCV3-associated disease in the United States swine herd. Emerg Microbes Infect, 8(1): 684-698. doi: 10.1080/22221751.2019.1613176

Baron, L., Atar, S., Zur, H., Roopin, M., Goz, E. and Tuller, T. (2021). Computational based design and tracking of synthetic variants of Porcine circovirus reveal relations between silent genomic information and viral fitness. Sci Rep, 11(1): 10620. doi: 10.1038/s41598-021-89918-6.

Bera, B.C., Choudhary, M., Anand, T., Virmani, N., Sundaram, K., Choudhary, B. and Tripathi, B.N. (2020). Detection and genetic characterization of porcine circovirus 3 (PCV3) in pigs in India. Transbound Emerg Dis, 67(3): 1062-1067. doi: 10.1111/tbed.13463.

Bhattacharjee, U., Sen, A. and Sharma, I. (2021). A retrospective study reveals the Porcine circovirus-2f genotype predominant in the indigenous pig population of North-eastern India. Infect Genet Evolut, 96: 105100. doi: 10.1016/j.meegid.2021.105100.

Brunborg, I.M., Jonassen, C.M., Moldal, T., Bratberg, B., Lium, B., Koenen, F. and Schönheit, J. (2007). Association of myocarditis with high viral load of porcine circovirus type 2 in several tissues in cases of fetal death and high mortality in piglets. A case study. J Vet Diagn Invest,19(4): 368-375. doi: 10.1177/104063870701900405.

Cadar, D., Csagola, A., Lorincz, M., Tombacz, K., Spinu, M. and Tuboly, T. (2012). Detection of natural inter- and intra-genotype recombination events revealed by cap gene analysis and decreasing prevalence of PCV2 in wild boars. Infect Genet Evolut, 12(2): 420-427. doi: 10.1016/j.meegid.2012.01.014.

Calsamiglia, M., Segalés, J., Quintana, J., Rosell, C. and Domingo, M. (2002). Detection of porcine circovirus types 1 and 2 in serum and tissue samples of pigs with and without postweaning multisystemic wasting syndrome. J Clin Microbiol, 40(5): 1848-1850. doi: 10.1128/JCM.40.5.1848-1850.2002.

Chang, H.W., Jeng, C.R., Lin, T.L., Liu, J.J. and Chiou, M.T. (2006). Immunopathological effects of porcine circovirus type 2 (PCV2) on swine alveolar macrophages by in vitro inoculation. Vet Immunol Immunopathol, 110(3-4): 207-219. doi: 10.1016/j.vetimm.2005.09.016.

Dorr, P.M., Baker, R.B., Almond, G.W., Wayne, S.R. and Gebreyes, W.A. (2007). Epidemiologic assessment of porcine circovirus type 2 coinfection with other pathogens in swine. J Am Vet Med Assoc, 230(2): 244-250. doi: 10.2460/javma.230.2.244.

Ellis, J.A, Bratanich, A., Clark, *E.G.*, Allan, G. and Meehan, B. (2000). Coinfection by porcine circoviruses and porcine parvovirus in pigs with naturally acquired post weaning multisystemic wasting syndrome. J Vet Diagn Invest, 12(1): 21-27. doi: 10.1177/104063870001200104.

Ellis, J., Hassard, L., Clark, E., Harding, J., Allan, G., Willson, P., Strokappe, J., Martin, K., McNeilly, F., Meehan, B., Todd, D. and Haines, D. (1998). Isolation of circovirus from lesions of pigs with postweaning multisystemic wasting syndrome. Can Vet J, 39(1): 44-51.

Ellis, J., Krakowka, S., Lairmore, M., Haines, D., Bratanich, A., Clark, E., Allan, G., Konoby, C., Hassard, L., Meehan, B., Martin, K., Harding, J., Kennedy, S. and McNeilly, F. (1999). Reproduction of lesions of postweaning multisystemic wasting syndrome in gnotobiotic piglets. J Vet Diagn Invest, 11(1): 3-14. doi: 10.1177/104063879901100101.

Fort, M., Olvera, A., Sibila, M., Segalés, J. and Mateu, E. (2007). Detection of neutralizing antibodies in post weaning multi systemic wasting syndrome (PMWS)-affected and non-PMWS-affected pigs. Vet Microbiol, 125(3-4): 244-255. doi: 10.1016/j.vetmic.2007.06.004.

Franzo, G. and Segalés, J. (2018). Porcine circovirus 2 (PCV-2) genotype update and proposal of a new genotyping methodology. PLoS One,13(12):e0208585. doi: 10.1371/journal.pone.0208585.

Ge, X., Wang, F., Guo, X. and Yang, H. (2012). Porcine circovirus type 2 and its associated diseases in China. Virus Res, 164(1-2): 100-106. doi: 10.1016/j.virusres.2011.10.005.

Guo, L.J., Lu, Y.H., Wei, Y.W., Huang, L.P. and Liu, C.M. (2010). Porcine Circovirus type 2 (PCV2). Genetic variation and newly emerging genotypes in China. Virol J, 7: 273-285. doi: 10.1186/1743-422X-7-273.

Holgado-Martín, R., Arnal, J.L., Sibila, M., Franzo, G., Martín-Jurado, D., Risco, D., Segalés, J. and Gómez, L. (2023). First detection of porcine circovirus 4 (PCV-4) in Europe. Virol J, 20(1): 230. doi: 10.1186/s12985-023-02181-1.

ICTV [International Committee on Nomenclature of Viruses] (2022): https://ictv.global/taxonomy. Accessed 20/10/2023

Kekarainen, T., McCullough, K., Fort, M., Fossum, C., Segales, J. and Allan, G.M. (2010). Immune responses and vaccine-induced immunity against Porcine circovirus type 2. Vet Immunol Immunopathol, 136(3-4): 185-193. doi: 10.1016/j.vetimm.2010.03.025.

Kim, H.R., Park, Y.R., Lim, D.R., Park, M.J., Park, J.Y., Kim, S.H., Lee, K.K., Lyoo, Y.S. and Park, C.K. (2017). Multiplex real-time polymerase chain reaction for the differential detection of porcine circovirus 2 and 3. J Virol Methods, 250: 11-16. doi: 10.1016/j.jviromet.2017.09.021.

Kumar, G.S., Sharma, R. and Paliwal, O.P. (2006). Occurrence and pathology of post weaning multisystemic wasting syndrome. In: Compendium of Invited Papers and Abstracts of XXIII Annual Conference of Indian Association of Veterinary Pathologists. p. 83.

Maclachlan, N.J. and Dubovi, E.J. (2016). Fenner's Veterinary Virology, 5thEdition, Elsevier, 602p

McIntosh, K.A., Tumber, A., Harding, J.C., Krakowka, S., Ellis, J.A. and Hill, J.E. (2009). Development and validation of a SYBR green real-time PCR for the quantification of porcine circovirus type 2 in serum, buffy coat, feces, and multiple tissues. Vete Microbiol, 133(1-2): 23-33. doi: 10.1016/j.vetmic.2008.06.010.

Meerts, P., Misinzo, G., Lefebvre, D., Nielsen, J., Botner, A., Kristensen, C.S. and Nauwynck, H.J. (2006). Correlation between the presence of neutralizing antibodies against porcine circovirus 2 (PCV2) and protection against replication of the virus and development of PCV2-associated disease. BMC Vet Res, 2: 6. doi: 10.1186/1746-6148-2-6.

Meng, X.J. (2013). Porcine circovirus type 2 (PCV2): pathogenesis and interaction with the immune system. Annu Rev Anim Bioscie, 1: 43-64. doi: 10.1146/annurev-animal-031412-103720.

Misinzo, G., Delputte, P.L., Meerts, P., Lefebvre, D.J. and Nauwynck, H.J. (2006). Porcine circovirus 2 uses heparan sulfate and chondroitin sulfate B glycosaminoglycans as receptors for its attachment to host cells. J Virol, 80(7): 3487-3494. doi: 10.1128/JVI.80.7.3487-3494.2006.

Olvera, A., Sibila, M., Calsamiglia, M., Segalés, J. and Domingo, M. (2004). Comparison of porcine circovirus type 2 load in serum quantified by a real time PCR in postweaning multisystemic wasting syndrome and porcine dermatitis and nephropathy syndrome naturally affected pigs. J Virol Methods,117(1): 75-80. doi: 10.1016/j.jviromet.2003.12.007.

Opriessnig, T. and Halbur, P.G. (2012). Concurrent infections are important for expression of porcine circovirus associated disease. Virus Res, 164(1-2): 20-32. doi: 10.1016/j.virusres.2011.09.014.

Opriessnig, T., Meng, X.J. and Halbur, P.G. (2007). Porcine circovirus type 2–associated disease: update on current terminology, clinical manifestations, pathogenesis, diagnosis, and intervention strategies. J Vet Diagn Invest, 19(6): 591-615. doi: 10.1177/104063870701900601.

Opriessnig, T., Thacker, E.L., Yu, S., Fenaux, M., Meng, X.J. and Halbur, P.G. (2004). Experimental reproduction of post weaning multisystemic wasting syndrome in pigs by dual infection with *Mycoplasma hyopneumoniae* and porcine circovirus type 2. Vet Pathol, 41(6): 624-640. doi: 10.1354/vp.41-6-624.

Palinski, R., Piñeyro, P., Shang, P., Yuan, F., Guo, R., Fang, Y., Byers, E. and Hause, B.M. (2017). A novel porcine circovirus distantly related to known circoviruses is associated with porcine dermatitis and nephropathy syndrome and reproductive failure. J Virol, 91(1): e01879-16. doi: 10.1128/JVI.01879-16.

Pallarés, F.J., Halbur, P.G., Opriessnig, T., Sorden, S.D. and Villar, D. (2002). Porcine circovirus type 2 (PCV-2) coinfections in US field cases of post weaning multisystemic wasting syndrome (PMWS). J Vet Diagn Invest, 14(6): 515-519. doi: 10.1177/104063870201400614.

Parthiban, S., Ramesh, A., Dhinakar Raj, G., Karuppannan, A.K., Hemalatha, S., Parthiban, M., Ravishankar, C., Senthilkumar, K. and Balasubramaniyam, D. (2022b). Molecular evidence of porcine circovirus 3 infection in swine: first report in southern India. Virus Dis, 33(3): 284-290. doi: 10.1007/s13337-022-00778-8.

Parthiban, S., Ramesh, A., Karuppannan, A. K., Dhinakar Raj, G., Hemalatha, S., Parthiban, M., Senthilkumar, K., Balasubramaniyam, D., Sumanth Kumar, R., Ranganatha, S. and Ravishankar, C. (2022a). Isolation and genetic analysis of Porcine circovirus 2 in southern India evidences high circulation of Porcine circovirus 2d genotype. Mol Biol Rep, 49(12): 11829-11846. doi: 10.1007/s11033-022-07946-5.

Phan, T.G., Giannitti, F., Rossow, S., Marthaler, D., Knutson, T.P., Li, L., Deng, X., Resende, T., Vannucci, F. and Delwart, E. (2016). Detection of a novel circovirus PCV3 in pigs with cardiac and multi-systemic inflammation. Virol J, 3(1): 184. https://doi.org/10.1186/s12985-016-0642-z

Pogranichniy, R.M., Yoon, K.J., Harms, P.A., Sorden, S.D. and Daniels, M. (2002). Case-control study on the association of porcine circovirus type 2 and other swine viral pathogens with post weaning multisystemic wasting syndrome. J Vet Diagno Invest, 14(6): 449-456. doi:10.1177/104063870201400601.

Ramamoorthy, S. and Meng, X.J. (2009). Porcine circoviruses: a minuscule yet mammoth paradox. Animal Health Research Reviews,10(1):1-20. doi: 10.1017/S1466252308001461.

Segales, J. (2012). Porcine circovirus type 2 (PCV2) infections: clinical signs, pathology and laboratory diagnosis. Virus Res, 164(1-2): 10-19. doi: 10.1016/j.virusres.2011.10.007.

Segales, J. (2021). Porcine Circovirus Disease. https://www.merckvetmanual.com/generalized-conditions/porcine-circovirus-diseases/porcine-circovirus-diseases. Accessed 20 October 2023.

Segales, J., Olvera, A., Grau-Roma, L., Charreyre, C., Nauwynck, H., Larsen, L., Dupont, K., McCullough, K., Ellis, J., Krakowka, S., Mankertz, A., Fredholm, M., Fossum, C., Timmusk, S., Stockhofe-Zurwieden, N., Beattie, V., Armstrong, D., Grassland, B., Baekbo, P. and Allan, G. (2008). PCV-2 genotype definition and nomenclature. Vet Rec, 162(26): 867-868. doi: 10.1136/vr.162.26.867.

Somashekara, S., Ravishankar, C., Ravindran, R., Rajappan, A., Kanjirakkuzhiyil, S., Paravalappil Muraleedharan, A., Janardhan, A. and John, K. (2023). Molecular characterization and phylogenetic analysis of porcine circovirus 2 from Kerala, India. Virus Dis, 34(2): 331-338. doi: 10.1007/s13337-023-00814-1.

Tian, R.B., Jin, Y., Xu, T., Zhao, Y., Wang, Z.Y. and Chen, H.Y. (2020). Development of a SYBR green I-based duplex real-time PCR assay for detection of pseudorabies virus and porcine circovirus 3. *Mol Cell* Prob, 53: 101593. doi: 10.1016/j.mcp.2020.101593.

Wellenberg, G.J., Stockhofe-Zurwieden, N., de Jong, M.F., Boersma, W.J. and Elbers, A.R. (2004). Excessive porcine circovirus type 2 antibody titres may trigger the development of porcine dermatitis and nephropathy syndrome: a case-control study. Vet Microbiol, 99(3-4): 203-214. doi: 10.1016/j.vetmic.2004.01.001.

Xu, X.G., Chen, G.D., Huang, Y., Ding, L., Li, Z.C., Chang, C.D., Wang, C.Y., Tong, D.W. and Liu, H.J. (2012). Development of multiplex PCR for simultaneous detection of six swine DNA and RNA viruses. J Virol Methods, 183(1): 69-74. doi: 10.1016/j.jviromet.2012.03.034.

Yang, Z., Habib, M., Shuai, J.B. and Fang, W.H. (2007). Detection of PCV2 DNA by SYBR Green I-based quantitative PCR. Journal of Zhejiang University Science B, 8(3): 162-169. doi: 10.1631/jzus.2007.B0162.

Zhao, K., Han, F., Zou, Y., Zhu, L., Li, C., Xu, Y., Zhang, C., Tan, F., Wang, J., Tao, S., He, X., Zhou, Z. and Tang, X. (2010). Rapid detection of porcine circovirus type 2 using a TaqMan-based real-time PCR. Virology J, 7: 374. doi: 10.1186/1743-422X-7-374.

Zheng, H.H., Zhang, S.J., Cui, J.T., Zhang, J., Wang, L., Liu, F. and Chen, H.Y. (2020). Simultaneous detection of classical swine fever virus and porcine circovirus 3 by SYBR green I-based duplex real-time fluorescence quantitative PCR. *Mol Cell* Prob, 50: 101524. doi: 10.1016/j.mcp.2020.101524.

Chapter 11

Bovine Papillomatosis

Geeta Devi Leishangthem[1]*, *Sonam Sarita Bal*[1] and *Ramesh Somvanshi*[2]

[1]*Department of Veterinary Pathology, Guru Angad Dev Veterinary and Animal Sciences University, Ludhiana, Punjab, India*
[2]*Division of Veterinary Pathology, Indian Veterinary Research Institute, Izatnagar, Bareilly, Uttar Pradesh, India*
**e-mail: drgeetapatho@gmail.com*

ABSTRACT

Bovine papillomatosis is a contagious viral disease caused by different types of bovine papillomaviruses (BPV-1 to BPV- 43) characterized by formation of warts on the skin as well as mucosa. These lesions caused economic loss to the farmers. The lesion may start with a benign nature however it may undergo to malignancy in immunocompromised animal or depending on the presence of certain synergistic factors such as bracken fern, types of viral oncoproteins or its predilection site. It can be transferred through direct contact or vertically. It can be diagnosed with the help of various ancillary techniques from gross to microscopical, immunological as well as molecular approaches. However there is no effective treatment and vaccine against BPV.

Keywords: *Bovine Papillomatosis, BPV and Diagnosis.*

Introduction

Bovine Papillomatosis is a contagious viral disease caused by heterogenous bovine papilloma virus affecting a wide range of animals from domestic to wild animals and characterized by the presence of benign proliferative lesionson the skin (warts or papillomas) or mucosal epithelium (condylomas) (Howley and Lowy, 2007). It can be either generalised (affecting the head, neck, back, abdomen, udder, perineum, and mucosal epithelium of the upper alimentary tract, including the oesophagus and rumen) or limited to the teat. These lesions can regress spontaneously or progress to malignant neoplasms. These lesions cause the animal's aesthetic and economic worth to decrease since they impair its ability of milking, suckling, and coitus, and raise the animal's risk of traumatic wounds, myiasis, necrotic dermatitis, and mastitis (Somvanshi, 2011).

The virus affects vertebrates, including humans, and can cause both benign and malignant epithelial lesions. It causes sarcoids in equines. In bovines it causes hyperproliferative lesions in teats, mastitis, bleeding, and economic depreciation of animal leather, bladder and gastrointestinal cancer. BPV-1 and 2 can also affect yaks, tapirs, zebras, giraffes, antelopes, buffaloes, and horses. Buffaloes and cattle, closely related to the Bovidae family, often share herds, posing a risk of cross-infection, which can spread without a cattle intermediary (Somvanshi, 2011).

Bovine papillomatosis is caused by various types of BPV. BPV is a cosmopolitan virus, worldwide distributed, with prevalence in both Asian countries and European countries like Turkey and Italy (Grindatto *et al.*, 2015; Timurkan and Alcigir, 2017). It has also been reported in African countries like Nigeria and in the Americas in Brazil and Mexico (Ugochukwu *et al.*, 2019). It is estimated that 60 percent of Brazilian cattle herd is infected by BPV (Stocco dos *Santos et al.*, 1998). Research shows that BPV can occur as a co-infection, as seen in Japan and other countries (Ogawa *et al.*, 2004). Papillomaviruses are specific to a single animal host species. However, only a few viral types can infect a second animal species, resulting in non-productive infections. Sarcoid, a fibroblastic skin tumor of equines, is a common example of heterologous PV infection because it is caused by BPV1 and 2 (Chambers *et al.*, 2003). Various studies indicate multiple infections in cattle with multiple cutaneous papillomas caused by different BPV types, and the possibility of viral co-infections in single lesions (Da Silva *et al.*, 2012; Carvalho *et al.*, 2012; Santos *et al.*, 2016).

The most frequently identified virus types of BPV are BPV-1 and 2, which are associated with both benign and malignant neoplasms. Teat warts (TWs) in Indian cattle have been reported to be infected with either BPV-1 or BPV-10, or their combined infections. However, in case of cutaneous warts (CWs) BPV-1, BPV-2, or their combined infections has been reported from India (Leishangthem *et al.*, 2006, 2008; Pangty *et al.*, 2010; Kumar *et al.*, 2013a; Kumar *et al.*, 2013b).

Bovine cutaneous papillomatosis is prevalent in young animals aged two to three years. The high incidence of papillomatosis is linked to the densification of animals, as confined populations are more susceptible to disease outbreaks. BPV have been identified in various tissues such as embryos, female reproductive tissues, blood, milk, urine, seminal fluid, and spermatozoa from infected cattle, suggesting potential involvement of these fluids and cell types in BPV transmission (Lindsey *et al.*, 2009). BPV has been detected in the placenta, bladder as well as lymphocytes that may act as reservoirs of this virus (Roperto *et al.*, 2011, 2012). The most often detected virus types are BPV-1 and -2. BPV-2 was detected in 46 percent of naturally occurring bladder malignancies and in 69 percent of experimental bladder lesions in animals raised on bracken fern (Campo *et al.*, 1992) and also reported BPV-4 as the major causative factor of bladder carcinoma in animals consuming bracken fern or animal with enzootic haematuria where the virus interacts with compounds present in bracken fern like quercetin, ptaquiloside leading to malignant transformation. Ptaquiloside is believed to be the carcinogenic principle, which alkylates DNA when activated to its unstable dienone formed under alkaline condition (Leishangthem *et al.*, 2008).

Unlike human papilloma virus (HPV) that exclusively infects human, BPV capability to infect various species, has been used as a prototype for studying papilloma virus biology and oncology. Its pathogenic and morphological characteristics are shared with HPV, contributing to understanding viral oncogenesis and its host interaction (Koller and Olson, 1972; Costa and Medeiros, 2014).

Bovine Papilloma Virus

Bovine Papillomaviruses (BPV) is a small non-enveloped, circular double-stranded DNA genome virus belonging to Papillomaviridae family. There are 43 types of *Bos taurus papillomavirus* (BPV) based on the diversity of nucleotide sequences of the L1 Open Reading Frame (ORF). They are classified into five known genera as shown in Table 11.1 such as *Deltapapillomavirus* (includes BPV-1, 2, 13, 14); *Epsilonpapillomavirus* (includes BPV-5, 8, 25, 32, 43); *Xipapillomavirus* (includes BPV-3, 4, 6, 9, 10, 11, 12, 15, 17, 20, 23, 24, 26, 28, 29, 30, 34, 35, 36, 37, 38, 39, 40, 41 and 42); *Dyoxipapillomavirus* (includes BPV-7, 31); *Dyokappapapillomavirus* (includes BPV-16, 18 and 22). In addition, BPV-19, 21, 27, 33 belongs to the unclassified genus(Sauthier *et al.*, 2021). *Xipapillomavirusesis* pure epitheliotropic BPVs that causes papillomas involving only the epithelium. While *Epsilonpapilloma* and *Deltapapillomavirus* viruses cause both pure papillomas and fibropapillomas (de Freitas and Almeida, 2012) as well as urinary bladder tumor (Borzacchiello and Roperto, 2008; Lunardi *et al.*, 2013, Roperto *et al.*, 2016).

Table 11.1: Types of BPV and its Genera

Genera of BPV	*Types of BPV*
Delta-papillomavirus	BPV-1, 2, 13, 14
Epsilon-papillomavirus	BPV-5, 8, 25, 32, 43
Xi-papillomavirus	BPV-3, 4, 6, 9, 10, 11, 12, 15, 17, 20, 23, 24, 26, 28, 29, 30, 34, 35, 36, 37, 38, 39, 40, 41 and 42
Dyoxi-papillomavirus	BPV-7, 31
Dyokappa-papillomavirus	BPV-16, 18 and 22
Unclassified	BPV-19, 21, 27, 33

BPV Structure

Papillomaviruses are a type of oncogenic virus that belongs to the Papillomaviridae family. Cottontail rabbit papillomavirus (CRPV), was discovered in warts on cottontail rabbits' skin in the 1930s, was the first DNA oncovirus being discovered (Shope and Hurst, 1933). These viruses have a double-stranded circular DNA structure and icosahedralcapsid. They are non-enveloped, approximately 8 kb long and 55-60 nm diameter. The viral capsid contains 360 copies of the L1 structural proteins and around 12 copies of the L2 structural protein. The viral genome is divided into early, late and long control regulatory regions by two polyadenylation sites (Zheng and Baker, 2006). The viral DNA is associated with histone-like proteins (Leto *et al.*, 2011).

The early control regulatory region (ECR) makes upto 50 percent of the viral genome and encodes non-structural and regulatory proteins, that includes E1, E2, E4, E5, E6, and E7 while the late control region (LCR) (occupying 40 percent of the genome) encodes L1 and L2 structural proteins. Expression of E1- E7 occurs in basal cells or during the intermediate stages of maturation and that of L1, L2 occurs in keratinocytes in the final stage of maturation (Zheng and Baker, 2006). The non coding region (NCR) occupies 10 percent of the genome and contains the origin of replication and binding sites for multiple transcription factors, important for RNA polymerase II regulation. BPVs of the *Xipa-pillomavirus* group lack E6 (Shah and Howley, 1996).

Different species within the same genus show 60 percent to 70 percent similarity in their L1 ORFs, while less than 60 percent similarity exists between genera and 71-89 percent similarity is seen between different viral types within the same species (de Villiers *et al.*, 2004).

The Functions of different genes of BPV are shown in Table 11.2.

Table 11.2: Functions of different Genes of BPV

Gene	*Functions*
E1	Encodes protein that binds with viral origin of replication, has helicase activity
E2	Master transcriptional regulator for early viral promoters, facilitates the binding of E1 to the viral origin of replication. E2 tethers viral genome to chromosomes (utilizes Bromodomain-4), ensuring genome distribution to daughter cells post-division, and regulates E6 and E7 expression in HPV-infected basal layer keratinocytes.
E3	Found in a few papillomavirus types and does not express any protein.
E4	E4 proteins, expressed at low levels in early viral infection, significantly increase in late infection and facilitate virion release by disrupting keratinocyte cytoskeleton filaments.
E5	Encodes for 44 amino acid protein, major BPV transforming oncoprotein. Act as oncogene by activating growth-promoting signals and it also down-regulates histocompatibility complex class I protein expression, potentially protecting cells from cytotoxic T cells.

Gene	Functions
E6	Encodes for 137 amino acids oncoprotein. Inactivate the tumor suppressor protein p53.
E7	Encodes a 127 amino acid zinc binding protein. Primarily inactivates retinoblastoma protein. E7, along with E5 and E6, promotes cell-cycle progression, priming cells for viral DNA replication and immortalizing infected cells by activating cellular telomerase.
E8	Few papillomaviruses express the E8 gene's short protein, like BPV-4, which may replace the absent E6 open reading frame in this genus.
L1	Major capsid proteins that interact with cell surface heparin sulphate. Stabilizes capsids by forming disulfide bonds between neighboring L1 molecules.
L2	L2 collaborates with L1 to package viral DNA into the virion, and help during the entry process. L2 induces virion assembly by binding to viral DNA.

Pathogenesis

Infectious Entry

BPV transmission can occur through direct or indirect contact with contaminated surfaces, as well as through flies, and ticks (Muro *et al.*, 2008; Cubie, 2013). Papillomaviruses are believed to enter keratinocyte stem cells via minor skin or mucosal surface injuries called microtraumas, exposing heparin sulphate peptidoglycans. The virus binds to heparin sulphate in the cytoplasm of the epithelium cell (McBride *et al.*, 2012). The micro-injury allows the virus to access basal keratinocytes, where the virus cycle begins. L1 binds to the integrin α6 and integrin 332 after tissue micro-injury (Sibbet *et al.*, 2000; Florin *et al.*, 2012). The binding of L1 to heparan sulphate causes changes in capsid icosahedral structure, exposing the L2 N-terminal to furin protein cleavage in the cell membrane, enabling L2 to bind to various receptors (Buck *et al.*, 2013). The virus is subsequently taken up from the cell surface and is internalized through a clathrin-dependent endocytosis mechanism to membrane-bound vesicles known as endosomes (Selinka, 2002, Day, 2003). Subsequently, the minor capsid protein L2 breaks the endosome's membrane, facilitating the release and transportation of the viral DNA, along with L2, to the cell nucleus (Day *et al.*, 2003).

Maintenance and Latency

Upon successfully infecting a keratinocyte, the virus exhibits very low levels of the early viral proteins E1 and E2. These proteins play a crucial role in the replication and maintenance of the viral DNA as a circular episome. The BPV genome is found in episomal form, while HPV can integrate into host genome sites. The viral oncogenes E6 and E7, which promote cell proliferation by inactivating the tumour suppressor proteins p53 and pRb, respectively, may also exhibit minimal expression. Keratinocyte stem cells located in the epithelial basement layer have the ability to preserve papillomavirus genomes in a dormant or latent condition for extended periods of time (Doorbar, 2005).

Production of Progeny Virus

The viral late genes, L1 and L2, are only expressed in differentiating keratinocytes located in the outermost layers of the skin or mucosal surface. The upregulation of L1 and L2 is commonly associated with a substantial rise in the quantity of viral genome copies. The limited immune system surveillance of stratified squamous epithelium's outer layers suggests that this restriction of viral late gene expression may be a form of immune evasion.

The BPV reproductive cycle involves the formation of a low-copy viral genome, maintenance of replication, and amplification of differentiated vegetative cells (McBride *et al.*, 2012). This cycle relies on epithelial cell differentiation, indicating the absence of an *in vitro* replication system. Papilloma virus lack polymerase and thus induces S-phase entry in BPV-infected cells through amplification (Moody and Laimins, 2010). This stimulates cell proliferation, causing mitotic stress and cytogenetic aberrations (Araldi *et al.*, 2013). The virus invades basal keratinocytes, resulting in the replication in the spinosum and granular layers of the skin, which leads to excessive growth (Hargis and Ginn, 2012). Neoplasia can manifest as either papilloma or fibropapilloma, depending on the relative proportions of epithelial and connective tissues present (Radostits *et al.*, 2006).

New infectious progeny virus particles are assembled in the cell nucleus. Desquamation is the process by which the surface layers of stratified squamous epithelia are lost into the environment through spontaneous cellular rupture. Papillomaviruses have adapted to utilise this regular process as a means of releasing virions into the surrounding environment. Papillomaviruses utilise desquamation as a covert and non-inflammatory method of releasing themselves (Hargis and Ginn, 2012).

Delta-papillomaviruses cause sub-epithelial fibroblast transformation, followed by plexiform acanthosis and papillomatosis. In contrast, *Xi-papillomaviruses* lead to the development of epithelial papillomas without fibroblast participation (Campo, 2006).The viral hyperproliferative action leads to exophytic mass development, acanthosis, and keratinization, characterized histologically by increased keratin granules in the granular layer. This process is a consequence of epithelium differentiation and virus assembly (Araldi *et al.*, 2015). Virus assembly occurs in differentiated epithelium layers, leading to virion release through cell degeneration and koilocyte formation (Araldi *et al.*, 2015).

The term "koilocyte" originates from the Greek word "koillos," which means "cavity" (Araldi *et al.*, 2015). Koilocytes are formed from cytopathic effect of E5 and E6 oncoproteins, contributing to cell fragility and virion release. Following the process of viral assembly, virions are discharged into the corneum layer, facilitating their penetration into the keratin matrix (Buck *et al.*, 2013).

Further, this keratin provides a form of physical safeguarding for virions due to their lack of an envelope.

Gross lesions

BPV-1 and BPV-2 cause various skin fibropapillomas, including those in the forehead, neck and back skin. BPV-1 causes penile and teat papilloma, while cauliflower-like fibropapillomas in the anal, genital, and ventral abdominal regions seen in BPV-2 infection. BPV-2 and 4 is also linked to urinary bladder cancer in cattle, especially when ingestion of bracken fern is associated (Radostits *et al.*, 2006). BPV-3 causes cutaneous papillomas, while BPV-4 causes upper digestive tract papillomas (Radostits *et al.*, 2006). BPV-5 causes rice grain fibropapillomas on the udder, and BPV-6 causes frond epithelial papillomas of udder and teats (Tsirimonaki *et al.*, 2003, Roperto *et al.*, 2015).

Gross lesions are either exophytic or endophytic in nature. These lesions consist of a hyperplastic epithelium, dermal tissues and dilated capillaries (Ginn *et al.*, 2007; Hargis and Ginn, 2012). Morphologically, as per Monteiro *et al.* (2008) they are classified as shown in Table 11.3.

Table 11.3: Gross Lesions Observed in BPV Infection

Sl.No.	*Classification of Lesions*
1	Classical-exophytic "cauliflower" masses with wide or narrow insertion base
2	Pedunculated - exophytic masses connected by a narrow base with a peduncle shape
3	Atypical or flat - dense and flat exophytic masses entirely connected with the tissue
4	Filamentous - exophytic masses with highly keratinized surface and a thin base, present in mammary glands
5	Rice-form - small papillomas with rice-like shape

Papilloma usually regresses within a year. But chronically immunosuppressed cattle have papillomas that do not regress after one year, spreading to other sites. This is often due to exposure to immunosuppressive chemicals, such as those found in bracken fern, or infection with bovine viral diarrhoea virus (Campo, 2006). Teat papillomatosis in cows can cause mastitis, prevent suckling, and make milking impossible. Teat papillomas also called as rice grain fibropapillomas, which appear pedunculated, can snap off and become infected, leading to milk canal distortions. Fibropapilloma of the penis can spread along the perineum causing sterility (Campo, 2006; Borzacchiello and Roperto, 2008).

Histopathology

Histopathological analysis of cutaneous papillomas show well-developed finger-like projections with overlying stratum corneum, acanthosis, and uniformly down-growing rete pegs, moderate orthokeratotic hyperkeratosis, prominent vacuolating (koilocytes) cells in stratum spinosum and negligible fibrocellular reaction (Leishangthem *et al.*, 2008; Jangir *et al.*, 2017).

Fibropapillomas are characterized by moderate to extensive hyperkeratosis, acanthosis, with a basket weave appearance. They have well developed finger like projections having connective tissue core and irregular rete pegs projecting deep into fibrous connective tissue. The stratum granulosum layer also shows marked thickening and numerous basophilic keratohyaline granules in the cytoplasm. Numerous koilocytes with condensed hyperchromatic nuclei are observed in the stratum spinosum and granulosum.

Fibrosarcoma shows excessive fibrous tissue proliferation with variable-sized fibrocytes and fibroblasts with hapazzard arrangement of pleomorphic nuclei (Jangir and Somvanshi, 2021). Fibroblastic proliferation can be seen in samples infected with BPV-2, a common symptom in papillomas infected with *Deltapapillomavirus* (Anjos *et al.*, 2010; Marins and Ferreira, 2011; Timurkan and Alcigir, 2017).

In buffalo, Cutaneous warts are microscopically characterized as exophytic fibropapilloma, endophytic papilloma, endophytic fibropapilloma and fibroma (Pangty *et al.*, 2010, Somvanshi, 2011, Jangir *et al.*, 2017).

Equine sarcoids are lesions with intense fibroblastic proliferation, presenting a fibropapilloma-like morphology (Martens *et al.*, 2000). They are characterized by anaplastic and pleomorphic fibroblasts perpendicularly oriented in relation to the basal membrane, found in the dermo-epidermal junction.

Diagnosis

Diagnosis of bovine papillomatosis can be achieved by gross and histopathology along with other ancillary techniques such as polymerase chain reaction, real time- polymerase chain reaction, immunohistochemistry, in situ hybridization, electron microscopy, comet assay and flow cytometry.

Virus Isolation

BPV can be isolated using density gradient separation *i.e.* either ultracentrifugation in cesium chloride (CsCl) gradients or sucrose gradients (Gillock *et al.*, 1997; Araldi *et al.*, 2014) and can be subjected to further confirmation by electron microscopy or molecular tecchniques.

Polymerase Chain Reaction (PCR)

It is the most sensitive method for viral detection. There is no "gold-standard" primer for PVs identification. PCR amplification targeting the L1, L2, E1-E7 primers gene of BPV-1, -2, BPV-5, and -10 was conducted using specific primers. Different primer setswere employed to amplify L1 ORF for detecting *Deltapapillomavirus* (BPV-1 and 2), *Epsilonpapillomavirus*(BPV-5), and *Xipapillomavirus* (BPV-3, 4, and 6), respectively (Araldi *et al.*, 2014). While real-time PCR is common for BPV identification, PCR followed by DNA sequencing is more reproducible for subsequent characterization of BPVs (Melo *et al.*, 2014; Araldi *et al.*, 2014). Genotype-specific primers were employed for the identification and characterization of BPV, and sequencing coupled with phylogenetic analysis was affirmed as the most suitable method for BPV characterization (Leishangthem *et al.*, 2007; Grindatto *et al.*, 2015, Ozmen and Kale, 2023).

Southern Blot

After PCR amplification, the material needs to be subjected to Southern blot, dot blot, reverse hybridization, restriction enzyme digestion, and sequencing for identifying papilloma virus types (Molijn *et al.*, 2005).Southern blot is a highly sensitive and accurate method for estimating the amount of DNA in a lesion. Dot blot and reverse blot are laborious techniques with similar sensitivity and good accuracy (Majewski andJablonska, 1997).

Immunohistochemistry

Melo *et al.* (2015) studied BPV protein expression in cutaneous papillomas using IHC using monoclonal antibodies against E2 and L1. L1 was found to be immunodetected in both epithelium and dermis, while E2 protein showed immunostaining in the granular layer and cytoplasmic labelling in the corneal layer (Nasir *et al.*, 1999; Roperto *et al.*, 2011; Melo *et al.*, 2015). BPV was immunolocalised in the epidermis, keratin layer, and dermis (Ozmen and Kale, 2023).

In Situ Hybridization

It uses radiolabelled probes to assess the distribution of papilloma virus in lesions and allows for viral localization using other markers. Non-radioactive hybrid capture is safe, easy to perform, and reproducible, providing good accuracy for mucosal lesions. Chromogenic in Situ Hybridization (CISH) is an additional method used to demonstrate viruses' physical state. The L1 gene, the most conserved in papillomaviruses, is used for BPV detection. Melo *et al.* (2015) utilized CISH to analyse BPV sequences in PBMCs, revealing viral DNA labelling in nuclear, perinuclear, and cytoplasmic regions, as an additional method to PCR.

Comet Assay

Recent studies confirm that peripheral blood mononuclear cells (PBMCs) carry viruses, with BPV found in CD4+ and CD8+ lymphocytes, and viral activity detected using comet assay and flow cytometry (Roperto *et al.*, 2011; Araldi *et al.*, 2013).

Electron Microscopy

Papilloma virus can be confirmed through electron microscopic studies in histopathologically confirmed cases of cutaneous warts. Transmission electron microscopic examination of warts revealed negatively stained, non-enveloped virus particles, approximately 60 nm in diameter, with icosahedral symmetry (Leishangthem *et al.*, 2008, Ata *et al.*, 2021). BPV particles were observed within intracytoplasmic and/or intranuclear areas in the epithelial cells (Melo *et al.*, 2015; Ozmen and Kale, 2023).

Enzyme-Linked Immunosorbent Assay (ELISA)

ELISA for detection of BPV antigen can be performed in blood sample (Ozmen and Kale, 2023).

Prevention and Treatment

Papilloma treatments are currently inefficient, with methods such as surgical excision, laser therapy, cryotherapy, and chemotherapy. However, surgical excision has a recurrence of 50-64 percent within six months, laser therapy has a 38 percent recurrence rate, cryotherapy is inefficient for lesions larger than 2 cm, and chemotherapy using cisplatin or 5′-fluouracyl (5-FU) can cause nephrotoxicity and hepatotoxicity (Araldi *et al.*, 2017). Homoepathic alternative medicine such as thuja has been used to treat papilloma. Saied (2021) reported that cutaneous papillomas were regressed with ivermectin. Controlling blood sucking insect populations in fields can effectively halt vector transmission of BPV, although this is not the main mechanism of viral spread.

Vaccine

Challenges in Vaccine Development

The development of vaccine strategies against papillomavirus (PV) is crucial due to the large national cattle herd and significant economic losses caused by BPV-related diseases (Mazzuchelli-Souza *et al.*, 2013). The main challenge is viral isolation of a particular viral type, as co-infection is common (Lindsey *et al.*, 2009; Araldi *et al.*, 2013). There is no *in vitro* PV replication system and PV is not cultivable, as amplification of virus only occurs in differentiated vegetative/epithelial cells (Araldi *et al.*, 2014). BPV cannot grow in conventional cell cultures, so protein expression systems like yeast and insect cells have been used to produce virus-like particles (VLPs) vaccines.

Preventive vaccines for cattle against BPV2 and 4 have been developed, targeting cutaneous and mucous BPVs associated with cancer in cattle (Campo *et al.*, 1993). A vaccine made from Escherichia coli's BPV2 L1 capsid protein produced neutralizing antibodies, preventing infection (Jarrett *et al.*, 1991). VLPs from BPV4 genes, specifically L1 and L2 genes, are highly immunogenic and produce potent prophylactic vaccines. L1 VLPs promote neutralizing antibodies and effectively prevent infection during BPV4 challenge, while L2 proteins in insect cells effectively prevent experimentally induced papillomas (Campo, 1997).A candidate vaccine against BPV1 using L1 VLPs produced in plant elicited a strong, specific immune response, demonstrating potential for a lower-cost vaccine (Love *et al.*, 2012).BPV1 L1 produced in a baculovirus expression system and purified as VLPs could be used for prophylactic immunization agent against BPV-1 infection (Vrablikova *et al.*, 2023). A binary ethylenimine inactivated and saponized cutaneous warts (BPV-2) therapeutic vaccine was assessed for enzootic bovine haematuria in hill cows, however no effective regression of the tumor was observed except for immunological response (Pathania *et al.*, 2011)

Conclusion

Bovine papillomavirus has been extensively studied in recent years due to its pathogenic and carcinogenic properties. However, a relationship between BPV types and anatomic location of lesion couldn't be established. Various ancillary techniques can be employed to confirmed the types of virus causing the disease. Further, there is a need for the development of various alternative medicine and vaccine for treatment of bovine papillomatosis.

REFERENCES

Anjos, B.L.D., Silva, M.S., Diefenbach, A., Brito, M.D.F., Seppa, G.D.S. and Brum, M.C.S. (2010). Sarcoideequinoassociadoao papillomavirus bovino BR-UEL-4.Ciência Rural, 40: 1456-1459.

Araldi, R.P., Assaf, S.M.R., Carvalho, R.F.D., Carvalho, M.A.C.R.D., Souza, J.M.D., Magnelli, R.F., Modolo, D.G., Roperto, F.P., Stocco R.C. and Beçak, W. (2017). Papillomaviruses: a systematic review. Genet Mol Biol, 40: 1-21.

Araldi, R.P., Carvalho, R.F.D., Melo, T.C.D., Diniz, N.S.P., Sant'Ana, T.A., Mazzuchelli-de-Souza, J., Spadacci-Morena, D.D., Beçak, W. and Stocco, R.D.C. (2014). Bovine papillomavirus in beef cattle: first description of BPV-12 and putative type BAPV8 in Brazil. Genet Mol Res, 13(3): 5644-5653.

Araldi, R.P., Melo, T.C.D., Diniz, N., Mazzuchelli-de-Souza, J., Carvalho, R.F.D., Beçak, W. and Stocco, R.D.C. (2013). Bovine papillomavirus clastogenic effect analyzed in comet assay. Bio Med Res Int, 2013, 630683.

Araldi, R.P., Melo, T.C., Neves, A.C., Spadacci-Morena, D.D., Magnelli, R.F., Modolo, D.G., de-Sá-Júnior, P.L., Mazucchelli-de-Souza, J., Carvalho, R.F., Beçak, W. and Stocco, R.C. (2015). Hyperproliferative action of bovine papillomavirus: genetic and histopathological aspects. Genet Mol Res, 14(4): 12942-12954.

Ata, E.B., Allam, A.M., Elbayoumy, M.K. and Mahmoud, M.A.E. (2021). Electron microscopy and phylogenetic analysis of Bovine papillomavirus infection in cattle from four Egyptian governorates. Trop Anim Health Prod, 53: 160.

Bogaert, L., Martens, A., Kast, W.M., Van Marck, E. and De Cock, H. (2010). Bovine papillomavirus DNA can be detected in keratinocytes of equine sarcoid tumors. Vet Microbiol, 146(3-4): 269-275.

Bogaert, L., Van Poucke, M., De Baere, C., Dewulf, J., Peelman, L., Ducatelle, R., Gasthuys, F. and Martens, A. (2007). Bovine papillomavirus load and mRNA expression, cell proliferation and p53 expression in four clinical types of equine sarcoid. J Gen Virol, 88(8): 2155-2161.

Borysiewicz, L.K., Fiander, A., Nimako, M., Man, S., Wilkinson, G.W.G., Westmoreland, D., Evans, A.S., Adams, M., Stacey, S.N., Boursnell, M.*E.G.* and Rutherford, E. (1996). A recombinant vaccinia virus encoding human papillomavirus types 16 and 18, E6 and E7 proteins as immunotherapy for cervical cancer. Lancet, 347(9014): 1523-1527.

Borzacchiello, G. and Roperto, F. (2008). Bovine papillomaviruses, papillomas and cancer in cattle. Vet Res, 39(5): 45.

Borzacchiello, G., Nasir, L., Roperto, F.P. and Campo, M.S. (2009). Human papillomavirus research: do we still need animal models? Int J Cancer, 123: 739-740.

Brandsma, J.L., Yang, Z.H., Barthold, S.W. and Johnson, E.A. (1991). Use of a rapid, efficient inoculation method to induce papillomas by cottontail rabbit papillomavirus DNA shows that the E7 gene is required. Proc Nat Acad Sci, 88(11): 4816-4820.

Buck, C.B., Day, P.M. and Trus, B.L. (2013). The papillomavirus major capsid protein L1. Virol, 445(1-2): 169-174.

Campo, M.S. (1997). Bovine papillomavirus and cancer. Vet J, 154(3): 175-188.

Campo, M.S. (2006). Bovine papillomavirus: old system, new lessons? In: Campo MS (ed) Papillomavirus research: from natural history to vaccine and beyond. Caister Academic Press, Wymondham, pp 1-31

Campo, M.S. and Roden, R.B. (2010). Papillomavirus prophylactic vaccines: established successes, new approaches. J Virol, 84(3): 1214-1220.

Campo, M.S., Grindlay, G.J., O'Neil, B.W., Chandrachud, L.M., McGarvie, G.M. and Jarrett, W.F. (1993). Prophylactic and therapeutic vaccination against a mucosal papillomavirus. J Gen Virol, 74(6): 945-953.

Carvalho, C.C.R., Batista, M.V.A., Silva, M.A.R., Balbino, V.Q. and Freitas, A.C. (2012). Detection of bovine papillomavirus types, coinfection and a putative new BPV11 subtype in cattle. Transbound Emerg Dis, 59(5): 441-447.

Chambers, G., Ellsmore, V.A., O'brien, P.M., Reid, S.W.J., Love, S., Campo, M.S. and Nasir, L. (2003). Association of bovine papillomavirus with the equine sarcoid. J Gen Virol, 84(5): 1055-1062.

Cubie, H.A. (2013). Diseases associated with human papillomavirus infection. Virol, 445(1-2), 21-34.

Da Silva, M.A.R., Carvalho, C.C.R., Coutinho, L.C.A., Reis, M.C., de Aragão Batista, M.V., De Castro, R.S., Dos Anjos, F.B.R. and De Freitas, A.C. (2012).

Coinfection of Bovine papillomavirus and felineassociated papillomavirus in bovine cutaneous warts. Transbound Emerg Dise, 59(6): 539-543.

Day, P.M., Lowy, D.R. and Schiller, J.T. (2003). Papillomaviruses infect cells via a clathrin-dependent pathway. Virol, 307(1): 1-11.

de Freitas, A.C. and Almeida, A.P. (2012). Genetic diversity of human and animal papillomaviruses.

De Villiers, E.M., Fauquet, C., Broker, T.R., Bernard, H.U. and ZurHausen, H. (2004). Classification of papillomaviruses. Virol, 324(1): 17-27.

Doorbar, J. (2005). The papillomavirus life cycle. J Clin Virol, 32: 7-15.

Embers, M.E., Budgeon, L.R., Pickel, M. and Christensen, N.D. (2002). Protective immunity to rabbit oral and cutaneous papillomaviruses by immunization with short peptides of L2, the minor capsid protein. J Virol, 76(19): 9798-9805.

Florin, L., Sapp, M. and Spoden, G.A. (2012). Host-cell factors involved in papillomavirus entry. Med Microbiol Immunol, 201: 437-448.

Gil da Costa, R.M. and Medeiros, R. (2014). Bovine papillomavirus: opening new trends for comparative pathology. Arch Virol, 159: 191-198.

Gillock, E.T., Rottinghaus, S., Chang, D., Cai, X., Smiley, S.A., An, K. and Consigli, R.A. (1997). Polyomavirus major capsid protein VP1 is capable of packaging cellular DNA when expressed in the baculovirus system. J Virol, 71(4): 2857-2865.

Ginn, P.E., Mansell, J.E.K.L. and Rakich, P.M. (2007). Skin and appendages. In: Maxie MG (ed) Jubb, Kennedy, and Palmer's pathology of domestic animals, vol 1. Saunders/Elsevier, Philadelphia, pp 748-751

Grindatto, A., Ferraro, G., Varello, K., Crescio, M. I., Miceli, I., Bozzetta, E., and Nappi, R. (2015). Molecular and histological characterization of bovine papillomavirus in North West Italy. Vet Microbiol, 180(1-2): 113-117.

Hamad, M.A., Al-Shammari, A.M., Odisho, S.M. and Yaseen, N.Y. (2018). Molecular epidemiology of bovine papillomatosis and identification of three genotypes in central Iraq. Intervirol, 60(4): 156-164.

Hargis, A.M. and Ginn, P.E. (2012). The integument. In: Zachary JF, McGavin MD (eds) Pathologic basis of veterinary disease, 5th ed. Elsevier/Mosby, St. Louis, pp. 1025-1027

Harro, C.D., Pang, Y.Y.S., Roden, R.B., Hildesheim, A., Wang, Z., Reynolds, M.J., Mast, T.C., Robinson, R., Murphy, B.R., Karron, R.A. and Dillner, J. (2001). Safety and immunogenicity trial in adult volunteers of a human papillomavirus 16 L1 virus-like particle vaccine. J Natl Cancer Inst, 93(4): 284-292.

Howley, P.M. and Lowy D.R. (2007). Papillomaviruses. In: Knipe DM, Howley PM. (Eds.) Fields' Virology, Philadelphia: Lippincott Williams and Wilkins. p 2300-2354.

Jangir, B.L. and Somvanshi, R. (2021).Pathological studies and detection of co-infection of bovine papilloma viruses in cattle cutaneous and teat warts. Vet Arh, 91(6): 589-602.

Jangir, B.L., Bind, R.B., Kumar, P. and Somvanshi, R. (2017). Pathologicalstudies and detection of different bovine papilloma virus types in buffalo cutaneous warts. Turk J Vet Anim, 41(2): 306-311.

Jarrett, W.F.H., Smith, K.T., O'neil, B.W., Gaukroger, J.M., Chandrachud, L.M., Grindlay, G.J., McGarvie, G.M. and Campo, M.S. (1991). Studies on vaccination against papillomaviruses: prophylactic and therapeutic vaccination with recombinant structural proteins. Virol, 184(1): 33-42.

Koller, L. D. and Olson, C. (1972). Attempted transmission of warts from man, cattle, and horses and deer fibroma to selected hosts. J Invest Dermatol, 58(6): 366-368.

Kumar, P., Jangir, B.L., Saikumar, G. and Somvanshi, R. (2013a). Co-infection of bovine papillomavirus type-1 and-10 in teat warts of dairy cattle. Vet Med-Czech J, 58(12): 605-608.

Kumar, P., Nagarajan, N., Kumar, D., Bind, R.B. and Somvanshi, R. (2013b). Detection and quantification of bovine papillomaviruses (BPVs) in cutaneous warts of cattle and buffaloes. Adv Anim Vet Sci, 1(2): 53-58.

Lehtinen, M., Kibur, M., Luostarinen, T., Anttila, A. and Pukkala, E. (2000). Prospects for phase III–IV HPV vaccination trials in the Nordic countries and in Estonia. J Clin Virol, 19(1-2): 113-122.

Leishangthem, G.D., Somvanshi, R. and Lauren, D.R. (2008a). Pathological studies on bovine papilloma virus-fern interaction in hamsters. Indian J Exp Biol, 46: 100-107.

Leishangthem, G.D., Somvanshi, R. and Tiwari, A.K. (2008b). Detection of bovine papillomaviruses in cutaneous warts/papillomas in cattle. Indian J Vet Pathol, 32(1): 15-20

Leto, M.D.G.P., Santos Júnior, G.F.D., Porro, A.M. and Tomimori, J. (2011). Human papillomavirus infection: etiopathogenesis, molecular biology and clinical manifestations. An Bras Dermatol, 86: 306-317.

Lindsey, C.J., Almeida, M.E.D., Vicari, C.F., Carvalho, C.D., Yaguiu, A., Freitas, A.C.D., Beçak, W. and Stocco, R.D.C. (2009). Bovine papillomavirus DNA in milk, blood, urine, semen, and spermatozoa of bovine papillomavirus-infected animals. Genet Mol Res, 8(1): 310-318.

Love, A.J., Chapman, S.N., Matic, S., Noris, E., Lomonossoff, G.P. and Taliansky, M. (2012). In planta production of a candidate vaccine against bovine papillomavirus type 1. Planta, 236: 1305-1313.

Majewski, S. and Jablonska, S. (1997). Human papillomavirus-associated tumors of the skin and mucosa. J Am Acad Dermatol, 36(5): 659-685.

Marins, R.D.Q. and Travassos, C.E.P.F. (2011). Clinicopathologic and epithelial regression study in cutaneous warts of bovines infected by papillomavirus. Int Res J Pharm Pharmacol, 1(1): 12-16.

Martens, A., Moor, A.D.E., Demeulemeester, J. and Ducatelle, R. (2000). Histopathologica characteristics of five clinical types of equine sarcoid. Res Vet Sci, 69(3): 295-300.

Mazzuchelli-de-Souza, J., Carvalho, R.F.D., Ruiz, R.D.M., Melo, T.C.D., Araldi, R.P., Carvalho, E., Thompson, C.E., Sircili, M.P., Beçak, W. and Stocco, R.D.C. (2013). Expression and in silico analysis of the recombinant bovine papillomavirus E6 protein as a model for viral oncoproteins studies. BioMed Res Int, 2013:421398. doi: 10.1155/2013/421398.

McBride, A.A., Sakakibara, N., Stepp, W.H. and Jang, M.K. (2012). Hitchhikingon host chromatin: how papillomaviruses persist. Biochim Biophys Acta, 1819(7): 820-825. doi: 10.1016/j.bbagrm.2012.01.011.

Melo, T.C., Araldi, R.P., Pessoa, N.S.D., de-Sá-Júnior, P.L., Carvalho, R.F., Beçak, W. and Stocco, R.C. (2015). Bostaurus papillomavirus activity in peripheral blood mononuclear cells: demonstrating a productive infection. Genet Mol Res, 14(4): 16712-16727.

Melo, T.C.D., Carvalho, R.F.D., Mazzucchelli-de-Souza, J., Diniz, N., Vasconcelos, S., Assaf, S.L.M.R., Araldi, R.P., Ruiz, R.D.M., Kerkis, I., Beçak, W. and Stocco, R.D.C. (2014). Phylogenetic classification and clinical aspects of a new putative Deltapapillomavirus associated with skin lesions in cattle. Genet Mol Res, 13(2): 2458-2469.

Molijn, A., Kleter, B., Quint, W. and Van Doorn, L.J. (2005). Molecular diagnosis of human papillomavirus (HPV) infections. J Clin Virol, 32: 43-51.

Monteiro, V.C., Coelho, M.C.O.C., da Silva Carneiro, A., de Araújo Silva, R.A., Teixeira, M.N., Wanderley, A.G., Wanderley, E.K. and Franco, E.D.S.F. (2008). Descriçãoclínica e histopatológica da papilomatosecutânea bovina (BPV). Ciênc Anim Bras, 9(4): 1079-1088.

Moody, C.A. and Laimins, L.A. (2010). Human papillomavirus oncoproteins: pathways to transformation. Nat. Rev. Cancer, 10(8): 550-560.

Mosseri, S., Hetzel, U., Hahn, S., Michaloupoulou, E., Sallabank, H.C., Knottenbelt, D.C. and Kipar, A. (2014). Equine sarcoid: In situ demonstration of matrix metalloproteinase expression. The Vet J, 202(2): 279-285.

Muro, L., Botteira, C. and Piccinin, A. (2008). Papilomatose bovina. RevistaCientíficaEletrônica de MedicinaVeterinária, 10: 1-7.

Nasir, L. and Reid, S.W.J. (1999). Bovine papillomaviral gene expression in equine sarcoid tumours. Virus Res, 61(2): 171-175.

Ogawa, T., Tomita, Y., Okada, M., Shinozaki, K., Kubonoya, H., Kaiho, I. and Shirasawa, H. (2004). Broad-spectrum detection of papillomaviruses in bovine teat papillomas and healthy teat skin. J Gen Virol, 85(8): 2191-2197.

O'Neil, B.W. (1988). Studies on Immunity to Bovine Papillomaviruses (Doctoral dissertation, University of Glasgow (United Kingdom)).

Özmen, G. and Kale, M. (2023). Searching bovine papillomavirus presence in lesions seen on teats of cows. Pesquisa Veterinária Brasileira Pesqui Vet Brasil, 43: e07150.

Pangty, K., Singh, S., Goswami, R., Saikumar, G. and Somvanshi, R. (2010). Detection of BPV1 and2 and quantification of BPV1 by realtime PCR in cutaneous warts in cattle and buffaloes. Transbound Emerg Dis, 57(3): 185-196.

Pathania, S., Kumar, P., Devi, L. G., Kumar, D., Dhama, K., and Somvanshi, R. (2011). Preliminary assessment of binary ethylenimine inactivated and saponized cutaneous warts (BPV-2) therapeutic vaccine for enzootic bovine haematuria in hill cows. Vaccine, 29(43): 7296-7302.

Pilacinski, W.P., Glassman, D.L., Glassman, K.F., Reed, D.E., Lum, M.A., Marshall, R.F., Muscoplat, C.C. and Faras, A.J. (2007). Immunization against bovine papillomavirus infection. In Ciba Foundation Symposium 120Papillomaviruses: Papillomaviruses: Ciba Foundation Symposium 120 (pp. 136-156). Chichester, UK: John Wiley and Sons, Ltd.

Radostits, O.M., Gay, C.C., Hinchcliff, H.W. and Constable, P.D. (2006). Viral diseases characterized by skin lesions. Veterinary medicine: a textbook of the diseases of cattle, horses, sheep, pigs and goats, 10th edn. Saunders Elsevier, Edinburgh, 1421-1422.

Rees, C.A. (2004). Disorders of the Skin.In: Reed, S.M., Bayly, W.M., Sellon, D.C. (Eds) Equine Internal Medicine, 2nd edn. Saunders, Elsevier, St. Louis, p 667-720.

Roperto, S., Brun, R., Paolini, F., Urraro, C., Russo, V., Borzacchiello, G., Pagnini, U., Raso, C., Rizzo, C., Roperto, F. and Venuti, A. (2008). Detection of bovine papillomavirus type 2 in the peripheral blood of cattle with urinary bladder tumours: possible biological role. J Gen Virol, 89(12): 3027-3033.

Roperto, S., Comazzi, S., Ciusani, E., Paolini, F., Borzacchiello, G., Esposito, I., Luca, R., Russo, V., Urraro, C., Venuti, A. and Roperto, F. (2011). PBMCs are additional sites of productive infection of bovine papillomavirus type 2. J Gen Virol, 92(8): 1787-1794.

Roperto, S., Russo, V., Leonardi, L., Martano, M., Corrado, F., Riccardi, M.G. and Roperto, F. (2016). Bovine papillomavirus type 13 expression in the urothelial bladder tumours of cattle. Transbound Emerg Dis, 63(6): 628-634.

Saied, A.A. (2021). Regression of bovine cutaneous papillomas via ivermectin-induced immunostimulant and oxidative stress. J Adv Vet Anim Res, 8(3): 370-377.

Santos, E.U.D., Silva, M.A.R., Pontes, N.E., Coutinho, L.C.A., Paiva, S.S.L., Castro, R.S. and Freitas, A.C. (2016). Detection of different bovine papillomavirus types and coinfection in bloodstream of cattle. Transbound Emergi Dis, 63(1): e103-e108.

Sauthier, J.T., Daudt, C., da Silva, F.R.C., Alves, C.D.B.T., Mayer, F.Q., Bianchi, R.M., Driemeier, D., Streit, R.S.A., Staats, C.C., Canal, C.W. and Weber, M.N. (2021). The genetic diversity of "papillomavirome" in bovine teat papilloma lesions. Anim Microbiome, 3(1): 1-13.

Shah, K.V. and Howley, P.M. (1996)."Papillomaviruses". In Fields BN, Knipe DM, Howley PM, *et al.* (eds.). Fields Virology (3rd ed.). Lippincott-Raven.

Shope, R.E. and Hurst, E.W. (1933). Infectious papillomatosis of rabbits: with a note on the histopathology. J Exp Med, 58(5): 607-624.

Sibbet, G., Romero-Graillet, C., Meneguzzi, G. and Campo, M. S. (2000). α6 integrin is not the obligatory cell receptor for bovine papillomavirus type 4. J Gen Virol, 81(2): 327-334.

Somvanshi, R. (2011). Papillomatosis in buffaloes: A lessknown disease. Transbound Emerg Dis, 58(4): 327-332.

Stocco dos Santos, R.C., Lindsey, C.J., Ferraz, O.P., Pinto, J.R., Mirandola, R.S., Benesi, F.J., Birgel, E.H., Bragança Pereira, C.A. and Beçak, W. (1998). Bovine papillomavirus transmission and chromosomal aberrations: an experimental model. J Gen Virol, 79(9): 2127-2135.

Timurkan, M.O. and Alcigir, M.E. (2017). Phylogenetic analysis of a partial L1 gene from bovine papillomavirus type 1 isolated from naturally occurring papilloma cases in the northwestern region of Turkey. Onderstepoort J Vet Res, 84(1): 1-6.

Tsirimonaki, E., O'neil, B.W., Williams, R. and Campo, M.S. (2003). Extensive papillomatosis of the bovine upper gastrointestinal tract. J Comp Pathol, 129(2-3): 93-99.

Ugochukwu, I.C.I., Aneke, C.I., Idoko, I.S., Sani, N.A., Amoche, A.J., Mshiela, W.P., Ede, R.E., Ibrahim, N.D.G., Njoku, C.I.O. and Sackey, A.K.B. (2019). Bovine papilloma: aetiology, pathology, immunology, disease status, diagnosis, control, prevention and treatment: a review. Comp Clin Pathol, 28: 737-745.

Vrablikova, A., Brezani, V., Psikal, I.,Fraiberk, M., Sebela, M., Fojtikova, M., Kulich, P., Hezova, R., Masek, J. (2023). Development of modern immunization agent against bovine papillomavirus type 1 infection based on BPV1 L1 recombinant protein. Front Vet Sci, 10: 1116661. doi: 10.3389/fvets.2023.1116661.

Zheng, Z.M. and Baker, C.C. (2006). Papillomavirus genome structure, expression, and post-transcriptional regulation. *Front Biosci*, 11: 2286.

Chapter 12

Bovine Viral Diarrhea - Mucosal Disease (BVD-MD)

Niranjan Mishra*, S. Kalaiyarasu and S.B. Sudhakar

Principal Scientist, Indian Council of Agricultural Research-National Institute of High Security Animal Diseases, Anand Nagar, Bhopal – 462 022, Madhya Pradesh, India
****e-mail: mishranir@rediffmail.com***

ABSTRACT

Bovine viral diarrhea (BVD) is prevalent worldwide and has high economic impact in cattle production due to a variety of disease syndromes, including reproductive disease, respiratory disease, gastroenteritis and the fatal mucosal disease. BVD is caused by three bovine pestiviruses, bovine viral diarrhea virus 1 (BVDV-1) of Pestivirus A species, BVDV-2 of Pestivirus B species and HoBi-like pestivirus (HoBiPeV) of Pestivirus H species, belonging to the Pestivirus genus within the Flaviviridae family. The term BVDV refers to all the three bovine pestiviruses in this chapter. BVD has a wide host range and bovine pestiviruses are highly successful to persist and spread in their host populations due to their unique ability to generate persistent infection through evasion of both innate and adaptive immune response. BVDV-1 and BVDV-2 are distributed in all continents, while HoBiPeV has been reported in South America, Europe and Asia. Bovine pestiviruses show high genetic and antigenic diversity, as demonstrated by identification of 21 BVDV-1 genotypes (BVDV-1a-1u), 4 BVDV-2 genotypes (BVDV-2a-2d) and 5 HoBiPeV genotypes (HoBiPeV-a-e) globally. All the three, BVDV-1, BVDV-2 and HoBiPeV are also prevalent in India. Providing acquired immune protection against BVD is challenging due to the antigenic diversity among BVDV strains and ability of BVDV to infect the fetus. Both killed and live attenuated vaccines have been reported to be effective in field and recent advancement in molecular studies may help towards future development of new generation vaccines against BVD. Successful BVD control programs are based on identification and removal of PI animals, movement controls, strict biosecurity, vaccination and surveillance. However, vaccination alone has not resulted in the elimination of BVD, highlighting importance of implementation of other associated control measures. Some of the developed countries have initiated successful BVD control programs at national level, while most developing and emerging economies live with BVD. This chapter focuses advances in knowledge involving all aspects of BVD-MD with special emphasis on molecular biology, epidemiology, genetic and antigenic diversity, diagnosis, prevention and control.

Keywords: *Pestivirus, Bovine viral diarrhea virus, Persistent infection, BVDV-1, BVDV-2, HoBiPeV, Epidemiology, Diagnosis, Control, Cattle.*

Introduction

Bovine viral diarrhea (BVD) is one of the major economically important viral diseases of cattle and is prevalent in cattle populations worldwide. It is one of the most complex infectious diseases with regard to its host spectrum, host virus interactions, virulence, immune response, pathogenesis, diagnosis, management and control. Bovine viral diarrhea (BVD) was first reported in USA in 1946 (Olafson *et al.*, 1946) in association with epizootics of acute disease characterized by fever, leucopenia, reduced milk yield, high rates of abortion, diarrhea and erosive lesions of the digestive tract of cattle. In the 1950s, a special form of viral diarrhea called mucosal disease (MD), with haemorrhages and intestinal erosions was reported in Iowa, USA (Ramsey *et al.*, 1953). However, the relationship between the two illnesses could be established after many years, when it was clear that both BVD and MD were caused by a viral agent, bovine viral diarrhea virus (BVDV).

In 1987, BVD associated with thrombocytopenia, haemorrhage and high mortality was reported in USA dairy herds (Perdrizet *et al.*, 1987), followed by other reports from USA and Canada (Bolin and Ridpath, 1992; Pellerin, 1994). The causative agent involved was identified as BVDV-2 and the classical BVDV strains identified earlier were termed as BVDV-1 (Ridpath *et al.*, 1994). The recently emerged HoBiPeV (strain D32/00-'HoBi'), was first detected in foetal bovine serum (FBS) originating from Brazil (Schirrmeier *et al.*, 2004) and has frequently been identified in commercial FBS, mostly of Brazilian origin but also originating from Argentina, Mexico, Canada and Australia (Xia *et al.*, 2011; Pecora *et al.*, 2019). Nevertheless, it was present since 1996 in Brazil, when it was detected and isolated from a diseased Brazilian buffalo in 1996 (Stalder *et al.*, 2005). HoBi-like pestivirus has since been recognized as one of the causative agents of BVD and is prevalent in Europe, South America and Asia with clinical symptoms akin to the classical BVDV-1 or BVDV-2 infections (Schirrmeir *et al.*, 2004; Liu *et al.*, 2009; Bauermann *et al.*, 2013, Mishra *et al.*, 2014). Among the bovine pestiviruses affecting cattle globally, BVDV-1 is predominantly prevalent, while BVDV-2 and HoBiPeV occurs sporadically.

BVD causes significant economic losses in cattle production and the economic impact varies within and between countries (Richter *et al.*, 2017). Direct losses due to BVDV occur on account of increased morbidity and mortality in adult cattle due to acute infection, reduced milk yield, respiratory disorders, extended calving interval, reproductive disorders such as repeat breeding and abortions, congenital defects, increased neonatal mortality, non-thriving and death among young stock besides the costs for treatment and prevention (Houe, 2003). Indirect losses of BVD arise due to implementation of control programs and trade restrictions. Devastating effects of BVD on the economy of dairy farmers have been seen in the form of severe acute BVD outbreaks with high mortalities in many countries in North America and Europe, specifically in USA, Canada and Germany, while it remains undetermined in other countries. A study on direct financial losses due to BVD in 15 countries around the world over the past 30 years has shown that direct financial losses were in the range of 0.50–687.80 USD per animal and the average direct losses were higher per dairy cow than per beef cow (Richter *et al.*, 2017). Based on the data of 31 published studies undertaken around the world during 1991-2015, it has been shown that the economic impact of BVD ranges from 0–552 GBP per cow per year and is based on outcome of the disease (Yarnal and Thrusfield, 2017). Despite variation, calculation of economic losses has been a significant motivator for considering implementation of BVD control programs and mitigation activities in many countries in Europe. In this chapter, unless specified, the term BVDV is used to refer all the three bovine pestiviruses.

The Disease

Bovine viral diarrhoea refers to the disease syndromes (reproductive, respiratory, gastrointestinal) following post natal infection with ncp bovine pestiviruses, whereas mucosal disease (MD) occurs mostly in cattle which had been persistently infected with a ncp BVDV followed by generation of cp BVDV by genetic alterations/mutations or super-infection with an antigenically related cp BVDV.

BVD

Acute bovine pestivirus infections in cattle develop when seronegative and immuno-competent cattle are infected with pestiviruses and the disease may be subclinical, severe acute or chronic.

The majority of postnatal BVDV infections are inapparent and virulence of BVDV strains is the key determinant of outcome of acute infection. Field outbreaks of acute BVD associated with thrombocytopenia, severe clinical signs and mortality in all age groups has been reported with death of about 40000 animals, due to high virulent strains of BVDV-2 (Perdriz et *et al.*, 1987; Carman *et al.*, 1998). However, subsequent studies showed that BVDV-1 is able to induce heamorrhagic disease and all the three bovine pestiviruses, BVDV-1,

BVDV-2 and HoBiPeV encompasses strains of high, moderate and low virulence (Ridpath *et al.*, 2000; Decaro *et al.*, 2012; Mishra *et al.*, 2014). Moreover, virulence is not correlated with the biotype, since all the three bovine pestiviruses have both ncp and cp strains. The differences in replication, have a great impact on the virulence of BVDV strains, since strains that produce highest degree of viremia result in the most severe clinical symptoms (Walz *et al.*, 2001). Acute BVDV infection also causes immunosuppression by depletion of both B and T lymphocytes, thereby leading to suppression of immune functions in the infected animal.

The outcome of BVDV fetal infection depends on the stage of development of the fetal immune system at the time of infection. Fetal infections during the first two trimesters of gestation may have severe reproductive consequences and may result in persistent infections, fetal death and abortion, or congenital anomalies (Brownlie *et al.*, 1998). Dual infection of the fetus with both BVDV-1 and BVDV-2 has also been reported. Although abortions due to BVD occur mostly in the early stages of gestation (<125 days), abortions in the late phase of gestation do occur.

Pestiviruses including BVDV, use novel strategies to persist and spread in the host population through persistent infection. PI animals develop following infection of fetus with ncp BVDV strains during the first trimester of gestation, before the development of lymphoid tissues and functional immune responses and the immunotolerance is specific to the persisting BVDV strain (McClurkin *et al.*, 1984). Primary acute BVDV infections of pregnant cows is the predominant cause of producing PI calves, although PI cows most often give birth to PI calves. Whereas, Trojan cow, a non-PI cow carrying a PI fetus is immune to BVDV and possesses significantly higher antibody titres during mid-late pregnancy. The ability to cause persistent infection is exclusively a biotype-specific feature of ncp BVDV. Immunotolerance develops through selective evasion of innate immunity in the fetus by inhibition of IFN I synthesis, following ncp BVDV infection in addition to complete avoidance of the adaptive immune system.

Mucosal Disease

Mucosal disease (MD) is a sporadic disease of cattle, in which mostly 6–24 months-old animals succumb, but it may arise in adult animals. MD is highly fatal and is associated with the presence of closely related ncp and cp biotypes of BVDV in these animals. The prominent characteristics of mucosal disease are bloody diarrhea along with fever, anorexia, ataxia and general debility. Mortality is often 100 percent and within 15 days after onset of clinical signs (Baker, 1987). During post mortem, extensive ulcerative lesions in the gastrointestinal tract, affecting especially the GALT in the mucosa are observed.

The first hypothesis on the mechanism of MD was that it develops only in PI animals. The second hypothesis was that animals suffering from MD always harbor a cp virus along with the persisting ncp strain (Bolin *et al.*, 1985). Then it was established that close antigenic relationship between ncp and cp strains from the same animal is obligatory for development of MD (Meyers *et al.*, 1991). Mucosal disease arises in PI animals by complex mutations of the ncp biotype to a cp biotype, or by superinfection of PI animals with a cp BVDV. In acute MD, genome analyses showed cp strain specific genomic alterations which are mainly due to recombination events resulting in insertion of cellular sequences, or duplications and deletions of viral sequences. The genomic alterations enhance NS3 protein production, which provides an apoptotic signal for the infected cells. Insertion of ubiquitin coding sequences upstream of the NS3 gene or insertion of host-cell origin Jiv sequences within NS2 upstream of the NS2–3 cleavage site are the most commonly observed genomic changes (Tautz *et al.*, 1993; Becher and Tautz, 2011).

Cp strains have been reported for all three bovine pestiviruses, BVDV-1, BVDV-2 and HoBiPeV. However, the clinical signs and postmortem findings in MD caused by HoBiPeV have been found to be less severe than those described for classical MD caused by BVDV-1/BVDV-2 (Decaro *et al.*, 2012). As clinical signs of both persistent and acute infections with bovine pestiviruses may be extremely diverse and MD can occur with different clinical pictures, it is likely that mucosal disease like signs in cattle can occur in field in certain cases without cytopathic biotype of HoBiPeV being detected (Kalaiyarasu *et al.*, 2022). MD in cattle associated with HoBiPeV with different clinical outcomes, including fatalities has recently been reported in India (Kalaiyarasu *et al.*, 2022) (Figure 12.1).

In animals suffering MD, strikingly higher numbers of cp BVDV-infected cells have been found compared to ncp-infected cells in PIs before outbreak of MD. In chronic MD, the antigenic properties of the persisting ncp and the super infecting exogenous cp viruses are more divergent and such animals suffer clinically for a prolonged period. MD can also develop several weeks after superinfection and is known as late-onset MD and cp viruses in them are recombinants between cp and ncp viruses, where structural genes are shared by ncp viruses and nonstructural genes are shared by cp viruses. Direct cell damage by cp BVDV is the major mechanism of disease in PI animals suffering from MD, while cp BVDV is mostly attenuated in acutely infected animals.

The Agent

Both BVD and MD are caused by three bovine pestiviruses, bovine viral diarrhea virus 1 (BVDV-1), BVDV-2 and the recently emerged HoBi-like pestivirus (HoBiPeV). In nature, bovine pestivirus strains exist as two biotypes, non-cytopathic (ncp) and cytopathic (cp) based on the cytopathogenicity of cultured cells. In most cases of mucosal disease, both ncp and cp biotypes occur as a pair, while BVD is mostly associated with ncp biotype, which is preponderant.

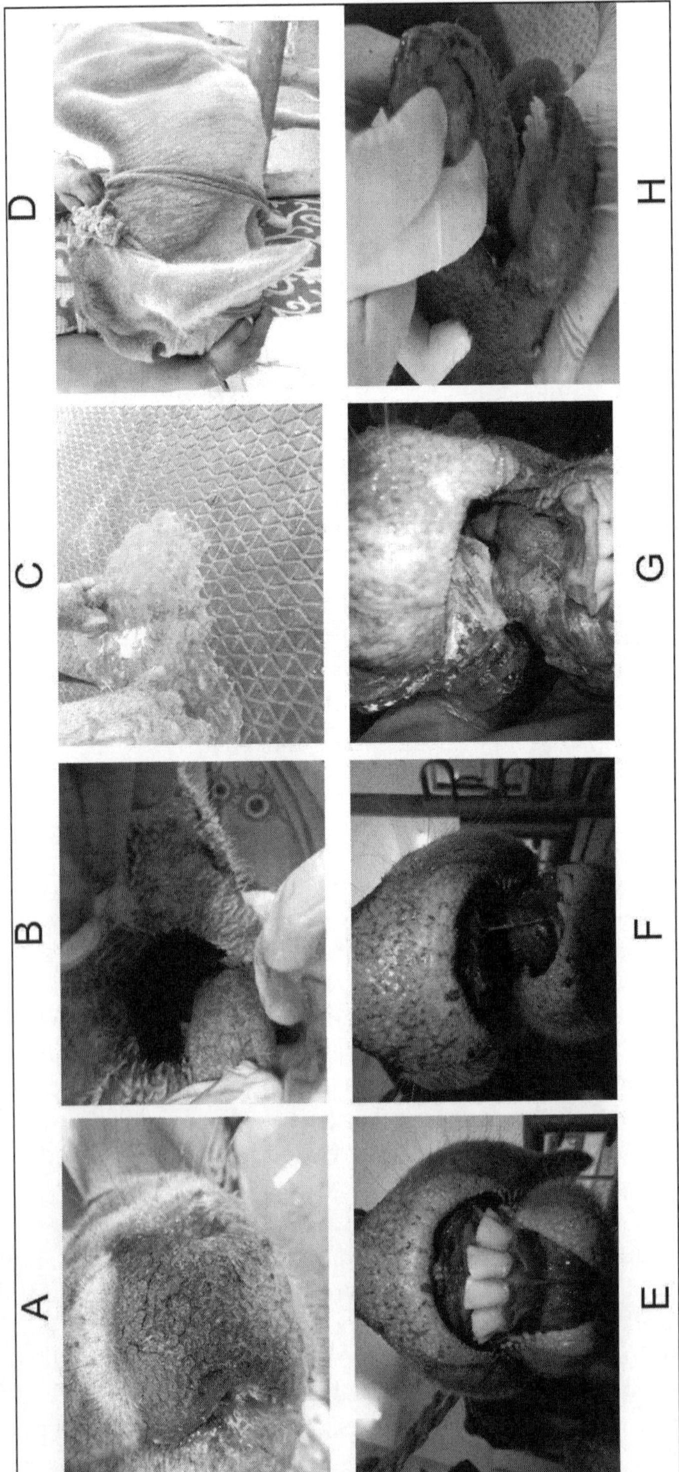

Figure 12.1: Clinical Signs of Mucosal Disease.

Taxonomy

Bovine pestiviruses belong to the *Pestivirus* genus within the *Flaviviridae* family that also comprises the genera *Flavivirus, Hepacivirus* and *Pegivirus*. The pestivirus taxonomy was problematic for long, since the earlier species names were derived from the names of virus isolates, based on host range and disease attributes. However, a new uniform naming system for *Pestivirus* species analogous to that used for species belonging to *Pegivirus* and *Hepacivirus* genera of *Flaviviridae*, has been approved by ICTV in 2017 (Smith *et al.*, 2017). The four existing species have been designated as *Pestivirus A*, which comprises of *Bovine viral diarrhea virus 1* (BVDV-1), *Pestivirus B*, which comprises of *Bovine viral diarrhea virus 2* (BVDV-2), *Pestivirus C*, which comprises of *Classical swine fever virus* (*CSFV*) and *Pestivirus D*, comprising of *Border disease virus* (*BDV*) along with seven additional species and some unassigned species. The HoBi-like pestivirus (HoBiPeV) (formerly atypical bovine pestivirus or BVDV-3) belongs to the species *Pestivirus H*.

Structure and Genomic Organization

The BVDV virions are enveloped, spherical particles of 40 to 60 nm diameter in size. The virion is composed of a core region, consisting of the genomic RNA coated with structural capsid or core protein which is surrounded by a lipid envelope. The capsid is about 30 nm diameter and appears as an electron-dense inner core (Horzinek *et al.*, 1971). However, cryo-electron microscopy of purified BVDV virions has demonstrated that viral particles display an electron-dense capsid surrounded by a phospholipid bilayer with no visible spikes and most BVDV particles are 50 nm diameter in size and about 2 percent are 65 nm in size, suggesting size flexibility during morphogenesis (Callens *et al.*, 2016). The lipid bilayer envelope is made up of three virus encoded glycoproteins, E^{rns}, E1, and E2. The heavily glycosylated E^{rns} glycoprotein is loosely associated with the virus particle and is secreted in soluble form by infected cells, while E1 and E2 glycoproteins are integral membrane proteins and E1-E2 heterodimer is essential for BVDV entry. Due to the pleomorphic nature of envelope and association of BVDV particles with host cells, it is extremely difficult to achieve highly purified infectious particles by ultracentrifugation. The buoyant density of virion in sucrose is 1.134 gm/ml and molecular weight of the virion is estimated as 6.0×10^7 (Lindenbach *et al.*, 2013). Similar to other enveloped viruses, BVDV is inactivated by organic solvents and detergents, but it is resistant to low pH unlike flaviviruses.

Bovine pestivirus genome consists of a positive sense single stranded RNA of about 12.3 kb size, but the size may vary up to 16.5 kb due to insertions, genomic duplications and genomic recombination events (Becher *et al.*, 1999). No subgenomic RNA is transcribed during BVDV replication and the plus strand genomic RNA only represents the viral mRNA and codes all viral proteins. The genomic organization of BVDV is similar to all the recognized pestiviruses

(Figure 12.2). The RNA genome codes for a polyprotein of about 3900 amino acids in one large open reading frame, that is flanked by 5′- and 3′-untranslated regions (UTR) of about 400 and 200 nucleotides respectively. The translated polyprotein is subsequently processed by viral and cellular proteases resulting into 12-13 proteins, N^{pro}, C (capsid), the envelope proteins E^{rns}, E1, and E2, and nonstructural proteins p7, NS2, NS3, (NS2–3), NS4A, NS4B, NS5A, and NS5B (Lindenbach *et al.,* 2013). The arrangement of proteins in the polyprotein is as NH2–N^{pro}/C/E^{rns}/E1/E2/p7/NS2/NS3/NS4A/NS4B/NS5A/NS5B–COOH.

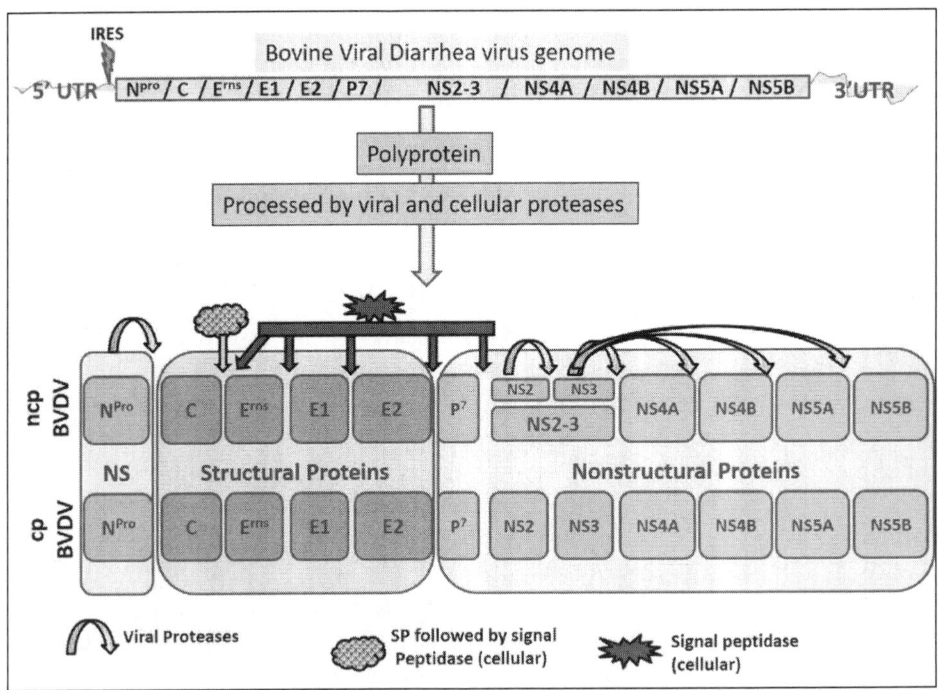

Figure 12.2: Genome Organization of Bovine Viral Diarrhea Virus.

Polyprotein processing occurs co-translationally and begins with the N-terminal autoprotease (N^{pro}) release. The N^{pro}, unique to pestiviruses is responsible for cleavage at its own carboxy-terminus thereby releasing amino-terminus of the C protein (Stark *et al.,* 1993). The C/E^{rns} cleavage occurs by cellular signal peptidase (SPase) and signal peptide peptidase (SPPase), while processing at the E^{rns}/E1, E1/E2, E2/p7, and p7/NS2 borders is carried out by SPase (Rumenapf *et al.,* 1993). The processing of NS2 and NS3 occurs by protease located within NS2, whereas all processing downstream of NS3 is done by NS3 protease. BVDV strains exist as two biotypes, non-cytopathic (ncp) and cytopathic (cp). Cytopathic viruses result following changes in the NS2/3 protein coding region leading to generation of NS2 and NS3 proteins. These genomic changes include point mutations, genomic duplications and insertion of cellular mRNA sequences (Meyers and Thiel, 1996; Kummerer *et al.*, 2000). The 53 kDa BVDV E2 glycoprotein is responsible for virus attachment, generation of neutralizing antibodies and host tropism (Weiland *et al.*, 1990; Liang *et al.*, 2003).

Entry and Replication

BVDV entry into bovine cells involves virion attachment to cellular receptors, internalization and membrane fusion. Bovine CD46 has been reported to act as a cellular receptor for BVDV (Maurer *et al.*, 2004), but it may not be sufficient and some unknown elements may be required for BVDV entry. For initiating the infection process, BVDV E1-E2 heterodimer binds the cellular receptor. BVDV entry into bovine cells occurs through clathrin dependent endocytosis and endosomal fusion in a pH dependent manner (Grummer *et al.*, 2004; Krey *et al.*, 2005) and a similar mechanism exists for entry into ovine cells (Mathapati *et al.*, 2010).

BVDV replication takes place in cellular cytoplasm. Following IRES mediated initiation of translation, BVDV replication process begins with a positive strand replicase complex consisting of viral and cellular components formed at the 3′-terminus of the genome. The replicase complex catalyzes transcription of positive sense RNA into full-length complementary negative sense RNA, which then acts as template for synthesizing additional positive sense RNA, using a semiconservative asymmetric replication model (Warrilow *et al.*, 2000). In comparison to negative sense RNA, a large amount of newly synthesized positive sense RNA is generated. In this model, three virus-specific RNAs, a double-stranded replicative form (RF), a partially single-stranded and partially double-stranded replicative intermediate (RI), and a single-stranded viral RNA are involved. The same positive sense genomic RNA acts as a template for both replication and translation. Regulation of this process is mediated by secondary structure of IRES in the 5′-UTR. The budding of BVDV takes place in endoplasmic reticulum, where the RNA-core complex is packed in envelopes (Schmeiser *et al.*, 2014). Following assembly, virions are released from the cell through the secretory pathway.

Host

Cattle of all ages are susceptible to bovine pestivirus infections and are the primary hosts. BVD has a wide host range and infects a variety of animals, both domesticated and wild. Among the domesticated ruminants, it infects cattle, buffalo, sheep, goats and yaks, while among wild ruminants it infects buffalo, eland, Canadian bison, alpaca, pudu, bongo, deer, roe deer, mouse deer, reindeer, giraffe, European bison, chamois, pronghorn antelope and mithun. BVDV has been isolated in over 40 species of ruminants and serological evidence indicates susceptibility of most free ranging ruminants to BVDV infection (Vilcek and Nettleton, 2006). Serological evidence of BVDV infection has also been reported in European rabbits. Although natural infection with BVDV occurs mainly in cattle, sheep and goats, it occurs also in pigs. In addition, natural or experimental BVDV persistent infection has been reported in mountain goats and domestic goats, domestic sheep, swine, alpaca, eland, mule deer, white-tailed deer and mouse deer.

Transmission

Transmission of bovine pestiviruses occurs through several modes. BVDV spreads horizontally within a herd while vertical transmission occurs from cow to calf. Persistently infected (PI) cattle results following infection of fetus with ncp BVDV in the first or second trimester (45-125 days) of pregnancy, before maturation of its immune system. The most common method of producing PI calves is through primary acute BVDV infections of pregnant cows. However, PI cows also invariably give birth to PI calves. Horizontal transmission occurs not only by viraemic transiently infected (TI) animals but also by PI animals, that shed virus lifelong in all secretions, such as nasal and ocular discharges, milk/colostrum, semen, urine and feces (Van Campen and Frolich, 2001). PI animals are the principal reservoirs of BVDV transmission, while TI animals transmit BVDV transiently and occasionally.

Semen from transiently infected bulls can transmit BVDV infection and virus can be detected up to 28 days in such bulls (Kirkland *et al.*, 1994). In contrast, in both raw and extended semen of PI bulls, BVDV concentration remains high and semen from PI bulls infects susceptible animals consistently. Persistent testicular infection (PTI) occurs following acute BVDV infection in bulls. BVDV persists in semen or testicular tissue of these non viremic and seropositive bulls and seronegative cows may be infected via artificial insemination. Environmental conditions favoring crowding and aerosol transmission enhance the chances of BVDV transmission from acutely infected calves having respiratory form of BVDV infection. BVDV can be transmitted indirectly from contaminated pens, rectal examination gloves, hypodermic needles, nose tongs, BVDV contaminated live vaccines and ambient air (Niskanen and Lindberg, 2003). Sufficient evidence exists regarding the spread of BVDV from domestic ruminants to wild ruminants. The four main factors which affect BVDV transmission are, infectiousness (virulence) of the virus strain, the number of adequate contacts per time period between infectious and susceptible animals, the prevalence of infectious animals in a herd and the presence of truly susceptible animals.

Pathogenesis

BVDV replicates in epithelial cells and lymphoid tissues of the oropharynx following infection via oronasal route. The phagocytic cells carry BVDV and/or BVDV infected cells to peripheral lymphoid tissues leading to viremia that occurs 2-4 days after exposure and spreads BVDV to internal organs. The pathogenesis of BVDV is a complex interaction between the agent, host and environmental factors and hence clinical signs are highly variable (Baker, 1995).

The virulence of BVDV strains and biotype of BVDV (ncp and cp) are the major determinants from the agent's side, while immune status, immune competence and stage of pregnancy are the major determinants from host's side.

In cultured cells infected by cp BVDV strains, there is rounding up and detachment of cells and cells die due to apoptosis, whereas in cells infected by ncp BVDV, no microscopically detectable alterations are seen. But both cp and ncp BVDV isolates induce apoptosis of T and B cells *in vivo*. Furthermore, ncp viruses are pathogenic to the host and establish persistent infection upon fetal infection. The host interactions with BVDV are highly variable, ranging from lack of immune response to a purifying immune response and from lack of clinical signs to highly lethal infection. Reproductive failure or immunosuppression are major consequences of BVDV infections because of its affinity for the fetus and for cells of lymphatic organs. Strong affinity of BVDV for lymphoreticular tissues causes necrosis in lymph nodes and spleen and destruction of Peyer's patches. The four major syndromes associated with BVDV pathogenesis are acute infection, transplacental infection, persistent infection and mucosal disease.

Genetic and Antigenic Diversity

Bovine pestiviruses show high genetic diversity, which results from accumulation of point mutations, homologous and non-homologous RNA recombination. Genetic typing of strains is important for BVD epidemiology and control. Accurate genetic typing of BVDV has been obtained from the application of advanced molecular techniques such as, reverse transcription polymerase chain reaction (RT-PCR), next generation sequencing of full-genome and phylogenetic analysis. Although sequence analysis of the highly conserved 5′-UTR is mostly used for the classification of pestivirus isolates into species level, segregation into genotypes within each species is more accurate by analysis of full genome sequence or complete N^{pro} and E2 genes (Becher *et al.*, 2003; Vilcek *et al.*, 2001, 2004; Yesilbag *et al.*, 2017). Additionally, antigenic similarity is detected by monoclonal antibody (MAb) binding and cross-neutralization assays with homologous and heterologous antisera is used for determining antigenic diversity of BVDV strains (Paton *et al.*, 1995; Becher *et al.*, 2003; Dias *et al.*, 2017).

Thus far, based on sequence analyses 21 genotypes of BVDV-1 (BVDV-1a to -1u), 4 genotypes of BVDV-2 (BVDV-2a to -2d) and 5 genotypes of HoBiPeV (HoBiPeV-a-e) have been identified (Mishra *et al.*, 2014; Yesilbag *et al.*, 2017; Kalaiyarasu *et al.*, 2022). Despite accurate distribution of BVDV genotypes in individual countries and continents is unknown, BVDV-1b is the predominant genotype worldwide, followed by BVDV-1a and 1c. BVDV-1b is the predominant genotype in the Europe, Asia and Americas, while BVDV-1c is predominantly prevalent in Australia and Mexico. Extensive genetic diversity of BVDV-1 has been reported from several countries in Europe and also in Turkey, China and Japan, but it is comparatively lesser in Americas, Africa and Australia as well as in India. Although BVDV-1 strains are antigenically closely related than BVDV-

2 or HoBiPeV, significant antigenic differences between some of the BVDV-1 genotypes have been reported.

BVDV-2 strains are genetically less diverse than BVDV-1 and occur less commonly than BVDV-1. Although BVDV-2 has been divided into four (2a–2d) genotypes, BVDV-2a is the most prevalent genotype in all continents, while BVDV-2b has been detected in Americas (Brazil, Argentina, Uruguay, USA) Europe (Portugal, Spain, Slovakia, Turkey) and Asia (India, China), and BVDV-2c has been detected only in Europe (Germany, Ireland) and North America (USA). BVDV-2 can occur also in sheep and goats. In sheep, BVDV-2a has been detected in USA and Italy, while BVDV-2b has been reported from India (Mishra *et al.*, 2008b) and Turkey (Yesilbag *et al.*, 2008). However, only BVDV-2a genotype has so far been reported in goats (Mishra *et al.*, 2007a) and significant antigenic variation between BVDV-2 genotypes has not yet been observed.

Till 2014, all the previously reported HoBiPeV strains were found to be closely related genetically until our previous work showed that HoBiPeV strains can be classified into 3 genotypes with two highly divergent HoBiPeV lineages co-circulating in Indian cattle (Mishra *et al.*, 2014). A recent study has shown that HoBiPeV strains can be classified into 5 genotypes (a-e) (Kalaiyarasu *et al.*, 2022). HoBiPeV-a has been detected in South America (Brazil) and Europe (Italy) and as contaminants of FBS, while HoBiPeV-b has only been reported from Bangladesh and HoBiPeV-c, HoBiPeV-d and HoBiPeV-e have been exclusively detected in India (Mishra *et al.*, 2014; Giammarioli *et al.*, 2015; Kalaiyarasu *et al.*, 2022). Nevertheless, full-genome sequence based Bayesian phylogenetic analyses confirmed the existence and independent evolution of four main HoBiPeV clades (a, c, d, and e) globally, with genetic divergence ranging from 13.0 percent to 18.2 percent (Kalaiyarasu *et al.*, 2023) (Figure 12.3). In addition, the Bayesian molecular clock estimates revealed that HoBiPeV most likely originated in India with a dated tMRCA of 1938 (1762–2000), evidencing a more recent origin of HoBiPeV. Marked antigenic differences exist between BVDV-1, BVDV-2 and HoBiPeV, while there is no significant antigenic variation among HoBiPeV strains.

Epidemiology

A large number of variables, such as clinical, pathological, virological, serological, and production measures can be used to quantify the occurrence of BVDV infections. Long presence of antibodies (often lifelong) in acutely infected animals and presence of virus in PI animals throughout their life make prevalence studies more suitable for BVD. Sensitivity and specificity of diagnostic tests play a key role in determining true prevalence of BVDV infection.

Global

BVD is prevalent worldwide (WOAH, 2022), except in a few Scandinavian

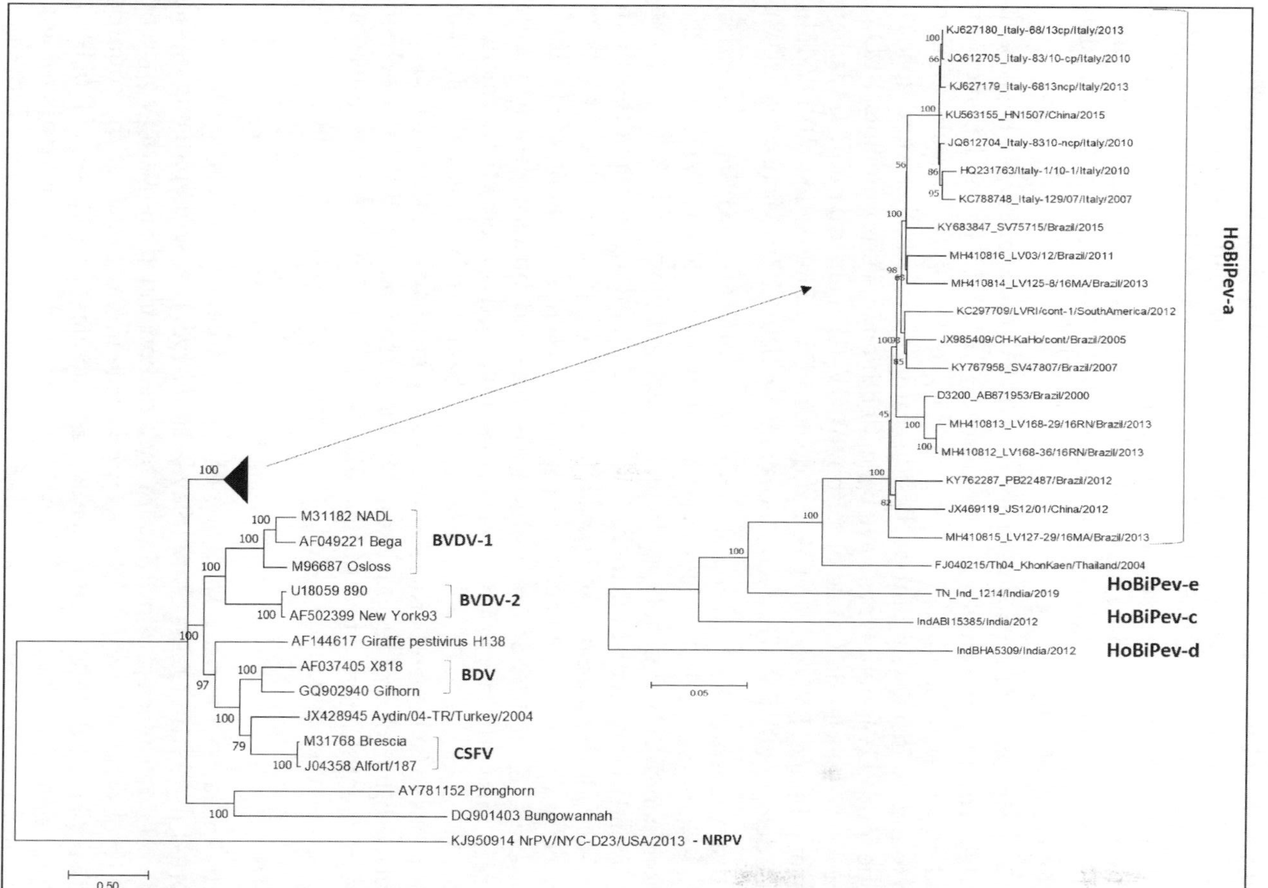

Figure 12.3: Bayesian Phylogenetic Analysis of HoBiPeV Full-Genome Sequences.

countries in Europe, where it has been eradicated or in the final stages of eradication. Despite variation in prevalence rates among surveys, BVDV infection is endemic in most cattle populations having 1-2 percent of the cattle being persistently infected (PI) and >90 percent of the cattle being antibody positive. Variations in BVDV prevalence rates among different countries or regions or introduction of virus into BVDV naïve herds are influenced by cattle population density, cattle trade and pasturing practices. BVDV-1 and BVDV-2 are distributed in all continents, while HoBiPeV has so far been reported in South America, Europe and Asia (Mishra *et al.*, 2014; Yesilbag *et al.*, 2017; Mishra and Kalaiyarasu, 2019).

Geographical patterns in distribution of BVDV-1 genotypes have been reported around the world. BVDV-1 genotypes, 1m, 1n, 1o, 1p and 1q have been found only in some countries in Asia, while 1f, 1g, 1h, 1k, 1l, 1r, 1s and 1t have been found exclusively in Europe. Besides cattle, buffaloes, sheep and goats, a number of BVDV-1 genotypes have been identified in wild ruminants. BVDV-1a has been detected in Canadian bison, BVDV-1b in alpaca, pudu, Canadian bison and bongo, BVDV-1c in yak and deer, BVDV-1d in roe deer, BVDV-1f in mouse deer and BVDV-1j in deer (Vilcek and Nettleton 2006; Mishra *et al.*, 2008).

BVDV-2 is prevalent in several countries of North America, South America, Europe and Asia and also in Australia (Vilcek *et al.*, 2005). BVDV-2a is the predominant BVDV-2 genotype circulating around the world. A very special case of fatal disease with high mortalities in cattle resulting from infection with BVDV-2c isolates consisting of 3 genomic variants (dup^+, dup_1^-, dup_2^-) has been reported in Germany (Jenckel *et al.*, 2014). The recent association of HoBiPeV with severe respiratory and reproductive disease and mucosal disease in cattle and respiratory disease in small ruminants has raised concerns (Bauermann *et al.*, 2013; Weber *et al.*, 2016; Kalaiyarasu *et al.*, 2022). Natural HoBiPeV infection in cattle has so far been reported sporadically in Brazil, Italy, Thailand, India and Bangladesh, but in some regions like in Northeastern Brazil, HoBiPeVs have been more frequently detected in cattle than BVDV-1 and BVDV-2. Natural infection has also been reported in buffalo in Brazil and in sheep and goats in China.

India

In India, serological evidence of BVDV infection in cattle was first reported in 1981 in Orissa State (Nayak *et al.*, 1981). A sero-epidemiological study on sera collected from 17 states then demonstrated an overall apparent BVD seroprevalence rate of 17 percent in cattle in most parts of the country (Sudharshana *et al.*, 1999). Subsequently, varying rates (up to 52 percent) of BVD seroprevalence in cattle and buffaloes in different parts of the country involving both commercial dairies and small holder units have been reported. Besides cattle and buffaloes, serological evidence of BVDV infection has been reported in sheep and goats (Mishra *et al.*, 2009), yaks (Mishra *et al.*, 2008a) and also in mithun (Singh *et al.*, 2017). A BVDV seroprevalence study involving sheep and goats from 13 States during 2004-2008 reported a true prevalence rate of 23.4 percent in sheep and 16.9 percent in goats and provided evidence of BVDV-1 infection predominantly and BVDV-2 occasionally (Mishra *et al.*, 2009).

Although earlier studies indicated serological evidence of BVDV infection, the conclusive virological evidence of BVDV infection in Indian cattle was provided by virus isolation and subsequent phylogenetic analysis of BVDV isolates in 2004 (Mishra *et al.*, 2004). The phylogenetic analysis of 13 BVDV isolates originating from cattle in eastern, northern and western India in 5′-UTR and N^{pro} genes revealed that they belong to BVDV-1b genotype and are closely related (Figure 12.4). In the same year, BVDV-1 was also reported in lambs showing RVF like symptoms (Yadav *et al.*, 2004). Subsequently, genetic typing of BVDV strains revealed that BVDV-1b is the predominant genotype circulating in cattle (Mishra *et al.*, 2007b, 2014; Behera *et al.*, 2011), while BVDV-1b and 1c genotypes have been reported in buffaloes (Mishra *et al.*, 2007b), and in sheep and goats (Mishra *et al.*, 2012). Besides domestic ruminants, BVDV-1 has been detected in yak species (*Bos grunniens*) for the first time and phylogenetic analysis revealed that they belong to BVDV-1c genotype (Mishra *et al.*, 2008a).

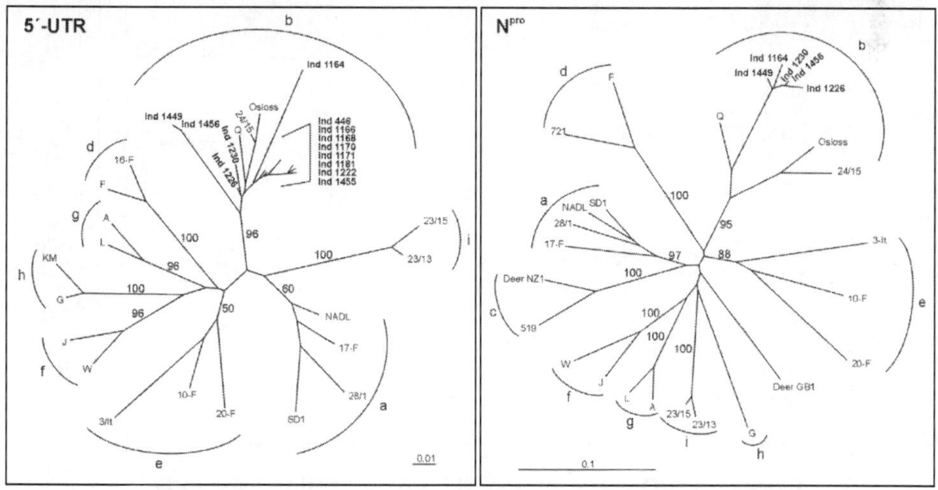

Figure 12.4: Genotyping of BVDV-1 Strains from India.

In India, BVDV-2a was first detected in cattle from Jammu and Kashmir in 2011 and then in bull semen from Tamil Nadu (Behera *et al.*, 2011; Mishra *et al.*, 2018). BVDV-2a has earlier been detected in goats from Northern India, while BVDV-2b has been reported in sheep from Western India (Mishra *et al.*, 2007a, 2008b). Molecular epidemiology studies on BVDV-2 have provided evidence of circulation of genetically divergent BVDV-2a strains in Southern India and in Northern India (Figure 12.5). India reported natural infection with HoBiPeV in cattle for the first time in 2014 in the States of Maharashtra, Punjab and Chhattisgarh, while conducting systematic surveillance in cattle from 21 dairy farms across India. Molecular characterization of HoBiPeV strains revealed co-circulation of two novel and divergent genotypes of HoBiPeV (c, d), highlighting the independent evolution of at least 3-4 genotypes/lineages of HoBiPeV strains globally (Mishra *et al.,* 2014). A recent study has confirmed the existence of another novel genotype, HoBiPeV-e in cattle showing mucosal disease like cases from Tamil Nadu and out of the four main HoBiPeV genotypes (a, c, d, and e), three occur exclusively in India (Kalaiyarasu *et al.,* 2022). These novel findings extended the knowledge on the epidemiology and genetic diversity of HoBiPeV strains globally, which is important for management and control of BVD.

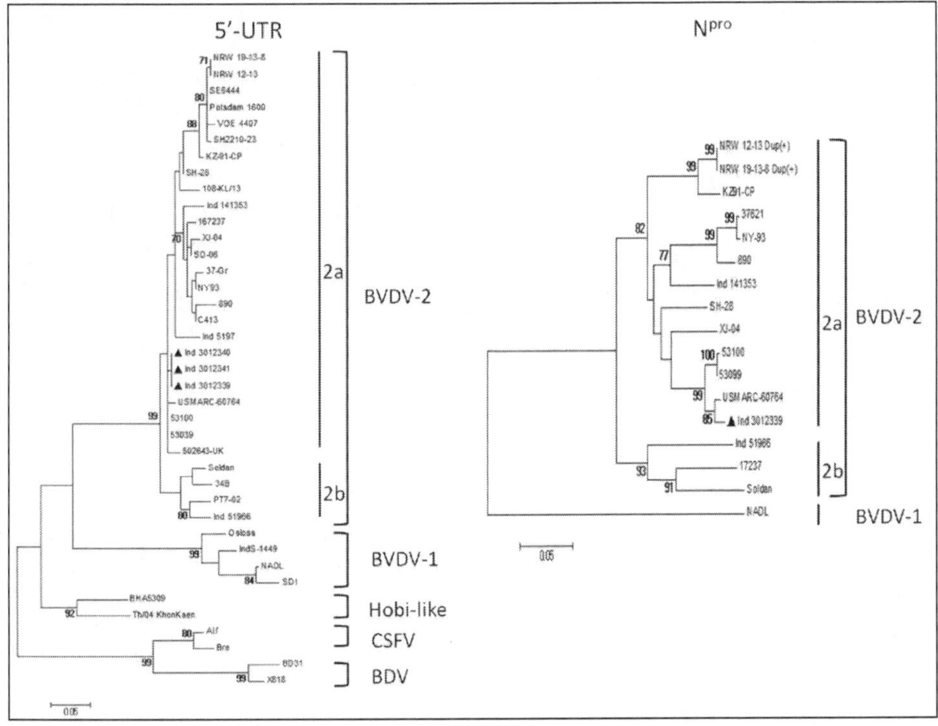

Figure 12.5: Genetic Typing of BVDV-2 Strains from India.

Diagnosis

The control of BVD is highly dependent on confirmed laboratory diagnosis that defines exposure of an individual animal or population to BVDV. WOAH recommended laboratory tests for BVD agent identification are virus isolation, real-time RT-PCR, pestivirus antigen ELISA and IPMA, while serological methods are virus neutralization test (VNT) and BVD antibody ELISA. For genetic typing, sequencing and phylogenetic analysis are the recommended tools. In spite of recent advances in BVDV diagnosis, virus isolation and identification still remains the gold standard test. BVDV free animals upon testing are negative for antibody, antigen and virus, while acutely infected animals or immunocompetent fetuses are antibody positive and generally antigen or virus negative. PI animals in contrast are positive for antigen or virus and negative for antibody. Since, acutely infected animals become positive for BVDV antibodies within 2–3 weeks post-infection, testing for antibody 4-8 weeks after initial testing can distinguish between acute and persistent infection in animals with positive antigen ELISA or RT-PCR results. Confirmatory diagnosis of mucosal disease relies on confirmation of PI status followed by isolation of both cp and ncp BVDV from the affected animal. However, it can also be confirmed based on identification of PI and characteristic pathological lesions.

Conventional Methods

Virus Isolation

Virus isolation (VI) is the gold standard test for BVDV diagnosis and is the WOAH recommended test for certifying individual freedom of infection during international trade. VI relies on the growth of BVDV in specific cell lines, such as MDBK or BT (bovine turbinate) followed by detection using specific antibodies or molecular methods. The cells and serum used for VI should be BVDV free and BVDV antibody free. As most of the BVDV isolates are of ncp biotype, detection by immunostaining or immunofluorescene using E^{rns} or NS3 specific monoclonal antibodies is recommended. In case of cp (cytopathic) strains, a characteristic cytopathic effect is observed. VI detects viremia in individual animals and is used to confirm the PI status of animals with positive ear notches, serum or whole blood buffy coats. Microplate immuno-peroxidase assay in 96-well plates is commonly used for PI animal detection. However, these methods suffer certain limitations such as varying sensitivity, slower test results and colostral antibodies can mask BVDV in PI animals and interfere their detection.

Antigen ELISA

Direct detection of viral antigen in leukocytes, serum or ear notch samples can be done by pestivirus antigen capture ELISA (PACE) and several commercial BVDV antigen ELISA kits are available which can be used in most of the veterinary diagnostic laboratories. PACE is usually recommended for PI screening, but it also detects some transiently infected animals, so a follow-up sample 4-6 weeks later is tested to confirm PI status. The test usually detects BVDV NS2-3 or E^{rns} antigen in peripheral blood leukocytes, plasma or serum. Skin biopsies or ear notches have become a popular sample for PI testing by PACE, since PI animals are consistently detected by PACE tests regardless of antibody status and the site of the biopsy. Although the test is easy to perform and rapid, it suffers from low sensitivity and specificity. Besides, false positives are not unusual. E^{rns} mAb based PACE is preferable since it can detect all the three species of bovine pestivirus, BVDV-1, BVDV-2 and HoBiPeV (Mishra *et al.*, 2014). Moreover, antigen capture ELISAs based on the detection of the p80 (NS3) protein of BVDV, have diagnostic gaps prior the age of 90 days both for serum and ear notches, instability of the p80/NS3 protein and the stronger inhibitory effect of colostral antibodies.

Immunohistochemistry (IHC)

Immunohistochemistry using ear notch tissue samples detects PI animals with 100 percent sensitivity and hence is one of the popular methods of BVDV Ag detection. While IHC is considered robust, it has many disadvantages in that it is subjective and restricted to tissue samples, is labor intensive, requiring experienced staff and its unreliability for use on samples stored in formalin for >15 days.

BVDV Specific Antibody Detection

Serological tests for BVDV can be used to determine previous exposure of animals to BVDV, colostral antibodies in calves, immune response in vaccinated animals and in confirmation of acute infection. However, antibody negative animals should be further tested for BVDV or Ag to rule out PI status. High prevalence of antibodies is indicative of current infection at a herd or region level. Although several antibody detection methods such as AGID, dot ELISA and microsphere based immunoassay have been reported, the virus neutralization test (VNT) and Ab ELISA are most commonly used.

Virus Neutralization Test (VNT)

VNT is the gold standard test to detect anti-BVDV antibodies, although antibody ELISA can alternatively be used. For demonstration of seroconversion, both acute and convalescent sera should simultaneously be tested. The test is based on determination of 50 percent neutralizing end point and both cytopathic and non-cytopathic strains of BVDV can be used. Most preferably, local isolates of both BVDV-1, BVDV-2 and HoBiPeV must be used as low levels of antibody to BVDV-1 may not be detectable by a VNT that uses only BVDV-2 and vice versa. A differential neutralization test against a BDV strain should also be carried out simultaneously for serological differential diagnosis of BVD, as BDV can also infect cattle naturally and serological cross reactivity occurs.

BVDV Antibody ELISA

BVD antibody ELISA is useful for screening large number of cattle herds for serological diagnosis of BVD in unvaccinated animals. Both mAb based competition ELISA and indirect ELISA are available commercially and used for detection of BVDV antibodies in serum, milk and bulk milk. Being rapid and cost effective, Ab ELISA is an efficient and economical alternative to VNT. But in BVDV vaccinated animals, the usefulness of ELISA tests is fairly limited, while in unvaccinated populations and in eradication phase, it has more utility. Moreover, VNT detects a rise in Abs following vaccination or infection, while Ab ELISAs fails in this regard. Caution should be taken in result interpretation, since some commercial BVDV antibody ELISA kits have been reported to yield false-negative results when serum samples of calves harboringHoBiPeV antibodies were tested.

Modern Methods

Newer technologies are constantly being developed and evaluated for their use in BVDV diagnostic testing through genome detection and/or amplification. Besides, nucleotide sequencing and sequence analysis has dramatically improved the molecular epidemiology of BVDV in detecting new and divergent strains and in tracing the origin of outbreaks.

RT-PCR

In BVDV infected cattle, viral RNA is detectable early and for a longer duration than virus isolation. RT-PCR employing pooled serum and milk samples has been found useful in identifying PI animals during BVDV surveillance. RT-PCR assays are able to detect acutely infected animals, PI animals, and animals vaccinated with modified-live vaccines but follow-up testing is necessary to define the status of positive animals. Several RT-PCR protocols have been developed, evaluated and are being used for BVDV diagnosis. The most widely used protocol utilizes primers 324/326, which is targeted at the highly conserved 5′-UTR and is pestivirus specific, but it fails to detect the highly divergent HoBiPeV strains. However, HoBiPeV specific RT-PCR has recently been reported. Differentiation between BVDV-1, BVDV-2 and BDV is also possible by nested RT-PCR. However, extreme precautions should be taken during nucleic acid based tests due to false positive cases arising from cross contamination and a positive RT-PCR does not define the clinical status of an animal in a single animal test or in a pooled sample.

Real-Time RT-PCR

As real time RT-PCR assay provides simultaneous quantitation and genotyping of BVDV, it is used more commonly now not only during BVD outbreaks but also for routine BVDV diagnosis, due to its rapidity in obtaining the results. Several real time PCR assays, both in uniplex, duplex or multiplex formats, are available commercially for diagnosis and genetic typing of BVDV and differentiation from other pestiviruses using primers and probes targeted at 5′-UTR (Hoffman *et al.*, 2006; Baxi *et al.*, 2006; Willoughby *et al.*, 2006; Liu *et al.*, 2008).

Sequencing and Next Generation Sequencing

Genetic typing of bovine pestiviruses provides useful information during BVD epidemiology and control. The more accurate genetic typing of BVDV strains is achieved from nucleotide sequencing or next generation sequencing data followed by phylogenetic analysis. Although sequence analysis of 5′-UTR can be used for pestivirus species assignment, sequence analysis of complete $N^{pro}$and E2 genes or combined datasets can classify them more accurately into genotypes/subtypes (Becher *et al.*, 2003; Vilcek *et al.*, 2001; Mishra *et al.*, 2014). Recently, analyses of whole genome sequencing data of BVDV strains are being more frequently used for molecular epidemiological studies.

Prevention and Control

Although initially it was thought that control of BVDV infection is not possible, a range of well-planned control strategies have been designed over the years, with successful implementation of BVDV control programs in many European countries. The common strategies for control programs are identification and removal of PI animals, movement control of infected animals, strict farm biosecurity and surveillance. BVD control programs without use of vaccination have been used successfully in Scandinavian countries, Austria, and Switzerland, whereas control with vaccination have been implemented in Germany, Belgium, Ireland and Scotland (Moennig and Becher, 2018). While compulsory and systematic control programs have been found most successful than voluntary control programs, there is a lack of official guidelines on BVD control in most of the countries.

Vaccination against BVD has been commonly used to prevent infection by enhancing immunity in cattle populations with an aim to prevent or reduce clinical disease and prevent the spread of infection within a herd by reducing BVDV viremia and preventing fetal infection and generation of PI calf. At present, both modified live vaccine (MLV) and killed vaccines are used to control BVD. Vaccination may reduce the incidence of acute and persistent infections, but may not prevent all infections in individual animals and vaccination failure is likely due to failure in broad protection arising from existence of three bovine pestivirus species (BVDV-1, BVDV-2 and HoBiPeV) and multiple subypes. Currently most of the commercially available BVD vaccines contain antigens of BVDV-1a, 1b, BVDV-2a or bivalent vaccines of different combinations but no vaccine is available against HoBiPeV.

As inactivated vaccines contain viral antigen(s) incapable of replication, the risk of adverse effects in vaccinates and the fetus in pregnant animals is minimal. Hence, it is safe and can be administered at any stage of gestation. But protective immunity is shorter and booster doses are required for its efficacy. Fetal protection varies from incomplete to satisfactory. Despite inactivated vaccines are predominantly used, several countries allow the use of inactivated BVD vaccine before breeding, while some countries implement use of inactivated vaccine first followed by modified live BVD vaccine (two-step vaccination). Although quite efficacious, the MLV vaccines have safety concerns due to possibility of reversion of virulence of attenuated virus, its ability to cause in-utero infections, ovarian lesions leading to infertility in cows and mucosal disease, risk of contamination with adventitious viruses and immunosuppressive effects. Hence, the MLVs are not recommended in first six months of unvaccinated pregnant animals, especially with ncp BVDV which may cross the placenta and infect the developing fetus. Most of the modern MLVs are prepared using cp BVDV vaccine strains. Vaccination with MLVs is not advised in calves younger than six weeks.

Recombinant and viral vectored vaccines against BVD have also been developed and protective efficacy of BVDV E2 antigen has been shown using various delivery platforms like viral vectors, DNA immunizations or as a recombinant protein produced in various expression systems. However, these vaccines are not available commercially either due to prohibitive cost or low protective efficacy. Although marker vaccines which allow differentiation of infected from vaccinated animals (DIVA) by serological tests have been successfully used in many other animal viral diseases, in case of BVD, the benefit of a marker vaccine is questionable. Because the PI animals, which are the main sources of infection do not produce antibodies against the homologous strain.

PI animals act as reservoirs of bovine pestiviruses and permanently shed large amounts of infectious virus. Hence, identification and removal of PI animals is the hall mark of BVD control. Virological screening using cost effective BVD antigen ELISAs and molecular tests have been found useful in identification and elimination of PI animals in successful control programs. All Scandinavian programs were successful following this method without vaccination. Additionally, strict farm biosecurity should be taken into account as a part of any BVD control program due to the vulnerability of susceptible herds to reinfection by PI animals. Purchase or trade with untested cattle is the predominant factor of BVDV introduction. Reports of high prevalence of BVDV-1 and BVDV-2 in sheep, and BVDV transmission from sheep to cattle pose hindrances to the success of BVD control and eradication programs.

Conclusions

Although significant advancements on bovine pestivirus/host interplay have been made, but many aspects of pestivirus biology including the genetic basis of attenuation are still obscure and await further work. Additionally, problems associated with the classification and nomenclature of pestiviruses, genotype assignment criteria and consistency in phylogenetic analyses methods and target genes still remain to be addressed. Furthermore, there is a need of assessing impact of recently emerged HoBiPeV and newly emerging pestiviruses on cattle health and production at a global level. Regarding diagnosis, virus isolation should be carefully undertaken by eliminating cross contamination from laboratory handling and identification of bovine pestiviruses in unusual hosts should be reported with caution. Moreover, no uniform approach exists in selection of correct virus strains for use in serological studies, selection of antigen ELISA kits, use of correct cells for virus isolation, use of appropriate primers and probes in real-time RT-PCR assays. Bovine pestivirus antigenic diversity, heterologous hosts and emergence of novel pestiviruses pose significant challenges in BVD prevention and control. As currently available

BVD vaccines, consisting of BVDV-1 or BVDV-2 strains provide only limited protection against HoBiPeV strains, future strategies should aim at development of efficient vaccines having ability of broad protection and complete protection from fetal infection.

REFERENCES

Baker, J.C. (1995). The clinical manifestation of bovine viral doarrhoea infection. Vet Clin N Am Food Anim Pract, 11(3): 425-445. doi: 10.1016/s0749-0720(15)30460-6.

Bauermann, F.V., Ridpath, J.F., Weiblen, R., Flores, E.F. (2013). HoBi-like viruses: an emerging group of pestiviruses. J Vet Diagn Invest, 25(1): 6-15. doi: 10.1177/1040638712473103

Baxi, M., McRae, D., Baxi, S., Greiser-Wilke, I., Vilcek, S., Amoako, K. and Deregt, D. (2006). A one-step multiplex real time RT-PCR for detection and typing of bovine viral diarrhea viruses. Vet Microbiol, 116(1-3): 37-44. doi: 10.1016/j.vetmic.2006.03.026

Becher, P. and Tautz, N. (2011). RNA recombination in pestiviruses: cellular RNA sequences in viral genomes highlight the role of host factors for viral persistence and lethal disease. RNA Biol, 8(2): 216-224. doi: 10.4161/rna.8.2.14514

Becher, P., Orlich, M., Kosmidon, A., Konig, M., Baroth, M. and Thiel, H.J. (1999). Genetic diversity of pestiviruses: identification of novel groups and implication for classification. Virol, 262(1): 64-71. doi: 10.1006/viro.1999.9872

Becher, P., Avalos Ramirez, R., Orlich, M., Cedillo Rosales, S., Konig, M., Schweizer, M., Stalder, H., Schirrmeier, H., Thiel, H.J. (2003). Genetic and antigenic characterization of novel pestivirus genotypes: implications for classification. Virol, 311(1): 96-104.doi: 10.1016/s0042-6822(03)00192-2.

Behera, S.P., Mishra, N., Vilcek, S., Rajukumar, K., Nema, R.K., Prakash, A., Kalaiyarasu, S., Dubey, S.C. (2011). Genetic and antigenic characterization of bovine viral diarrhoea virus type 2 isolated from cattle in India. Comp Immunol Microbiol Infect Dis, 34(2): 189-196. doi: 10.1016/j.cimid.2010.11.002.

Bolin, S.R. and Ridpath, J.F. (1992). Differences in virulence between two noncytopathic bovine viral diarrhea viruses in calves. Am J Vet Res, 53(11): 2157-2163. PMID: 1334641.

Bolin, S.R., McClurkin, A.W., Cutlip, R.C. and Coria, M.F. (1985). Severe clinical disease induced in cattle persistently infected with noncytopathogenic bovine viral diarrhea virus by superinfection with cytopathogenic bovine viral diarrhea virus. Am J Vet Res, 46(3): 573-576.PMID: 2986494.

Brownlie, J., Hooper, L.B., Thompson, I. and Collins, M.E. (1998). Maternal recognition of foetal infection with bovine viral diarrhea virus (BVDV) -

the bovine pestivirus. Clin Diag Virol, 10(2-3): 141-150.doi: 10.1016/s0928-0197(98)00030-0

Callens, N., Brügger, B., Bonnafous, P., Drobecq, H., Gerl, M.J., Krey, T., Roman-Sosa, G., Rümenapf, T., Lambert, O., Dubuisson, J. and Rouillé, Y. (2016). Morphology and molecular composition of purified bovine viral diarrhea virus envelope. Plo Pathog, 12: e1005476. doi.10.1371/journal.ppat.1005476

Carman, S., van Dreumel, T., Ridpath, J., Hazlett, M., Alves, D., Dubovi, E., Tremblay, R., Bolin, S., Godkin, A. and Anderson, N. (1998). Severe acute bovine viral diarrhea in Ontario, 1993-1995. J Vet Diagn Invest, 10(1): 27-35. doi: 10.1177/104063879801000106

Decaro, N., Mari, V., Pinto, P., Lucente, M.S., Sciarretta, R. and Cirone, F. (2012). Hobi-like pestivirus: both biotypes isolated from a diseased animal. J Gen Virol, 93(9): 1976-1983.doi: 10.1099/vir.0.044552-0.

Dias, R.K., Cargnelutti, J.F., Weber, M.N., Canal, C.W., Bauermann, F.V., Ridpath, J.F., Weiblen, R., Flores, E.F. (2017). Antigenic diversity of Brazilian isolates of HoBi-like pestiviruses. Vet Microbiol, 203: 221-228.doi: 10.1016/j.vetmic.2017.03.021.

Giammarioli, M., Ridpath, J.F., Rossi, E., Bazzucchi, M., Casciari, C., De Mia, G.M. (2015). Genetic detection and characterization of emerging HoBi-like viruses in archival foetal bovine serum batches. Biologicals, 43(4): 220-224. doi: 10.1016/j.biologicals.2015.05.009.

Grummer, B., Grotha, S. and Greiser-Wilke, I. (2004). Bovine viral diarrhoea virus is internalized by clathrin-dependent receptor-mediated endocytosis. J Vet Med Ser B 51(10): 427-432.doi: 10.1111/j.1439-0450.2004.00798.x.

Hoffmann, B., Depner, K., Schirrmeier, H. and Beer, M. (2006). A universal heterologous internal control system for duplex real-time RT-PCR assays used in a detection system for pestiviruses. J Virol Methods, 136(1-2): 200-209.doi: 10.1016/j.jviromet.2006.05.020.

Horzinek, M., Maess, J. and Laufs, R. (1971). Studies on the substructure of togaviruses. Arch Gesamte Virusforsch, 33: 306-318. https://doi.org/10.1007/BF01254687.

Houe, H. (2003). Economic impact of BVDV infection in dairies. Biologicals, 31(2): 137-143. doi: 10.1016/s1045-1056(03)00030-7.

Jenckel, M., Hoper, D., Schirrmeier, H., Reimann, I., Goller, K.V., Hoffmann, B. and Beer, M. (2014). Mixed triple: allied viruses in unique recent isolates of highly virulent type 2 bovine viral diarrhea virus detected by deep sequencing. J Virol, 88(12): 6983-6992.doi: 10.1128/JVI.00620-14.

Kalaiyarasu, S., Mishra, N., Jayalakshmi, K., Selvaraj, P., Sudhakar, S.B., Jhade, S.K., Sood, R., Premalatha, N. and Singh, V.P. (2022). Molecular characterization of recent HoBi-like pestivirus isolates from cattle showing mucosal disease-like signs in India reveals emergence of a novel genetic lineage. Transbound Emerg Dis, 69(2): 308-326. doi: 10.1111/tbed.13981

Kalaiyarasu, S., Mishra, N., Subramaniam, S., Moorthy, D., Sudhakar, S.B., Singh, V.P. and Sanyal, A. (2023). Whole-genome-sequence-based evolutionary analyses of HoBi-like pestiviruses reveal insights into their origin and evolutionary history. Viruses, 15(3): 733. doi: 10.3390/v15030733.

Kirkland, P.D., MacIntos,h S.G. and Moyle, A. (1994). The outcome of widespread use of semen from a bull persistently infected with pestivirus. Vet Rec, 135(2): 527-529.doi: 10.1136/vr.135.22.527.

Krey, T., Thiel, H.J. and Rumenapf, T. (2005). Acid-resistant bovine pestivirus requires activation for pH-triggered fusion during entry. J Virol, 79(7): 4191-4200.doi: 10.1128/JVI.79.7.4191-4200.2005.

Kummerer, B.M. and Meyers, G. (2000). Correlation between point mutations in NS2 and the viability and cytopathogenicity of bovine viral diarrhea virus strain Oregon analyzed with an infectious cDNA clone. J Virol, 74(1): 390-400.doi: 10.1128/jvi.74.1.390-400.2000.

Liang, D., Sainz, I.F., Ansari, I.H., Gill, L.H., Vassilev, V. and Donis, R.O. (2003). The envelope glycoprotein E2 is a determinant of cell culture tropism in ruminant pestiviruses. J Gen Virol, 84(5): 1269-1274.doi: 10.1099/vir.0.18557-0

Lindenbach, B.D., Murray, C.L., Thiel, H.J., Rice, C.M. (2013). Flaviviridae. In: Knipe DM, Howley PM (eds) Fields virology, vol 6. Lippincott Williams and Wilkins, Philadelphia, pp 712-746.

Liu, L., Xia, H., Belak, S. and Baule, C. (2008). A TaqMan real-time RT-PCR assay for selective detection of atypical bovine pestiviruses in clinical samples and biological products. J Virol Methods, 154(1-2): 82-85. doi: 10.1016/j.jviromet.2008.09.001.

Liu, L., Xia, H., Wahlberg, N., Belák, S. and Baule, C. (2009). Phylogeny, classification and evolutionary insights into pestiviruses. Virol, 385(2): 351-357. doi: 10.1016/j.virol.2008.12.004.

Mathapati, B.S., Mishra, N., Rajukumar, K., Nema, R.K., Behera, S.P. and Dubey, S.C. (2010). Entry of bovine viral diarrhoea virus into ovine cells occurs through clathrin-dependent endocytosis and low pH-dependent fusion. In Vitro Cell Dev Biol Anim, 46(5): 403-407.doi: 10.1007/s11626-009-9263-9.

Maurer, K., Krey, T., Moennig, V., Thiel, H.J. and Rumenapf, T. (2004). CD46 is a cellular receptor for bovine viral diarrhea virus. J Virol, 78(4): 1792-1799. doi: 10.1128/jvi.78.4.1792-1799.2004.

McClurkin, A.W., Littledike, E.T., Cutlip, R.C., Frank, G.H., Coria, M.F. and Bolin, S.R. (1984). Production of cattle immunotolerant to bovine viral diarrhea virus. Can J Comp Med, 48(2): 156-161.PMCID: PMC1236029

Meyers, G. and Thiel, H.J. (1996). Molecular characterization of pestiviruses. Adv Virus Res, 47: 53-118.doi: 10.1016/s0065-3527(08)60734-4.

Meyers, G., Tautz, N., Dubovi, E.J. and Thiel, H.J. (1991). Viral cytopathogenicity correlated with integration of ubiquitin-coding sequences. Virol, 180(2): 602-616.doi: 10.1016/0042-6822(91)90074-l.

Mishra, N. and Kalaiyarasu, S. (2019). Bovine Viral Diarrhea Virus. In: Malik, Y., Singh, R., Yadav, M. (eds) Recent Advances in Animal Virology. Springer, Singapore. https://doi.org/10.1007/978-981-13-9073-9_14

Mishra, N., Pattnaik, B., Vilcek, S., Patil, S.S., Jain, P., Swamy, N., Bhatia, S. and Pradhan, H.K. (2004). Genetic typing of bovine viral diarrhoea virus isolates from India. Vet Microbiol, 104(3-4): 207-212.doi: 10.1016/j.vetmic.2004.08.003.

Mishra, N., Dubey, R., Rajukumar, K., Tosh, C., Tiwari, A., Pitale, S.S. and Pradhan, H.K. (2007a). Genetic and antigenic characterization of bovine viral diarrhea virus type 2 isolated from Indian goats (*Capra hircus*). Vet Microbiol, 124(3-4): 340-347.doi: 10.1016/j.vetmic.2007.04.023.

Mishra, N., Dubey, R., Galav, V., Tosh, C., Rajukumar, K., Pitale, S.S. and Pradhan, H.K. (2007b). Identification of bovine viral diarrhea virus 1 in Indian buffaloes and their genetic relationship with cattle strains in 5′ UTR. Curr Sci, 93(1): 97-100. http://www.jstor.org/stable/24099435.

Mishra, N., Vilcek, S., Rajukumar, K., Dubey, R., Tiwari, A., Galav, V. and Pradhan, H.K. (2008a). Identification of bovine viral diarrhea virus type 1 in yaks (*Bos poepaghus grunniens*) in Himalayan region. Res Vet Sci, 84(3): 507-510.doi: 10.1016/j.rvsc.2007.05.019.

Mishra, N., Rajukumar, K., Vilcek, S., Tiwari, A., Satav, J.S. and Dubey, S.C. (2008b). Molecular characterization of bovine viral diarrhea virus type 2 isolate originating from a native Indian sheep (*Ovies aries*). Vet Microbiol, 130(1-2): 88-98.doi: 10.1016/j.vetmic.2008.01.005.

Mishra, N., Rajukumar, K., Tiwari, A., Nema, R.K., Behera, S.P., Satav, J.S., Dubey, S.C. (2009). Prevalence of bovine viral diarrhoea virus antibodies among sheep and goats in India. Trop Anim Health Prod, 41(7): 1231-1239. doi: 10.1007/s11250-009-9305-z.

Mishra, N., Rajukumar, K., Pateriya, A., Kumar, M., Dubey, P., Behera, S.P., Verma, A., Bhardwaj, P., Kulkarni, D.D., Vijaykrishna, D. and Reddy, N.D. (2014). Identification and molecular characterization of novel and divergent HoBi-like pestiviruses from naturally infected cattle in India. Vet Microbiol, 174(1-2): 239-246.doi: 10.1016/j.vetmic.2014.09.017.

Mishra, N., Pitale, S.S., Rajukumar, K., Prakash, A., Behera, S.P., Nema, R.K. and Dubey, S.C. (2012). Genetic variety of bovine viral diarrhoea virus 1 strains isolated from sheep and goats in India. Acta Virol, 56(3): 209-215. doi: 10.4149/av_2012_03_209.

Mishra, N., Kalaiyarasu, S, Mallinath, K.C., Rajukumar, K., Khetan, R.K., Gautam, S., Venkatesha, M.D. and Byregowda, S.M. (2018). Identification of bovine viral diarrhoea virus type 2 (BVDV-2) in cattle bull semen from southern India and its genetic characterization. Curr Sci, 114(3): 666-670. http://www.jstor.org/stable/26495122

Moennig, V. and Becher, P. (2018). Control of bovine viral diarrhea. Pathogens, 7: 29.doi: 10.3390/pathogens7010029.

Nayak, B.C., Panda, S.N., Misra, D.B., Kar, B.C. and Das, B.C. (1981). Note on serological evidence of viral abortion in cattle in Orissa. Indian J Anim Sci, 52: 102-103.

Niskanen, R. and Lindberg, A. (2003). Transmission of bovine viral diarrhoea virus by unhygienic vaccination procedures, ambient air, and from contaminated pens. Vet J, 165(2): 125-130.doi: 10.1016/s1090-0233(02)00161-2.

Olafson, P., McCallum, A.D. and Fox, F.H. (1946). An apparently new transmissible disease of cattle. Cornell Vet, 36: 205-213. PMID: 20995890

Paton, D.J., Sands, J.J., Lowings, J.P., Smith,.JE.,Ibata, G. and Edwards, S. (1995). A proposed division of the pestivirus genus using monoclonal antibodies, supported by cross neutralization assays and genetic sequencing. Vet Res, 26(2): 92-109. PMID: 7735309

Pecora, A., Perez Aguirreburualde, M.S., Ridpath, J.F. and Dus Santos, M.J. (2019). Molecular Characterization of pestiviruses in fetal bovine sera originating from Argentina: evidence of circulation of HoBi-like viruses. Front Vet Sci, 6: 359. doi: 10.3389/fvets.2019.00359.

Pellerin, C., Van den Hurk, J., Lecomte, J. and Tijssen, P. (1994). Identification of a new group of bovine viral diarrhea virus strains associated with severe outbreaks and high mortality. Virol, 203(2): 260-268.doi: 10.1006/viro.1994.1483.

Perdrizet, J.A., Rebhun, W.C., Dubovi, E.J. and Donis, R.O. (1987). Bovine virus diarrhea-clinical syndromes in dairy herds. Cornell Vet, 77(1): 46-74. PMID: 3802830

Ramsey, F.K. and Chivers, W.H. (1953). Mucosal disease of cattle. North. Am. Vet. 34:629–633 Richter, V., Lebi, K., Baumgartner, W., Obritzhauser, W., Käsbohrer, A., Pinior, B. (2017). A systematic worldwide review of the direct monetary losses in cattle due to bovine viral diarrhoea virus infection. Vet J, 220: 80-87.doi: 10.1016/j.tvjl.2017.01.005.

Ridpath, J., Bolin, S.R. and Dubovi, E.J. (1994). Segregation of bovine viral diarrhea virus into genotypes. Virol, 205(1): 66-74.doi: 10.1006/viro.1994.1620.

Ridpath, J.F., Neill, J.D., Frey, M., Landgraf, J.G. (2000). Phylogenetic, antigenic and clinical characterization of type 2 BVDV from North America. Vet Microbiol, 77(1-2): 145-155.doi: 10.1016/s0378-1135(00)00271-6.

Rumenapf, T., Unger, G., Strauss, J.H. and Thiel, H.J. (1993). Processing of the envelope glycoproteins of pestiviruses. J Virol, 67(6): 3288-3294.doi: 10.1128/JVI.67.6.3288-3294.1993.

Schirrmeier, H., Strebelow, G., Depner, K., Hoffmann, B. and Beer, M. (2004). Genetic and antigenic characterization of an atypical pestivirus isolate, a putative member of a novel pestivirus species. J Gen Virol, 85(2): 3647-3652. doi: 10.1099/vir.0.80238-0.

Schmeiser, S., Mast, J., Thiel, H.J. and Konig, M. (2014). Morphogenesis of pestiviruses: new insights from ultrastructural studies of strain Giraffe-1. J Virol, 88(5): 2717-2724.doi: 10.1128/JVI.03237-13.

Singh, V., Mishra, N., Kalaiyarasu, S., Khetan, R.K., Hemadri, D., Singh, R.K., Rajukumar, K., Chamuah, J., Suresh, K.P., Patil, S.S. and Singh, V.P. (2017). First report on serological evidence of bovine viral diarrhea virus (BVDV) infection in farmed and free ranging mithuns (*Bos frontalis*). Trop Anim Health Prod, 49(6): 1149-1156.doi: 10.1007/s11250-017-1310-z.

Smith, D.B., Meyers, G., Bukh, J., Gould, E.A., Monath, T., Muerhoff, A.S., Pletnev, A., Rico-Hesse, R., Stapleton, J.T., Simmonds, P. and Becher, P. (2017). Proposed revision to the taxonomy of the genus Pestivirus, family Flaviviridae. J Gen Virol, 98(8): 2106-2112.doi: 10.1099/jgv.0.000873.

Stalder, H.P., Meier, P., Pfaffen, G., Wageck-Canal, C., Rüfenacht, J., Schaller, P., Bachofen, C., Marti, S., Vogt, H.R. and Peterhans, E. (2005). Genetic heterogeneity of pestiviruses of ruminants in Switzerland. Prev Vet Med, 72(1-2): 37-41; discussion 215-9. doi: 10.1016/j.prevetmed.2005.01.020.

Stark, R., Meyers, G., Rumenapf, T. and Thiel, H.J. (1993). Processing of pestiviruspolyprotein: cleavage site between autoprotease and nucleocapsid protein of classical swine fever virus. Virol, 67(12): 7088-7095.doi: 10.1128/JVI.67.12.7088-7095.

Sudharsana, K.J., Suresh, K.B. and Rajasekhar, M. (1999). Prevalence of bovine viral diarrhea virus antibodies in India. Rev Sci Tech, 18(3): 667-671.doi: 10.20506/rst.18.3.1189.

Tautz, N., Meyers, G. and Thiel, H.J. (1993). Processing of poly-ubiquitin in the polyprotein of an RNA virus. Virol, 197(1): 74-85.doi: 10.1006/viro.1993.1568.

Van Campen, H. and Frolich, K. (2001). Pestivirus infections In: Williams, E.S., Barker, I.K. (eds) Infectious diseases of wild mammals. Iowa State University Press, Iowa City, pp 232244.

Vilcek, S. and Nettleton, P.F. (2006). Pestiviruses in wild animals. Vet Microbiol, 116(1-3): 1-12.doi: 10.1016/j.vetmic.2006.06.003.

Vilcek, S., Durkovic, B., Kolesarova, M., Greiser-Wilke, I. and Paton, D.J. (2004). Genetic diversity of international bovine viral diarrhoea virus (BVDV) isolates: identification of a new BVDV-1 genetic group. Vet Res, 35: 609-615. doi: 10.1051/vetres:2004036

Vilcek, S., Paton, D.J., Durkovic, B., Strojny, L., Ibata, G., Moussa, A., Loitsch, A., Rossmanith, W., Vega, S., Scicluna, M.T. and Palfi, V. (2001). Bovine viral diarrhea virus genotype 1 can be separated into at least eleven genetic groups. Arch Virol, 146(1): 99-115.doi: 10.1007/s007050170194.

Vilèek, Š., Ridpath, J.F., Van Campen, H., Cavender, J.L. and Warg, J. (2005). Characterization of a novel pestivirus originating from a pronghorn antelope. Virus Res, 108(1-2): 187-193.doi: 10.1016/j.virusres.2004.09.010.

Walz, P.H., Bell, T.G., Wells, J.L., Grooms, D.L., Kaiser, L., Maes, R.K. and Baker, J.C. (2001). Relationship between degree of viremia and disease manifestation in calves with experimentally induced bovine viral diarrhea virus infection. Am J Vet Res, 62(7): 1095-1103.doi: 10.2460/ajvr.2001.62.1095.

Warrilow, D., Lott, W.B., Greive, S. and Gowans, E.J. (2000). Properties of the bovine viral diarrhoea virus replicase in extracts of infected MDBK cells. Arch Virol, 145(10): 2163-2171.doi: 10.1007/s007050070046.

Weber, M.N., Streck, A.F., Silveira, S., Mósena, A.C., Silva, M.S. and Canal, C.W. (2015). Homologous recombination in pestiviruses: identification of three putative novel events between different subtypes/genogroups. Infect Genet Evol, 30: 219-224.doi: 10.1016/j.meegid.2014.12.032.

Weiland, E., Stark, R., Haas, B., Rumenapf, T., Meyers, G., Thiel, H.J. (1990). Pestivirus glycoprotein which induces neutralizing antibodies forms part of a disulfide linked heterodimer. J Virol, 64(8): 3563-3569.doi: 10.1128/JVI.64.8.3563-3569.

Willoughby, K., Valdazo-Gonzalez, B., Maley, M., Gilray, J. and Nettleton, P.F. (2006). Development of a real time RT-PCR to detect and type ovine pestiviruses. J Virol Methods, 132(1-2): 187-194.doi: 10.1016/j.jviromet.2005.10.007.

WOAH. (2017). Manual of diagnostic tests and vaccines for terrestrial animals. Chapter 3.04.7, Bovine viral diarrhea. WOAH, Paris, pp 1-23

Xia, H., Vijayaraghavan, B., Belák, S. and Liu, L. (2011). Detection and identification of the atypical bovine pestiviruses in commercial foetal fovine serum batches. PLoS One, 6(12): e28553. doi: 10.1371/journal.pone.0028553.

Yadav, P., Barde, P.V., Jadi, R., Gokhale, M.D., Basu, A., Joshi, M.V., Mehla, R., Kumar, S.R., Athavale, S.S. and Mourya, D.T. (2004). Isolation of bovine viral diarrhea virus 1, a pestivirus from autopsied lamb specimen from Tamil Nadu, India. Acta Virol, 48(4): 223-227. PMID: 15745045.

Yarnall, M.J. and Thrusfield, M.V. (2017). Engaging veterinarians and farmers in eradicating bovine viral diarrhoea: a systematic review of economic impact. Vet Rec, 181(13): 347.doi: 10.1136/vr.104370.

Yesilbag, K., Forster, C., Bank-Wolf, B., Yilmaz, Z., Alkan, F., Ozkul, F.A., Burgu, I., Rosales, S.C., Theil, H.J. and Konig, M. (2008). Genetic heterogeneity of bovine viral diarrhoea virus (BVDV) isolates from Turkey: identification of a new subgroup in BVDV-1. Vet Microbiol, 130(3-4): 258-267.doi: 10.1016/j.vetmic.2008.01.016.

Yesilbag, K., Alpay, G. and Becher, P. (2017). Variability and global distribution of subgenotypes of bovine viral diarrhea virus. Viruses, 9(6): 128. doi: 10.3390/v9060128.

Chapter 13

Canine Distemper: An Insight into Current Indian Scenario Perspectives

*H.R. Jogi[1], A.K. Yadav[1], M. Bhatt[2], C. Einstein[1], A. Revathi[1], R.K. Thakur[1], V. Rai[3], R.P. Singh[4] and K.K. Rajak[1]**

[1]Division of Biological Products, ICAR-IVRI, Izatnagar, Bareilly, Uttar Pradesh
[2]CADRAD, ICAR-IVRI, Izatnagar, Bareilly, Uttar Pradesh
[3]ICAR-NRC on Pig, Rani, Guwahati, Assam
[4]ICAR-NIFMD, Arugul, Bhubaneswar, Odisha
**e-mail: kaushalvirol@gmail.com*

ABSTRACT

Canine distemper virus (CDV) is one of the important, contagious and widely distributed pathogens amongst various animal species. The virus belongs to genus Morbillivirus and family Paramyxoviridae. CDV has only one serotype with wide host range which makes it a constant threat to various conserved endangered wild animal species. It is a virus causing multi-systemic infection and thus, can damage the nervous, respiratory and gastro-intestinal systems in affected animal. Till date, total seventeen lineages of CDV have been reported throughout the world. Recently, India has reported the eighteenth lineage of this virus with marked genomic difference. The ability of virus to cross the species barrier and infect a variety of mammals including primates has raised the concern regarding its future emergence as a zoonotic disease in humans. Amongst all the genes of virus, H gene is more likely to undergo rapid mutations and is solely responsible for the expanded host range. Furthermore, due to recurrent emergence of this virus, it is necessary to conduct advanced molecular studies regarding the strains circulating within different geographical regions around the world. For effective virus control and disease prevention in our valuable animals, it is of utmost importance to develop novel rapid diagnostic tools which can be needful in early screening. Additionally, a suitably planned vaccination strategy is quite essential along with the development of advanced generation vaccines.

Keywords: *Canine distemper virus (CDV), Emergence, Lineages, Diagnostic tools, Vaccines.*

Introduction

Canine distemper (CD) is an extremely contagious and fatal viral disease of dogs as well as wild animals. For the first time, CD was reported in Spain during the year 1761 and subsequently, it spread throughout the world (Bieringer *et al.*, 2013). In the year 1809, Edward Jenner initially gave the description of disease but later, as an etiological agent it was first studied by Henri Carre, a French veterinarian in 1905 (Appel *et al.*, 1973). The disease is caused by pantropic canine morbillivirus or canine distemper virus (CDV). It is a single stranded (ss) RNA virus, belonging to genus *Morbillivirus,* family *Paramyxoviridae* and order *Mononegavirales* (ICTV). The genus is comprised of other members like Rinderpest virus (RPV), Peste des petits ruminants virus (PPRV) and Measles virus which are responsible for causing various devastating diseases in both animals and humans (Taques *et al.*, 2018). Formerly, it was described as an infectious agent of dogs but now it has become a multi-host pathogen causing infection along with mortality in a wide range of carnivores. Approximately more than 19 families of animals are known to be affected by this virus including both carnivores as well as non-carnivores. The CDV host range contains members of several animal families such as Canidae, Ailruidae, Procyonidae, Mustelidae, Ursidae, Viverridae, Felidae, Ursidae, Hyaenidae, Tayassuidae, and Cercopithecidae (Loot *et al.*, 2018). The incubation period of this disease varies from 7 to 28 days and is manifested in two forms such as acute and chronic nervous form. The virus can cause the severe suppression in the immune system of infected animals that lead to development of clinical signs such as anorexia, coughing, conjunctivitis, biphasic fever and diarrhoea. The second phase of fever is seen after 6-9 days of the first peak (Pope *et al.*, 2015). Moreover, the chronic form of disease is known as "Old dog encephalitis" which is characterized by tremors, seizures, twitching movements and hard foot pads (Rendon-Marin *et al.*, 2019). In the last decade, many CDV outbreaks have been reported in wild animal populations showing the expansion of virus's host range. These outbreaks were responsible for the advancement of CD as an emerging threat in endangered wildlife species due to host switching (Beineke *et al.*, 2015). Additionally, the lethal infections in non-carnivores and non-human primates have been demonstrated to study regarding the remarkable capacity of virus to jump species barriers. Generally, this disease can be prevented by using the available attenuated vaccines. However, lately several outbreaks of disease have been reported in fully vaccinated animals which provided insight towards the probable emergence of novel CDV field strains that further helped in the immune escape (Bhatt *et al.*, 2019). This chapter provides an overview on the canine distemper virus including the virion properties, epidemiology, recent diagnostic tools and vaccines followed by conclusions.

Virion Properties and Transmission

CDV is a non-segmented, negative-sense, enveloped and ss-RNA virus with a diameter of 150-300 nm. The viral genome is comprised of 15,960 nucleotides and encodes six structural proteins such as nucleocapsid protein (N), phosphoprotein (P), matrix protein (M), fusion protein (F), haemagglutinin protein (H) and polymerase protein (L). Additionally, the gene sequence of P protein encodes non-structural protein (C) (Panzera *et al.*, 2015). The haemagglutinin (H) and fusion (F) proteins are the major antigenic determinants of CDV which are present on viral envelope. Amongst them, H protein is highly diversified in nature with approximately $5.4X10^{-4}$ to $1.8X10^{-3}$ nucleotide substitution rate per site per year that makes it suitable for lineage determination and genetic assessment of CDV. For the entry of CDV inside the host cell, interaction between H protein and signaling lymphocyte activation molecule (SLAM/CD150) receptor is necessary while, for the cell to cell spread it utilizes Nectin-4 receptor (Figure 13.1). Additionally, it can invade the central nervous system by use of putative receptor GliaR, present in glial cells (Anis *et al.*, 2018). Currently, due to the advancement in molecular studies, identification of virus genetic lineages along with its geographical origin is possible. Previously, based on the sequence of H protein there were nine reported genetic lineages of CDV. However, at present the number of lineages have reached seventeen which includes America 1 to America 5, Asia 1 to Asia 4, Europe/South America 1, South America 2 and 3, Europe wildlife, Arctic, Rockborn-like, Africa 1 and Africa 2 (Duque-Valencia *et al.*, 2019).Moreover, the eighteenth lineage (India-1/ Asia-5) was proposed by Bhatt *et al.* (2019) that represents the circulating strains of CDV in India.

As a known fact, CDV is highly contagious and can be easily transmitted to susceptible hosts through direct mode and various body secretions such as respiratory, oral and ocular. However, during the acute phase of infection other excretory substances like faeces and urine also contain the virus. Usually, the viral shedding may occur up to 90 days of infection even if the animal is sub-clinically infected (Carvalho *et al.*, 2012). Due to the presence of lipid envelope, CDV is highly sensitive to oxidizing agents, UV radiation, heat desiccation and lipid solvents. Therefore, close association is the prime requirement for the transmission between infected and naïve animal. The presence of unvaccinated population of dogs around the protected wildlife sanctuaries or national parks is highly responsible for CDV transmission to wildlife (Weckworth *et al.*, 2020).

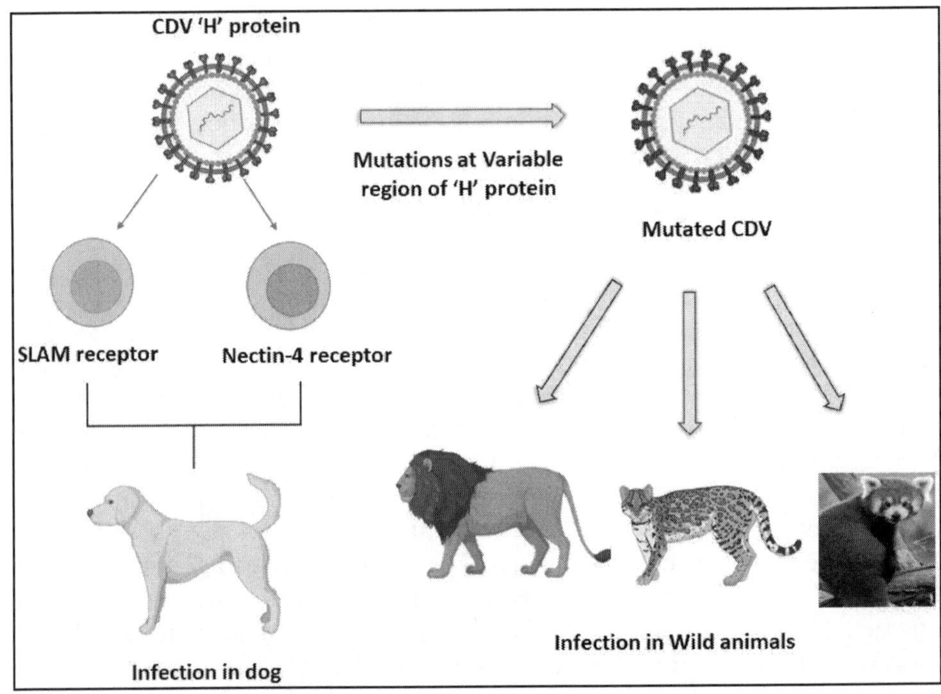

Figure 13.1: Canine Distemper Virus (CDV) Tropism and Receptor Fidelity in Domestic - Wild Animals (Created with biorender.com).

Epidemiology

CD has now become an endemic viral infection and high degree of disease outbreaks have been reported from various parts of the world. According to serological surveys in several countries, high rate of seropositivity to CDV was recorded in free ranging dog population (Acosta-Jamett *et al.*, 2015; Dorji *et al.*, 2020; McDermott *et al.*, 2023). Moreover, in India several reports regarding isolation and characterization of CDV derived from dogs as well as wild animals belonging to different geographical regions have been published (Swati *et al.*, 2015; Ashmi *et al.*, 2017; Kadam *et al.*, 2022). Dogs have been considered as the most common source for the transmission of this virus to the wild animal population. Severe impact of CDV infection has been observed in various wildlife species such as Silver fox, Siberian tigers, Ethiopian wolves, Red Panda, Cheetah and Lions (Yuan *et al.*, 2017; Mourya *et al.*, 2019). Additionally, the infection of CDV in primates has raised the concern about its transmission to humans. However, there is no confirmed report regarding the presence of this virus in human beings. Recently, in South Africa, due to fatal outbreak of CDV in wild animal population about 95 percent mortality was recorded which included various wild carnivores such as hyena, wild dogs and lions (Loots *et al.*, 2018).

As a result of species jumping, selection pressure and new host adaptation, it was concluded that the phocine distemper virus (PDV) has been derived from CDV (Kennedy *et al.*, 2019).

Genetic Diversity

As a matter of proven fact, the H protein of CDV determines its cellular tropism, host range and mainly responsible for interactions with host cell receptors. Additionally, it also possesses the property of undergoing high mutation rate and undergoes positive selection to adapt in any novel host. Consequently, the H gene is widely studied for the genetic and phylogenetic assessment of CDV (Fischer *et al.*, 2013). Till date, there are total 18 lineages of CDV available across the world and all are based on the nucleotide sequences of H gene. It includes America-1 (vaccine strains), America-2, America-3, Asia-1, Asia-2, Asia-3, Rockborn-like, European Wildlife, Europe/South America-1, South America-2, South America-3 and India-1/Asia-5 (Bhatt *et al.*, 2019). The detailed information regarding well known strains of CDV is compiled in Table 13.1. Comparative genetic analysis of H gene has revealed that occurrence of specific substitutions at 530 and 549 positions are important for the host adaption of CDV (McCarthy *et al.*, 2007). Moreover, analysis of the CDV strains obtained during the outbreak of Serengeti National Park in 1990 has demonstrated that there was a 530D substitution in all the affected species of animals. However, Y549H mutation occurred due to positive selection pressure and it might be the key determinant for adaptation of CDV in wildlife population (Nikolin *et al.*, 2017). Analysis of the antigenic characteristics by neutralization test has given two classes of CDV: old CDV and new CDV. It has been validated that mutations at the intergenic regions of CDV genome were responsible for the increased virulence of field strains (Sarute *et al.*, 2014). Majority of the currently used vaccines belong to America-1 strain and have shown significant genetic variation as compared to the circulating CDV strains of other lineages (Bhatt *et al.*, 2019). Therefore, co-circulation of different lineages might be responsible for various recombination events and the emergence of novel viral variants or sub-lineages.

Table 13.1: A List of Important and well Characterized Strains of CDV

Sl.No.	*Name of strain*	*Lineage*	*Gene*	*Accession Number*
1.	Convac	America-I	H	Z35493
2.	Synder Hill	America-I	H	AF259552
3.	Lederle	America-I	H	EF418782
4.	Rockborn	America-I	H	GU810819
5.	Ondersteepoort	America-I	H	AF014953
6.	Dog/Bly/Ind/2018	India-1/Asia 5	H	MF964178

Diagnosis

Ante-mortem diagnosis is the most preferable method in case of CDV due to its fast progression and highly infectious potential. However, the diagnosis of CDV based on its clinical signs is often difficult and confusing because it shows high similarity in symptoms with that of many other microbial infections such as rabies, parvovirus, coronavirus and bacterial enteritis. Therefore, effective laboratory diagnosis should be done for the correct detection of CDV in domestic as well as wild animals. Currently, there are a number of serological, immunological and molecular diagnostic assays which are available for the detection of CDV showing high sensitivity and specificity rate. It includes virus isolation, enzyme linked immunosorbent assay (ELISA), indirect fluorescent antibody test (IFAT), lateral flow assay (LFA), reverse transcription-polymerase chain reaction (RT-PCR) and reverse transcription loop mediated isothermal amplificationassay (RT-LAMP) (Joshi *et al.*, 2022; Menezes *et al.*, 2023). Here, we have provided some brief information regarding available diagnostic tools for the detection of CDV as below.

Virus Isolation

Virus isolation is considered as a gold standard for the diagnosis of viral diseases having animal as well as human importance. However, in case of CDV there is low success rate during virus isolation due to high sensitivity towards temperature and light. The isolation often require canine or ferret pulmonary macrophages and mitogen-stimulated healthy lymphocytes. Recently, vero cell lines showing stable expression of several animal species origin SLAM such as dog, sheep or goat have been used for effective isolation of CDV field strains in our laboratory (Yadav *et al.*, 2021). In addition to vero cells, Madin-Darby canine kidney (MDCK), Mink lung epithelial cells (MV1 Lu) and B95a cell lines also can be used for the primary isolation of CDV (Lednicky *et al.*, 2004; Romanutti *et al.*, 2020). Currently, our laboratory has isolated and recognized a novel lineage of CDV virus (CDV (Dog)/Bly/Ind/2018) (Bhatt *et al.*, 2019). It has projected as attenuated vaccine candidate. (Indian Patent application number 202011057169, dated 30/12/2020). The CDV shows a typical feature of multinucleated giant-cells (syncytia) like cytopathic effect (CPE) which is similar to other morbilliviruses. However, the vaccine and wild type strains of CDV show different types of CPE that is directly proportional to the level of attenuation and H protein fusion efficiency (von Messling *et al.*, 2001).

Serological Assay

Several serological tests are available for the detection of antibodies to CDV such as ELISA, IFAT and virus neutralization test (VNT). Amongst them, VNT is the gold standard and OIE recommended test for the detection of neutralizing antibodies. Usually, in the case of CDV, infected animal shows high titre (1:100) of neutralizing antibodies just after the infection. According to certain studies, the titre of 1:32 is an indicator for borderline protection against CDV infection (Bergmann *et al.*, 2021). The procedure of VNT is time taking and too much skill is required to perform test and handle the virus. Therefore, for the effective detection of IgM and IgG, ELISA is widely used till three months in infected dog. Furthermore, plaque reduction neutralization test can be also used to check the pre and post vaccination status of animals. There are some cross-reactive epitopes among CDV, MV, PPRV and RPV that must be considered beforehand during serological testing.

Monoclonal Antibody-Based Assay

In the year 1985, firstly monoclonal antibodies (mAbs) have been developed against N, F, H and P proteins of CDV (Orvell *et al.*, 1985). For the detection of CDV in nasal swabs, conjunctiva and lungs, mAb based sandwich ELISA has been used. With the help of two mAb directed against different epitopes of F protein, a sandwich ELISA was developed for the detection of CDV in faecal and serum samples (An *et al.*, 2008). However, to conduct any large-scale epidemiological surveillance, a rapid method of detection is required. Therefore, a novel mAb based sandwich dot ELISA has been developed (Li *et al.*, 2013). Furthermore, our laboratory has developed highly reactive mAbs against the representative isolate of Indian CDV strains (CDV/dog/bly/Ind/2018) (Fayaz *et al.*, 2022). These mAbs have been utilized for the development of a highly sensitive competitive ELISA assay to detect the CDV (Personal information). The monoclonal antibodies have provided high rate of specificity in diagnostic assays. As a future perspect, it will be highly helpful for the development of CDV diagnostics in various ways.

Lateral Flow Assay

Nowadays, several types of novel diagnostic tests are commercially available for the CDV detection with high sensitivity and specificity. However, certain problems like difficulty in wildlife sample collection and transport related damages have raised demand for the development of specific, easy and on-site diagnostic assay like lateral flow assay (LFA). Till date, very limited studies have been conducted regarding the formulation of pen-side diagnostic LFA assay for CDV (An *et al.*, 2008). Currently, some LFA kits are available commercially like Lilif™ CDV Rapid Antigen kit and Rapigen Inc. CDV Antigen test (Karki *et al.*, 2022). But the presently available kits are not based on the CDV strains circulating in India and it has raised the concern of false positive results. In order to overcome this problem and to conduct the rapid-accurate diagnosis of CDV, recently out laboratory has developed competitive LFA kit based on monoclonal antibodies. This assay has been optimized and it has high sensitivity as well as specificity (Karki *et al.*, 2023).

Molecular Assay

In order to provide a rapid diagnosis and avoid the complex procedure of virus isolation, various PCR based molecular methods have been developed. These methods are solely based on viral nucleic acid detection. Nowadays, molecular methods have proven to be more advantageous in terms of both diagnostic sensitivity and specificity. Amongst all the commonly used molecular assays, RT-PCR is a widely used test for the detection of CDV in cerebrospinal fluid (CSF), blood and serum (Freitas *et al.*, 2019). However, one-step nested RT-PCR and double step real time-RT-PCR have 100 fold more sensitivity than conventional RT-PCR. It usually targets the N gene of CDV and can be used for the quantitative estimation of viral load in the clinical samples (Duque-Valencia *et al.*, 2019). Furthermore, RT-LAMP assay targeting H gene is to be considered as 100 times more sensitive than RT-PCR and it gives result in 60 minutes (Liu *et al.*, 2015). To differentiate the vaccine and wild-type strains of CDV, multiplex RT-nested PCR targeting H gene can be used (Si *et al.*, 2010). Additionally, for the accurate identification of virus, whole genome sequencing is to be considered as the most important tool.

Treatment and Control

As a well-known fact, the treatment and control of infectious viral diseases is often difficult, especially in population of wild animals. In case of domestic dogs, the treatment for CDV is solely based on symptoms. There is no effective antiviral drug available for therapeutic use in any species of animals including dogs to combat against CDV infection. However, one such compound named fucoidan (sulphated polysaccharide found in brown algae) has been evaluated as an antiviral against CDV. During in-vitro experiment, it was observed that fucoidan inhibited the initial steps of viral replication (Trejo-Avila *et al.*, 2014). Additionally, other therapies such as mesenchymal stem cell therapy and silver nanoparticles have been explored extensively. Effective and timely vaccination of domestic as well as wild animals is the only available way for prevention of further viral spread and emergence of novel strains.

Conclusions

Canine distemper is an extremely contagious disease of domestic and wild carnivores along with some non-carnivores that is distributed worldwide. Although, vaccines having high level of safety and efficacy are available for use in domestic dogs. However, it was observed that rarely mild form of disease has been reported in vaccinated dogs and wildlife species which has raised the concern towards disease control and animal protection. Commonly used strains for the preparation of CDV vaccines are Onderstepoort, Convac, Synder Hill and Rockborn. The specific mutation at 549th position of H gene is responsible for expanded host range of virus. The continuous emergence of novel lineages has raised demand for more advanced studies on recombination and evolutionary events of CDV. However, for an effective control of CDV, advanced, rapid and accurate diagnostics are required along with highly potent vaccines. Currently, a novel lineage (CDV (Dog)/Bly/Ind/2018) of CDV has been isolated and identified in India (Bhatt *et al.*, 2019). It has been projected as an attenuated vaccine candidate. Additionally, a panel of mAb's directed against this strain have been developed and fully characterized. Occasionally, the conventional vaccines retain virulence and do not provide complete protection. Therefore, in order to overcome these difficulties, new generation vaccines such as recombinant and oral vaccines can be developed. To prevent the transmission of virus from dogs to wildlife, specific root level control strategies should be planned. As we know that, broad host range of virus is creating major difficulties in eradication but by planned strategic research and its application at field level can be more helpful for the control of future outbreaks.

REFERENCES

Acosta-Jamett, G., Cunningham, A.A., Bronsvoort, B.D., and Cleaveland, S. (2015). Serosurvey of canine distemper virus and canine parvovirus in wild canids and domestic dogs at the rural interface in the Coquimbo Region, Chile. Eur J Wildl Res, 61: 329-332. doi 10.1007/s10344-014-0886-0.

An, D.J., Kim, T.Y., Song, D.S., Kang, B.K., and Park, B.K. (2008). An immunochromatography assay for rapid antemortem diagnosis of dogs suspected to have canine distemper. J Virol Methods, 147(2): 244-249. doi 10.1016/j.jviromet.2007.09.006

Anis, E., Newell, T.K., Dyer, N., and Wilkes, R.P. (2018). Phylogenetic analysis of the wild-type strains of canine distemper virus circulating in the United States. Virol J, 15(1): 118. doi 10.1186/s12985-018-1027-2

Appel, M. and Robson, D.S. (1973). A microneutralization test for canine distemper virus. Am J Vet Res, 34(11): 1459-1463.

Ashmi, J.M., Thangavelu, A., Senthilkumar, T.M.A., and Manimaran, K. (2017). Molecular characterization of canine distemper virus from Tamil Nadu, India. Indian J Anim Sci, 87(9): 1062-1067.

Beineke, A., Baumgärtner, W. and Wohlsein, P. (2015). Cross-species transmission of canine distemper virus-an update. One health (Amsterdam, Netherlands), 1, 49-59. DOI 10.1016/j.onehlt.2015.09.002

Bergmann, M., Freisl, M., Zablotski, Y., Khan, M. A. A., Speck, S., Truyen, U. and Hartmann, K. (2021). Prevalence of Neutralizing Antibodies to Canine Distemper Virus and Response to Vaccination in Client-Owned Adult Healthy Dogs. Viruses, 13(5): 945. doi 10.3390/v13050945

Bhatt, M., Rajak, K.K., Chakravarti, S., Yadav, A.K., Kumar, A., Gupta, V., Chander, V., Mathesh, K., Chandramohan, S., Sharma, A.K., Mahendran, K., Sankar, M., Muthuchelvan, D., Gandham, R.K., Baig, M., Singh, R.P. and Singh, R.K. (2019). Phylogenetic analysis of haemagglutinin gene deciphering a new genetically distinct lineage of canine distemper virus circulating among domestic dogs in India. Transbound Emerg Dis, 66(3): 1252-1267. doi 10.1111/tbed.13142

Bieringer, M., Han, J.W., Kendl, S., Khosravi, M., Plattet, P. and Schneider-Schaulies J. (2013). Experimental adaptation of wild-type canine distemper virus (CDV) to the human entry receptor CD150. PloS one, 8(3): 57488. doi 10.1371/journal.pone.0057488

Carvalho, O.V., Botelho, C.V., Ferreira, C.G., Scherer, P.O., Soares-Martins, J.A., Almeida, M.R. and Silva Júnior, A. (2012). Immunopathogenic and neurological mechanisms of canine distemper virus. Adv Virol, 163860. doi 10.1155/2012/163860

Dorji, T., Tenzin, T., Tenzin, K., Tshering, D., Rinzin, K., Phimpraphai, W. and de Garine-Wichatitsky M. (2020). Seroprevalence and risk factors of canine distemper virus in the pet and stray dogs in Haa, western Bhutan. BMC Vet Res, 16(1): 135. doi 10.1186/s12917-020-02355-x

Duque-Valencia, J., Forero-Muñoz, N.R., Díaz, F.J., Martins, E., Barato, P. and Ruiz-Saenz, J. (2019). Phylogenetic evidence of the intercontinental circulation of a Canine distemper virus lineage in the Americas. Sci Rep, 9(1): 15747. Doi 98-019-52345-9

Fayaz, A., Rajak, K.K., Kumar, A., Karki, M., Kiran, Rai, V., Bhatt, M. and Singh, R.P. (2022). Development and characterization of mouse monoclonal antibodies to canine morbillivirus. Biologicals, 79: 19-26. doi 10.1016/j.biologicals.2022.08.005

Fischer, C.D., Ikuta, N., Canal, C.W., Makiejczuk, A., Allgayer, M. daC, Cardoso, C.H., Lehmann, F.K., Fonseca, A.S. and Lunge, V.R. (2013). Detection and differentiation of field and vaccine strains of canine distemper virus using reverse transcription followed by nested real time PCR (RT-nqPCR) and RFLP analysis. J Virol Methods, 194(1-2): 39-45. doi 10.1016/j.jviromet.2013.08.002

Freitas, L.A., Leme, R.A., Saporiti, V., Alfieri, A.A., and Alfieri, A.F. (2019). Molecular analysis of the full-length F gene of Brazilian strains of canine distemper virus shows lineage co-circulation and variability between field and vaccine strains. Virus Res, 264: 8-15. doi 10.1016/j.virusres.2019.02.009

International Committee on Taxonomy of Viruses (ICTV):http://ictv.global/taxonomy/"

Joshi, V.R., Bhanderi, B.B., Nimavat, V.R., Jhala, M.K., and Desai, D.N. (2022). Comparison of lateral flow assay and RT-PCR for detection of canine distemper virus in dogs. Indian J Vet Sci Biotechnol, 18(3): 79-83.

Kadam, R.G., Karikalan, M., Siddappa, C.M., Mahendran, K., Srivastava, G., Rajak, K.K., Bhardwaj, Y., Varshney, R., War, Z.A., Singh, R., Ghosh, M., Beena, V., Pawde, A.M., Singh, K.P., and Sharma, A.K. (2022). Molecular and pathological screening of canine distemper virus in Asiatic lions, tigers, leopards, snow leopards, clouded leopards, leopard cats, jungle cats, civet cats, fishing cat, and jaguar of different states, India. Infect Genet Evol, 98: 105211. doi: 10.1016/j.meegid.2022.105211.

Karki, M., Rajak, K.K., and Singh, R.P. (2022). Canine morbillivirus (CDV): a review on current status, emergence and the diagnostics. Virus Dis, 33(3), 309-321. doi 10.1007/s13337-022-00779-7

Karki, M., Rajak, K.K., Singh, P., Fayaz, A., Kiran, Kumar, A., Bhatt, M., Rai, V., Einstein, C., Yadav, A.K., and Singh, R.P. (2023). Optimization of lateral flow assay for Canine morbillivirus detection and the application of the strip as sample substitute. J Immunol Methods, 514: 113438. doi 10.1016/j.jim.2023.113438

Kennedy, J.M., Earle, J.A.P., Omar, S., Abdullah, H., Nielsen, O., Roelke-Parker, M.E., and Cosby, S.L. (2019). Canine and phocine distemper viruses: global spread and genetic basis of jumping species barriers. Viruses, 11(10): 944. doi 10.3390/v11100944

Lednicky, J.A., Meehan, T.P., Kinsel, M.J., Dubach, J., Hungerford, L.L., Sarich, N.A., Witecki, K.E., Braid, M.D., Pedrak, C. and Houde, C.M. (2004). Effective primary isolation of wild-type canine distemper virus in MDCK, MV1 Lu and Vero cells without nucleotide sequence changes within the entire haemagglutinin protein gene and in subgenomic sections of the fusion and phospho protein genes. J Virol Methods, 118(2): 147-157. doi 10.1016/j.jviromet.2004.02.004.

Li, Z., Zhang, Y., Wang, H., Jin, J. and Li, W. (2013). Sandwich-dot enzyme-linked immunosorbent assay for the detection of canine distemper virus. Can J Vet Res, 77(4), 303-308.

Liu, D.F., Liu, C.G., Tian, J., Jiang, Y.T., Zhang, X.Z., Chai, H.L., Yang, T.K., Yin, X.C., Zhang, H.Y., Liu, M., Hua, Y.P., and Qu, L.D. (2015). Establishment of reverse transcription loop-mediated isothermal amplification for rapid detection and differentiation of canine distemper virus infected and vaccinated animals. Infection, genetics and evolution. Infect Genet Evol, 32: 102-106. doi 10.1016/j.meegid.2015.03.002

Loots, A.K., Mokgokong, P.S., Mitchell, E., Venter, E.H., Kotze, A. and Dalton, D.L. (2018). Phylogenetic analysis of canine distemper virus in South African wildlife. PloS one, 13(7): 0199993. doi 10.1371/journal.pone.0199993

McCarthy, A.J., Shaw, M.A. and Goodman, S.J. (2007). Pathogen evolution and disease emergence in carnivores. Proc Biol Sci, 274(1629): 3165-3174. doi 10.1098/rspb.2007.0884

McDermott, I., Gilbert, M., Shah, M.K., Sadaula, A. and Anderson, N.E. (2023). Seroprevalence of canine distemper virus (CDV) in the free-roaming dog (*Canis familiaris*) population surrounding Chitwan National Park, Nepal. PloS one, 18(2): 0281542. doi 10.1371/journal.pone.0281542

Menezes, K.M.F., Damasceno, A.D., Souza, M. and Torres, B.B.J. (2023). Epizootiology of canine distemper in naturally infected dogs in Goiania, Brazil. Cienc Rural, 53.

Mourya, D.T., Yadav, P.D., Mohandas, S., Kadiwar, R.F., Vala, M.K., Saxena, A.K., Shete-Aich, A., Gupta, N., Purushothama, P., Sahay, R.R., Gangakhedkar, R.R., Mishra, S.C.K., and Bhargava, B. (2019). Canine distemper virus in Asiatic lions of Gujarat state, India. Emerg Infect Dis, 25(11): 2128-2130. doi 10.3201/eid2511.190120

Nikolin, V.M., Olarte-Castillo, X.A., Osterrieder, N., Hofer, H., Dubovi, E., Mazzoni, C.J., Brunner, E., Goller, K.V., Fyumagwa, R.D., Moehlman, P.D., Thierer, D. and East, M.L. (2017). Canine distemper virus in the Serengeti ecosystem: molecular adaptation to different carnivore species. Mol Ecol, 26(7): 2111-2130. doi 10.1111/mec.13902

Orvell, C., Sheshberadaran, H. and Norrby, E. (1985). Preparation and characterization of monoclonal antibodies directed against four structural components of canine distemper virus. J Gen Virol, 66(Pt 3), 443-456. doi 10.1099/0022-1317-66-3-443

Panzera, Y., Sarute, N., Iraola, G., Hernández, M. and Pérez, R. (2015). Molecular phylogeography of canine distemper virus: Geographic origin and global spreading. Mol. Phylogenetics Evol, 92: 147-154. doi 10.1016/j.ympev.2015.06.015

Pope, J.P., Miller, D.L., Riley, M.C., Anis, E. and Wilkes, R.P. (2016). Characterization of a novel Canine distemper virus causing disease in wildlife. J Vet Diagn Invest, 28(5): 506-513. doi 10.1177/1040638716656025

Rendon-Marin, S., da Fontoura Budaszewski, R., Canal, C.W., and Ruiz-Saenz, J. (2019). Tropism and molecular pathogenesis of canine distemper virus. Virol J, 16(1): 30. doi 10.1186/s12985-019-1136-6

Romanutti, C., Keller, L., La Torre, J., Panzera, Y., Fuques, E., Pérez, R. and Gallo Calderon, M. (2020). Virus isolation and full-length genome sequencing of a representative canine distemper virus wild type strain of the South America 2 clade. J Virol Methods, 279: 113857. doi 10.1016/j.jviromet.2020.113857

Sarute, N., Pérez, R., Aldaz, J., Alfieri, A.A., Alfieri, A.F., Name, D., Llanes, J., Hernández, M., Francia, L. and Panzera, Y. (2014). Molecular typing of canine distemper virus strains reveals the presence of a new genetic variant in South America. Virus genes, 48(3): 474-478. doi 10.1007/s11262-014-1054-z

Si, W., Zhou, S., Wang, Z. and Cui S.J. (2010). A multiplex reverse transcription-nested polymerase chain reaction for detection and differentiation of wild-type and vaccine strains of canine distemper virus. Virol J, 7: 86. doi 10.1186/1743-422X-7-86.

Swati, Deka, D., Uppal, S.K., and Verma, R. (2015). Isolation and phylogenetic characterization of Canine distemper virus from India. Virus Dis, 26(3): 133-140. doi 10.1007/s13337-015-0256-x

Taques, I.I., Morgado, T.O., Braga, Í.A., Paz, R.C., Corrêa, S.H., Fritzen, J.T. and Aguiar, D.M. (2018). Antibodies against canine distemper virus, parvovirus and Ehrlichia spp. in wild captive carnivores in midwestern Brazil. Pesqui Vet Bras, 38: 1681-1684.

Trejo-Avila, L.M., Morales-Martínez, M.E., Ricque-Marie, D., Cruz-Suarez, L.E., Zapata-Benavides, P., Moran-Santibañez, K. and Rodríguez-Padilla, C. (2014). In vitro anti-canine distemper virus activity of fucoidan extracted from the brown alga Cladosiphonokamuranus. Virus Dis, 25(4): 474-480. doi 10.1007/s13337-014-0228-6

von Messling, V., Zimmer, G., Herrler, G., Haas, L. and Cattaneo, R. (2001). The hemagglutinin of canine distemper virus determines tropism and cytopathogenicity. J Virol, 75(14): 6418-6427. doi 10.1128/JVI.75.14.6418-6427.2001

Weckworth, J.K., Davis, B.W., Dubovi, E., Fountain-Jones, N., Packer, C., Cleaveland, S., Craft, M.E., Eblate, E., Schwartz, M., Mills, L.S., and Roelke-Parker, M. (2020). Cross-species transmission and evolutionary dynamics of canine distemper virus during a spillover in African lions of Serengeti National Park. Mol Ecol, 29(22): 4308-4321. doi 10.1111/mec.15449

Yadav, A.K., Rajak, K.K., Kumar, A., Bhatt, M., Chakravarti, S., Muthu, S., Dubal, Z.B., Khulape, S., Yousuf, R.W., Rai, V., Kumar, B., Muthuchelvan, D., Gupta, P.K., Singh, R.P. and Singh, R. (2021). Replication competence of canine distemper virus in cell lines expressing signaling lymphocyte activation molecule (SLAM) of goat, sheep and dog origin. Microb Pathog, 156: 104940. doi 10.1016/j.micpath.2021.104940

Yuan, C., Liu, W., Wang, Y., Hou, J., Zhang, L. and Wang, G. (2017). Homologous recombination is a force in the evolution of canine distemper virus. PloS one, 12(4): 0175416. doi 10.1371/journal.pone.0175416.

Chapter 14

African Swine Fever

Pankaj Deka[1] *and Sangeeta Das*[2]

[1]*Assistant Professor cum Virologist, Department of Veterinary Microbiology, College of Veterinary Science, Assam Agricultural University, Khanapara, Guwahati – 781 022, Assam, India*

[2]*Veterinary Officer, Institute of Veterinary Biologicals, A.H. and Veterinary Department, Assam, Khanapara, Guwahati – 781022, Assam, India*

e-mail: [1]*drpankajaau@gmail.com,* [2]*sangeetadas31280@gmail.com*

ABSTRACT

African swine fever (ASF) is a World Organization for Animal Health (OIE) notifiable, highly contagious, haemorrhagic viral disease of both domestic pigs and wild boars. It is caused by ASF virus (ASFV), the only known DNA arbovirus, of the family Asfarviridae and genus Asfivirus. ASF is endemic in Africa and has spread throughout Europe and Asia. It has a high fatality rate that challenges the pig industry worldwide and negatively impacts the regional and national trade of live pigs and pork products. To date, there is no effective vaccine to control ASF outbreaks. The ASF control and eradication measures are based on early detection and strict biosecurity measures. In this chapter, we have described the etiology, the epidemiological cycles including the potential vectors, the global distribution and current situation, the disease, and the control methods of ASF.

Keywords: *African swine fever (ASF), DNA arbovirus, Asfarviridae, Domestic pigs, Wild boars.*

Introduction

African swine fever (ASF) is a highly contagious and devastating viral disease affecting both domestic and wild swine of all ages. The causative agent, ASF virus (ASFV) is the only large double-stranded DNA virus belonging to the family *Asfarviridae* and genus *Asfivirus*, and the only known DNA arbovirus (Alonso *et al.*, 2018; Dixon *et al.*, 2019). The ASFV genome contains many open reading frames (ORFs), ranging from 151 to 167, and encodes more than 50 functional proteins (Yanej *et al.*, 1995; Andres *et al.*, 2002; Dixon *et al.*, 2013). ASF was first identified in 1921 in Kenya and since then it has remained endemic in Africa and has spread throughout Europe and Asia (Montgomery, 1921; Penrith *et al.*, 2013). Domestic pigs (*Sus scrofa domesticus*), Eurasian wild boars (*Sus scrofa*), warthogs (*Phacochoerus africanus*), bushpigs (*Potamochoerus larvatus*), and giant forest hogs (*Hylochoerusmeinert zhageni*) are all susceptible to ASFV infections. The virus is highly stable and transmission occurs *via* direct contact with infected animals, indirect contact *via* contaminated fomites pork products, and by the biological vector soft ticks (*Ornithodoros* spp) (Jori and Bastos, 2009; Jori *et al.*, 2013). Currently, ASFV can be classified into 24 genotypes based on the B646L gene, which encodes the viral capsid protein p72. All 24 ASFV genotypes have been detected in Africa, and only genotypes I and II are reported outside Africa (Boshoff *et al.*, 2007). Depending on the virulence of the ASFV strain, clinical forms vary from per-acute or acute form (highly virulent strains), sub-acute form (moderately virulent strains), and chronic form (low virulent strains) (Nurmoja *et al.*, 2017; Pikalo *et al.*, 2020). There is no approved ASFV vaccine available, and current control measures rely on early detection, strict quarantine, biosecurity measures, and culling procedures (Blome *et al.*, 2020).

Etiology

The causative agent of ASF is ASF virus (ASFV), a large, enveloped, double-stranded DNA (dsDNA) of the genus Asfivirus, the only member of the *Asfarviridae* family and the only DNA virus transmitted by arthropod vectors (Alonso *et al.*, 2018). Recently, the International Committee on Taxonomy of Viruses (ICTV), 2019 grouped ASFV with nucleocytoplasmic DNA viruses (NCLDVs), a distinct monophyletic group of viruses. This tentative NCLDVs group contains Poxviridae, Iridoviridae, Asfarviridae, Phycodnaviridae, Mimiviridae, Ascoviridae, and Marseilleviridae. The ASFV has unique features from other NCLDVs while sharing similarities with them regarding structure, genomic organization, and replication traits (Andres *et al.*, 2020).

Structure

The virion has a complex multi-layer structure with an interior nucleoprotein core (70-100 nm in diameter), surrounded by an internal lipid layer and an icosahedral capsid with 1892–2172 capsomers and an outer lipid envelope. The ASFV genome is a linear ds DNA molecule with a length of 170 to 195 kb. The

ASFV genome contains 151 to 167 open reading frames (ORF) and includes inverted repeat sequences in terminal position and a hairpin structure sharing a similar genome organization with that of Poxvirus and Iridovirus. The ORFs encode structural proteins and multiple enzymes essential for virus replication and play important roles in host-cell interaction. The virus encodes 50 different structural proteins, such as pp220, pp62, p72, p54, p30, p10, p12, p14.5, p17 and CD2v, along with a number of non-structural proteins. The inner protein layer comprises proteins derived from the viral polyproteins pp220 (p5, p14, p34, p37, and p150), pp62 (p35, p15, and p8) and pS273R. The ASFV polyprotein pp220 plays a key role in core assembly and p15 is the major capsid protein of the inner layer. The outer capsid forms a hexagonal lattice and is composed of 8280 copies of major capsid protein p72, arranged in trimers, and 60 copies of a penton protein at the vertices. PS273R is an enzyme that processes viral polyproteins and plays a role in virus assembly. The p12 protein is the viral attachment protein associated with cellular protein receptors, thus promoting virus attachment to host cells. The p17 protein is essential for capsid assembly. The p54 and p30 proteins are important antigenic structural proteins. The CD2v protein is responsible for the haemadsorbing property of the viral particle (Salas and Andres, 2012; Alonso *et al.*, 2018; Wang *et al.*, 2019; Liu *et al.*, 2019; Karger *et al.*, 2019; Andres *et al.*, 2020). A schematic presentation of the ASFV virion structure is depicted in Figure 14.1.

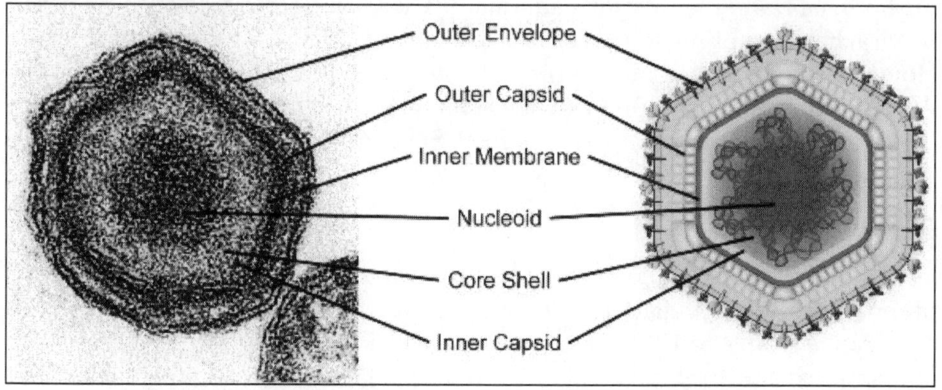

Figure 14.1: Electron Micrograph Structure (Left) and Schematic Overview (Right) of AFSV Particle.
(*Source*: https://doi.org/10.1016/j.virusres.2020.198099)

Antigenicity

The ASFV shows high genetic and antigenic diversity based on its genome size and enzymatic restriction profile. The multigene family (MGF) proteins are extensively distributed in the genome of ASFV and are the principal sites of insertions and deletions. The MGF proteins are generally classified into 5 different families including MGF 110, MGF 360, MGF 530/505, MGF 300, and MGF 100. Homologous gene recombination events change the number of genes in the MGF leading to variations between ASFV isolates (Teklue *et al.*, 2020). A large diversity of ASFV isolates is also attributed to a change in the number of amino acids in tandem repeats detected in 14 ASFV proteins. The inhibition of the haemadsorption properties of the virus using sera of infected animals also distinguishes ASFV isolates. Several studies report that haemadsorption inhibitory antibodies can neutralize ASFV infection in vitro and partially protect against ASFV challenge in vivo.

Genetic Diversity and Typing

Genotyping of ASFV strains has been based on the nucleotide sequence of p54, p72, and, Pb602L genes. Currently, based on the major coat protein p72 (B646L), ASFV has been classified into 24 different genotypes (I to XXIV), which fall into three main lineages. However, strain virulence cannot be predicted by p72-based genetic typing. All of the 24 genotypes have been identified in Africa, and ASFV genotypes I and II have spread to other continents. The first emergence of ASFV outside Africa occurred in an outbreak in Portugal which consisted of Genotype I ASFVs. Since then, it has been prevalent in Portugal, Spain, Cuba, Brazil, and other countries. Later, in 2007, genotype II ASFV was introduced in Georgia, and remains the current ASFV genotype circulating in Europe, Asia, and the Russian Federation. There are 8 serotypes of ASFV based on viral hemagglutinins(HAI) CD2-like protein (CD2v) as well as C-type lectin. However, p72-based genetic typing is not related to HAI serogroups; *e.g.*, the ASFV genotype I can simultaneously be classified as serogroup 1, 2, and 4 strains (Boshoff *et al.*, 2007; Gallardo *et al.*, 2009; Sanchez-Vizcaino *et al.*, 2013; Brown *et al.*, 2018; Wade *et al.*, 2019).

Stability

ASFV is highly stable in a wide range of physical and chemical factors in the environment as well as in raw pork products. The virus can remain stable in a pH range from 4 to 10. The virus can survive in blood for 18 months at room temperature and for 6 years at 4 °C, in skin fat for 300 days and, even a few years in frozen carcasses. ASFV can remain infectious for 18 days in salami, 60 days in pork belly, 83 days in tenderloin, 15 weeks in frozen meat, 6 months in cured ham, and 399 days in Parma ham. ASFV also remains viable for 4 days in urine and 3 days in feces at 37°C. Experimental studies have shown that ASFV can survive up to 112 days in forest soil and for 50-176 days in stagnant water. Due to the sturdy nature of ASFV, it can survive under extreme conditions (Mebus *et al.*, 1993; Penrith and Vosloo, 2009; Oganesyan *et al.*, 2013; Bellini *et al.*, 2016; Davies *et al.*, 2017; Alonso *et al.*, 2018; Petrini *et al.*, 2019; Probst *et al.*, 2019; Carlson *et al.*, 2020; Mazur-Panasiuk and WoŸniakowski, 2020).

Virus Inactivation

ASFV is inactivated by heating at 56°C for 70 min and 60°C for at least 20 min. Chemical compounds such as formaldehyde (1 per cent), sodium hypochlorite (0.5 per cent), sodium or calcium hydroxide (0.5-1 per cent), sodium hydroxide (0.2-1 per cent), caustic soda solution (2 per cent), and quaternary ammonium compounds (0.003 per cent) are shown to be effective in inactivation of ASFV. Amongst all, 2 percent caustic soda solution is the strongest virucidal agent to deactivate ASFV. ASFV is sensitive to glutaraldehyde, formic acid, phenols-lysol, lysephoform, and creolin. Multi-constituent compounds such as Virkon (1:100), lysoformin, and desoform are effective at ASFV inactivation. The OIE has recommended 8/1000 sodium hydroxide (30 min), 2.3 percent hypochlorite (3 min), 3/1000 formalin (30 min), and 3 percent n-phenyl phenol and iodine compound (30 min) for the inactivation of ASFV (EFSA, 2009; Juszkiewicz *et al.*, 2020; Tanneberger *et al.*, 2021; Pan *et al.*, 2021).

Virus Replication

ASFV infects the mononuclear-phagocytic system and primarily replicates in monocytes and macrophages. Virus entry takes place through clathrin-mediated and dynamin-dependent endocytosis and micropinocytosis. The clathrin-mediated endocytosis is a low pH- and temperature-dependent process. No specific cellular receptors and attachment factors for ASFV entry into cells have been identified to date. However, several macrophage receptors including CD163, CD45, and MHC II are assumed to play a possible role in virus entry. Upon entry, virus uncoating occurs by disassembling the capsid at the acidic pH of the endosomal lumen. After capsid degradation and membrane fusion, the pE248R enzyme mediates the release of the viral core into the cytoplasm. The viral core uses the ubiquitin-protease system to release viral genes. ASFV encodes its DNA replication-associated enzymes for transcription and replication. Viral genome replication and assembly take place in a specific region of the cytoplasm near the nucleus, called the virus factories. Icosahedral particles can be observed in the virus factory. After the formation of the capsid, the viral core is assembled under the inner envelope of the virus. The mature virus factory inside the cell is then transferred to the cell surface and is released from the infected cells by budding from the plasma membrane (Gomez-Villamandos *et al.*, 2013; Munoz-Monero *et al.*, 2015; Sanchez *et al.*, 2017; Simoes *et al.*, 2019). A schematic presentation of the ASFV replication cycle is depicted in Figure 14.2.

Transmission

Domestic pigs (*Sus scrofa domesticus*), Eurasian wild boars (*Sus scrofa*), warthogs (*Phacochoerus africanus*), bushpigs (*Potamochoerus larvatus*), and giant forest hogs (*Hylochoerus meinertzhageni*) are all susceptible to ASFV infections. The transmission of ASFV occurs *via* direct contact with infected animals, and indirect contact *via* contaminated fomites or environment, uncooked meat or

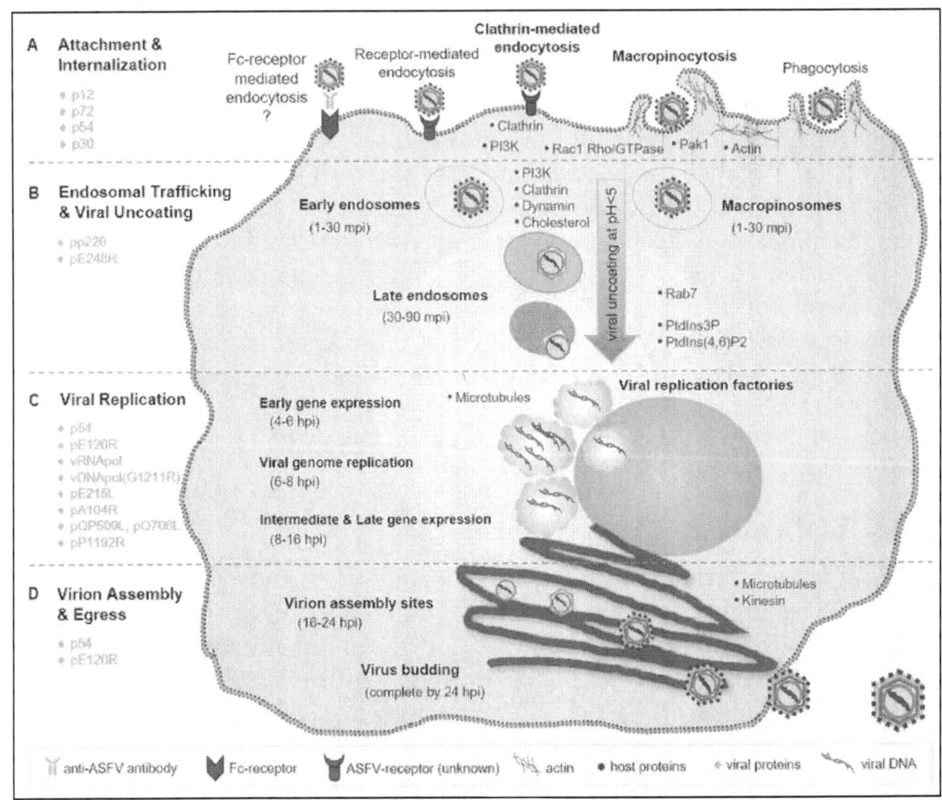

Figure 14.2: Replication of ASFV.
(*Source*: https://doi.org/10.3389/fvets.2020.00215**).**

pig-derived products from infected animals, and potential vectors, particularly soft ticks in the genus *Ornithodoros*, which serve as the biological vector and reservoir hosts for ASFV in both domestic and wild pigs (Figure 14.3) (Chenais *et al.*, 2019). Eight *Ornithodoros* species including *O. moubata* (southern and eastern Africa, as well as in Madagascar), *O. erraticus* (Mediterranean countries), *O. sonrai* (Western Africa), and *O. coriaceus*, *O. turicata*, *O. parkeri*, *O. puertoricensis* (Central and North America), and *O. savignyi* (North African desert) have been shown as competent ASFV vectors (Kleiboeker *et al.*, 1998; Boinas *et al.*, 2011; Golnar *et al.*, 2019). ASFV has been detected in several blood-feeding arthropods such as the swine lice (*Haematopinus suis*), the mite *Triatoma gerstaaekeri*, the stable flies (*Stomoxys calcitrans*), horse flies (*Tabanidae*) and the leeches; however, they are not associated with ASFV transmission (Sanchez Botija and Badiola, 1966; Petrov *et al.*, 2018; Karalyan *et al.*, 2019; Fila and WoŸniakowski, 2020).

Domestic pigs have been shown to shed infectious virus particles in all their secretions and excretions. Domestic pig-to-pig transmission occurs primarily through the oronasal route. The infected pigs remain carriers and shed the

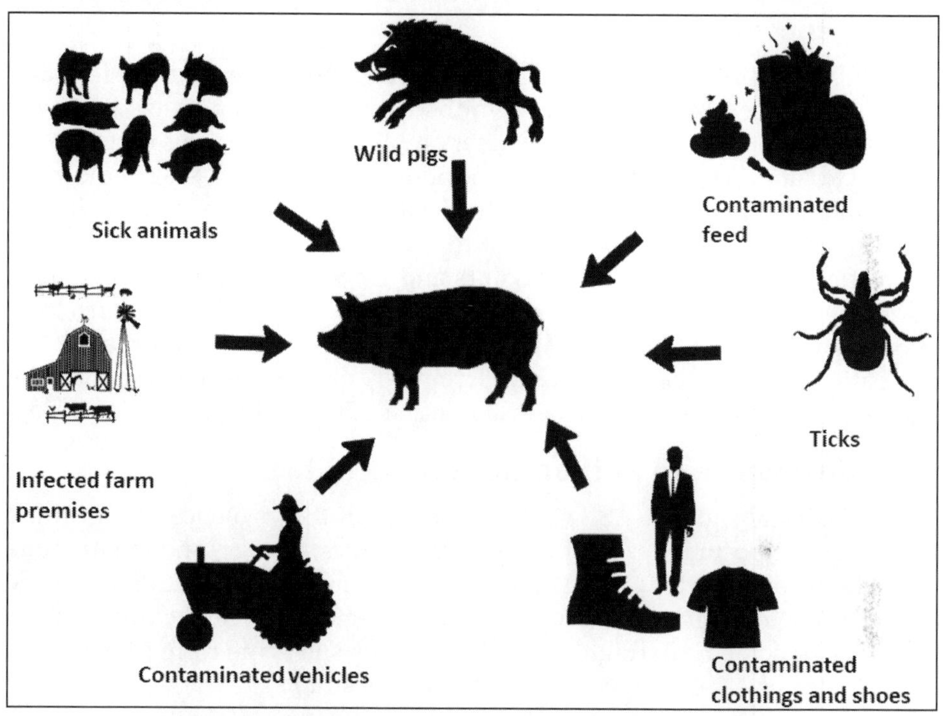

Figure 14.3: Potential Routes of ASFV Transmission.

virus into the environment. Since ASFV is persistent in blood and tissues for longer periods, infected carcasses can contribute to virus dissemination in the environment. Domestic pigs manifest the clinical disease in multiple ways from death with no signs to asymptomatic infection; however, acute haemorrhagic fever is common in domestic pigs. Pigs of all age groups are equally susceptible to ASFV infection (Blome *et al.*, 2013). Swill feeding is a widespread practice in traditional pig production systems with backyard and free-ranging pigs worldwide that contributes significantly to the spread of ASFV in domestic pigs (Wen *et al.*, 2019). The local wild boar population may acquire the infection from the outbreaks in domestic pigs through incidental spillover of ASFV. Further, wild boars play a significant epidemiological role in the transboundary dissemination of ASFV because of their natural dispersal ecology in quest of new territory. In wild boar habitat, carcasses maintain the infection cycles (Depner *et al.*, 2016). Generally, warthogs and bushpigs develop asymptomatic infections and serve as reservoirs of ASFV (Petrov *et al.*, 2018; Stahl *et al.*, 2019). On the other hand, importing ASFV-contaminated pork products is thought to be the most possible source and cause of infection across ASF-free countries. Although ASFV is shed in the genital secretions, no evidence of sexual transmission in pigs has been found. Transmission of ASFV from pregnant sows to foetus has also not been documented (Stoain *et al.*, 2019; Gebhardt *et al.*, 2022).

Pathogenesis

ASFV enters the body through the tonsils and respiratory system and replicates in the nasopharyngeal lymphoid tissues. A generalized viremia occurs 48-72 hours post-infection. Primary replication of the virus occurs in the monocytes and macrophages of the lymph nodes that are closest to the virus entry site and the virus spreads systemically through viremia. The viremia usually lasts for a longer period due to the lack of neutralizing antibodies. Thereafter, secondary replication occurs and the virus is detectable in almost all organs, including the lymph nodes, bone marrow, spleen, kidney, lungs, and liver. The virus affects the vascular endothelium and causes haemorrhages. The incubation period of the disease ranges from 3-19 days. Extensive haemorrhage and oedema cause death to ensue (Gabriel *et al.*, 2011; Sehl-Ewert *et al.*, 2022).

Clinical Signs and Pathomorphological Lesions

The clinical signs of ASF vary depending on the virulence of the strain, the route of exposure, dose of the virus, the species, the age, and the immune status of the pig infected. Usually, wild boars are more resistant than domestic pigs. The clinical signs observed in the ASF-infected pigs can be described as per-acute or acute form (highly virulent strains), sub-acute form (moderately virulent strains), and chronic form (low virulent strains).

Per-Acute or Acute Form

Highly virulent ASFV strains cause acute to per-acute forms of the disease with up to 100 percent mortality. Clinical signs are often non-specific and its clinical course is characterized by high fever, loss of appetite, reluctance to move, huddling, respiratory and gastrointestinal signs (vomiting and diarrhoea), haemorrhages in the skin and epistaxis (Figure 14.4). Sudden death can occasionally be seen without any symptoms of the disease. Some animals may develop conjunctivitis and nasal discharge. Pregnant sows may abort. Gross lesions may be absent in per-acute form. The acute form is characterizedby severe haemorrhages in lymph nodes, spleen, kidney, and occasionally in the heart. The spleen becomes enlarged (splenomegaly), infarcted, and friable. Excess of pleural, pericardial, or peritoneal fluid and petechiae in the epicardium and endocardium may be observed. There may be edema in the mesenteric structures of the colon and gall bladder.

Sub-Acute Form

Moderately virulent ASFV strains cause a sub-acute form of the disease with 70-80 percent mortality in young and 20-40 percent mortality in adults. Clinical signs and lesions are milder in sub-acute form. Affected pigs show moderate to high fever. The vascular changes mostly haemorrhage and odema can be more intense in the sub-acute form than in the acute form of the disease. Lymph nodes and kidneys show extensive hemorrhages. Spleen becomes haemorrhagic and enlarged. The lung may be congested, edematous, and present interstitial pneumonia. Hydropericardium, ascites, and multifocal oedema, are very characteristic in the wall of the gall bladder or the perirenal fat.

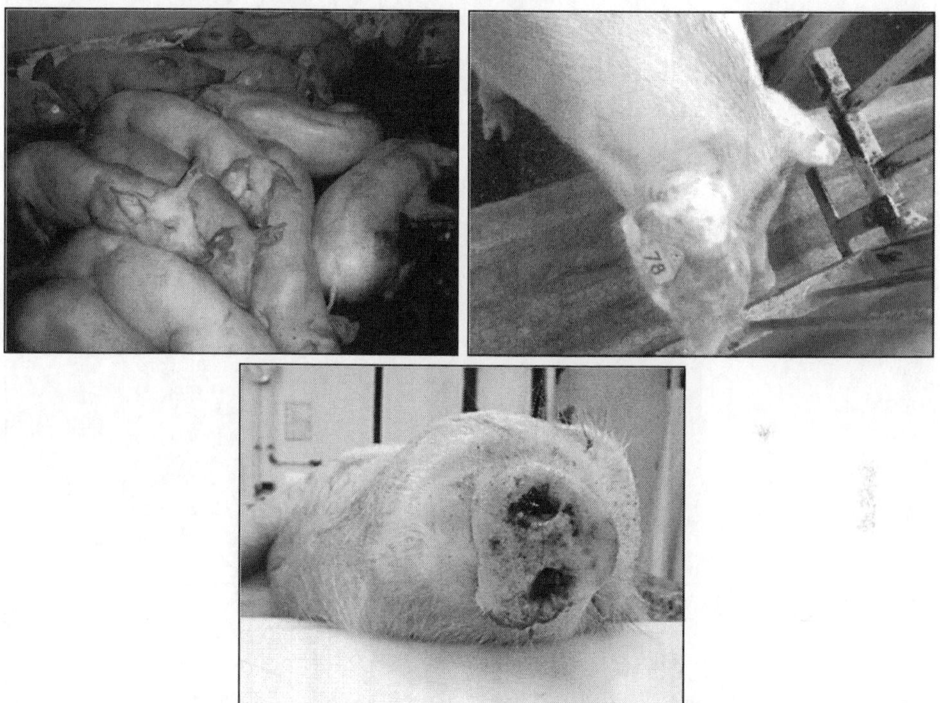

Figure 14.4: Clinical Signs of ASFV – Huddling, Skin Haemorrhages and Epistaxis. (*Source*: https://doi.org/10.1016/j.virusres.2020.198099).

Chronic Form

Low virulent ASFV strains cause a chronic form of the disease, which is characterized by the absence of vascular lesions and low mortality rates (10-30 per cent). The clinical manifestations of chronic ASF are retarded growth, weight loss, emaciation, arthritis, and skin ulcers associated with secondary bacterial infection. The chronic form is characterized by enlarged lymph nodes and spleen. Focal caseous necrosis may be observed in the lungs. Survivors or chronically infected pigs may play a role in the persistence of the disease (Sanchez-Vizcaino *et al.*, 2015; OIE, 2019; Pikalo *et al.*, 2019; Sanchez-Cordon *et al.*, 2019).

Epidemiological Cycles of ASFV

The epidemiology of ASF varies greatly throughout continents, countries, and areas depending on thepresence of wild reservoirs and competent tick vectors. The distinct epidemiologic cycles - the sylvatic, tick-pig, and wild-boar habitat cycles involve soft ticks, especially the *Ornithodoros* species, wild African pigs (mostly warthogs), domestic pigs, and pig-derived products or pork, and contaminated habitats (Figure 14.5.)

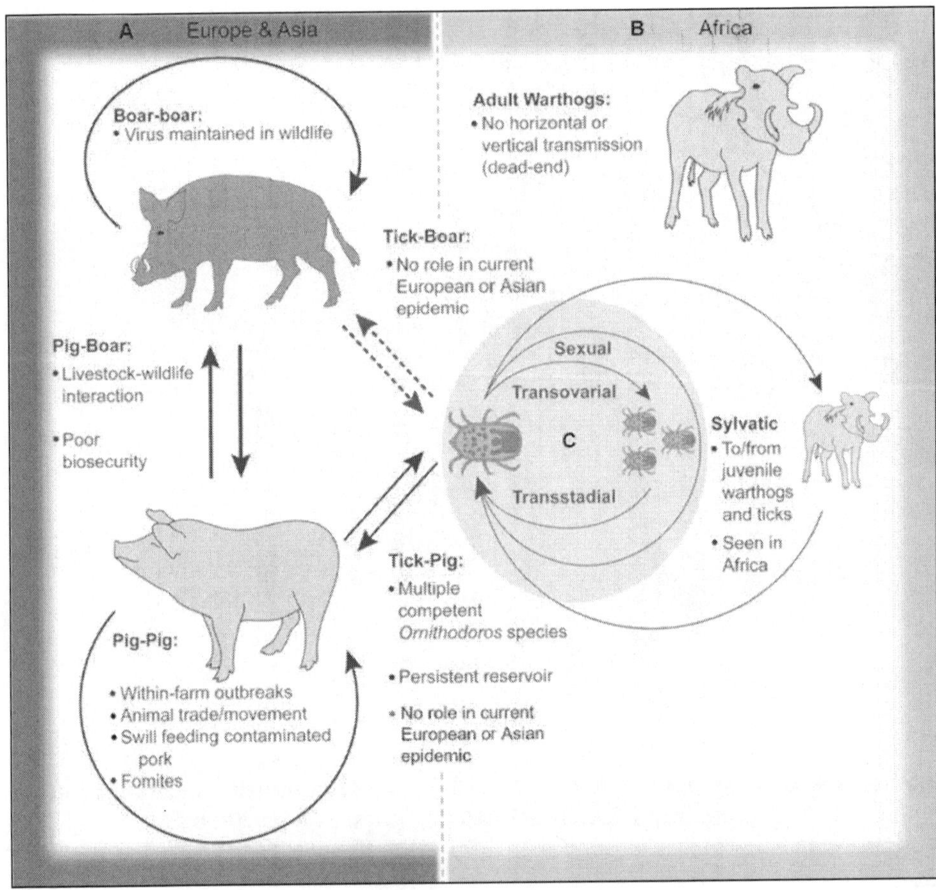

Figure 14.5: Epidemiological Cycles of ASFV
(*Source*: https://doi.org/10.3389/fvets.2020.00215).

Sylvatic Epidemiologic Cycle

ASF is beleived to have evolved in eastern and southern Africa in a sylvatic cycle between warthogs (*Phacochorus africanus*) and *Ornithodoros* soft argasid ticks that live in their burrows. Juvenile warthogs acquire infection from infected ticks. Warthogs become viremic but remain asymptomatic carriers of ASFV throughout their whole lives and cannot transmit the virus either horizontally or vertically among representative species. Adult warthogs usually do not maintain high levels of viremia and are dead-end hosts. Transmission to naive ticks occurs when they feed on the blood meal of viremic warthogs. The tick population maintains the virus through transtadial, transovarial, and sexual transmission even in the absence of viremic warthogs. Domestic pigs may get the infection from infected tick bites or warthog meat. Bushpigs and giant forest hogs play a minor role compared to warthogs in the sylvatic cycle (Thomson *et al.*, 1980; Jori and Bastos, 2009; Costard *et al.*, 2013; Penrith *et al.*, 2019).

Tick-Pig Epidemiologic Cycle

ASFV is transmitted to domestic pigs through *Ornithodoros* soft tick species - *O. moubata* complex in Africa and *O. erraticus* in Europe. Transmission to ticks occurs after feeding on viremic pigs. Soft ticks play an important role in maintenance of ASFV, as infected ticks of the *O. moubata* complex have been isolated from ASF outbreak pig farms in Africa and Madagascar, where no contact between domestic and wild pigs is reported. A similar pattern of ASFV transmission was reported in the Iberian Peninsula, where *O. erraticus* ticks were associated with the persistence of ASFV. The virus is maintained among domestic pigs, with the ticks serving as a reservoir without involving the warthogs (Wilkinson, 1984; Basto *et al.*, 2006; Boinas *et al.*, 2014; Diaz *et al.*, 2012; Costard *et al.*, 2013).

Wild Boar Habitat Epidemiologic Cycle

This cycle, reported in Africa, Europe, and Asia involves the Eurasian wild boar and their contaminated habitat with ASFV-infected wild boar carcasses. ASFV transmission occurs primarily *via* direct contact between infected and susceptible domestic and wild pigs, indirect contact via contaminated environment with wild boar, and indirect contact via the habitat. Wild boars were reported to maintain the virus without reintroduction from infected domestic pigs and capable of transboundary spread of ASFV (Laddomada *et al.*, 1994; Chenais *et al.*, 2019; Probst *et al.*, 2019).

Global Distribution

Historical Overview

ASF was first described by Montgomery in 1921 in Kenya and since then it has remained endemic in Africa, affecting up to 35 African countries. In 1957, ASF reached Europe in a Portugal pig farm, then was reintroduced in 1960 and rapidly spread to Spain, Italy, France, Sardinia, Malta, Belgium, and the Netherlands. In 1977, ASF was first reported in Russia and subsequently, during the 1970-80s, ASF emerged in Brazil, Cuba, the Caribbean Islands, and the Americas. The feeding of contaminated pork products is thought to be the cause of these outbreaks. However, by the mid-1990s, ASF was eradicated outside of Africa, except for an isolated outbreak in Portugal in 1999 and in the island of Sardinia, where it has been still endemic, since 1978 (Montgomery, 1921; Costard *et al.*, 2009; Sánchez-Vizcaíno *et al.*, 2012; Gallardo *et al.*, 2015).

Current Situation

In 2007, ASF re-emerged in the European continent, reporting the first outbreak in the Caucasus region of Georgia and subsequently spread into the Russian Federation and Eastern Europe, most likely through the movements of infected boars and illegal import of contaminated meat. In 2014, it reached the European Union, when the outbreak was reported in wild boars. Recently, in 2018, ASF spread to northwestern Europe in Belgium in wild boar. In August 2018, ASF was reported for the first time in Asia, causing devastating outbreaks in the People's Republic of China, and by the end of September 2019, ASFV was detected in neighboring countries including Cambodia, Laos, Thailand, Vietnam, Mongolia, the Philippines, North Korea, South Korea, Myanmar, Timor-Leste, and Indonesia. Twenty-four African countries have notified the OIE of the presence of ASF between 2018-2019, including Benin, Burkina Faso, Burundi, Cabo Verde, Central African Republic, Democratic Republic of the Congo, Republic of the Congo, the Gambia, Ghana, Guinea Bissau, Madagascar, Malawi, Mozambique, Namibia, Nigeria, Rwanda, Senegal, Sierra Leone, South Africa, Tanzania, Togo, Uganda, Zambia, and Zimbabwe (Bosch *et al.*, 2007; Gogin *et al.*, 2013; Bellini *et al.*, 2016; Frant *et al.*, 2017; Ge *et al.*, 2018; Le *et al.*, 2019; Kim *et al.*, 2019; Patil *et al.*, 2020). In May 2020, India reported the first occurrence of ASF. The global ASF situation for the period 2021-2023 is depicted in Figure 14.6. As of 2023, ASF reported countries in Asia and Pacific are China, Mongolia, Vietnam, Cambodia, Democratic People's Republic of Korea, Lao People's Democratic Republic, Myanmar, the Philippines, Republic of Korea, Timor-Leste, Indonesia, Papua New Guinea, India, Malaysia, Bhutan, Thailand, Nepal and Singapore (Figure 14.7).

Diagnosis

Anamnesis, clinical signs, and pathomorphological lesions are the basis fora provisional diagnosis of ASF, however, laboratory tests are essential to establish a definitive diagnosis (Arias *et al.*, 2018).

Detection of Virus/Viral Antigen

Methods for antigen detection include virus isolation, haemadsorption (HAD) test, fluorescent antibody test (FAT), and virus detection by Sandwich ELISA or lateral flow assays. Virus isolation and a positive HAD test are considered definitive for ASFV diagnosis. ASFV can be isolated in primary porcine cells (pig leukocyte cells, porcine alveolar macrophages, bone marrow cultures, porcine blood monocytes), monkey-derived established cell lines (Vero, COS.1 cells), and porcine established cell lines (IPAM, ZMAC, WSL cells). However, both assays are time-consuming and expensive. Furthermore, not all ASFV strains are hemadsorbing. The antigen-based ELISA or lateral flow assays and FAT have lower sensitivity and specificity (de Leon *et al.*, 2013; Beltran *et al.*, 2017; Sánchez-Vizcaíno and Heath, 2019; OIE, 2019). Due to these limitations, the molecular detection of antigen using polymerase chain reaction (PCR) is considered the 'gold standard' test for early detection of ASFV in clinical samples. Conventional and real-time PCR methods are rapid diagnostic tools

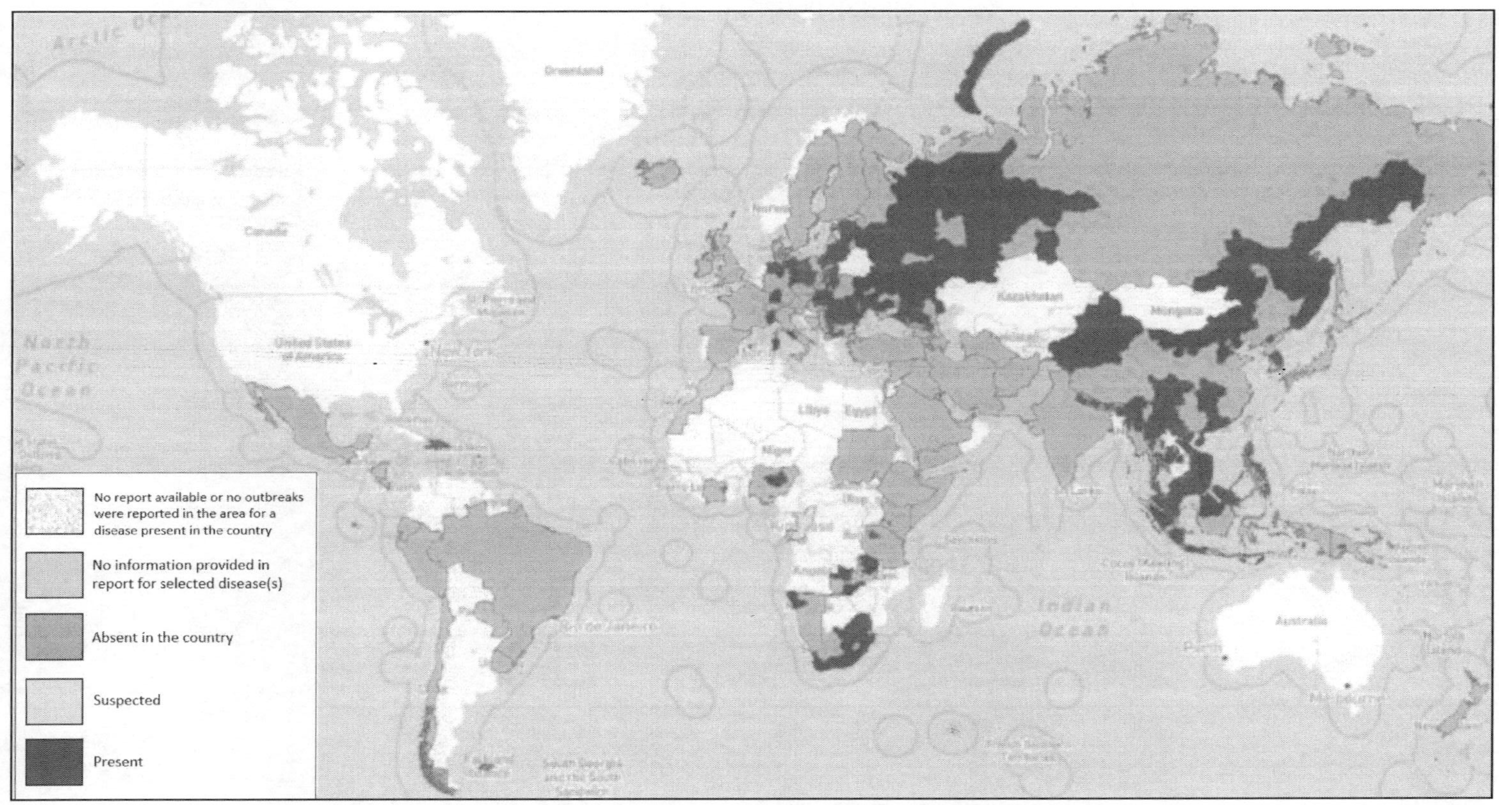

Figure 14.6: Global ASF Epidemic Situation (2021-2023)
[*Source*: asf-report25.pdf (woah.org)].

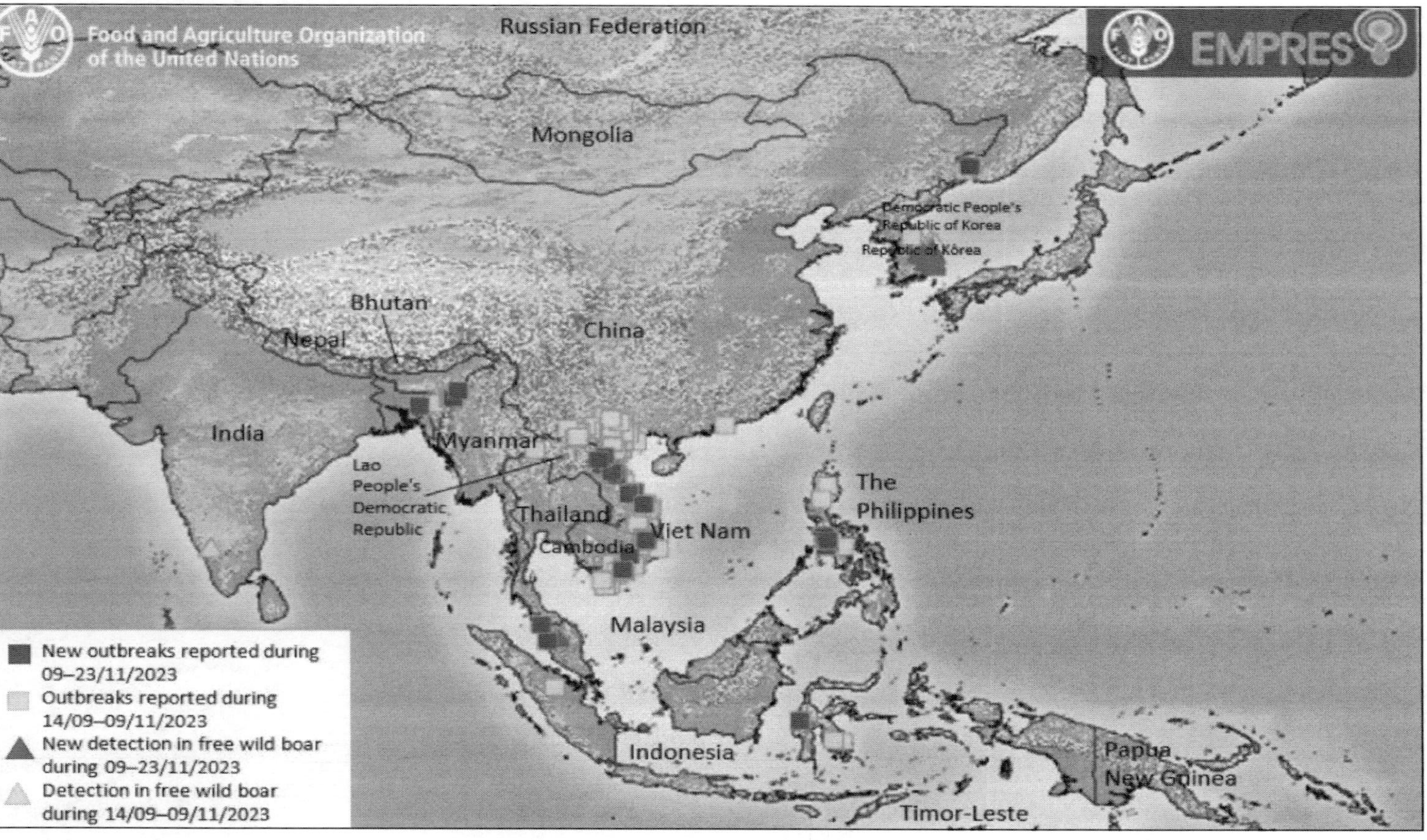

Figure 14.7: ASF Situation in Asia (2023)

(*Source*: https://www.fao.org/animal-health/situation-updates/asf-in-asia-pacific/en).

with superior sensitivity and specificity, and sample throughput. For these reasons, r-time quantitative PCR (qPCR) is recommended for screening and confirmatory testing during active ASF infection. Additionally, conventional and real-time PCR can be used as diagnostic methods to detect ASFV formats in *Ornithodoros* ticks. Furthermore, for simultaneous and differential detection of other swine pathogens such as classical swine fever virus (CSFV) multiplex PCR techniques have also been developed. Other alternative molecular tests, loop-mediated isothermal amplification (LAMP) are also described as ASFV diagnostic methods (Aguero *et al.*, 2003; Zsak *et al.*, 2005; Basto *et al.*, 2006; Fernandez-Pinero *et al.*, 2013).

Detection of Antibodies

There is no effective ASF vaccine to date, therefore, the presence of anti-ASFV antibodies indicates current or previous infection. ASFV-specific antibodies appear early in infection and in convalescent animals persist for months or years; the viremia also persists for long periods due to the lack of neutralizing antibodies (Beltran *et al.*, 2017; Sánchez-Vizcaíno and Heath, 2019). Methods for serological detection of ASFV-specific antibodies include indirect ELISAs, immunoblots (IBTs), indirect fluorescent antibody tests (IFATs), indirect immunoperoxidase tests (IPTs), and lateral flow tests (LFTs). Several of these tests are recommended by the OIE for disease surveillance. ELISA followed by immunoblotting is a recommended standard serological test often used for international trade purposes. However, these tests are limited by a low sensitivity in detecting early-stage infections (<7 days post-infection) and those infected by highly virulent strains where per-acute death occurs before induction of antibodies, necessitating the improvement of existing tools, to detect and diagnose the disease rapidly. Multiple commercial and in-house indirect ELISA formats are validated as fit-for-purpose by the OIE for high throughput ASF serological testing. For ELISA-positive serum, confirmatory testing using IFAT is recommended. IFAT provides a high level of specificity by allowing direct visualization of antibody reactivity with intracellular ASFV antigens. The ASFV IPT is similar to IFAT and has comparable sensitivity and specificity. Both IFAT and IPT are confirmatory serological tests for ELISA-positive samples recommended by the OIE (Pan *et al.*, 1982; Pastor *et al.*, 1990; Arias *et al.*, 1993; Perez-Filgueira *et al.*, 2006; Gallardo *et al.*, 2009; Cubillos *et al.*, 2013; Nieto-Pelegrin *et al.*, 2015; Sastre *et al.*, 2016; Kazakova *et al.*, 2017).

It is a foundational principle for any laboratory test procedure to collect proper specimens for accurate test results. Whole blood, serum, and tissues, mainly spleen, lymph nodes, bone marrow, lung, tonsil, and kidney are appropriate samples for laboratory testing of ASFV. In wild boar ASF surveillance programs, non-invasive sampling strategies - unpreserved field fecal samples, dried blood spots (DBS) on filter papers and swabs, and use of bait (ropes) for collection of oral fluid have been introduced for ASFV detection (OIE, 2019).

Control

Currently, neither vaccine nor treatment is available to control ASF. In the absence of an effective ASF vaccine, successful prevention and mitigation of ASF outbreaks becomes difficult. The broad geographic distribution of wild and feral pig populations, potential arthropod vectors capable of maintaining the virus, and the increasingly globalized nature of animal husbandry are further factors impeding disease control. In this context, ASF control strategies primarily focus on early detection, timely and reliable diagnosis, and implementation of strict quarantine and biosecurity measures, restriction on the movement of pigs, meat, or feeds, including surveillance, epidemiological investigations, and culling of infected herds. Trade restrictions also play a key role in minimizing a potential outbreak (Dixon *et al.*, 2020; Zhang *et al.*, 2023).

Government

The control effort of ASF requires a comprehensive set of measures covering the entire pork chain "from farm to fork". At the government level, territorial management and support policy are crucial for compliance with measures and to encourage all actors to participate in the prevention and control of ASF. An extensive ASF surveillance and monitoring network, blocking of transmission routes including slaughter orders, the establishment of restricted areas, movement restriction of pig and their products,and registration and inspection of pig transport vehicles are the goals of disease control initiatives. The establishment of a binding legal framework for control programs as well as a timely and appropriate compensation scheme are the most vital components of a disease control measure. Without a financial compensation policy to compensate farmers for the slaughter of infected pigs, they would sell all their sick pigs, resulting in the circulation of ASFV in slaughter houses and markets and increasing the risk of disease transmission, which is not conducive to the prevention and control of ASF. The lack of awareness of ASF control measures among farmers also hinders the control programs.

The establishment of veterinary teams and the development of policies by animal health authorities – regional biosafety regulations, trade regulations, test and slaughter policy, carcass disposal guidelines, and zonation determining an infected zone and surrounding buffer and control zones can reduce ASF outbreaks. Numerous national and/or governmental authorities have issued guidelines and documentation for the prevention and control of ASF. The FAO has recommended comprehensive contingency plans, strategies for ASFV detection and diagnosis, and action plans in the event of an ASF outbreak (FAO, 2011; Hoinville *et al.*, 2013; Beltran-Alcrudo *et al.*, 2017; EFSA, 2017).

Domestic Pigs and Breeding Farms

Pig farmers shoulder the most important responsibilities for controlling ASF. A sound biosecurity system including both on-farm and off-farm biosecurity practices is crucial to keep ASFV away from their farm premises. For this, it is recommended that farms should have adequate fencing to avoid entry of outside wild animals, avoid swill feeding (garbage feeding), comply strictly with biosecurity measures – proper disinfection of clothing and boots, feeding utensils, premises, *etc.*, controlled animal and human movements, disinfection of vehicles and equipment before entering the premises, proper disposal of carcasses and disinfection of carcass-hauling trucks. Additionally, pig farms should maintain a rigorous health monitoring program. A key component for controlling the disease involving a domestic cycle depends on early detection and the ability to respond to ASF outbreaks efficiently so that virus spread can be prevented. A veterinarian should be immediately notifiedupon spotting signs of disease. Strict quarantine measures should be practiced in both diseased and disease-free zones to prevent further spread of infections. Further, financial compensation policies will aid the farmers in overcoming the fear of economic losses and encourage them to participate in disease control programs (Dixon *et al.*, 2020).

Wild Pigs

The control and eradication measures for ASF in the wild boar cycle were implemented after the successful eradication of ASF from the wild boar population in the Czech Republic. The European Commission recommended a control strategy involving zonation determining the infected zone, surrounding buffer, and control zones. These zones are to be established immediately based on the ecological, epidemiological, and social context of the outbreak area. Carcasses should be searched and disposed of under strict biosecurity measures. Fencing should be applied to prevent the movement of wild boars and delineate the restricted area. Hunting activities should be banned in the infected and buffer zones to minimize the spread of ASFV. Strict wild boar depopulation strategies should be implemented in the control zone to reduce the wild boar densities (EFSA, 2018; Guberti *et al.*, 2018).

Vector Control

The control and eradication of softticks, which inhabit the warthog's burrows is difficult with the application of acaricides since these ticks are the biological vector and reservoir hosts for ASFV (Arias *et al.*, 2018). The soft ticks transmit ASFV through all stages of their life cycle. However, immunological control using anti-tick vaccines can be used as an alternative to the traditional application of acaricides (Manzano-Román *et al.*, 2007; Díaz-Martín *et al.*, 2015).

Vaccination

Despite decades of research, a broadly protective, safe, and efficacious vaccine against ASF is still not available. Different vaccine development strategies have been employed, but production of most vaccine candidates against ASFV is massively hampered by the genetic complexity of the virus, limited understanding of virus-host cell interactions, including mechanisms of immune evasion, identification of protective immunogens and virulence-related genes of ASFV, lack of development of neutralizing antibodies, ability of the virus to modulate host immune responses, and technical difficulties such as the lack of stable cell lines which is sufficiently susceptible to ASFV and does not permit further genetic adaptations within the virus genome. Three main approaches have been employed in the development of ASFV vaccine candidates - whole virus inactivated vaccines, live virus-vectored recombinant, subunit, and DNA vaccines, and live attenuated virus vaccines (LAVs) (Dixon *et al.*, 2020).

Studies have shown whole virus inactivated vaccines does not induce protective immunity (Blome *et al.*, 2014; Revilla *et al.*, 2018). Based on the immunoprotection capacity, live virus-vectored recombinant, subunit, and DNA vaccines target individual or cocktails of specific viral structural proteins such as p30 (encoded by CP204 L), p54 (E183L), p72 (B646L), pp62 (CP530R) and CD2v (EP402R) (Argilaguet *et al.*, 2012; Jancovich *et al.*, 2018; Sunwoo *et al.*, 2019; Goatley *et al.*, 2020). However, the results were not always conclusive. The inconsistency could be attributed to limited understanding of protective immunogens and virulence-related genes of ASFV which leads to variable protection to enhancement of disease and even accelerated mortality (Gaudreault and Richt, 2019). However, virus-vectored and subunit vaccines have a better safety profile than live attenuated vaccines and allows differentiation of infected from vaccinated animals (DIVA). However, most approaches yielded partial or no protection. Currently, there is no specific viral protein that can trigger complete immune protection. It is necessary to identify highly efficient epitopes that could stimulate stronger immune responses against ASFV (Dixon *et al.*, 2013; Reis *et al.*, 2017).

LAVs can confer great protection but virulence, antigen diversity, and the lack of strain cross-protective immunity, continue to affect these vaccines. LAVs are derived from naturally occurring ASFV strains of reduced virulence (*e.g.*, Non-HAD genotype I ASFV strain NH/P68 and OURT88/3) or virulent strains attenuated by extensive viral passaging in cell lines or by deleting multiple virulence-related genes (Dixon *et al.*, 2013; Gallardo *et al.*, 2018; Reis *et al.*, 2020; Chen *et al.*, 2022; Zhang *et al.*, 2023). The safety of LAVs was improved by deleting ASFV virulence-associated genes - 9GL (B119L), UK (DP96R), CD2v (EP402R), DP148R, and different members of the multigene family (MGF) (O'Donnell *et al.*, 2016; Reis *et al.*, 2017; Monteagudo *et al.*, 2017; Borca *et al.*, 2020; Teklue *et al.*, 2020). However, modified LAVs have side effects causing arthritis, skin necrosis, and chronic infections. Studies have shown LAVs can confer protection against homologous parental virus challenge, but provide little to no cross-protection against heterologous virulent ASFV strains (Gallardo *et al.*, 2018; Sang *et al.*, 2020; Sereda *et al.*, 2022).

Conclusion

ASF is a highly contagious disease of domestic pigs and wild boars causing 100 percent mortality leading to significant health and socioeconomic consequences, especially in the absence of any available treatment or effective vaccine. Moreover, trade restrictions imposed on the affected countries can dramatically disrupt the global swine industry. Therefore, epidemiological tracing and early detection, stringent on-farm biosecurity measures, ban on swill feeding, vector control and wild boar control programs continue to be of significant priority. Furthermore, expanding studies on ASFV biology, gene functions, virus and cellular factors and protective immunity, will be critical for development of a rationally-designed safe and effective vaccine.

REFERENCES

Aguero, M., Fernandez, J., Romero, L., Sanchez Mascaraque, C., Arias, M. and Sanchez-Vizcaino, J.M. (2003). Highly sensitive PCR assay for routine diagnosis of African swine fever virus in clinical samples. J Clin Microbiol, 41: 4431-4434. doi: 10.1128/JCM.41.9.4431-4434.2003.

Alonso, C., Borca, M., Dixon, L., Revilla, Y., Rodriguez, F. and Escribano, J.M. (2018). ICTV Report Consortium. ICTV Virus Taxonomy Profile: ASFarviridae. J Gen Virol, 99: 613-614.

Andrés, G., Charro, D., Matamoros, T., Dillard, R.S., and Abrescia, N.G.A. (2020). The cryo-EM structure of African swine fever virus unravels a unique architecture comprising two icosahedral protein capsids and two lipoprotein membranes. J Biol Chem, 295(1): 1-12. doi: https://doi.org/10.1074/jbc.AC119.011196.

Andres, G., Garcia-Escudero, R., Salas, M.L. and Rodriguez, J.M. (2002). Repression of African Swine Fever Virus Polyprotein Pp220-Encoding Gene Leads to the Assembly of Icosahedral Core-Less Particles. J Virol, 76: 2654-2666.

Argilaguet, J.M., Pérez-Martín, E., Nofrarías, M., Gallardo, C., Accensi, F., Lacasta, A., Mora, M., Ballester, M., Galindo-Cardiel, I., López-Soria, S., Escribano, J.M., Reche, P.A. and Rodríguez, F. (2012). DNA vaccination partially protects against African swine fever virus lethal challenge in the absence of antibodies. PLoS One, 7(9): e40942. doi: 10.1371/journal.pone.0040942.

Arias, M., Escribano, J.M. and Sanchez-Vizcaino, J.M. (1993). Persistence of African swine fever antibody reactivity on ELISA and immunoblotting assays. Vet Rec, 133: 189-190. doi: 10.1136/vr.133.8.189

Arias, M., Jurado, C., Gallardo, C., Fernández-Pinero, J. and Sánchez-Vizcaíno, J.M. (2018). Gaps in African swine fever: Analysis and priorities. Transbound Emerg Dis, 65 (Suppl 1): 235-247. doi: https://doi.org/10.1111/tbed.12695

Basto, A.P., Nix, R.J., Boinas, F., Mendes, S., Silva, M.J., Cartaxeiro, C., Portugal, R.S., Leitão, A., Dixon, L.K. and Martins, C. (2006). Kinetics of African swine fever virus infection in *Ornithodoros erraticus* ticks. J Gen Virol, 87(Pt 7): 1863-1871. doi: 10.1099/vir.0.81765-0.

Basto, A.P., Portugal, R.S., Nix, R.J., Cartaxeiro, C., Boinas, F., Dixon, L.K., Leitão, A. and Martins, C. (2006). Development of a nested PCR and its internal control for the detection of African swine fever virus (ASFV) in *Ornithodoros erraticus*. Arch Virol, 151(4): 819-826. doi: 10.1007/s00705-005-0654-2.

Bellini, S., Rutili, D. and Guberti, V. (2016). Preventive measures aimed at minimizing the risk of African swine fever virus spread in pig farming systems. Acta Vet Scand, 58(1): 82. doi:10.1186/s13028-016- 0264-x.

Beltran-Alcrudo, D., Arias, M., Gallardo, C., Kramer, S. and Penrith, M.L. (2017). African Swine Fever: Detection and Diagnosis - A Manual for Veterinarians. FAO Animal Production and Health Manual No. 19. FAO, Rome, Italy 2017, 88.

Beltrán-Alcrudo, D.A.M., Gallardo, C., Kramer, S., *et al.* (2017). African Swine Fever: Detection and Diagnosis - A Manual for Veterinarians. 19th ed. Rome: Food and Agriculture Organization of the United Nations.

Blome, S., Franzke, K. and Beer, M. (2020). African swine fever - A review of current knowledge. Virus Res, 287: 198099. doi: https://doi.org/10.1016/j.virusres.2020.198099.

Blome, S., Gabriel, C. and Beer, M. (2013). Pathogenesis of African swine fever in domestic pigs and European wild boar. Virus Res, 173: 122-130. doi: 10.1016/j.virusres.2012.10.026.

Boinas, F., Ribeiro, R., Madeira, S., Palma, M., de Carvalho, I.L., Núncio, S. and Wilson, A.J. (2014). The medical and veterinary role of *Ornithodoros erraticus* complex ticks (Acari: Ixodida) on the Iberian Peninsula. J Vector Ecol, 39(2): 238-248. doi: 10.1111/jvec.12098. PMID: 25424252.

Boinas, F.S., Wilson, A.J., Hutchings, G.H., Martins, C. and Dixon, L.J. (2011). The persistence of African swine fever virus in field infected *Ornithodoros erraticus* during the ASF endemic period in Portugal. PLoS One, 6(5): e20383. doi: 10.1371/journal.pone.0020383.

Borca, M.V., Ramirez-Medina, E., Silva, E., Vuono, E., Rai, A., Pruitt, S., Holinka, L.G., Velazquez-Salinas, L., Zhu, J. and Gladue, D.P. (2020). Development of a highly effective African swine fever virus vaccine by deletion of the I177L gene results in sterile immunity against the current epidemic Eurasia strain. J Virol, 94(7): e02017-19. doi: 10.1128/JVI.02017-19.

Bosch, J., Rodríguez, A., Iglesias, I., Muñoz, M.J., Jurado, C., Sánchez-Vizcaíno, J.M., de la Torre, A. (2017). Update on the risk of introduction of African swine fever by wild boar into disease-free European union countries. Transbound Emerg Dis, 64(5): 1424-1432. doi: 10.1111/tbed.12527.

Boshoff, C.I., Bastos, A.D., Gerber, L.J. and Vosloo, W. (2007). Genetic characterisation of African swine fever viruses from outbreaks in southern Africa (1973-1999). Vet Microbiol, 121(1, 2): 45-55. doi: 10.1016/j.vetmic.2006.11.007

Brown, V.R. and Bevins, S.N. (2018). A Review of African swine fever and the potential for Introduction into the United states and the possibility of subsequent establishment in feral swine and native ticks. Front Vet Sci, 5: 11. doi: 10.3389/fvets.2018.00011.

Carlson, J., Fischer, M., Zani, L., Eschbaumer, M., Fuchs, W., Mettenleiter, T., Beer, M. and Blome S. (2020). Stability of African swine fever virus in soil and options to mitigate the potential transmission risk. Pathogens, 9: 977. doi: 10.3390/pathogens9110977.

Chen, L., Zhang, X., Shao, G., Shao, Y., Hu, Z., Feng, K., Xie, Z., Li, H., Chen, W., Lin, W., Yuan, H., Wang, H., Fu, J. and Xie, Q. (2022). Construction and evaluation of recombinant pseudorabies virus expressing African swine fever virus antigen genes. Front Vet Sci, 9: 832255. doi: 10.3389/fvets.2022.832255.

Chenais, E., Ståhl, K., Guberti, V. and Depner, K. (2019). Identification of wild boar–habitat epidemiologic cycle in African swine fever epizootic. Emerg Infect Dis, 24(4): 810-812. doi:10.3201/eid2404.172127.

Costard, S., Mur, L., Lubroth, J., Sanchez-Vizcaino, J.M. and Pfeiffer, D.U. (2013). Epidemiology of African swine fever virus. Virus Res, 173: 191-197. doi: 10.1016/j.virusres.2012.10.030.

Costard, S., Wieland, B., de Glanville, W., Jori, F., Rowlands, R., Vosloo, W., Roger, F., Pfeiffer, D.U. Dixon, L.K. (2009). African swine fever: how can global spread be prevented? Philos Trans R Soc Lond B Biol Sci, 364(1530): 2683-2696. doi: 10.1098/rstb.2009.0098.

Cubillos, C., Gómez-Sebastian, S., Moreno, N., Nuñez, M.C., Mulumba-Mfumu, L.K., Quembo, C.J., Heath, L., Etter, E.M., Jori, F., Escribano, J.M. and Blanco, E. (2013). African swine fever virus serodiagnosis: a general review with a focus on the analyses of African serum samples. Virus Res, 173(1):159-167. doi: 10.1016/j.virusres.2012.10.021.

Davies, K., Goatley, L.C., Guinat, C., Netherton, C.L., Gubbins, S., Dixon, L.K. and Reis, A.L. (2017). Survival of African swine fever virus in excretions from pigs experimentally infected with the Georgia 2007/1 isolate. Transbound Emerg Dis, 64(2): 425-431. doi: 10.1111/tbed.12381.

de Leon, P., Bustos, M.J. and Carrascosa, A.L. (2013). Laboratory methods to study African swine fever virus. Virus Res, 173: 168-179. doi: 10.1016/j.virusres.2012.09.013

Depner, K., Staubach, C., Probst, C., Globig, A., Blome, S., Dietze, K., Sauter Louis, C. and Conraths F.J. (2016). African swine fever - epidemiological considerations and consequences for disease control. Tierarztl Umschau, 71: 72-78.

Diaz, A.V., Netherton, C.L., Dixon, L.K. and Wilson, A.J. (2012). African swine fever virus strain Georgia 2007/1 in *Ornithodoros erraticus* ticks. Emerg Infect Dis, 18: 1026-1028. doi: 10.3201/eid1806.111728.

Díaz-Martín, V., Manzano-Román, R., Obolo-Mvoulouga, P., Oleaga, A. and Pérez-Sánchez, R. (2015). Development of vaccines against Ornithodoros soft ticks: An update. Ticks Tick Borne Dis, 6(3): 211-220. doi:10.1016/j.ttbdis.2015.03.006.

Dixon, L.K., Chapman, D.A., Netherton, C.L. and Upton, C. (2013). African swine fever virus replication and genomics. Virus Res, 173: 3-14.

Dixon, L.K., Stahl, K., Jori, F., Vial, L. and Pfeiffer, D.U. (2020). African swine fever epidemiology and control. Annu Rev Anim Biosci, 8: 221-246. doi: https://doi.org/10.1146/annurev-animal-021419-083741.

Dixon, L.K., Sun, H. and Roberts, H. (2019). African swine fever. Antivir Res, 165: 34-41.

EFSA (European Food Safety Authority). (2009). Scientific report prepared by Sánchez-Vizcaíno, J.M., Martínez-López, B., Martínez-Avilés, M., Martins, C., Boinas, F., Vial, L., Michaud, V., Jori, F., Etter, E., Albina, E. and Roger, F. on African Swine Fever. 1-141.

EFSA (European Food Safety Authority) (2017). Scientific report on epidemiological analyses on African swine fever in the Baltic countries and Poland prepared by Cortiñas Abrahantes J, Gogin A, Richardson J, Gervelmeyer A. EFSA Journal 15(4732):1-73. doi: 10.2903/j.efsa.2017.4732.

EFSA (2018). Scientific Report on the Epidemiological Analyses of African Swine Fever in the European Union (November 2017 Until November 2018) prepared by Boklund, A., Cay, K., Depner, Z., Földi, V., Guberti, M., Masiulis, A. et.al., European Food Safety Authority), p. 106. doi: https://doi.org/10.2903/j.efsa.2018.5494

FAO (Food and Agricultural Organization of the United Nations). (2011). Good Emergency Management Practices: The essentials. Honhold N, Douglas I, Geering W, Shimsoni A, Lubroth J (editors). FAO Animal Production and Health Manual No. 11. Rome, Italy 2011.

Fernández-Pinero, J., Gallardo, C., Elizalde, M., Robles, A., Gómez, C., Bishop, R., Heath, L., Couacy-Hymann, E., Fasina, F.O., Pelayo, V., Soler, A. and Arias, M. (2013). Molecular diagnosis of African swine fever by a new real-time PCR using universal probe library. Transbound Emerg Dis, 60(1): 48-58. doi: 10.1111/j.1865-1682.2012.01317. x.

Fila, M. and WoŸniakowski, G. (2020). African swine fever virus - the possible role of flies and other insects in virus transmission. J Vet Res, 64(1): 1-7. doi: https://doi.org/10.2478/jvetres-2020- 0001.

Frant, M., WoŸniakowski, G. and Pejsak, Z. (2017). African swine fever (ASF) and ticks. No risk of tick-mediated ASF spread in Poland and Baltic states. J Vet Res, 61(4): 375-380. doi:10.1515/jvetres-2017-0055

Gabriel, C., Blome, S., Malogolovkin, A., Parilov, S., Kolbasov, D., Teifke, J.P. and Beer, M. (2011). Characterization of African swine fever virus Caucasus isolate in European wild boars. Emerg Infect Dis, 17(12): 2342-2345. doi: https://doi.org/10.3201/eid1712.110430.

Gallardo, C., Mwaengo, D.M., Macharia, J.M., Arias, M., Taracha, E.A., Soler, A., Okoth, E., Martín, E., Kasiti, J. and Bishop, R.P. (2009). Enhanced discrimination of African swine fever virus isolates through nucleotide sequencing of the p54, p72, and pB602L (CVR) genes. Virus genes, 38(1): 85-95. doi: https://doi.org/10.1007/s11262-008-0293-2

Gallardo, C., Reis, A.L., Kalema-Zikusoka, G., Malta, J., Soler, A., Blanco, E., Parkhouse, R.M. and Leitão, A. (2009). Recombinant antigen targets for serodiagnosis of African swine fever. Clin Vaccine Immunol, 16(7): 1012-1020. doi: 10.1128/CVI.00408-08.

Gallardo, C., Sánchez, *E.G.*, Pérez-Núñez, D., Nogal, M., de León, P., Carrascosa, Á.L., Nieto, R., Soler, A., Arias, M.L. and Revilla, Y. (2018). African swine fever virus (ASFV) protection mediated by NH/P68 and NH/P68 recombinant live-attenuated viruses. Vaccine. 36(19): 2694-2704. doi: 10.1016/j.vaccine.2018.03.040.

Gallardo, M.C., Reoyo, A.T., Fernández-Pinero, J., Iglesias, I., Muñoz, M.J. and Arias, M.L. (2015). African swine fever: a global view of the current challenge. Porcine Health Manag, 1: 21. doi: 10.1186/s40813-015-0013-y.

Gaudreault, N.N. and Richt, J.A. (2019). Subunit vaccine approaches for African swine fever virus. Vaccines, 7(2): 56. doi: https://doi.org/10.3390/vaccines7020056.

Ge, S., Li, J., Fan, X., Liu, F., Li, L., Wang, Q., Ren, W., Bao, J., Liu, C., Wang, H., Liu, Y., Zhang, Y., Xu, T., Wu, X. and Wang, Z. (2018). Molecular characterization of African swine fever virus, China, Emerg Infect Dis, 24(11): 2131-2133. doi: 10.3201/eid2411.181274.

Gebhardt, J.T., Dritz, S.S., Elijah, C.G., Jones, C.K., Paulk, C.B. and Woodworth, J.C. (2022). Sampling and detection of African swine fever virus within a feed manufacturing and swine production system. Transbound Emerg Dis, 69: 103-114. doi: 10.1111/tbed.14335.

Goatley, L.C., Reis, A.L., Portugal, R., Goldswain, H., Shimmon, G.L., Hargreaves, Z., Ho, C. S., Montoya, M., Sánchez-Cordón, P.J., Taylor, G., Dixon, L.K. and Netherton, C.L. (2020). A Pool of eight virally vectored African swine fever antigens protect pigs against fatal disease. Vaccines, 8(2): 234. doi: https://doi.org/10.3390/vaccines8020234.

Gogin, A., Gerasimov, V., Malogolovkin, A. and Kolbasov, D. (2013). African swine fever in the North Caucasus region and the Russian Federation in years 2007-2012. Virus Res, 173(1): 198-203. doi:10.1016/j.virusres.2012.12.007

Golnar, A.J., Martin, E., Wormington, J.D., Kading, R.C., Teel, P.D., Hamer, S.A. and Hamer, G.L. (2019). Reviewing the Potential Vectors and Hosts of African Swine Fever Virus Transmission in the United States. Vector Borne Zoonotic Dis, 19(7): 512-524. doi: 10.1089/vbz.2018.2387.

Gomez-Villamandos, J.C., Bautista, M.J., Sanchez-Cordon, P.J. and Carrasco, L. (2013). Pathology of African swine fever: the role of monocyte-macrophage. Virus Res, 173:140-149. doi: 10.1016/j.virusres.2013.01.017

Guberti, V., Khomenko, S., Masiulis, M. and Kerba S. (2018). Handbook on African swine fever in wild boar and biosecurity during hunting. Standing Group of Experts on African swine fever in Europe under the GF-TADs umbrella: OIE 2018.

Hoinville, L.J., Alban, L., Drewe, J.A., Gibbens, J.C., Gustafson, L., Häsler, B., Saegerman, C., Salman, M. And Stärk, K.D. (2013). Proposed terms and concepts for describing and evaluating animal-health surveillance systems. Prev Vet Med, 112(1-2): 1-12. doi: 10.1016/j.prevetmed.2013.06.006.

Jancovich, J.K., Chapman, D., Hansen, D.T., Robida, M.D., Loskutov, A., Craciunescu, F., Borovkov, A., Kibler, K., Goatley, L., King, K., Netherton, C.L., Taylor, G., Jacobs, B., Sykes, K., Dixon, L.K. (2018). Immunization of pigs by DNA prime and recombinant vaccinia virus boost to identify and rank African swine fever virus immunogenic and protective proteins. J Virol, 92(8): e02219-17. doi: 10.1128/JVI.02219-17.

Jori, F. and Bastos, A.D. (2009). Role of wild suids in the epidemiology of African swine fever. Ecohealth, 6: 296-310. doi: 10.1007/s10393-009-0248-7.

Jori, F., Vial, L., Penrith, M.L., Pérez-Sánchez, R., Etter, E., Albina, E., Michaud, V. and Roger, F. (2013). Review of the sylvatic cycle of African swine fever in sub-Saharan Africa and the Indian ocean. Virus Res, 173(1): 212-227. doi: 10.1016/j.virusres.2012.10.005.

Juszkiewicz, M., Walczak, M., Mazur-Panasiuk, N. and WoŸniakowski, G. (2020). Effectiveness of chemical compounds used against African swine fever virus in commercial available disinfectants. Pathogens, 9: 878. doi: 10.3390/pathogens9110878.

Karalyan, Z., Avetisyan, A., Avagyan, H., Ghazaryan, H., Vardanyan, T., Manukyan, A., Semerjyan, A. and Voskanyan, H. (2019). Presence and survival of African swine fever virus in leeches. Vet Microbiol, 237: 108421. doi: 10.1016/j.vetmic.2019.108421.

Kazakova, A.S., Imatdinov, I.R., Dubrovskaya, O.A., Imatdinov, A.R., Sidlik, M.V., Balyshev, V.M., Krasochko, P.A. and Sereda, A.D. (2017). Recombinant protein p30 for serological diagnosis of African swine fever by immunoblotting assay. Transbound Emerg Dis, 64(5): 1479-1492. doi: 10.1111/tbed.12539.

Kim, H.J., Cho, K.H., Lee, S.K., Kim, D.Y., Nah, J.J., Kim, H.J., Kim, H.J., Hwang, J.Y., Sohn, H.J., Choi, J.G., Kang, H.E. and Kim, Y.J. (2020). Outbreak of African swine fever in South Korea, 2019. Transbound Emerg Dis, 67(2): 473-475. doi: 10.1111/tbed.13483.

Kleiboeker, S.B., Burrage, T.G., Scoles, G.A., Fish, D. and Rock, D.L. (1998). African swine fever virus infection in the argasid host, *Ornithodoros porcinus porcinus*. J Virol, 72: 1711-1724. doi: 10.1128/JVI.72.3.1711-1724.1998.

Kolbasov, D., Titov, I., Tsybanov, S., Gogin, A. and Malogolovkin, A. (2018). African swine fever virus, Siberia, Russia, 2017. Emerg Infect Dis, 24(4): 796-798. doi:10.3201/eid2404.171238.

Laddomada, A., Patta, C., Oggiano, A., Caccia, A., Ruiu, A., Cossu, P. and Firinu, A. (1994). Epidemiology of classical swine fever in Sardinia: a serological survey of wild boar and comparison with African swine fever. Vet Rec, 134(8): 183-187. doi: 10.1136/vr.134.8.183.

Le, V.P., Jeong, D.G., Yoon, S.W., Kwon, H.M., Trinh, T.B.N., Nguyen, T.L., Bui, T.T.N., Oh, J., Kim, J.B., Cheong, K.M., Van Tuyen, N., Bae, E., Vu, T.T.H., Yeom, M., Na, W. and Song, D. (2019). Outbreak of African swine fever, Vietnam, 2019. Emerg Infect Dis, 25(7): 1433-1435. doi: 10.3201/eid2507.190303.

Manzano-Román, R., García-Varas, S., Encinas-Grandes, A. and Pérez-Sánchez, R. (2007). Purification and characterization of a 45-kDa concealed antigen from the midgut membranes of *Ornithodoros erraticus* that induces lethal anti-tick immune responses in pigs. Vet Parasitol, 145(3-4): 314-325. doi:10.1016/j.vetpar.2007.01.011.

Mazur-Panasiuk, N. and WoŸniakowski, G. (2020). Natural inactivation of African swine fever virus in tissues: Influence of temperature and environmental conditions on virus survival. Vet Microbiol, 242: 108609. doi: 10.1016/j.vetmic.2020.108609.

Mebus, C.A., House, C., Ruiz-Gonzalvo, F.R., Pineda, J.M., Tapiador, J.T., Pire, J.J. Bergada, J., Yedloutschnig, R.J., Sahu, S., Becerra, V. and Sánchez-Vizcaíno, J.M. (1993). Survival of foot-and-mouth disease, African swine fever, and hog cholera viruses in Spanish Serrano cured hams and iberian cured hams, shoulders and loins. Food Microbiol, 10(2): 133-143. doi:10.1006/fmic.1993.1014.

Monteagudo, P.L., Lacasta, A., López, E., Bosch, L., Collado, J., Pina-Pedrero, S., Correa-Fiz, F., Accensi, F., Navas, M.J., Vidal, E., Bustos, M.J., Rodríguez, J.M, Gallei, A., Nikolin, V., Salas, M.L. and Rodríguez, F. (2017). BA71ΔCD2: a new recombinant live attenuated African swine fever virus with cross-protective capabilities. J Virol, 91(21): e01058-17. doi: 10.1128/JVI.01058-17.

Montgomery, R.E. (1921). On a form of swine fever occurring in British East Africa. J Comp Pathol, 34: 59-191. doi: 10.1016/S0368-1742(21)80031-4

Munoz-Moreno, R. Galindo, I., Cuesta-Geijo, M.A., Barrado-Gil, L. and Alonso, C. (2015). Host cell targets for African swine fever virus. Virus Res, 209: 118-27. doi: 10.1016/j.virusres.2015.05.026.

Nieto-Pelegrin, E., Rivera-Arroyo, B. and Sanchez-Vizcaino, J.M. (2015). First detection of antibodies against African swine fever virus in faeces samples. Transbound Emerg Dis, 62: 594-602. doi: 10.1111/tbed.12429

Nurmoja, I., Petrov, A., Breidenstein, C., Zani, L., Forth, J.H., Beer, M., Kristian, M., Viltrop, A. and Blome, S. (2017). Biological characterization of African swine fever virus genotype II strains from north-eastern Estonia in European wild boar. Transbound Emerg Dis, 64(6): 2034-2041. doi: https://doi.org/10.1111/tbed.12614

O'Donnell, V., Risatti, G.R., Holinka, L.G., Krug, P.W., Carlson, J., Velazquez-Salinas, L., Azzinaro, P.A., Gladue, D.P. and Borca, M.V. (2016). Simultaneous deletion of the 9GL and UK genes from the African swine fever virus Georgia 2007 isolate offers increased safety and protection against homologous challenge. J Virol, 91(1): e01760-716. doi: 10.1128/JVI.01760-16.

Oganesyan, A.S., Petrova, O.N., Korennoy, F.I., Bardina, N.S., Gogin, A.E., Dudnikov, S.A. (2013). African swine fever in the Russian Federation: spatio-temporal analysis and epidemiological overview. Virus Res, 173(1): 204-211. doi: 10.1016/j.virusres.2012.12.009

OIE. (2019). African Swine Fever (Infection with African Swine Fever Virus) Manual of Diagnostic Tests and Vaccines for Terrestrial Animals. 8th ed. Terrestrial Manual, 1-3: 3.8.1.

Pan, I.C., Huang, T.S. and Hess, W.R. (1982). New method of antibody detection by indirect immunoperoxidase plaque staining for serodiagnosis of African swine fever. J Clin Microbiol, 16: 650-655. doi: 10.1128/JCM.16.4.650-655.1982

Pan, L., Luo, R., Wang, T., Qi, M., Wang, B., Sun, M., Luo, Y., Ji, C., Sun, Y., Qiu, H.-J. (2021). Efficient inactivation of African swine fever virus by a highly complexed iodine. Vet Microbiol, 263: 109245. doi: 10.1016/j.vetmic.2021.109245.

Pastor, M.J., Arias, M. and Escribano, J.M. (1990). Comparison of two antigens for use in an enzyme-linked immunosorbent assay to detect African swine fever antibody. Am J Vet Res, 51: 1540-1543.

Patil, S.S., Suresh, K.P., Vashist, V., Prajapati, A., Pattnaik, B. and Roy, P. (2020). African swine fever: A permanent threat to Indian pigs. Vet world, 13(10): 2275-2285. doi: https://doi.org/10.14202/vetworld.2020.2275-2285.

Penrith, M. and Vosloo, W. (2009). Review of African swine fever: Transmission, spread and control: review article. J S Afr Vet Assoc, 80(2): 58-62. doi:10.4102/jsava.v80i2.172.

Penrith, M.L., Bastos, A.D., Etter, E.M.C. and Beltran-Alcrudo, D. (2019). Epidemiology of African swine fever in Africa today: sylvatic cycle versus socio-economic imperatives. Transbound Emerg Dis, 66: 672-686. doi: 10.1111/tbed.13117.

Penrith, M.L., Vosloo, W., Jori, F. and Bastos, A.D. (2013). African swine fever virus eradication in Africa. Virus Res, 173: 228-246.

Pérez-Filgueira, D.M., González-Camacho, F., Gallardo, C., Resino-Talaván, P., Blanco, E., Gómez-Casado, E., Alonso, C. and Escribano, J.M. (2006). Optimization and validation of recombinant serological tests for African Swine Fever diagnosis based on detection of the p30 protein produced in Trichoplusia ni larvae. J Clin Microbiol, 44(9): 3114-3121. doi: 10.1128/JCM.00406-06.

Petrini, S., Feliziani, F., Casciari, C., Giammarioli, M., Torresi, C. and De Mia, G.M. (2019). Survival of African swine fever virus (ASFV) in various traditional Italian dry-cured meat products. Prev Vet Med, 162: 126-130. doi: 10.1016/j.prevetmed.2018.11.013.

Petrov, A., Forth, J.H., Zani, L., Beer, M. and Blome, S. (2018). No evidence for long-term carrier status of pigs after African swine fever virus infection. Transbound Emerg Dis, 65(5): 1318-1328. doi:10.1111/tbed.12881.

Pikalo, J., Schoder, M.E., Sehl, J., Breithaupt, A., Tignon, M., Cay, A.B., Gager, A.M., Fischer, M., Beer, M. and Blome, S. (2020). The African swine fever virus isolate Belgium 2018/1 shows high virulence in European wild boar. Transbound Emerg Dis, 67(4): 1654-1659. doi: https://doi.org/10.1111/tbed.13503

Pikalo, J., Zani, L., Hühr, J., Beer, M. and Blome, S. (2019). Pathogenesis of African swine fever in domestic pigs and European wild boar - Lessons learned from recent animal trials. Virus research, 271: 197614. doi: https://doi.org/10.1016/j.virusres.2019.04.001.

Probst, C., Gethmann, J., Amler, S., Globig, A., Knoll, B. and Conraths, F.J. (2019). The potential role of scavengers in spreading African swine fever among wild boar. Sci Rep, 9: 11450. doi: https://doi.org/10.1038/s41598-019-47623-5.

Reis, A.L., Goatley, L.C., Jabbar, T., Lopez, E., Rathakrishnan, A., Dixon, L.K. (2020). Deletion of the gene for the type I interferon inhibitor I329L from the attenuated African swine fever virus OURT88/3 strain reduces protection induced in pigs. Vaccine, 8: 262. doi: 10.3390/vaccines8020262.

Reis, A.L., Goatley, L.C., Jabbar, T., Sanchez-Cordon, P.J., Netherton, C.L., Chapman, D.A.G. and Dixon, L.K. (2017). Deletion of the African swine fever virus gene DP148R does not reduce virus replication in culture but reduces virus virulence in pigs and induces high levels of protection against challenge. J Virol, 91(24): e01428-17. doi: https://doi.org/10.1128/JVI.01428-17.

Revilla, Y., Pérez-Núñez, D. and Richt, J.A. (2018). African swine fever virus biology and vaccine approaches. Adv Virus Res, 100: 41-74. doi: https://doi.org/10.1016/bs.aivir.2017.10.002.

Sanchez Botija, C. and Badiola, C. (1996). Presencia del virus de la pesteporcina africanaenel *Haematopinus suis* [Presencie of the African swine pest virus in Haematopinus suis]. Bull Off Int Epiz, 66(1): 699-705.

Sánchez, *E.G.*, Riera, E., Nogal, M., Gallardo, C., Fernández, P., Bello-Morales, R., López-Guerrero, J.A., Chitko-McKown, C.G., Richt, J.A. and Revilla, Y. (2017). Phenotyping and susceptibility of established porcine cells lines to African swine fever virus infection and viral production. Sci Rep, 7(1): 10369. doi: 10.1038/s41598-017-09948-x.

Sánchez-Cordón, P. J., Nunez, A., Neimanis, A., Wikström-Lassa, E., Montoya, M., Crooke, H. and Gavier-Widén, D. (2019). African swine fever: Disease dynamics in wild boar experimentally infected with ASFV isolates belonging to genotype I and II. Viruses, 11(9): 852. doi: https://doi.org/10.3390/v11090852.

Sánchez-Vizcaíno, J. M., Mur, L. and Martínez-López, B. (2013). African swine fever (ASF): five years around Europe. Veterinary Microbiology, 165(1-2): 45–50. DOI: https://doi.org/10.1016/j.vetmic.2012.11.030

Sánchez-Vizcaíno, J.M. and Heath, D.L. (2019). African swine fever (infection with African swine fever virus). In: Manual of Diagnostic Tests and Vaccines for Terrestrial Animals. 8th ed. Paris: Office International des Epizooties. Available online at: https://www.oie.int/fileadmin/Home/eng/Health_standards/tahm/3.08.01_ASF.pdf

Sánchez-Vizcaíno, J.M., Mur, L. and Martínez-López, B. (2012). African swine fever: An epidemiological update. Transbound Emerg Dis, 59: 27-35. doi:10.1111/j.1865-1682.2011.01293.x

Sánchez-Vizcaíno, J.M., Mur, L., Gomez-Villamandos, J.C. and Carrasco, L. (2015). An update on the epidemiology and pathology of African swine fever. J Comp Pathol,152(1): 9-21. doi:10.1016/j.jcpa.2014.09.003.

Sang, H., Miller, G., Lokhandwala, S., Sangewar, N., Waghela, D., Bishop, R.P. and Mwangi, W. (2020). Progress toward development of effective and safe African swine fever virus vaccines. Front Vet Sci, 7: 84. doi: 10.3389/fvets.2020.00084.

Sastre, P., Pérez, T., Costa, S., Yang, X., Räber, A., Blome, S., Goller, K.V., Gallardo, C., Tapia, I., García, J., Sanz, A. and Rueda, P. (2016). Development of a duplex lateral flow assay for simultaneous detection of antibodies against African and Classical swine fever viruses. J Vet Diagn Invest, 28(5): 543-549. doi: 10.1177/1040638716654942.

Sehl-Ewert, J., Deutschmann, P., Breithaupt, A. and Blome, S. (2022). Pathology of African swine fever in wild boar carcasses naturally infected with German virus variants. Pathogens (Basel, Switzerland), 11(11): 1386. doi: https://doi.org/10.3390/pathogens11111386.

Sereda, A.D., Kazakova, A.S., Namsrayn, S.G., Vlasov, M.E. and Kolbasov D.V. (2022). The attenuated ASFV strains MK-200 and FK-32/135 as possible models for investigation of protective immunity by ASFV infection. PLoS One, 17: e0270641. doi: 10.1371/journal.pone.0270641.

Simoes, M., Freitas, F.B., Leitao, A., Martins, C. and Ferreira, F. (2019). African swine fever virus replication events and cell nucleus: new insights and perspectives. Virus Res, 270: 197667. doi: 10.1016/j.virusres.2019.197667

Ståhl, K., Sternberg-Lewerin, S., Blome, S., Viltrop, A., Penrith, M. L. and Chenais, E. (2019). Lack of evidence for long term carriers of African swine fever virus - a systematic review. Virus Res, 272: 197725. doi: https://doi.org/10.1016/j.virusres.2019.197725.

Stoian, A.M.M., Zimmerman, J., Ji, J., Hefley, T.J., Dee, S., Diel, D.G., Rowland, R.R.R. and Niederwerder, M.C. (2019). Half-Life of African swine fever virus in shipped feed. Emerg Infect Dis, 25(12): 2261-2263. doi: 10.3201/eid2512.191002.

Sunwoo, S.Y., Pérez-Núñez, D., Morozov, I., Sánchez, *E.G.*, Gaudreault, N.N., Trujillo, J.D., Mur, L., Nogal, M., Madden, D., Urbaniak, K., Kim, I.J., Ma, W., Revilla, Y. and Richt, J.A. (2019). DNA-protein vaccination strategy does not protect from challenge with African swine fever virus Armenia 2007 strain. Vaccines (Basel), 7(1): 12. doi: 10.3390/vaccines7010012.

Tanneberger, F., Abd El Wahed, A., Fischer, M., Blome, S. and Truyen U. (2021). The efficacy of disinfection on modified Vaccinia Ankara and African swine fever virus in various forest soil types. Viruses, 13: 2173. doi: 10.3390/v13112173.

Teklue, T., Wang, T., Luo, Y., Hu, R., Sun, Y. and Qiu, H.J. (2020). Generation and evaluation of an African swine fever virus mutant with deletion of the CD2v and UK genes. Vaccines, 8: 763. doi: https://doi.org/10.3390/vaccines8040763

Thomson, G.R., Gainaru, M.D. and Van Dellen, A.F. (1980). Experimental infection of warthogs (*Phacochoerus aethiopicus*) with African swine fever virus. Onderstepoort J Vet Res, 47: 19-22.

Wade, A., Achenbach, J.E., Gallardo, C., Settypalli, T.B.K., Souley, A., Djonwe, G., Loitsch, A., Dauphin, G., Ngang, J.J.E., Boyomo, O., Cattoli, G., Diallo, A. and Lamien, C.E. (2019). Genetic characterization of African swine fever virus in Cameroon, 2010-2018. J Microbiol, 57(4): 316-324. doi: 10.1007/s12275-019-8457-4.

Wen, X., He, X., Zhang, X., Zhang, X., Liu, L., Guan, Y., Zhang, Y. and Bu, Z. (2019). Genome sequences derived from pig and dried blood pig feed samples provide important insights into the transmission of African swine fever virus in China in 2018. Emerging microbes and infections, 8(1): 303-306. doi: https://doi.org/10.1080/22221751.2019.1565915.

Wilkinson, P. (1984). The persistence of African swine fever in Africa and the Mediterranean. Prev Vete Med, 2(1-4): 71-82. doi:10.1016/0167-5877(84)90050-3.

Yáñez, R.J., Rodríguez, J.M., Nogal, M.L., Yuste, L., Enríquez, C., Rodriguez, J.F. and Viñuela, E. (1995). Analysis of the complete nucleotide sequence of African swine fever virus. Virol, 208: 249-278.

Zhang, H., Zhao, S., Zhang, H., Qin, Z., Shan, H. and Cai, X. (2023). Vaccines for African swine fever: an update. Front Microbiol, 14: 1139494. doi: https://doi.org/10.3389/fmicb.2023.1139494.

Zsak, L., Borca, M.V., Risatti, G.R., Zsak, A., French, R.A., Lu, Z., Kutish, G.F., Neilan, J.G., Callahan, J.D., Nelson, W.M. and Rock, D.L. (2005). Preclinical diagnosis of African swine fever in contact-exposed swine by a real-time PCR assay. J Clin Microbiol, 43(1): 112-119. doi: 10.1128/JCM.43.1.112-119.2005.

Chapter 15

Understanding Nipah Virus Disease in Pigs: Transmission Dynamics, Impact, Control and Future Directions

*Deepak Y. Patil, Rima R. Sahay, Sreelekshmy Mohandas, Anita M. Shete and Pragya D. Yadav**

Indian Council of Medical Research-National Institute of Virology, Pune – 411 021, Maharashtra, India
**e-mail: hellopragya22@gmail.com*

ABSTRACT

This chapter delves into various aspects of Nipah virus (NiV) disease in pig including its epidemiology and the various presentation of the illness, and many protracted efforts needed to manage this problem. This zoonotic virus that mainly resides in Pteropus bats creates several risks to the health of animals and humans. Beginning with background information on the disease; the first outbreak occurred in the Malaysia during 1998-1999. It not only affected animal and human health but also brought vast economic problems and over a million pigs had to be culled in order to contain the virus. There is more on the transmission, where NiV spreads through direct contact with affected tissues and secretions of pigs and through contact with contaminated surfaces. These also includes housing arrangements that involve close contact as well as inter-farm pig movements and international trade which enables the virus spread. The Malaysian outbreak alone caused the deaths of almost one million pigs while unleashing severe impact on people's sources of income. In response to such outbreaks, the post-outbreak measures focused on the control and prevention as the major concepts that included, collaborative approach involving multiple stakeholders, increased biosafety, allowing pig farming in specified regions only, promoting public awareness, and stringent sanitary control. Close monitoring and active surveillance in areas where they are adjacent is imperative to prevent similar future epidemic occurrences. In conclusion, this chapter lowers down all the details regarding the interaction of different factors in the transmission of nipah virus stressing on the combination of multifaceted strategies in controlling this disease and preventing it from spreading in through pig population.

Keywords: *Nipah, Outbreak, Pig, Transmission, Prevention, Control.*

Brief History of Nipah Virus

Nipah virus (NiV) is a naturally occurring *Paramyxovirus* of bats and causes fatal encephalitis/respiratory disease outbreak in pigs and humans (Singh *et al.*, 2019). In humans, NiV infection has a broad range of clinical presentations, from asymptomatic infection to acute respiratory syndrome and fatal encephalitis, with case-fatality rates in excess of 70 percent (Tan *et al.*, 1999; Chan *et al.*, 2002; Wong *et al.*, 2002). Fruit bats of the genus *Pteropus*, family *Pteropodidae* are considered as natural reservoir host for nipah virus. They have wide geographical distribution that encompasses north, east and south-east areas of Australia, Indonesia, Malaysia, the Philippines and some of the Pacific Islands (Deka *et al.*, 2018).

The virus was first reported in 1999 in an epidemic of encephalitis among pig farmers in Perak State of Malaysia resulting in a fatality rate of 39.6 percent with 265 human cases of acute encephalitis and 105 deaths (Chua *et al.*, 1999; Chua *et al.*, 2000; Chua *et al.*, 2002; Chua *et al.*, 2003a; Chua *et al.*, 2003b; Chua *et al.*, 2010). In the Malaysian outbreak, NiV was introduced into the pig population, and most of the human cases resulted from exposure to ill pigs (Chua *et al.*, 1999). The virus was isolated from *Pteropus vampyrus* bats, which were considered as the source of infection for Malaysia outbreak (Chua *et al.*, 2002). In 1999, Singapore experienced a brief but concerning outbreak of NiV linked to infected pigs imported from Malaysia. Though limited to 11 cases and one tragic death, the outbreak highlighted the potential for zoonotic transmission of NiV through direct contact with infected pigs and their bodily fluids. Notably, unlike outbreaks in Malaysia, Singapore saw no evidence of human-to-human transmission (Chua *et al.*, 1999, Chua *et al.*, 2000; Chua *et al.*, 2002; Chua *et al.*, 2003a; Chua *et al.*, 2003b; Chua *et al.*, 2010; Epstein *et al.*, 2016; Paton *et al.*, 2016).

The first NiV outbreak in Bangladesh was identified in 2001, and since then, the country has reported 79 spill over events from April 2001 to April 2014 (Hsu *et al.*, 2004; Islam *et al.*, 2016). These outbreaks are typically seasonal, occurring between December and May, coinciding with the date harvest season when their droppings contaminate date palm sap, a popular beverage in Bangladesh. The virus mainly spreads through consuming contaminated date palm sap. However, human-to-human transmission through close contact with infected individuals has also been documented. Bangladesh stands out as the only nation with consistent reporting of Nipah virus outbreaks (Hossain *et al.*, 2008; Hegde *et al.*, 2016).

In India, Siliguri district in West Bengal State first witnessed the emergence and re-emergence of the NiV during years 2001 and 2007 (Chadha *et al.*, 2006; Arankalle *et al.*, 2011). In Siliguri based on the case definition, 66 cases of encephalitis were identified, and the case-fatality ratio was ≤74 per cent. Subsequently, NiV caused an intra-familial outbreak with a 100 percent case-fatality rate in Nadia District during year 2007, which confirmed person-to-person transmission. Both outbreaks (2001 and 2007) occurred in the state of West Bengal bordering Bangladesh (Chadha *et al.*, 2006; Arankalle *et al.*, 2011). During May 2018, the first outbreak of NiV occurred in Kozhikode and Malappuram districts of Kerala. Twenty-three cases (18 laboratory confirmed and 5 probable) of NiV infection were reported in two generations. Most patients presented with acute neurological syndrome and respiratory symptoms with a case fatality rate of 91.3 percent (Arunkumar*et al.*, 2018; Yadav *et al.*, 2019). While the source of infection of index case is not known, all the remaining cases are linked to close contact with confirmed/probable case at six hospitals. Kerala reported the second and third outbreak of Nipah virus diseases with single cases from Ernakulum district in June 2019 and from Kozhikode district in 2021(Sahay *et al.*, 2020; Sudeep *et al.*, 2021; Yadav *et al.*, 2022). The case from Ernakulam survived with supportive care (Sahay *et al.*, 2020). The single case from 2021 presented with fever and late-onset encephalitis and subsequently died (Yadav *et al.*, 2022). The current NiV outbreak 2023 Kozhikode Kerala was declared on September 13, 2023. Nipah virus was confirmed amongst the six cases with two fatalities (case fatality ratio of 33.33 per cent). Rest all the close contacts and other suspected non-epidemiologically linked cases were found negative for NiV infection (AS *et al.*, 2024). In all the outbreaks in Kerala, the *Pteropus medius* bats were tested positive for nipah viral RNA or IgG antibodies near the affected area, pointing the probable source of infection (Yadav *et al.*, 2019; Sudeep *et al.*, 2021; Balasubramanian *et al.*, 2024).

The presence of nipah viral RNA in fruit bats from Dubri region of Assam, Myanaguri and Cooch Behar area of West Bengal, Kozhikode and Wayanad states, Kerala state (Yadav *et al.*, 2015; Sudeep *et al.*, 2021; Gokhale *et al.*, 2022) is a significant concern and highlights the need for proactive measures to prevent future outbreaks. Although pigs have not been identified as source of infection in nipah outbreak in India, the lack of surveillance data on NiV presence in pigs, particularly in the northeast bordering Bangladesh and Kerala state, indicates a potential surveillance gap.

Transmission Dynamics

Routes of Transmission in Livestock

Nipah outbreaks in pigs and other domestic animals (horses, goats, sheep, cats and dogs) were first reported during the initial Malaysian outbreak in 1999 (Chua *et al.*, 1999). Many pigs had no symptoms, but others developed acute feverish illness, laboured breathing, and neurological symptoms such as trembling, twitching and muscle spasms (Mohd Nor *et al.*, 2000). Generally, mortality was low except in young piglets. These symptoms are not dramatically different from other respiratory and neurological illnesses of pigs. Nipah should be suspected if pigs also have an unusual barking cough or if human cases of encephalitis are associated with pig handlers. Nipah virus is highly contagious in pigs. Pigs are infectious during the incubation period, which lasts from 4 to 14 days (Goh *et al.*, 2000).

The current scenario suggests that, when huge population of pigs and bats comes in close contact with the human beings, possible risk of spread of this virus to humans exists. Currently, there is no information available for the presence and seasonality of this virus in pigs in India and Bangladesh.

Epidemiology of Nipah Virus in Pigs

Nipah Virus Outbreak in Pigs

In 1998-1999, a major disease outbreak affected pigs and humans in Malaysia. To control the outbreak, about 1.1 million pigs were culled (Chua *et al.*, 2000). Pigs of all ages were infected with a highly contagious disease. The disease caused acute fever, respiratory problems, and sometimes nervous signs. In some cases, pigs died very quickly (Mohd Nor *et al.*, 2000). The disease spread mainly through the movement of pigs between farms and regions. The pigs were most likely infected through the oro-nasal route. The outbreak is believed to have begun in the northern Malaysian state of Perak. Pigs sold from there during the outbreak helped spread the disease throughout the country. The disease may have also spread through infected dogs, cats, and shared boar semen, although the virus has not been found in boar semen. Trucks transporting pigs may have also played a role in spreading the disease. The transmission of the virus between pig farms in peninsular Malaysia was primarily attributed to the emergency sale of sick pigs and the movement of pigs from infected farms in Perak to Bukit Pelandok, Negri Sembilan. Notably, the pig farms in the Kinta district were spatially isolated from each other by a distance of at least half to one kilometre, whereas those in Bukit Pelandok, the largest pig growing area in Southeast Asia, were densely clustered with neighbouring farms within proximity of less than 100 meters. Moreover, the latter farms lacked substantial bio-security measures. Consequently, the disease spread rapidly among pigs on infected farms and also propagated due to the close proximity of one farm to another. The transmission of the virus within a farm possibly occurred through direct contact with the infected pigs' excretory and secretary fluids, including urine, saliva, and pharyngeal and bronchial secretions, particularly in pigs that were kept in close confinement. The origin of the disease in Perak and how it first infected pigs is still unknown (Goh *et al.*, 2000; Muniandy *et al.*, 2004; Mohd Nor *et al.*, 2020). The disease was not recognized as new because the symptoms were similar to other pig diseases, and the number of sick and dead pigs was not unusual (Aziz *et al.*, 1999; Bunning *et al.*, 2001).Researchers investigated potential secondary reservoirs for nipah virus, focusing on common animals living near humans. They tested rodents, insectivores, and birds in Malaysia, finding no evidence of the virus in any of them. This suggests these animals didn't play a role in spreading the virus further. While dogs could catch nipah virus from infected pigs, further investigation revealed the virus didn't spread between dogs horizontally (Mill *et al.*, 2009).

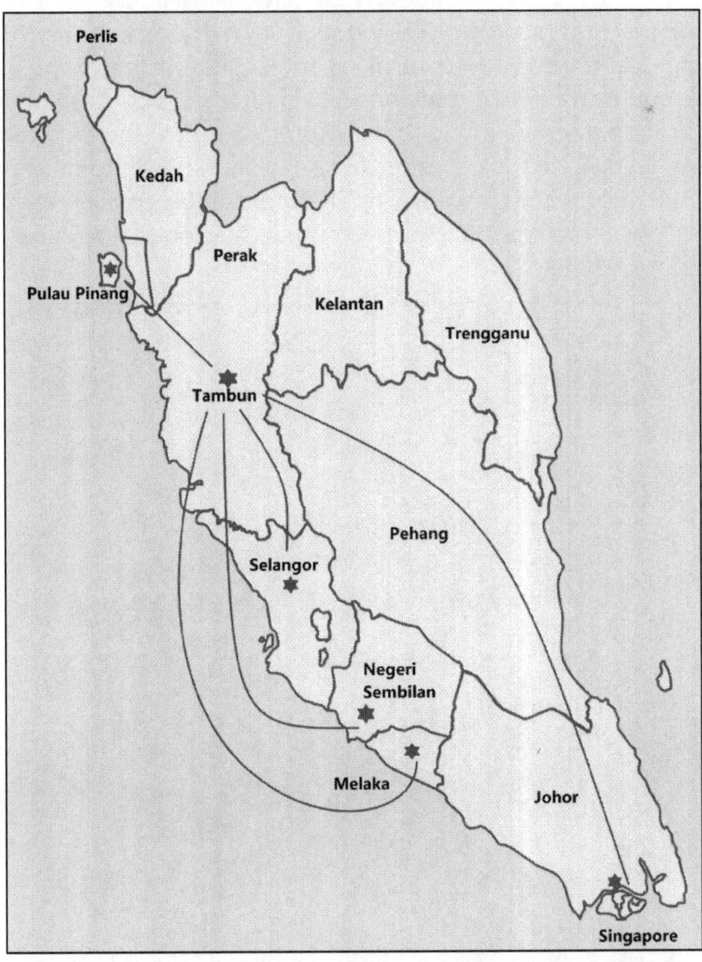

Figure 15.1: Spread of Nipah Virus amongst Pig Population during Malaysia 1998-1999 Outbreak.

Factors Contributing to the Spread of Nipah Virus within Pig Populations

Understanding the factors that contribute to the spread of the virus within pig populations is crucial for implementing effective control and prevention measures. Pigs are social animals that often live in close proximity to one another, creating an environment conducive to the spread of infectious diseases such as NiV. The close contact and social behaviour within pig herds facilitate the transmission of the virus from infected to susceptible animals. Additionally, the movement of pigs between farms can contribute to the dissemination of the virus, leading to widespread outbreaks within pig populations (Figure 15.2). Environmental contamination, particularly through contaminated water sources and feed, can contribute to the spread of NiV within pig populations. Pigs may come into contact with virus-contaminated materials, leading to the ingestion or inhalation of infectious particles. The persistence of the virus in the environment further complicates efforts to control its spread within pig populations. The global trade and movement of pigs present challenges in controlling the spread of NiV. International trade in pigs can facilitate the introduction of the virus to new regions, leading to cross-border transmission and outbreaks. Regulating the movement of pigs and implementing stringent biosecurity measures across international borders is essential for preventing the global dissemination of NiV within pig populations.

Figure 15.2: Pictorial Representation of Nipah Outbreak in Pig Population in Malaysia.

The nipah outbreak in Malaysia started when pig farms encroached on bat habitats, allowing the virus to jump from bats to pigs. Packed pig farms acted like breeding grounds for the virus, fuelling its spread within pig communities. Transporting infected pigs to southern Malaysia and Singapore sparked a wider outbreak, putting many more pigs at risk. These densely populated pig farms became hubs for human-to-human transmission, posing a serious public health threat (Augustyniak *et al.*, 2023).

Impact of NiV Outbreak on the Pig Farming Industry

The 1999 nipah outbreak in Malaysia forced the evacuation of nearly 618 homes and 111 crucial facilities, impacting lives and businesses beyond farms. Over 36,000 people lost jobs, including farmers, suppliers, and support services, dealing a heavy blow to livelihoods (Looi *et al.*, 2007). The pig population, once standing at 2.4 million, suffered a devastating loss. Nearly 1 million pigs across almost 900 farms were culled to control the outbreak (Looi *et al.*, 2007). This drastic response marked a turning point for the industry. Pig farming is now restricted to designated areas with stricter biosecurity measures, pushing many farmers to transition to other agricultural activities.

Nipah outbreaks pose a significant threat to pig farmers' livelihoods and the broader economy. Addressing this challenge requires a multi-pronged approach involving individual farmers, industry stakeholders, and government intervention, focusing on prevention, early detection, and economic support for affected communities.

Clinical Manifestations and Pathogenesis in Pigs

Clinical Manifestations

There are no pathognomonic features ascribable to nipah virus disease in pigs. Differential diagnoses should include the following classical swine fever, porcine reproductive and respiratory syndrome, Aujeszky's disease (Pseudorabies) - Swine enzootic pneumonia (*Mycoplasma hyopneumoniae*), Porcine pleuropneumonia (*Actinobacillus pleuropneumoniae*), Pasteurellosis (Aziz *et al.*, 1999). While not a definitive sign, illness in sows can point towards a possible NiV infection. This is because other diseases typically don't cause severe breathing problems, neurological issues, or high death rates in sows.

The NiV infection is highly contagious and affects the respiratory and nervous systems of pigs. It is also known as porcine respiratory and neurologic syndrome, porcine respiratory and encephalitic syndrome, barking pig syndrome, or one-mile cough. The disease can be asymptomatic, mild or severe. The clinical signs vary based on the age and response of the individual animal. Generally, mortality rates are low ranging from less than 1 percent to 5 per cent, except in piglets, but morbidity rates are high in all age groups (Nor *et al.*, 2000).

In the Malaysian outbreak, a mortality rate of about 40 percent was observed in suckling pigs. However, it is unclear whether piglet mortality is caused by disease or the sow's inability to nurse. The majority of the infected piglets showed open-mouth breathing, limb weakness, muscle tremors, and twitching (Aziz *et al.*, 1999; Nor *et al.*, 2000; Middleton *et al.*, 2002; Muniandy *et al.*, 2004). In the weaners (1 to 6 months of age), mortality was less than 5 percent and animals showed fever, difficulty in breathing, and severe non-productive cough (Nor *et al.*, 2000; Muniandy *et al.*, 2004). Severe cases can cause hemoptysis, whereas less severe cases may lead to open-mouth breathing. Respiratory signs may be followed by neurological signs like twitching, trembling, muscle weakness, spasms, convulsions, lameness, uncoordinated gait, and even death. The adult animals may display similar symptoms when infected. Other symptoms include laboured breathing, increased salivation, and nasal discharge. In the case of sows, early abortion within the first three months is also possible (Nor *et al.*, 2000; Muniandy *et al.*, 2004). Neurological symptoms such as agitation, head pressing, rapid eye movements, seizures, difficulty swallowing, excessive drooling, and protrusion of the tongue may also be present. Experimental infection of pigs with the nipah virus resulted in virus shedding through oral and nasal cavities with or without any clinical signs (Middleton *et al.*, 2002).

Pathogenesis

The nipah virus specifically infects the vascular, nervous, and lymphoreticular systems. The dissemination of the virus by viremia as well as central nervous system invasion has been experimentally demonstrated in pigs (Kasloff *et al.*, 2019). These studies suggest that the central nervous system can be infected through cranial nerves or by crossing the blood-brain barrier after initial virus replication in the upper respiratory tract (Weingartl *et al.*, 2005). Most of the pigs examined in the Malaysian outbreak had lung lesions, which varied in severity from mild to severe. The observed lesions include frothy fluid in the bronchi/trachea, emphysema, consolidation, and hemorrhages in the pulmonary parenchyma. Brain tissue may exhibit generalized congestion and edema, while kidney tissue may display both surface and cortical congestion. Histologically, generalized vasculitis in the lung, kidney, and brain tissues as well as interstitial pneumonia with extensive haemorrhages and syncytial cell formation in the vascular endothelial cells of the lung could be seen. The other significant finding in the brain was non-suppurative meningitis with gliosis. Viral antigens can be demonstrated in the endothelial cells by immunohistochemistry (Stachowiak *et al.*, 2012).

Diagnosis and Surveillance

Overview of Diagnostic Methods

Diagnosis of NiV infection is primarily done by detecting viral RNA in clinical or post-mortem specimens. Virus isolation or demonstration of viral antigen in tissue samples can also be used. In pigs, the detection of specific antibodies can be useful as NiV infection may be subclinical. Enzyme-linked immunosorbent assays, virus-neutralization tests, and bead-based assays are available for serology. Demonstration of specific antibodies to NiV is of diagnostic significance due to the rarity of the infection and the serious zoonotic implications of transmission. The samples for diagnosis should be sent to the designated laboratories with containment facilities appropriate to handle the virus by following the suitable transportation standards for biological specimens. All laboratory manipulations of suspected samples, potentially contaminated materials, or virus cultures must be performed at an appropriate containment level (Biosafety level 3 or 4) after the bio-risk analysis (MSD Vet Manual, 2012; WOAH, 2024).

PCR-based Methods

These techniques can be used to identify the presence of viral genome in samples. PCR-based methods are commonly used for diagnosis and are highly sensitive and specific. Conventional RT-PCR followed by Sanger's sequencing and Real-time reverse-transcription polymerase chain reaction is available. Swab samples from the nasal or oral cavities, urine, blood in EDTA, placenta, fetal tissues, and visceral organs such as lungs, kidneys, spleen, uterus, and brain are all suitable samples to use for diagnosis (WOAH, 2024).

Histopathology

While histopathology is valuable, it's important to remember that other pig diseases like Aujeszky's disease, swine fever, and enzootic pneumonia can cause similar lung and brain damage, potentially leading to misdiagnosis. In pigs specifically, the most telling histopathological signs of nipah appear in the lungs. The histopathology of infected pig typically shows widespread inflammation of blood vessels with tissue necrosis, bleeding, and white blood cell infiltration, especially in the lungs, kidneys, and lymph nodes. The lungs also show moderate to severe inflammation between air sacs, with widespread bleeding. Airways exhibit thickened lining, white blood cell build up around them, and dead cells clogging them. Additionally, numerous white blood cells fill the air sacs and airways themselves. In some cases, those distinctive clumps of infected cells are seen in blood vessel walls and air spaces. The brain may also show signs of inflammation, with swelling and white blood cell infiltration around the membranes and inflamed blood vessel walls (WOAH, 2024).

Immunohistochemistry (IHC)

Nipah virus antigen can be visualized within the formalin-fixed tissues by this method. It's a safe and reliable test performed on tissues preserved with formalin, making it readily available. The antisera against NiV nucleoprotein or phosphoprotein are generally used as the primary antibody for detection by this method. The test procedures like incubation time, antibody dilutions, *etc.*

need to be standardized at each laboratory. While Nipah primarily affects the lining of blood vessels (endothelium), its reach extends beyond the lungs. To effectively capture the virus, it's crucial to submit a wider range of tissues, not just the lungs, for testing. Besides lungs, include porcine meninges (membranes surrounding the brain), trachea (windpipe), and kidneys. For pregnant animals, uterus, placenta, and fetal tissues are equally important. By adopting this comprehensive approach, we can significantly improve the chances of detecting Nipah virus early and preventing its spread (WOAH, 2024).

Virus Isolation

This should be performed only in the appropriate containment facility ensuring all biosafety and biosecurity measures. Virus isolation is crucial in new outbreaks or areas without Nipah virus disease. The transportation of samples in a cold chain to the laboratory is critical for virus isolation. Swab samples from the nasal or oral cavities, urine, blood in EDTA, placenta, fetal tissues, and visceral organs such as lungs, kidneys, spleen, uterus, and brain are all suitable samples to use for isolation. African green monkey kidney cells, rabbit kidney cells, and suckling mouse brains are susceptible models for virus isolation and propagation. In cultured cells, the Nipah virus induces typical cytopathic changes like syncytia formation. PCR, immunostaining, or electron microscopy can be used for the confirmation of virus isolation (WOAH, 2024).

Enzyme-Linked Immunosorbent Assay

Indirect ELISA using tissue culture grown and gamma irradiated viral antigen is being used. The relative reactivity of the sera is assessed with nipah viral antigen vs. control cell antigen. A recombinant nucleocapsid antigen ELISA is also reported. For the confirmation of the serological assay virus neutralization is recommended as an accepted reference standard. The samples should be inactivated by gamma-irradiation and heat-inactivated at 56°C for 30 minutes before handling the tests with biosafety precautions (WOAH, 2024).

Virus Neutralization Test

The live virus neutralization test should be performed with strict containment protocols after the risk assessment at a BSL-4 facility. The diluted sera samples are mixed with the nipah virus of known titre and the virus serum mixture after incubation is added onto Vero cell monolayers. The serum sample dilution which shows a complete absence of the cytopathic effect after 3-5 days of incubation is termed as neutralizing antibody positive. Pseudo virus neutralization test is also described for nipah virus (WOAH, 2024).

Magnetic Bead-Based Serological Assays

These assays measure antibodies against the recombinant glycoproteins coupled to magnetic beads. The bound antibodies are detected using a fluorescent detection system by employing a biotinylated protein conjugate and a streptavidin fluorochrome system. The fluorescence median intensity per 100 beads is measured. For the confirmation of this assay virus neutralization is recommended as an accepted reference standard (WOAH, 2024).

Importance of Surveillance

Nipah virus is listed among the priority pathogens listed by the World Health Organisation, which has the potential to cause outbreaks in the future. The presence of the virus in reservoir hosts, *i.e.*, fruit bats across a wide geographic range highlights the need for robust disease surveillance and biosecurity measures to enable early detection and prevent reintroduction. The virus introduced into pig farms can spread rapidly and infect almost all the herd within the premises. The infections in pigs can be asymptomatic. Poor biosecurity measures and the movement of infected animals can transmit the virus between premises. An active surveillance program, strict quarantine measures, and the slaughtering of pigs resulted in the containment of the outbreak in Malaysia. It is crucial to follow the right biosecurity and quarantine measures in facilities to prevent the spread of nipah virus disease. The absence of any approved vaccine or treatment for the infection points to the importance of surveillance of the disease in the pig population in the disease hotspots (Aziz *et al.*, 1999; Bunning *et al.*, 2001; WOAH, 2004; Mills *et al.*, 2009; Singh *et al.*, 2019; Goh *et al.*, 2024).

Prevention and Control Measures

Overview of Existing Vaccines for Pigs

A devastating pig outbreak in Malaysia sparked the urgent need for a vaccine to shield both animals and humans from Henipaviruses. This aligns with the "One Health" concept promoting interconnected animal and human health. While some vaccines showed promise in other animals, they fell short in pigs (Aziz *et al.*, 1999; Bunning *et al.*, 2001; WOAH, 2004; Mills *et al.*, 2009; Singh *et al.*, 2019; Goh *et al.*, 2024). HeV-sG, for example, offered only partial protection against HeV and none against NiV. Other available options are canarypox and viral vectors generated antibodies, but their effectiveness needs further testing. An oral RABV vector could be a cost-effective solution for mass vaccination (Degeling *et al.*, 2015). Experts recently reviewed the unique challenges and current progress in developing this crucial vaccine. Studies like HeV-sG vaccination, which generated immunity against HeV but not NiV, highlighted the need for a different approach, potentially focusing on cellular immunity instead of just antibodies (Amaya *et al.*, 2020).

Public Health Interventions

Controlling a major outbreak like nipah demands a highly organized response across various levels: from legislation and management to logistics, technical expertise, and established procedures. This includes plans for:

- Management: Setting up control centers, coordinating efforts, managing information flow, and ensuring laboratories are prepared.
- Control: Safely culling and disposing of infected animals, compensating farmers, and decontaminating affected areas.
- Livestock management: Addressing the needs of various livestock enterprises involved.

Two zones were established around infected farms: a 2km infected zone and a 10km buffer zone. All pigs within the buffer zone were humanely culled over two months, totalling 901,228 animals from 896 farms (Suit-B *et al.*, 2021). This involved collaboration between various government agencies, NGOs, and the army. Culling methods involved shooting, and disposal involved burying animals in deep pits within the infected area. Disinfection of premises and burial sites was ensured using chlorinated lime and detergents. Once Nipah was declared a controlled disease, all pig and pork movement within the country was banned. Public awareness campaigns informed people about the restrictions. Increased patrols on roads ensured enforcement. Limited movement permits allowed transporting pigs outside designated zones to government slaughterhouses under strict supervision (Suit-B *et al.*, 2021).

Malaysia launched a three-phase program to eradicate the virus from its pig population. Every pig farm was tested, leading to the identification and culling of 50 additional infected farms. Pigs sent for slaughter were carefully tracked back to their farms using ear-notch identification. Young pigs, known as "porkers," were specifically tested for antibodies, indicating recent infection on their farms (Daszak *et al.*, 2006). This comprehensive program succeeded in eradicating Nipah virus from Malaysian pigs by mid-2001. By demonstrating the absence of the virus, the program also aimed to rebuild public confidence in pork consumption (WOAH, 2004).

Zoonotic Implications

Several factors amplify the zoonotic potential of Nipah from pigs. Pigs are susceptible to the virus, readily shedding it through saliva and respiratory secretions. Additionally, intensive pig farming practices, characterized by high animal density and poor hygiene, provide fertile ground for viral amplification and further transmission between animals (Bruno *et al.*, 2002). These factors, coupled with the lack of a readily available vaccine or treatment for humans, raise significant concerns.

Understanding the transmission pathways offers opportunities for intervention. Malaysia's Nipah virus outbreak highlighted the critical role of pig movement in spreading the disease. Farms that avoided receiving animals from outside generally stayed infection-free, even with infected neighbours (Espinosa *et al.*, 2020). This suggests that biosecurity inspections should be a top priority when facing nipah concerns. Implementing biosecurity measures in pig farms, such as restricting human-animal contact and preventing bat access to feed and water sources, can significantly reduce the risk of spill over (Alarcón *et al.*, 2021). Public awareness campaigns educating communities about safe handling practices for pigs and avoiding consumption of contaminated food sources are equally crucial. Combating this zoonotic threat requires a multi-pronged approach. Strengthening veterinary surveillance to detect outbreaks in animals early, coupled with rapid diagnostic tools, is essential for prompt containment. Research efforts directed towards developing vaccines for both animals and humans hold immense promise for long-term prevention. In conclusion, nipah virus represents a significant zoonotic threat with pigs acting as its amplifier. Recognizing the transmission pathways and implementing preventive measures across various sectors is vital to protect human health and prevent future outbreaks. By remaining vigilant and investing in research and public health initiatives, we can move towards a future where the shadow of Nipah virus no longer looms over humans and animals alike.

Future Directions and Research Needs

Future research should focus on identifying the reservoir hosts of nipah virus, particularly in pig populations. Understanding the natural hosts and vectors of the virus is crucial for developing targeted control measures. Additionally, the development of effective vaccines for pigs can significantly reduce the risk of nipah virus outbreaks in swine populations. Improving surveillance and early detection methods is also a priority to enable rapid response to potential outbreaks. Implementing robust biosecurity measures on pig farms is critical for preventing the spread of nipah virus. This includes controlling access to farms, proper waste management, and regular disinfection protocols. Educating pig farmers about best biosecurity practices can significantly reduce the risk of virus transmission within pig populations. International collaboration and the sharing of research findings are essential for advancing our understanding of nipah virus in pigs. By fostering global partnerships, researchers can exchange data, insights, and best practices, leading to more effective control and prevention strategies. Adopting a One Health approach is essential for understanding the complex dynamics of nipah virus transmission. This approach emphasizes the interconnectedness of human, animal, and environmental health, highlighting the need for collaborative research efforts to address zoonotic diseases such as nipah virus.

Nipah virus outbreaks in pig populations can have devastating economic consequences for the farming industry. The cost of implementing control measures, loss of livestock, and trade restrictions all contribute to the economic burden of nipah virus. Understanding the economic impact is crucial for advocating investment in research and control efforts. The potential for zoonotic transmission of nipah virus from pigs to humans underscores the importance of protecting public health. Strategies such as promoting awareness, implementing surveillance systems, and developing rapid response protocols are essential for mitigating the risk of human infections.

Ethical treatment of animals in nipah virus research is paramount. Balancing the need for scientific advancement with animal welfare is a critical consideration in conducting studies related to nipah virus in pigs. Engaging with local communities and educating pig farmers about nipah virus awareness is crucial for early detection and prevention. Empowering communities with knowledge about the virus and its transmission pathways can contribute to effective control measures. Government policies play a significant role in nipah virus control, particularly in regulating pig farming practices and trade. Establishing clear regulatory frameworks and guidelines can help mitigate the risk of virus transmission and outbreaks. The impact of climate change on nipah virus transmission patterns is an emerging area of concern.

Understanding how environmental changes influence the spread of the virus is essential for developing adaptive strategies to mitigate the effects of climate change on Nipah virus in pigs. Incorporating technological innovations, such as advanced diagnostic tools and surveillance systems, can enhance our ability to detect and monitor nipah virus in pig populations. Leveraging technology can lead to more efficient control and prevention strategies. Drawing insights from previous nipah virus outbreaks is invaluable for informing future research and control efforts. Analyzing past experiences can provide valuable lessons that can be applied to improve prevention and response strategies.

In conclusion, addressing the future research needs and directions regarding nipah virus in pigs requires a multidisciplinary and collaborative approach. By prioritizing research areas such as reservoir host identification, vaccine development, and global knowledge sharing, we can work towards effective control and prevention of nipah virus in pig populations. It is imperative to emphasize the importance of biosecurity, public health protection, and ethical considerations in research to mitigate the impact of nipah virus on both animal and human health. Through concerted efforts and strategic investments, we can pave the way for a future where Nipah virus poses minimal threat to pig populations and public health.

REFERENCES

Alarcón, L.V., Allepuz, A. and Mateu, E. (2021). Biosecurity in pig farms: a review. Porcine Health Manag, 7(1): 5. Erratum in: Porcine Health Manag. 2021 Mar 5; 7(1): 24

Amaya, M. and Broder, C.C. (2020). Vaccines to emerging viruses: Nipah and Hendra. Annu Rev Virol, 7(1): 447-473.

Arankalle, V.A., Bandyopadhyay, B.T., Ramdasi, A.Y., Jadi, R., Patil, D.R., Rahman, M., Majumdar, M., Banerjee, P.S., Hati, A.K., Goswami, R.P., Neogi, D.K. and Mishra, A.C. (2011). Genomic characterization of Nipah virus, West Bengal, India. Emerg Infect Dis, 17(5): 907-909. doi: 10.3201/eid1705.100968.

Arunkumar, G., Chandni, R., Mourya, D.T., Singh, S.K., Sadanandan, R., Sudan, P. and Bhargava, B. (2018). Outbreak investigation of Nipah Virus disease in Kerala, India. J Infect Dis, 219(12): 1867-1878.

As, A.K., Sahay, R.R., Radhakrishnan, C., P.S., Kandath, S., Patil, D.Y., Shete, A.M.M.S., Ramakrishnan, G., Moorkoth, A.P., Gupta, N., Yadav, P.D., Godbole, S., Ramakrishnan, L.V., Vadekkandiyil, S., Ekkalayil, D.V.N., Balakrishnan, A., Pullor, N.K., Asokan, N., Joseph, R.K., Nair, P.R., Purayil, S.M., Mathew, T., Kizhakkekandiyil, R., Poovullathil, J.K., Ps, K.S., Pt, U., George, K., Rahim, A., Kumar, S.S.S., Mohandas, S., Rajan, L.S., Ramachandran, S.P., Thampi, S.P., Ashadevi, Anish, T.S., Chandran, P., Mohan, A., Vadakkayil, B., Koroth, S.C., Hafeez, N., Sasi, R.R. and Abraham, M. (2024). Clinico-epidemiological presentations and management of Nipah virus infection during the outbreak in Kozhikode district, Kerala state, India 2023. J Med Virol, 96(3): e29559. doi: 10.1002/jmv.29559.

Augustyniak, A. and Pomorska-Mól, M. (2023). An update in knowledge of pigs as the source of zoonotic pathogens. Animals (Basel), 13(20): 3281.

Aziz, A.J., Mahendran, R., Daniels, P., Shahiruddin, S., Narasiman, M., Azizah, D. and Johara, M.Y. (1999). The status, public response and challenges in overcoming emerging and exotic diseases-Nipah virus disease experience. In National Congress on Animal Health and Production: Environmental Care in Animal Production. Alor Gajah, 1999.

Balasubramanian, R., Mohandas, S., Thankappan, U.P., Shete, A., Patil, D., Sabarinath, K., Mathapati, B., Sahay, R., Patil, D. and Yadav, P.D. (2024). Surveillance of Nipah virus in *Pteropus medius* of Kerala state, India, 2023. Front Microbiol, 5(15):1342170.

Bruno, L., Nappo, M.A., Ferrari, L., Di Lecce, R., Guarnieri, C., Cantoni, A.M. and Corradi, A. (2022). Nipah virus disease: epidemiological, clinical, diagnostic and legislative aspects of this unpredictable emerging zoonosis. Animals (Basel), 13(1): 159.

Bunning, M. (2001). Nipah virus outbreak in Malaysia, 1998-1999. J Swine Health Prod, 9(6): 295-299.

Chadha, M.S., Comer, J.A., Lowe, L., Rota, P.A., Rollin, P.E., Bellini, W.J., Ksiazek, T.G. and Mishra, A.C. (2006). Nipah-virus associated encephalitis outbreak, Siliguri, India. Emerg Infect Dis, 12(2): 235-240.

Chan, K.P., Rollin, P.E., Ksiazek, T.G., Leo, Y.S., Goh, K.T., Paton, N.I., Sng, E.H. and Ling, A.E. (2002). A survey of Nipah virus infection among various risk groups in Singapore. Epidemiol Infect, 128(1): 93-98. doi: 10.1017/s0950268801006422.

Chua, K.B., Bellini, W.J., Rota, P.A., Harcourt, B.H., Tamin, A., Lam, S.K., Ksiazek, T.G., Rollin, P.E., Zaki, S.R., Shieh, W.J., Goldsmith, C.S., Gubler, D.J., Roehrig, J.T., Eaton, B., Gould, A.R., Olson, J., Field, H., Daniels, P., Ling, A.E., Peters, C.J., Anderson, L.J. and Mahy, B.W.J. (2000). Nipah virus: a recently emergent deadly paramyxovirus. Science, 288: 1432-1435.

Chua, K.B., Goh, K.J., Wong, K.T., Kamarulzaman, A., Tan, P.S., Ksiazek, T.G., Zaki, S.R., Paul, G., Lam, S.K. and Tan, C.T. (1999). Fatal encephalitis due to Nipah virus among pig-farmers in Malaysia. Lancet, 354(9186): 1257-1259.

Chua, K.B., Koh, C.L., Hooi, P.S., Wee, K.F., Khong, J.H., Chua, B.H., Chan, Y.P., Lim, M.E. and Lam, S.K. (2002). Isolation of Nipah virus from Malaysian Island flying-foxes. Microbes Infect, 4(2): 145-151. doi: 10.1016/s1286-4579(01)01522-2.

Chua, K.B. (2003a). Nipah virus outbreak in Malaysia. J Clin Virol, 26: 265-275.

Chua, K.B. (2010). Epidemiology, surveillance and control of Nipah virus infections in Malaysia. Malays J Pathol, 32(2): 69-73.

Chua KB. Nipah virus outbreak in Malaysia. JClinVirol. 2003b Apr 1;26(3):265-75.

Daszak, P., Plowright, R.K., Epstein, J.H., Pulliam, J.H., Rahman and Abdul, S. (2006). The emergence of Nipah and Hendra virus: pathogen dynamics across a wildlife-livestock-human continuum. Disease Ecology: Community structure and pathogendynamics.Oxford, 2006. https://doi.org/10.1093/acprof:oso/9780198567080.003.0013, accessed 23 Feb. 2024.

Degeling, C., Johnson, J., Kerridge, I., Wilson, A., Ward, M., Stewart, C. and Gilbert, G. (2015). Implementing a one health approach to emerging infectious disease: reflections on the socio-political, ethical and legal dimensions. BMC Pub Health, 15: 1307.

Deka, M.A. and Morshed, N. (2018). Mapping disease transmission risk of Nipah virus in South and Southeast Asia. Trop Med Infect, 3(2): 57.

Epstein, J.H., Anthony, S.J., Islam, A., Kilpatrick, A.M., Khan, S.A., Ross, N. (2016). Nipah virus ecology and infection dynamics in its bat reservoir, *Pteropus medius*, in Bangladesh. Int J Infect Dis, 53: 20-21.

Espinosa, R., Tago, D. and Treich, N. (2020). Infectious diseases and meat Production. Environ Resour Econ (Dordr), 76(4): 1019-1044.

Goh, K.J., Tan, C.T., Chew, N.K., Tan, P.S., Kamarulzaman, A., Sarji, S.A., Wong, K.T., Abdullah, B.J., Chua, K.B. and Lam, S.K. (2000). Clinical features of Nipah virus encephalitis among pig farmers in Malaysia. N Engl J Med, 342(17): 1229-35. doi: 10.1056/NEJM200004273421701.

Gokhale, M., Sudeep, A.B., Mathapati, B., Balasubramanian, R., Ullas, P.T., Mohandas, S., Patil, D.R., Shete, A.M., Gopale, S., Sawant, P., Jain, R., Holeppanavar, M., Suryawanshi, A.T., Chopade, G., Dhaigude, S., Patil, D.Y., Mourya, D.T., Yadav, P.D. (2022). Serosurvey for Nipah virus in bat population of southern part of India. Comp Immunol Microbiol Infect Dis, 85: 101800. doi: 10.1016/j.cimid.2022.101800.

Hegde, S.T., Sazzad, H.M., Hossain, M.J., Alam, M.U., Kenah, E., Daszak, P., Rollin, P., Rahman, M., Luby, S.P. and Gurley, E.S. (2016). Investigating rare risk factors for Nipah virus in Bangladesh: 2001-2012. Ecohealth, 13(4): 720-728.

Hossain, M.J., Gurley, E.S., Montgomery, J.M., Bell, M., Carroll, D.S., Hsu, V.P., Formenty, P., Croisier, A., Bertherat, E., Faiz, M.A., Azad, A.K., Islam, R., Molla, M.A.R., Ksiazek, T.G., Rota, P.A., Comer, J.A., Rollin, P.E., Luby, S.P. and Breiman, R.F. (2008). Clinical presentation of nipah virus infection in Bangladesh. Clin Infect Dis, 46: 977-984.

Hsu, V.P., Hossain, M.J., Parashar, U.D., Ali, M.M., Ksiazek, T.G., Kuzmin, I., Niezgoda, M., Rupprecht, C., Bresee, J. and Breiman, R.F. (2004). Nipah virus encephalitis reemergence, Bangladesh. Emerg Infect Dis, 10(12): 2082-2087. doi: 10.3201/eid1012.040701.

Islam, M.S., Sazzad, H.M., Satter, S.M., Sultana, S., Hossain, M.J., Hasan, M., Rahman, M., Campbel,l S., Cannon, D.L., Ströher, U., Daszak, P., Luby, S.P. and Gurley, E.S. (2016). Nipah virus transmission from bats to humans associated with drinking traditional liquor made from date palm sap, Bangladesh, 2011-2014. Emerg Infect Dis, 22(4): 664-670. doi: 10.3201/eid2204.151747.

Kasloff, S.B., Leung, A., Pickering, B.S., Smith, G., Moffat, E., Collignon, B., Embury-Hyatt, C., Kobasa, D. and Weingartl, H.M. (2019). Pathogenicity of Nipah henipavirus Bangladesh in a swine host. Sci Rep, 9(1): 5230. doi: 10.1038/s41598-019-40476-y.

Looi, L.M. and Chua, KB. (2007). Lessons from the Nipah virus outbreak in Malaysia. Malays J Pathol, 29(2): 63-67.

Middleton, D.J., Westbury, H.A., Morrissy, C.J., Van der Heide, B.M., Russell, G.M., Braun, M.A. and Hyatt, A.D. (2002). Experimental Nipah virus infection in pigs and cats. J Comp Patho, 126(2-3): 124-136.

Mills, J.N., Alim, A.N., Bunning, M.L., Lee, O.B., Wagoner, K.D., Amman, B.R., Stockton, P.C. and Ksiazek, T.G. (1999). Nipah virus infection in dogs, Malaysia, 1999. Emerg Infect Dis, 15(6): 950-952. doi: 10.3201/eid1506.080453.

MohdNor, M.N., Gan, C.H. and Ong, B.L. (2000). Nipah virus infection of pigs in peninsular Malaysia. Rev Sci Tech, 19(1): 160-165.

MSD Vet Manual. Nipah Virus Infection in Swine - Generalized Conditions. MSD Veterinary Manual. https://www.msdvetmanual.com/generalized-conditions/nipah-virus-infection/nipah-virus-infection-in-swine. Accessed February 8, 2024.

Muniandy, N. and Aziz, J.A. (2004). Effects of intensification of the traditional farming system on the environment and bio-safety of the human population: Nipah virus outbreak in Malaysia. Kyoto Area Studies on Asia, 8: 303-318.

Paton, N.I., Leo, Y.S., Zaki, S.R., Auchus, A.P., Lee, K.E., Ling, A.E., Chew, S.K., Ang, B., Rollin, P.E., Umapathi, T., Sng, I., Lee, C.C., Lim, E., Ksiazek, T.G. (1999). Outbreak of Nipah-virus infection among abattoir workers in Singapore. Lancet, 354(9186): 1253-1256. doi: 10.1016/S0140-6736(99)04379-2.

Sahay, R.R., Yadav, P.D., Gupta, N., Shete, A.M., Radhakrishnan, C., Mohan, G., Menon, N., Bhatnagar, T., Suma, K., Kadam, A.V and Ullas, P.T. (2020). Experiential learnings from the Nipah virus outbreaks in Kerala towards containment of infectious public health emergencies in India. Epidemiol Infect, 2020; 148.

Singh, R.K., Dhama, K., Chakraborty, S., Tiwari, R., Natesan, S., Khandia, R., Munjal, A., Vora, K.S., Latheef, S.K., Karthik, K., Singh Malik, Y., Singh, R., Chaicumpa, W. and Mourya, D.T. (2019). Nipah virus: epidemiology, pathology, immunobiology and advances in diagnosis, vaccine designing and control strategies - a comprehensive review. Vet Q. 39(1): 26-55. doi: 10.1080/01652176.2019.1580827.

Stachowiak, B. and Weingartl, H.M. (2012). Nipah virus infects specific subsets of porcine peripheral blood mononuclear cells. PloS one, 7(1): e30855.

Sudeep, A.B., Yadav, P.D., Gokhale, M.D., Balasubramanian, R., Gupta, N., Shete, A., Jain, R., Patil, S., Sahay, R.R., Nyayanit, D.A., Gopale, S., Pardeshi, P.G., Majumdar, T.D., Patil, D.R., Sugunan, A.P. and Mourya, D.T. (2012). Detection of Nipah virus in Pteropus medius in 2019 outbreak from Ernakulam district, Kerala, India. BMC Infect Dis. 21(1): 162. doi: 10.1186/s12879-021-05865-7.

Suit, B.Y., Hassan, L., Krauss, S.E., Ooi, P.T., Ramanoon, S.Z., Yasmin, A.R. and Epstein, J.H. (2021). Mental model of Malaysian pig farmers in implementing disease prevention and control practices. Front Vet Sci, 8: 695702. doi: 10.3389/fvets.2021.695702.

Tan, K.S., Tan, C.T. and Goh, K.J. (1999). Epidemiological aspects of Nipah virus infection. Neurol J Southeast Asia, 4(1): 77-81.

Weingartl, H., Czub, S., Copps, J., Berhane, Y., Middleton, D., Marszal, P., Gren, J., Smith, G., Ganske, S., Manning, L. and Czub, M. (2005). Invasion of the central nervous system in a porcine host by Nipah virus. J Virol, 79(12): 7528-7534.

WOAH Manual of diagnostic Tests and Vaccines for Terrestrial Animals. Chapter 3.1.15. Nipah and Hendra virus diseases. https://www.woah.org/fileadmin/Home/eng/Health_standards/tahm/3.01.15_NIPAH_HENDRA.pdf. Accessed February 8, 2024.

Wong, K.T., Shieh, W.J., Kumar, S., Norain, K., Abdullah, W., Guarner, J., Goldsmith, C.S., Chua, K.B., Lam, S.K., Tan, C.T., Goh, K.J., Chong, H.T., Jusoh, R., Rollin, P.E., Ksiazek, T.G. and Zaki, S.R. (2002). Nipah virus pathology working group. Nipah virus infection: pathology and pathogenesis of an emerging paramyxoviral zoonosis. Am J Pathol, 161(6): 2153-2167. doi: 10.1016/S0002-9440(10)64493-8.

Yadav, P., Sudeep, A., Gokhale, M., Pawar, S., Shete, A., Patil, D., Kumar, V., Lakra, R., Sarkale, P., Nichol, S. and Mourya, D. (2018). Circulation of Nipah virus in *Pteropus giganteus* bats in northeast region of India, 2015. Indian J Med Res, 147(3): 318-320. doi: 10.4103/ijmr.IJMR_1488_16.

Yadav, P.D., Sahay, R.R., Balakrishnan, A., Mohandas, S., Radhakrishnan, C., Gokhale, M.D., Balasubramanian, R., Abraham, P., Gupta, N., Sugunan, A.P., Khobragade, R., George, K., Shete, A., Patil, S., Thankappan, U.P., Dighe, H., Koshy, J., Vijay, V., Gayathri, R., Kumar, P.J., Rahim, A., Naveen, A., Nair, S., Rajendran, V.R., Jayasree, V., Majumdar, T., Jain, R., Viswanathan, P., Patil, D.Y., Kumar, A., Nyayanit, D.A., Sarkale, P., Waghmare, A., Baradkar, S., Gawande, P., Bodke, P., Kalele, K., Yemul, J., Dhaigude, S., Holepannawar, M., Gopale, S., Chopade, G., Ray, S., Waghmare, P., Narayan, J., Mathapati, B., Kadam, M., Kumar, A., Suryawanshi, A., Jose, B.P., Sivadas, S., Akash, N.P., Vimisha, T.V., Keerthi, K.V. (2022). Nipah virus outbreak in Kerala state, India amidst of COVID-19 pandemic. Front Public Health, 10: 818545. doi: 10.3389/fpubh.2022.818545.

Yadav, P.D., Shete, A.M., Kumar, G.A., Sarkale, P., Sahay, R.R., Radhakrishnan, C., Lakra, R., Pardeshi, P., Gupta, N., Gangakhedkar, R.R., Rajendran, V.R., Sadanandan, R. and Mourya, D.T. (2019). Nipah virus sequences from humans and bats during Nipah outbreak, Kerala, India, 2018. Emerg Infect Dis, 25(5): 1003-1006. doi: 10.3201/eid2505.181076.

Chapter 16

Overview of Classical Swine Fever

Rahul Singh Arya

Associate Professor, Department of Veterinary Pathology, College of Veterinary Science, D.U.V.A.S.U., Mathura – 281 001, Uttar Pradesh
e-mail: aryarahulsingh@gmail.com

ABSTRACT

Classical swine fever (CSF) is an important infectious, viral, transboundary disease of domestic and wild pigs. In the following chapter the basic biology, epidemiology, pathology, prophylaxis and diagnosis of CSF is discussed as per standard review. The CSFV belongs to pestivirus genus. It is an enveloped virus with positive sense single stranded RNA. There are three groups of genotype-1, 2, and 3. The disease had been eradicated from some parts of the world while in some region it is still endemic. It manifests multiple clinico-pathological syndromes depending on multiple factors like virulence, host susceptibility and secondary infection. Consequently, differential diagnosis from other swine diseases requires laboratory confirmation. There are different vaccines available for prophylaxis. Vaccination, quarantine, strict biosecurity and efficient animal movement laws can be helpful in control of the disease.

Keywords: *Classical swine fever, Genotypes, Pathology, Vaccination, Diagnosis.*

Introduction

Classical swine fever (CSF) is an important infectious viral disease of domestic and wild pigs caused by classical swine fever virus (CSFV), a small (40-60 nm) enveloped virus with a single stranded RNA genome of positive polarity, glycosylated membrane proteins and an icosahedral nucleocapsid.

Other names of the disease include Hog cholera, *Schweinepest* (German), *Peste Du Porc* (French) and *Peste Suina* (Italian) (Dunn, 1958). CSF virus belongs to the Pestivirus genus of the *Flaviviridae* family. It is antigenically related to the bovine viral diarrhea (BVD) virus and the border disease (BD) virus of sheep, Bungowannah virus, and HoBi-like virus, with total 11 member species of the genus (Moennig, 2000; Blome *et al.,* 2017). The genome is of about 12,300 base pairs and comprises one large Open Reading Frame (ORF) between the 5′UTR (untranslated region) and 3′UTR (Ruggli *et al.,* 1996). The ORF codes for a polyprotein of 3,898 amino acids, which is processed co- and post-translationally by both host and virus encoded proteases to produce 13 products. Based on sequence analysis of 5′ UTR, E2 and NS5B gene, Paton *et al.* (2000) described phylogenetic classification of CSFV into three main groups and a number of subgroups *viz.*, group-1 and subgroups- 1.1, 1.2, 1.3; group-2 and sub-group- 2.1, 2.2, 2.3; group-3 sub-group-3.1, 3.2, 3.3, 3.4.Van Oirschot (1986) had described that the disease occurs in three different forms: acute, chronic and late onset. The disease is endemic in many parts of the world while some parts are free of CSF. Different types of vaccines are being used all over the world for control of this disease like lapinized Chinese C Strain which is one of the most widely used ones. In India Lapinized Weybridge strain imported from UK is in use for many decades (Coronado *et al.,* 2021). The virus has potential for transboundary spread (Brown and Bevins, 2018). Further we discuss some important aspects of this virus to get an overview of the disease.

Classical Swine Fever Virus

CSFV comes under the *Pestivirus* genus of the *Flaviviridae* family. The closely related important members include BVDV-1 and 2 and, Border disease virus (Heinz *et al.,* 2000; Blome *et al.,* 2017). There are number of relatively recently described atypical porcine pestiviruses, HoBi like pestivirus and Bungowannah virus. It is an enveloped virus with a single stranded RNA genome of positive polarity of about 12.3 kb in size. It does not possess 5′ and 3′ poly-A sequences. Instead, it has 5′UTR and 3′UTR. Between the UTR there is a long open reading frame (ORF) which encodes a polyprotein of about 3,898 amino acids. It encodes for 13 viral proteins arranged in the order NH2-Npro-C-E^{rns}-E1-E2-p7-NS2-3-NS2-NS3-NS4A-NS4B-NS5A-NS5B-COOH. The proteins produced are products of co- and post- translational processing involving cellular and viral proteases (Lindenbach and Rice, 2001; Blome *et al.,* 2017).

The CSFV is a temperature sensitive virus. It can survive for few days at room temperature. However, the survival is better in cold and humid conditions. At 5°C it can remain infective for few weeks. The virus remains alive for longer period in the cellular and proteinaceous medium. Therefore, it can survive in the excretions, body fluids and stored meat and meat products. This property of virus raises important issues for biosecurity and international spread via tourist carrying pork products. CSFV had been reported to survive for years in frozen meat and meat products. It can also survive curing and smoking (Morley, 1997; Gale, 2004; Wijnker *et al.*, 2008; Farez and Weesendorp *et al.*, 2008a) Temperatures above 35°C are detrimental for virus and its survivability drops down. Above 50°C the virus is inactivated within minutes (Depner, *et al.*, 1992, Panina *et al.*, 1992, Haas *et al.*, 1995, Turner *et al.*, 2000; Botner and Belsham, 2012). Properly cooked pork can be therefore safe (Edwards, 2000). During winters and generally colder geographic regions the virus can survive for several days to a month in infected pens. Freeze thaw cycle reduces viable virus titres. Consequently, stabilizers like dimethyl sulfoxide are needed to prevent loss of infectivity of virus in laboratories (Tessler *et al.*, 1975). The virus can survive a wide pH range of 5 to 10. Low pH survivability is temperature dependent and at refrigeration temperature the virus can survive longer as compared to room temperature (Kramer *et al.*, 2009).

Infection and Replication of the Virus in Cells

The virus enters the target cell by attachment to receptors which is followed by endocytosis. The receptors involved include porcine complement regulatory protein CD-46 and heparin sulphates. The heterodimer of envelope glycoproteins E1 and E2 is important for binding with receptors. There is fusion of virus envelope with the membrane of the endosome and thus the viral core reaches the cytoplasm (Donis, 1995; Wang *et al.*, 2004; Krey *et al.*, 2005; Lecot *et al.*, 2005). The translation of viral genome occurs in the cytoplasm.

There is no 5′cap in the viral genome. Instead, there is an internal ribosomal entry site (IRES) at the 5′NTR of the viral nucleic acid. This initiates the translation or also referred as cap independent translation (Poole *et al.*, 1995; Rijnbrand *et al.*, 1997; Pestova and Hellen, 1999). Both, the viral and cellular proteases take part in the post translational processing of the polyprotein precursor. Later in the replication process the positive sense RNA strand is synthesized (Tautz *et al.*, 1997; Lackner *et al.*, 2004; Lackner *et al.*, 2006). The assembly of the virion and the maturation take place in the endoplasmic reticulum and Golgi apparatus. The progeny particles are exocytosed from the cell by budding through the cell membrane (Gray and Nettleton, 1987; Ohmann, 1990).

Epidemiology

Classical swine fever is endemic in many regions of the world. It is found both in domesticated as well as the wild hogs. It is a notifiable disease and is continuously monitored worldwide by OIE. (OIE reference) Many countries like Canada, USA, Australia, New Zealand and some parts of Europe and Africa are considered free from the virus. This number is increasing. However, in many parts of the world it continues to be endemic. The worst affected parts of the world include Asian countries particularly South East Asia and Indian sub-continent. There is prevalence of largest diversity of viral strains in this region (Blome *et al.*, 2017; Brown and Bevins, 2018; Coronado *et al.*, 2021). In India it was first reported in 1962 (Sapre *et al.*, 1962). It has the status of being endemic in the country and has been reported from almost every part of the country. Genogroup 1 (1.1) and 2 (2.1 and 2.2) have been reported time and again (Patil *et al.*, 2010; Patil *et al.*, 2012; Roychoudhury *et al.*, 2014; Rajkhowa *et al.*,2014)

Transmission

The CSF virus is contagious. Susceptible pigs contract infection by oral and oronasal routes via direct or indirect contact. Secretions and excretions from infected animals easily spread the virus. The virus also shows iatrogenic transmission. 'Neighborhood effect' during outbreaks in areas of high pig farm density has been found with airborne transmission over short distances of up to 1 km (Van Oirschot, 1986; Terpstra, 1987; Weesendorp *et al.*, 2008b). Poor biosecurity practices lead to spread of virus by farm visitors, vehicles, fomites (farm implements, vehicles, clothes, instruments). Cannibal ration with improperly cooked pork is a mode of transmission within a region and across the international borders as well (Farez and Morley, 1997). Transplacental infection gives rise to inapparent carrier piglets. Wild boar populations may harbor the virus and act as reservoirs. Introduction of new infected animals in farm, transportation vehicles, farm workers, carcass of dead pigs, artificial insemination by contaminated semen, slurry of effluents from farm, *etc.*, all can spread the virus. The slurry used in biogas plants can remain infective for longer period during winters (Dunne, 1958; Dunne, 1975, Van Oirschot, 1986).

Pathogenesis

Most commonly CSFV infection occurs via oro-nasal rout, however congenital and iatrogenic transmissions can also occur. The primary site of viral replication has been described as palatine tonsils, particularly in oronasal transmission. (Ressang, 1973b; Van Oirschot, 1986; Liess, 1987, Trautwein and Leiss, 1988). Next target of virus are the regional lymph nodes where it reaches via lymphatic system. Here again there is replication of virus. Further there is viraemia (free and cell associated) and virus spreads by peripheral circulation to other lymphoid and hemopoietic tissues like spleen, mesenteric lymph nodes, gut associated lymphoid tissues,thymusandbone marrow (Ressang, 1973a; Ressang, 1973b; Dunne 1975; Trautwein and Leiss 1988; Summerfield *et al.*, 2000). These are sites of secondary viral replication. The primary cellular targets of CSFV include, mononuclear cells particularly monocyte-macrophage system, dendritic cells and endothelium (Trautwein and Leiss, 1988; Knoetig *et al.*, 1999; Summerfield *et al.*, 2000; Gomez-Villamandos *et al.*, 2001; Jamin *et al.*, 2008; Liu *et al.*, 2011). The virus can replicate in them and with them spread to lymphoid and other tissues. The virus has been demonstrated in numerous other organs, predominantly associated with macrophages and endothelial cells (Trautwein and Leiss, 1988; Liu *et al.*, 2011).

The pathogenesis of CSFV involves both viral replication in target organs and aberrant immunological response. The latter is involved, especially, in those tissues of the affected animal where the lesions do not develop due to direct virus replication. The major cells of immunopathogenesis are the macrophages. During the course of disease many macrophage populations show activation with increased phagocytic activity and secretory responses. Many will also show apoptosis, especially in the animals which are infected with strains of higher virulence. Further there is damage to dendritic cells and very high INF-α in serum. In-fact high INF-α has been considered an important observation in acute phase of disease caused by virulent strains. (Pauly *et al.*,1998; Summerfield *et al.*, 2000; Carrasco *et al.*, 2004; Bauhofer *et al.*, 2005; Summerfield *et al.*, 2006; Jamin *et al.*, 2008; Renson *et al.*, 2010; Fiebach *et al.*, 2011; Summerfield *et al.*, 2015).

The virulent CSFV strains induce severe lymphopenia which is contributed by extensive lymphocytic apoptosis. The lymph nodes and mucosal lymphoid tissues also show depletion. There is also granulocytopenia and thrombocytopenia. The thymus shows apoptosis of thymocytes. Bone marrow of affected pigs shows depletion with poor myelopoiesis and megakaryocytopoiesis. There is disseminated intravascular coagulopathy, petechiation and hemoconcentration. The death is resultant of hypotension and circulatory failure (Trautwein and Leiss, 1988; Susa *et al.*, 1992; Pauly *et al.*,1998; Summerfield *et al.*, 1998; Summerfield *et al.*, 2000; Bautista *et al.*, 2002; Sanchez-Cordon *et al.*, 2002; Gomez-Villamandos *et al.*, 2003; Sanchez-Cordon *et al.*, 2003). The haemorrhages develop due to diapedesis of RBCs, capillary dilatation with increased permeability, endothelial damage and thrombocytopenia (Hoffmann *et al.*, 1971; Heene *et al.*,1971; Weiss *et al.*, 1973; Trautwein, 1988; Gomez-Villamandos, 2000). The role of DIC is pathogenesis of CSF has been controversial (Blome *et al.*, 2013).

Clinico-Pathology of CSF

Dunne (1958) has mentioned that classical swine fever suffering pigs may show generalized hemorrhages, 95 to 100 percent morbidity and high mortality. He also classified it into different forms according to clinico-pathological manifestations and course of disease as - per acute (4-7 days), acute (8-19 days), sub-acute (20-29 days) and chronic (30-95 days). Van Oirschot (1986) described the disease course as acute (10-20 days), chronic (1-3 months) and late onset (2-11 months). This classification was related to the variation in virulence of the virus, time of infection (pre- and postnatal) and the animal's immune response. Likewise, Moennig *et al.* (2003) classified its different clinical manifestations as- acute (transient or lethal), chronic and persistent course (congenital).The clinical picture of disease is variable and has been observed to change with time. The milder sub-acute forms of the disease with atypical signs were observed becoming prominent in some parts of the world towards the end of the last century. It is milder, appears for a longer duration and show lower morbidity and mortality. Younger age groups are more affected and sometimes vaccinated pigs may also show such manifestations (Changchun *et al.*, 2001; Suradhat *et al.*, 2007).

Acute CSF

Dunne (1970) has described the gross lesions of acute CSF which include erythema of ear, tail and extremities. The lymph nodes are enlarged, edematous, swollen and hemorrhagic. Inflammation of tonsils which may turn necrotic due to bacterial invasion and petechial hemorrhages on epiglottis and larynx are typical. Kidneys may reveal petechial or ecchymotic subcapsular hemorrhages. Van Oirschot (1986) has described multiple hemorrhages especially in kidneys and lymph nodes with infarction in spleen.

Chronic CSF

Anthony and Lewis (1961) described chronic form of CSF which is characterized by emaciation, splenic atrophy, button ulcers in large intestines, pneumonia and pleurisy. Dunne (1970) has described atrophy of thymus and exostosis of costo-chondral junctions of the ribs of young pigs. In chronic CSF, a single organ system usually is involved more and secondary bacterial infection may be involved. The lymph nodes may show hyperplasia instead of typical hemorrhages (Trautwein, 1988; Terpstra, 1991). Kamolsiriprichaiporn *et al.* (1992) observed that, pigs infected with low virulent strain (NSW strain) of CSFV, does not showed distinct lesions and were not suggestive of CSF. Lesions were characterized by scattered petechial hemorrhages with swollen, congested lymph nodes and multiple abscesses, variable atelectasis or consolidation in the antero-ventral portions of the lung's indicative of chronic bronchopneumonia.

Late Onset CSF

Congenital CSF infection can result in mummification, still birth and malformations such as hypoplasia of cerebellum, lungs and hypomyelogenesis. Generalized subcutaneous edema and hydrothorax are most pronounced lesions in stillborn piglets. Those dying shortly after birth may reveal petechial hemorrhages of skin and internal organs (Van Oirschot, 1979). Van Oirschot (1986) mentioned swollen lymph nodes and thymic atrophy while Moennig *et al.* (2003) mentioned poor growth, wasting and occasionally congenital tremors in congenitally infected piglets.

Blood

Dunne (1958) reported leucopenia (TLC may drop to 9000/mm^3 and as low as 3000/mm^3) in CSF infected pigs. Further, Dunne (1975) described neutropenia, eosinopenia and relative lymphocytosis and severe thrombocytopenia from a normal value of 250 to 850 thousand/mm^3 to values below 50,000/mm^3. Virulent virus can cause the platelets to drop below 20,000/mm^3 and some may reach zero level before death. Susa *et al.* (1992) have found severe leukopenia and increase in band neutrophils in CSF infected pigs. Gomez-Villamandos *et al.* (2003) noted marked and significant decrease in TLC with lymphopenia, monocytopenia, neutropenia and increase in band neutrophils. They also described that platelets decreased from $475 x 10^3/\mu l$ to $342 x 10^3/\mu l$ in the first 48 hours post infection, coinciding with pyrexia and becoming more severe in later stages.

Skin

Dunne (1958) mentioned active hyperemia of skin coinciding with pyrexia in early part of disease, especially noticeable in white hogs. It later turns to purplish discoloration which extends over the abdomen, the snout, the ears, and the medial sides of legs. A peculiar "blotching" effect on the ears may occur in acute infections but is more often seen in chronic cases. Chronic cases reveal partial alopecia characterized by thinning of the bristles. Kaden *et al.* (2007) studied the skin punch biopsies of experimentally infected pigs. They have detected viral antigen in endothelial cells of vascular plexus in the upper dermis, spindle shaped cells of the dermis, keratinocytes of the epidermis, inner and outer root sheath epithelium, hair bulb, and intravascular leukocytes.

Kidneys

One of the characteristic findings in classical swine fever are the cortical and subcapsular hemorrhage on kidneys which give the typical 'turkey egg kidney' appearance. Dunne (1958) described it as numerous ecchymotic "turkey egg" hemorrhages ranging in size up to 2mm in diameter. Van Oirschot (1986) reported petechial cortical hemorrhages which seldom extended to the medulla and has also described that glomerulonephritis was observed in chronic CSF but not in "late onset" CSF. Gomez-Villamandos *et al.* (2000) experimentally studied the pathological changes in kidneys by Quillota strain of CSFV. They have recorded renal petechial hemorrhages which were due to erythrodiapedesis, increased capillary area, mononuclear cell infiltration and increase in number of degranulated mast cells leading to capillary vasodilatation and increased vascular permeability. Infection of endothelial cells did not play a major role in the vascular changes. There was no evidence of micro-thrombosis in the blood vessels and activated platelets revealed no signs of aggregation or attachment to the endothelium. Calderon *et al.* (2000) experimentally studied the renal pathology caused by CSFV. They have observed a segmental thickening of the glomerular basement membrane and a nuclear swelling of endothelial cells. Several areas of the urinary space contained eosinophilic homogeneous material. Some renal tubules were showing vacuolization of epithelial cells and presence of eosinophilic homogeneous material in the tubular lumen.

Nervous System

Van Oirschot (1986) reported convulsions and posterior paralysis in terminally ill pigs infected with CSF. Gomez-Villamandos *et al.* (2006) have described the pathological changes in brain caused by the Alfort strain of CSF.

They observed the typical development of disease and the nervous signs were appeared 12 days post infection comprising in coordination, staggering walk and paralysis of the posterior third of the body. Initially hyperemia and one cell thick perivascular cuffing were observed. Later the cellular infiltration increased and non-purulent meningitis started developing. Ultrastructural examination revealed micro-hemorrhages which were associated to diapedesis of erythrocytes. The satellitosis and apoptosis of infiltrating cells were also observed. Immunohistochemical assay could demonstrate viral antigen in the perivascular and interstitial infiltrating cells with the capillary endothelial cells. However, the infection of capillary endothelial cells could not be associated to the micro-hemorrhages.

Lymphoid System

Dunne (1958) has considered lymph nodes as the first tissue to show microscopic pathological changes and among the tissues which are most constant in the development of lesion. In early stages, they are enlarged, edematous and show hyperplasia, congestion and hemorrhages. An increased number of polymorphonuclear cells observed as early as 5h after infection. The parotid, submaxillary, cervical, bronchial, iliac and superficial inguinal lymph nodes generally manifest a diffuse type of hemorrhage appearing moderately red to totally black in color. However, other lymph nodes like mesenteric and colic are characterized by peripheral hemorrhages giving a mottled or 'straw berry' like appearance. Occasionally all lymph nodes may show 'straw berry' appearance. Dunne (1970) has described suppurative tonsillitis in CSF infected pigs. It begins with mild inflammation which soon spreads bilaterally into necrotic tonsillitis. Infarction was thought to be the reason behind necrosis and rough feed can aggravate it. The tonsillitis turns suppurative due to secondary bacterial invasion. Van Oirschot (1986) reported thymic atrophy in persistent hog cholera as the most outstanding lesion. Susa *et al.* (1992) have found that B-lymphocytes are depleted from the peripheral circulation as well from the spleen and various lymph nodes in the pigs suffering from CSF. They could locate CSFV genome in the germinal center of lymph nodes and concluded that CSFV replication in B-follicles was a specific and early event in pathogenesis of the disease and preceded generalized infection. On the same lines Bruschke *et al.* (1997) found that E^{rns} glycoprotein of CSFV could cause inhibition of lymphocyte proliferation as well as trigger the apoptosis in lymphocytes.

Gastro Intestinal System

Dunne (1958) reported 'button ulcers' occurring in first part of colon which appears encrusted, circular, raised and having concentric lines. It starts with an infarction as a small indiscernible necrotic area to which fecal plaques adhere. Later, the secondary infection sets in and the circular lesion progressively becomes larger as exuding mucus, combine with cellular and fecal debris becomes encrusted over the eroded surface. Thus, the lesion is actually a combined effect of the viral and bacterial action.

Diagnosis of CSF

Diagnostic Samples

Sample of choice for early detection of infection are whole blood and tissues from multiple febrile or recently dead animals. The samples should be refrigerated and shipped to laboratory as quickly as possible. The samples which can reveal diagnostic load of virus include: tonsil, lymph nodes (pharyngeal, mesenteric), spleen, kidney, distal ileum, blood in EDTA or heparin (live cases).

Virus Isolation

Virus isolation is the 'gold standard' test for the diagnosis of CSF (OIE, 2000). However, it is labor intensive and takes 3 days to perform (Moening, 2000). Paton *et al.* (2000) have emphasized that even when much progress had been made in the development of new methods for the direct detection of CSFV, the `gold standard' is still the isolation of the virus in cell culture. CSFV can be isolated from buffy coat cells or organ suspensions of viremic animals. Suitable organs are spleen, tonsils, lymph nodes and kidneys. The samples can be incubated on susceptible cell cultures of porcine origin. Since CSFV is non-cytopathogenic, CSFV specific antibodies are used for detecting the virus in cell culture. Differentiation of the CSFV from ruminant pestiviruses is usually done using monoclonal antibodies (mAbs). Choi and Chae (2003) successfully isolated CSFV from semen of experimentally infected boars on PK-15 cell line.

Fluorescent Antibody Test (FAT)

Fluorescent antibody test involves use of antibodies conjugated with fluorescent dye (FITC) to detect the target antigen. It is used extensively to detect viral antigen in cell cultures. Dunne (1970) had mentioned that FAT was one of the most important tests for herd screening in USA. Paton *et al.* (2000) considered it as a rapid, though less sensitive test for CSFV diagnosis. Choi and Chae (2003) demonstrated viral antigen in PK-15 cell culture. After acetone fixation the wells were stained with an anti-CSFV monoclonal antibody and FITC-conjugated anti-mouse immunoglobulin. The fluorescence was detected under UV microscope. The utility of FAT in research still remains under the control conditions of laboratory but its diagnostic potential is now considered conditional. Blome *et al.* (2006) have mentioned that the FAT has been replaced in many laboratories by RT-PCR. They emphasized that the test is valid only when the herd size to be monitored is large enough and the sample analyzed is proper like the tonsils.

Immunohistochemistry

Immunohistochemistry is a sensitive test to demonstrate the viral antigen in tissues, smears and cell cultures. It is similar to FAT but differs in use of peroxidase enzyme as the detection system. Mulas *et al.* (1997) demonstrated CSFV in formalin fixed and paraffin embedded tissues of experimentally infected pigs with help of monoclonal antibodies against Gp55 of CSFV and avidin-biotin-peroxidase complex (ABC) as the detection system. The test was able to differentiate CSFV infected tissues from control pig tissues, African swine fever virus infected tissues, or bovine viral diarrhea virus infected porcine and bovine tissues. Narita *et al.* (1999) proved the utility of immunohistochemistry by demonstrating the CSFV antigen in tissues preserved in formalin for 18 years using avidin-biotin-peroxidase complex as detection system. Choi and Chae (2003) used immunohistochemistry to demonstrate that CSFV is present in non-sperm cells of semen and not the sperms themselves. Gomez-Villamandos *et al.* (2006) could demonstrate Gp55 of CSFV in perivascular infiltrate, interstitial infiltrate and microglial cells of brain of pigs infected with CSFV.

ELISA

Dunne (1975) has mentioned various serodiagnostic tests for CSFV for large scale surveillance like hemagglutination test, compliment fixation test and most recommended serological test in those times the AGPT. Paton *et al.* (2000) recommend ELISA for the screening of large numbers of animals in herds suspected of being recently infected by CSFV. Virus antigen capture enzyme-linked immunosorbent assays (AgC-ELISAs) are also being used because of its high accuracy. Lin *et al.* (2005) proved that chimeric protein of fragments of E^{rns} and E2 gives better diagnostic efficiency as compared to the same peptides alone. Lin *et al.* (2010) designed an ELISA with quadruple antigenic epitope gene region of the classical swine fever virus (CSFV) E2 glycoprotein expressed in *E. coli.* The region is highly conserved and the epitope is not found in BVDV. The ELISA was highly efficient and did not react with the sera of pigs infected with BVD.

Reverse Transcription-Polymerase Chain Reaction (RT-PCR)

Polymerase chain reaction is one of the most important diagnostic tools of the modern times. Harding *et al.* (1996) investigated the capability of the RT-PCR test to discriminate between CSFV, bovine viral diarrhea virus (BVDV), African swine fever virus (ASFV) and pseudorabies virus (PRV). They found it to be 100 percent efficient and specific in differential diagnosis. Paton *et al.* (2000) have described PCR as a better and efficient diagnostic tool as compared to FAT and ELISA. They have mentioned the RT-PCR of 5′nontranslated region of the CSFV genome to be used more for diagnosis. The importance of PCR for phylogenetic analysis of various genotypes of the virus strains was also emphasized by them. Choi and Chae (2003) tested a semi-nested RT-PCR for the detection of classical swine fever virus (CSFV) in semen of experimentally infected boars. The seminal fluid and non-sperm cells revealed the presence of virus but not the sperm cells. The tremendous potential of RT-PCR in molecular diagnosis was once again proven when Singh *et al.* (2005) successfully standardized RT-PCR amplification of a 172-bp fragment from E2/NS2 region of CSFV genome from archival formalin-fixed CSF samples.

Real Time PCR

It is a modification of conventional PCR with the added advantage of monitoring the amplification while the reaction is in process. Risatti *et al.* (2003) tested the efficiency of fluorogenic-probe hydrolysis (TaqMan) - reverse transcriptase PCR assay for CSFV. They found that the test could detect the virus in the oronasal secretions 3 to 5 days and in blood 0 to 3 days before the appearance of clinical disease. The test was negative for border disease virus and BVDV used as negative control. Ophuis *et al.* (2006) used real time PCR (TaqMan assay) to detect the sequential spread of CSFV virus in various body tissues of experimentally infected pigs. The viral RNA could be detected as early as 1 dpi in tonsils and 3 dpi in lymph nodes. Brain and pancreas revealed the virus last. They also found that tonsil, nasal and rectal swabs could be used to detect the virus. Weesendorp *et al.* (2008a) conducted a study to estimate survival times of CSFV in feces and urine for different environmental temperatures with the help of real time PCR and cell culture. The real time PCR was used to estimate the rate of viral RNA decay. They found that viral survivability varied according to strains in feces but not in urine.

Vaccination

Live Attenuated Vaccines

Live vaccines against CSF had been used for more than half century duration. Such vaccines are developed by inducing adaptive mutations in the live virus by serial passages in either rabbits (Lapinized) or in permissive cell cultures. Some examples include Chinese C strain vaccine, Lapinized Philippines Coronel (LPC) strain, South Korean LOM, PAV-250, Thiverval strain, Russian LK-VNIVViM, Japanese guinea-pig exaltation-negative (GPE-) strain and the French cell culture adapted. The live attenuated vaccines provide early protection and the immunity can last up to 10 months (Qiu *et al.*, 2006; Blome *et al.*, 2017; Coronado, 2021)

Marker Vaccines

These are DIVA vaccines. It means they can differentiate between animals infected with virulent strain of virus and a vaccinated animal. The types of marker vaccines include subunit vaccines (*e.g.*, E2 Subunit Vaccine), viral vectors (chimera vaccines and replicons), live marker vaccines, immunogenic CSFV peptides, and DNA vaccines (Qiu *et al.*, 2006; Blome *et al.*, 2017; Coronado, 2021).

Conclusion

CSF causes severe morbidity and mortality. It is a resistant virus outside the host and therefore spread is easy. Even the countries free of it continuously implement measures to prevent its entry and monitor it worldwide. International movement of pigs, pork products, pig semen and other swine based biological are a continuous threat for even the endemic countries for introduction of newer genotypes. In endemic countries the existence of virus in feral swine makes its management and eradication programs weaker. Emergence of new CSFV genotypes in endemic countries has also been due to inefficient vaccination. CSF can be easily detected by most conventional laboratory diagnostic techniques. Emergence of ASF in CSF endemic countries like India has made it important to efficiently differentially diagnose CSF. There is urgent need of DIVA vaccine in endemic countries to implement better control measures.

REFERENCES

Anthony, D.J. and Lewis, E.F. (1961). Swine fever. In: Disease of the pig, 5th edn, BTC, pp.118-129.

Bauhofer, O., Summerfield, A., McCullough, K.C. and Ruggli, N. (2005). Role of double-stranded RNA and N^{pro} of classical swine fever virus in the activation of monocyte-derived dendritic cells. Virol, 343: 93-105.

Bautista, M.J., Ruiz-Villamor, E., Salguero, F.J., Sanchez-Cordon, P.J., Carrasco, L. and Gomez-Villamandos, J.C. (2002). Early platelet aggregation as a cause of thrombocytopenia in classical swine fever. Vet Pathol, 39: 84-91.

Blome, S., Meindl-Böhmer, A., Loeffen, W., Thuer, B. and Moennig, V. (2006). Assessment of classical swine fever diagnostics and vaccine performance. Rev Sci Tech Off Int Epiz, 25(3): 1025-1038.

Blome, S., Staubach, C., Henke, J., Carlson, J. and Beer, M. (2017). Classical Swine Fever - An updated review. Viruses, 9(86): doi:10.3390/v9040086.

Blome, S., Meindl-Böhmer, A., Nowak, G., and Moennig, V. (2013). Disseminated intravascular coagulation does not play a major role in the pathogenesis of classical swine fever. Vet Microbiol, 162: 360-368.

Blome, S., Wernike, K., Reimann, I., König, P., Moβ, C. and Beer, M.A. (2017). Decade of research into classical swine fever marker vaccine CP7_E2alf (Suvaxyn® CSF Marker): a review of vaccine properties. Vet Res, 48(1): 51. doi: 10.1186/s13567-017-0457-y.

Botner, A. and Belsham, G.J. (2012). Virus survival in slurry: Analysis of the stability of foot-and-mouth disease, classical swine fever, bovine viral diarrhoea and swine influenza viruses. Vet Microbiol, 157: 41-49.

Brown, V.R. and Bevins, S.N. (2018). A Review of Classical swine fever virus and routes of introduction into the United States and the potential for virus establishment. Front Vet Sci, 5: 31. doi: 10.3389/fvets.2018.00031

Bruschke, C.J.M., Hulst, M.M., Moormann, R.J.M., Van Rijn, P.A. and Van Oirschot, J.T. (1997). Glycoprotein E^{rns} of pestiviruses induces apoptosis in lymphocytes of several species. J Virol, 71(9): 6692-6696.

Calderon, N.L., Paasch, L.H. and Fortoul, T.I. (2000). Ultrastructural glomerular changes in experimental infection with the classical swine fever virus. Vet Res, 31: 447-453.

Carrasco, C.P., Rigden, R.C., Vincent, I.E., Balmelli, C., Ceppi, M., Bauhofer, O., Tâche, V., Hjertner, B., McNeilly, F., van Gennip, H.G., McCullough, K.C. and Summerfield, A. (2004). Interaction of classical swine fever virus with dendritic cells. J Gen Virol, 85(6): 1633-1641. doi: 10.1099/vir.0.19716-0. PMID: 15166448.

Changchun, T., Zongji, L., Hongwei, L., Xinglong, Y., Xiangtao, L., Yuehong, L., Hongyong, Z. and Zhen, Y. (2001). Phylogenetic comparison of classical swine fever virus in China. Virus Res, 81: 29-37.

Choi, C. and Chae, C. (2003). Detection of classical swine fever virus in boar semen by reverse transcription-polymerase chain reaction. J Vet Diagn Invest, 15: 35-41.

Coronado, L., Perera, C.L., Rios, L., Frías, M.T. and Pérez, L.J. (2021) A critical review about different vaccines against classical swine fever virus and their repercussions in endemic regions. Vaccines, 9: 154. https://doi.org/10.3390/vaccines9020154

Depner, K., Bauer, T. and Liess, B. (1992). Thermal and pH stability of pestiviruses. Rev Sci Tech, 11: 885-893.

Donis, R.O. (1995). Molecular biology of bovine viral diarrhea virus and its interactions with the host. Vet Clin N Am Food Anim Pract, 11: 393-423.

Dunne, H.W. (1958). Hog Cholera. In: Diseases of Swine (edited by H.W. Dunne) 1st Ed. Iowa State Univ. Press. pp 111-144.

Dunne, H.W. (1970). Hog Cholera. In: Diseases of Swine (edited by H.W. Dunne and A.D. Leman) 3rd Ed. Iowa State Univ. Press. pp 177- 239.

Dunne, H.W. (1975). Hog Cholera. In: Diseases of Swine (edited by H.W. Dunne and A.D. Leman) 4th Ed. Iowa State Univ. Press. pp 189-255.

Edwards, S. (2000). Survival and inactivation of classical swine fever virus. Vet Microbiol, 73: 175-181.

Farez, S. and Morley, R.S. (1997). Potential animal health hazards of pork and pork products. Rev Sci Tech, 16: 65-78.

Fiebach, A.R., Guzylack-Piriou, L., Python, S., Summerfield, A. and Ruggli, N. (2011). Classical swine fever virus N^{pro} limits type I interferon induction in plasmacytoid dendritic cells by interacting with interferon regulatory factor 7. J Virol, 85: 8002-8011.

Gale, P. (2004). Risks to farm animals from pathogens in composted catering waste containing meat. Vet Rec, 155: 77-82.

Gomez -Villamandos, J.C., Garcia, D.I., Nunez, A., Salguero, F.J., Ruiz-Villamor, E., Romero-Trevejo, J.L. and Sanchez-Cordon, P.J. (2006). Neuropathologic study of experimental classical swine fever. Vet Pathol, 43: 530-540.

Gomez-Villamandos, J.C., Ruiz-Villamor, E., Bautista, M.J., Sanchez, C.P., Sanchez-Cordon, P.J. and Salguero, F.J. (2001). Morphological and immunohistochemical changes in splenic macrophages of pigs infected with classical swine fever. J Comp Pathol, 125: 98-109. doi:10.1053/jcpa.2001.0487.

Gomez-Villamandos, J.C., Ruiz-Villamor, E., Bautista, M.J., Quezada, M., Sanchez, C.P., Salguero, F.J., and Sierra, M.A. (2000). Pathogenesis of classical swine fever: renal hemorrhages and erythrodiapedesis. J Comp Path, 123: 47-54.

Gomez-Villamandos, J.C., Salguero, F.J., Ruiz-Villamor, E., Sanchez-Cordon, P.J., Bautista, M.J., and Sierra, M.A. (2003). Classical swine fever: pathology of bone marrow. Vet. Pathol, 40: 157-163.

Gray, E.W. and Nettleton, P.F. (1987). The ultrastructure of cell cultures infected with border disease and bovine virus diarrhoea viruses. J Gen Virol, 68: 2339-2346.

Haas, B., Ahl, R., Böhm, R. and Strauch, D. (1995). Inactivation of viruses in liquid manure. Rev Sci Tech, 14: 435-445.

Harding, M.J., Prud'homme, I., Gradil, C.M., Heckert, R.A., Riva, J., McLaurin, R., Dulac, G.C. and Vydelingum, S. (1996). Evaluation of nucleic acid amplification methods for the detection of hog cholera virus. J Vet Diagn Invest, 8: 414-419.

Heene, D., Hoffmann-Fezer, G., Müller-Berghaus, G., Hoffmann, R., Weiss, E. and Lasch, H.G. (1971). Coagulation disorders in acute hog cholera. Beitr Pathol, 144: 259-271.

Heinz, F.X., Collett, M.S., Purcell, R.H., Gould, E.A., Howard, C.R., Houghton, M., Moormann, R.J.M., Rice, C.M. and Thiel, H.J. (2000). Family Flaviviridae. In: Virus Taxonomy. Seventh Report of the International Committee on Taxonomy of Viruses. San Diego, CA: Academic Press. pp. 859-878.

Hoffmann, R., Hoffmann-Fezer, G., Kimeto, B. and Weiss, E. (1971). Microthrombi as morphological evidence of consumption coagulopathy in acute hog cholera. Zentralbl Veterinarmed B, 18: 710-718.

Jamin, A., Gorin, S., Cariolet, R., LePotier, M.F. and Kuntz-Simon, G. (2008). Classical swine Fever virus induces activation of plasmacytoid and conventional dendritic cells in tonsil, blood, and spleen of infected pigs. Vet Res, 39: 7. doi:10.1051/vetres:2007045

Kaden, V., Lange, E., Faust, A. and Teifke J.P. (2007). Value of skin punch biopsies for the diagnosis of acute classical swine fever. J Vet Diagn Invest, 19: 697-701.

Kamolsiriprichaiporn, S., Hooper, P.T., Morrissy, C.J. and Westbury, H.A. (1992). A comparison of the pathogenecity of two strains of hog cholera virus. 1. Clinical and pathological studies. Aust Vet J, 69(10): 240-244.

Knoetig, S.M., Summerfield, A., Spagnuolo-Weaver, M. and McCullough, K.C. (1999). Immunopathogenesis of classical swine fever: role of monocytic cells. Immunol, 97: 359-366. doi:10.1046/j.1365-2567.1999.00775.x

Kramer, M., Staubach, C., Koenen, F., Haegeman, A., Pol, F., Le Potier, M.F., and Greiser-Wilke, I. (2009). Scientific review on Classical swine fever. EFSA Support, Publ, 6.

Krey, T., Thiel, H.J. and Rümenapf, T. (2005). Acid-resistant bovine pestivirus requires activation for pH-triggered fusion during entry. J Virol, 79: 4191-4200.

Lackner, T., Müller, A., Pankraz, A., Becher, P., Thiel, H.J., Gorbalenya, A.E. and Tautz, N. (2004). Temporal modulation of an autoprotease is crucial for replication and pathogenicity of an RNA virus. J Virol, 78: 10765-10775.

Lackner, T., Thiel, H.J. and Tautz, N. (2006). Dissection of a viral autoprotease elucidates a function of a cellular chaperone in proteolysis. Proc Natl Acad Sci USA, 103: 1510-1515.

Lecot, S., Belouzard, S., Dubuisson, J. and Rouille, Y. (2005). Bovine viral diarrhea virus entry is dependent on clathrin-mediated endocytosis. J Virol, 79: 10826-10829.

Liess, B. (1987). Pathogenesis and epidemiology of hog cholera. Ann Rech Vet, 18: 139-145.

Lin, M., Trottier, E. and Mallory, M. (2005). Enzyme-linked immunosorbent assay based on a chimeric antigen bearing antigenic regions of structural proteins E^{rns} and E2 for serodiagnosis of classical swine fever virus infection. Clin Diag Lab Immunol, 7: 877-881.

Lin, G., Zheng, F., Zhou, J., Cao, X., Gong, X., Wang, G. and Qiu, C. (2010). An indirect ELISA of classical swine fever virus based on quadruple antigenic epitope peptide expressed in *E. coli.* Virol Sinica, 25: 71-76.

Lin, M., Trottier, E. and Mallory, M. (2005). Enzyme-linked immunosorbent assay based ona chimeric antigen bearing antigenic regions of structural proteins E^{rns} and E2 for serodiagnosis of classical swine fever virus infection. Clin Diag Lab. Immunol, 7: 877-881.

Lindenbach, B.D. and Rice, C.M. (2001). Flaviviridae: The viruses and their replication. In Field's Virology, 4th Edn., vol. 1, pp. 991-1041, Lippincott Williams and Wilkins. Philadelphia, USA.

Liu, J., Fan, X.Z., Wang, Q., Xu, L., Zhao, Q.Z., Huang, W., Zhou, Y.C., Tang, B., Chen, L., Zou, X.Q., Sha, S. and Zhu, Y.Y. (2011). Dynamic distribution and tissue tropism of classical swine fever virus in experimentally infected pigs. Virol J, 8: 201. doi: 10.1186/1743-422X-8-201.

Moennig, V. (2000). Introduction to classical swine fever: virus, disease and control policy, Vet Microbiol, 73(2-3): 93-102.

Moennig, V., Floegel-Niesmann, G. and Greiser-Wilke, I. (2003). Clinical signs and epidemiology of classical swine fever: A review of new knowledge. Vet J, 165: 11-20.

Mulas, J.M.D.L., Ruiz-Villamor, E., Donoso, S., Quezada, M., Lecocq, C. and Sierra, M.A. (1997). Immunohistochemical detection of hog cholera viral glycoprotein 55 in paraffin-embedded tissues. J Vet Diagn Invest, 9: 10-16.

Narita, M., Kimura, K., Tanimura, N. and Ozaki, H. (1999). Immunohistochemical detection of hog cholera virus antigen in paraffin wax-embedded tissues from naturally infected pigs. J Comp Path, 121: 283-286.

Ohmann, H.B. (1990). Electron microscopy of bovine virus diarrhoea virus. Rev Sci Tech, 9: 61-73.

OIE (Office International des Epizootics/World Organization for Animal Health) (2000). In: Santéanimalemondial (SAM) - India, pp. 195-196.

Ophuis, R.J.A.O., Morrissy, C.J. and Boyle, D.B. (2006). Detection and quantitative pathogenesis study of classical swine fever virus using a real time RT-PCR assay. J Virol Methods, 131: 78-85.

Panina, G.F., Civardi, A., Cordioli, P., Massirio, I., Scatozza, F., Baldini, P. and Palmia, F. (1992). Survival of hog cholera virus (HCV) in sausage meat products (italian salami). Int J Food Microbiol, 17: 19-25.

Patil, S.S., Hemadri, D., Shankar, B.P., Raghavendra, A.G., Veeresh, H., Sindhoora, B., Chandan, S., Sreekala, K., Gajendragad, M.R. and Prabhudas, K. (2010). Genetic typing of recent classical swine fever isolates from India. Vet Microbiol, 141: 367-373.

Patil, S.S., Hemadri, D., Veeresh, H., Sreekala, K., Gajendragad, M.R. and Prabhudas, K. (2012). Phylogenetic analysis of NS5B gene of classical swine fever virus isolates indicates plausible Chinese origin of Indian subgroup 2.2 viruses. Virus Genes, 44: 104-108.

Paton, D.J., McGoldrick, A., Bensaude, E., Belak, S., Mittelholzer, C., Koenenc, F., Vanderhallen, H., Greiser-Wilke. I., Scheibner, H., Stadejek, T., Hofmann, M. and Thuer, B. (2000) Classical swine fever virus: a second ring test to evaluate RT-PCR detection methods. Vet Microbiol, 77: 71-81.

Pauly, T., König, M., Thiel, H.J. and Saalmüller, A. (1998). Infection with classical swine fever virus: Effects on phenotype and immune responsiveness of porcine T lymphocytes. J Gen Virol, 79: 31-40.

Pestova, T.V. and Hellen, C.U. (1999). Internal initiation of translation of bovine viral diarrhea virus RNA. Virol, 258: 249-256.

Poole, T.L., Wang, C., Popp, R.A., Potgieter, L.N., Siddiqui, A. and Collett, M.S. (1995). Pestivirus translation initiation occurs by internal ribosome entry. Virol, 206: 750-754.

Qiu, H.J., Shen, R.X., and Tong, G.Z. (2006). The Lapinized chinese strain vaccine against classical swine fever virus: a retrospective review spanning half a century. Agric Sci China, 5: 1-14.

Rajkhowa, T.K., Hauhnar, L., Lalrohlua, I. and Mohanarao, G.J. (2014). Emergence of 2.1. Subgenotype of classical swine fever virus in pig population of India in 2011. Vet Q, 34: 224-228.

Renson, P., Blanchard, Y., Le Dimna, M., Felix, H., Cariolet, R., Jestin, A. and Le Potier, M.F. (2010). Acute induction of cell death-related IFN stimulated genes (ISG) differentiates highly from moderately virulent CSFV strains. Vet Res, 41: 7.

Ressang, A.A. (1973). Studies on the pathogenesis of hog cholera. I. Demonstration of hog cholera virus subsequent to oral exposure. Zentralbl Veterinarmed B, 20: 256-271.

Ressang, A.A. (1973). Studies on the pathogenesis of hog cholera. II. Virus distribution in tissue and the morphology of the immune response. Zentralbl Veterinarmed B, 20: 272-288.

Rijnbrand, R., Van der Straaten, T., Van Rijn, P.A., Spaan, W.J. and Bredenbeek, P.J. (1997) Internal entry of ribosomes is directed by the 50 noncoding region of classical swine fever virus and is dependent on the presence of an RNA pseudoknot upstream of the initiation codon. J Virol, 71: 451-457.

Risatti, G.R., Callahan, J.D., Nelson, W.M. and Borca, M.V. (2003). Rapid detection of classical swine fever virus by a portable real-time reverse transcriptase PCR assay. J Clin Microbiol, 41(1): 500-505.

Roychoudhury, P., Sarma, D.K., Rajkhowa, S., Munir, M. and Kuchipudi, S.V. (2014). Predominance of genotype 1.1 and emergence of genotype 2.2 classical swine fever viruses in north-eastern region of India. Transbound Emerg Dis, 61(1): 69-77.

Ruggli, N., Tratschin, J.D., Mittelholzer, C. and Hofmann, M.A. (1996). Nucleotide sequence of classical swine fever virus strain Alfort/187 and transcription of infectious RNA from stably cloned full-length cDNA. J Virol, 70: 3478-3487.

Sánchez-Cordón, P.J., Romanini, S., Salguero, F.J., Núñez, A., Bautista, M.J., Jover, A. and Gómez-Villamos, J.C. (2002). Apoptosis of thymocytes related to cytokine expression in experimental classical swine fever. J Comp Pathol, 127(4): 239-248. doi: 10.1053/jcpa.2002.0587.

Sanchez-Cordon, P.J., Romanini, S., Salguero, F.J., Ruiz-Villamor, E., Carrasco, L. and Gomez-Villamandos, J.C. (2003). A histopathologic, immunohistochemical, and ultra-structural study of the intestine in pigs inoculated with classical swine fever virus. Vet Pathol, 40: 254-262. doi:10.1354/vp.40-3-254

Singh, V.K., Saikumar, G. and Paliwal, O. P. (2005). Detection of classical swine fever virus in archival formalin-fixed tissues by reverse transcription polymerase chain reaction. Res Vet Sci, 79: 81-84.

Summerfield, A., Alves, M., Ruggli, N., de Bruin, M.G. and McCullough, K.C. (2006). High IFN- responses associated with depletion of lymphocytes and natural IFN-producing cells during classical swine fever. J Interferon Cytokine Res, 26: 248-255.

Summerfield, A., Hofmann, M.A. and McCullough, K.C. (1998). Low density blood granulocytic cells induced during classical swine fever are targets for virus infection. Vet Immunol Immunopathol, 63: 289-301. doi:10.1016/S0165-2427(98)00108-1

Summerfield, A., Knoetig, S.M., Tschudin, R. and McCullough, K.C. (2000). Pathogenesis of granulocytopenia and bone marrow atrophy during classical swine fever involves apoptosis and necrosis of uninfected cells. Virol, 272: 50-60.

Summerfield, A. and Ruggli, N. (2015). Immune responses against classical swine fever virus: Between ignorance and lunacy. Front Vet Sci, 2: 10.

Suradhat, S., Damrongwatanapokin, S. and Thanawongnuwech, R. (2007). Factors critical for successful vaccination against classical swine fever in endemic areas. Vet Microbiol, 119: 19.

Susa, M., König, M., Saalmüller, A., Reddehase, M.J. and Thiel, H.J. (1992). Pathogenesis of classical swine fever: B-lymphocyte deficiency caused by hog cholera virus. J Virol, 66: 1171-1175.

Tautz, N., Elbers, K., Stoll, D., Meyers, G. and Thiel, H.J. (1997). Serine protease of pestiviruses: Determination of cleavage sites. J Virol, 71: 5415-5422.

Terpstra, C. (1987). Epizootiology of swine fever. Vet Q, 9 (Suppl. 1): 50S-60S.

Terpstra, C. (1991). Hog cholera: An update of present knowledge (special review series). Br Vet J, 147: 397-406.

Tessler, J., Stewart, W.C. and Kresse, J.I. (1975). Stabilization of hog cholera virus by dimethyl sulfoxide. Can J Comp Med, 39: 472-473.

Trautwein, G. and Leiss, B. (1988). Pathology and Pathogenesis of the Disease, Classical Swine Fever and Related Infections. Boston, MA: Martinus Nijhoff Publishing. pp.27-54.

Trautwein, G. (1988). Classical swine fever and related infections. In: Pathology and Pathogenesis of the Disease, Martinus Nijhoff: Boston, MA, USA, pp. 27-54.

Turner, C., Williams, S.M. and Cumby, T.R. (2000). The inactivation of foot and mouth disease, aujeszky's disease and classical swine fever viruses in pig slurry. J Appl Microbiol, 89: 760-767.

Van Oirschot, J.T. (1979). Experimental production of congenital persistent swine fever infections: Clinical, pathological and virological observations. Vet Microbiol, 4: 117-132.

Van Oirschot, J.T. (1986). Hog Cholera. In: Diseases of Swine, (edited by A.D. Leman, Barbara Straw, Robert D. Glock, William L. Mengeling, R.H.C. Penny and Erwin Scholl) 6th Ed. Iowa State Univ. Press. pp 289-300.

Wang, Z., Nie, Y., Wang, P., Ding, M. and Deng, H. (2004). Characterization of classical swine fever virus entry by using pseudotyped viruses: E1 and E2 are sufficient to mediate viral entry. Virol, 330: 332-341.

Weesendorp, E., Stegeman, A. and Loeffen, W.L. (2008). Survival of classical swine fever virus at various temperatures in faeces and urine derived from experimentally infected pigs. Vet Microbiol, 132: 249-259.

Weesendorp, E., Landman, W.J., Stegeman, A. and Loeffen, W.L. (2008) Detection and quantification of classical swine fever virus in air samples originating from infected pigs and experimentally produced aerosols. Vet Microbiol, 127: 50-62.

Weiss, E., Teredesai, A., Hoffmann, R. and Hoffmann-Fezer, G. (1973). Volume distribution and ultrastructure of platelets in acute hog cholera. Thromb Diath Haemorrh, 30: 371-380.

Wijnker, J.J., Depner, K.R. and Berends, B.R. (2008). Inactivation of classical swine fever virus in porcine casing preserved in salt. Int J Food Microbiol, 128: 411-413.

Chapter 17

Rabies

S. Isloor*, L. Dilip and R. Sharada

KVAFSU-CVA Rabies Diagnostic Laboratory, Department of Veterinary Microbiology, Veterinary College, KVAFSU, Hebbal, Bengaluru – 560 024, Karantaka, India
****e-mail: kisloor@gmail.com***

ABSTRACT

Rabies is one of the most important and fatal zoonotic diseases in the world. This viral disease is endemic in various parts of Asia, Africa, Western Europe and North and South America. Although rabies is transmitted through rabid animal bites, dogs are the main source of transmission of Asia, Africa and Latin America. Rabies is endemic in all countries in the Indian subcontinent. The disease often affects the central nervous system, causing behavioural abnormalities and paralysis in most of the hosts it afflicts. Rabies virus belongs to the Lyssavirus genus of Rhabdoviridae family that generally affects the nervous system and causes abnormal behaviour and/or paralysis in the bitten victims including animals and human beings. Direct Fluorescent Antibody test (DFA) is often used for postmortem diagnosis of rabies. The development of direct rapid immunohistochemistry test (dRIT) is one of the most important advances in rabies diagnosis. Additionally, Lateral Flow assay (LFA) is an immunochromatography-based tool and is a rapid field test for diagnosing rabies without the need for field diagnostic equipment. Recently, Reverse transcriptase PCR (RT-PCR) and real-time PCR (including LN 34 real-time PCR) methods have become popular in molecular analysis and epidemiological studies. Furthermore, the rabies virus neutralization test (FAVN or RFFIT) is considered the gold standard for assessing the efficacy of rabies vaccinations. Alternatively, ELISA (Enzyme Linked immunosorbent assay) can be used. Controlling rabies in animals, especially dogs, requires a coordinated, multi-agency, multi-pronged approach. Programs emphasizing promotion of mass dog immunization and seromonitoring of vaccination should be developed for rabies endemic countries.

Keywords: *Rabies, Dogs, Diagnosis, Vaccine, Control.*

Introduction

Rabies is a fatal progressive neurologocal infection that affects all warm blooded animals including man. The word rabies is etymologically derived from the sanskrit word "Rabhas" which means "to do violence' describing it's furious form. Rabies infects warm blooded animals leading to fatality (WHO, 2018). The disease was first described in the Eshnunna code in the 23rd century BC. Aristotle (4th century BC) had quoted that, "Dogs suffer from madness and become irritable, and all animals they bite become diseased". Cornelius Celsus coined the term "hydrophobia" (fear of water) in the first century AD, as he described a case of human rabies. The illness was known as lytta or lyssa, derived from the belief that the illness was caused by a worm under the tongue (lytta), or hydrophobia, which depicts the fear of water associated with the illness. For the first time in 1885, Louis Pasteur (Figure 17.1) developed, tested and used the anti rabies vaccine without knowing the nature of the virus and this led to the prevention of rabies by vaccination. Remlinger first demonstrated the filterability of this infectious agent in 1904 and in the same year, Negri first demonstrated the intracytoplasmic inclusion bodies that were subsequently known as 'Negri bodies' in brain impressions of rabid animals.

Figure 17.1: Louis Pasteur.

Globally, dogs are the major source of exposure of human beings and livestock to rabies virus. Furthermore, biting animals, especially cats, wild carnivores and bats also pose a threat as they are considered to be the potential reservoirs of the virus (OIE, 2013; Birhane *et al.*, 2017; Rupprecht *et al.*, 2017).

Wild life including wolf, jackals, foxes, mongoose, racoon, skunk and vampire bats are the natural reservoirs.

Classification

Rabies is caused by negative strand RNA-viruses belonging to the genus *Lyssavirus*, family *Rhabdoviridae* of the order *Mononegavirales*. Lyssa virus species are divided into three phylogroups based on their genetic distance and serological cross-reactivity. The rabies virus belongs to phylogroup 1 (ICTV 2017; Rupprecht *et al.*, 2002). The other lyssaviruses produce diseases similar to rabies, and bats are an important reservoir host to several of these viruses (Zee and MacLachlan, 2005). The Phylogroup I includes Rabies lyssavirus, Duvenhage lyssavirus, European bat 1 lyssavirus, European bat 2 lyssavirus, Bokeloh bat lyssavirus, Kotalahti bat lyssavirus, Aravan lyssavirus, Irkut lyssavirus, Khujand lyssavirus, Australian bat lyssavirus, Gannoruwa bat lyssavirus, Taiwan bat lyssavirus, Lagos bat lyssavirus. The Phylogroup II includes the Mokola lyssavirus, Shimoni bat lyssavirus and Ikoma lyssavirus. The Phylogroup III has the Lleida bat lyssavirus and the West Caucasian bat lyssavirus. Monoclonal antibody analysis has demonstrated considerable antigenic variation among virus isolates from outbreaks associated with different wildlife vectors or from different geographical areas (Debbie and Trimarchi, 1992). Using specific monoclonal antibodies, researchers determined whether a viral species was a rabies virus or a rabies-related virus by recognising different antigenic sites. Viruses currently assigned to the Lyssavirus genus have been isolated from various mammals all over the world. All phylogroup I genotypes have caused fatal rabies-like encephalitis in humans, whereas Mokola virus is thought to have caused three known human infections, one of which was a fatal encephalitis with no rabies-like symptoms. Phylogroup II viruses are less pathogenic in experiments, and there is little, if any, cross-neutralisation with phylogroup I lyssaviruses. Almost all infections are due to classical rabies (genotype 1) viruses. The remainders are due to the six genotypes of rabies-related viruses (Greene and Rupprecht, 2006). Some lyssaviruses have a more restricted geographic distribution: In Africa the Lagos bat, Shimoni bat, Mokola, and Duvenhage viruses have been identified, European lyssaviruses are found exclusively in Europe. Aravan, Khujand, Irkut, and West Caucasian bat viruses are limited to Asia and all Australian bat viruses are found only in Australia.

Highly virulent strains of rabies virus isolated from naturally occurring cases are referred to as "Street viruses". The attenuated laboratory strains are referred to as "Fixed viruses". These are street viruses adapted to the laboratory by passaging them in unnatural hosts like rabbits or in cell culture. These strains differ from each other in their biological properties in laboratory animals, for example, virulence, incubation period and distribution and nature of lesions in target tissues. The street viruses have a long incubation period, an affinity for salivary glands and produce intracytoplasmic inclusion bodies but do not produce any cytopathic effect in cell cultures. The fixed strains have short and defined incubation period of 5–7 days and are more neurotropic for rabbits but have no affinity for the salivary glands; they multiply faster and are stable viruses. They are comparatively less pathogenic and do not produce inclusion bodies, thus are used for vaccine production. These can infect only by the CNS route. Further, these are not seen in the saliva and other peripheral secretions. For example, Flury strain, Street Alabama Dufferin (SAD), Vnukovo, Kelev. The antigenic variation between the strains can be distinguished by their reaction with monoclonal antibodies.

Epidemiology

Rabies occurs in all countries except Japan, the United Kingdom, Ireland, Cyprus, New Zealand, Scandinavia, the Hawaiian Islands, the Caribbean Islands, Australia and Switzerland. The disease is Endemic in many countries of Africa, Asia, North and South America and Western Europe. Rabies has two transmission cycles: urban cycle and sylvatic cycle. The urban cycle is retained in dogs and transmitted to other species through dog bite. Dogs are still the most important source of infection in developing countries of Africa, Asia and Latin America (Zee and MacLachlan, 2005). Sylvatic cycles are maintained in the wild and can cause spillover infections in domestic animals and humans. Based on susceptibility, the hosts are broadly classified into four categories; most susceptible are the fox, coyote, wolf, jackal and voles. Skunk, bat, mongoose and cattle are considered increasingly susceptible. The moderately susceptible hosts are dogs, sheep, goat and horses; birds are said to be least susceptible. Foxes are important reservoirs in Western Europe, portions of Canada, Alaska and the desert south-western regions of the USA, whereas skunks and raccoons act as reservoirs in regions of North America. In Asia and Africa, mongoose is one of the reservoirs. The first case report of rabies in a wolf (*Canis lupus pallipes*) from India was documented based on laboratory evidence (Isloor *et al.*, 2014). The first confirmed case of rabies in a sloth bear (*Melursus ursinus*) from India was reported by Patel *et al.* (2018) from Gujarat state. Cattle and equines are considered dead-end hosts since cattle do not evince biting tendency and equines very rarely. Rats and bandicoots are naturally susceptible. However, there are no documented evidences of transmission of rabies through rat bites. Laboratory animals such as mice, rabbits and guinea pigs can be infected experimentally (Sarma, 2009).

Global Scenario

Rabies is reported from all the continents except Antarctica, with over 95 percent of human deaths occurring in Asia and Africa. Based on the prevalence of rabies, the countries have been classified as (a) countries with enzootic canine rabies - Asia, Latin America, Africa, (b) countries where canine rabies is under control and wildlife rabies is prevalent - Western Europe, Canada, the USA, (c) rabies free countries - most islands, Australia, England, Japan (De Serres *et al.*,

2008). In America, various species of insectivores and vampire bats overlap the rabies virus and there are 30 different variants of the virus identified in various bat species in North America (Nadin-Davis *et al.*, 2001; Nadin-Davis and Loza Rubio, 2006) that can be transmitted to other animal species in the wild. The rabies virus is maintained in domestic dogs, vampires and insectivorous bats in sylvatic cycles in South America (Favi *et al.*, 2002; Kobayashi *et al.*, 2005). The virus is also reported in monkeys, wolves, coyotes, skunks, and foxes and mongooses (Belotto *et al.*, 2005; Everard and Everard, 1988) in southern and Central America. With the discovery of Australian bat lyssavirus (ABLV) in 1996 Australia lost its 'rabies-free' status. ABLV was isolated from an insectivorous bat and four species of flying fox bats after the death of a ten-year-old girl with clinical signs of rabies (Gould *et al.*, 2002). The last case of dog transmitted rabies was reported in 1867, and since then Australia was considered "rabies-free" though incidences of dog rabies were occasionally reported which were attributed to importation of the dogs (Warrilow, 2005). In Europe, intense vaccination campaigns were successful in controlling dog rabies in 1940s, but wild canines escalated the incidences of rabies since then. Terrestrial rabies is under control in countries such as the United Kingdom, Finland and the Netherlands through strict vaccination campaigns (Bourhy *et al.*, 2005). There was paucity of information regarding rabies in the African continent until late twentieth century. Rabies occurs as scattered foci and is spread by dogs in sparsely populated countries spanning the Saharan desert; occasional cases in camels have been reported (Swanepoel, 2004). The rabies-related viruses, *viz.* Mokola, Lagos bat and Duvenhage virus have been presumed to be in circulation since hundreds of years in the African continent (Nel and Rupprecht, 2007). The World Health Organization (WHO) has reported 23,700 human rabies deaths per annum in Africa (WHO, 2013) which is reported mostly in poor rural communities and children and is attributed to the inadequate and costly necessary resources for rabies prevention and treatment.

Indian Scenario

In India, the disease is widespread in all the states and union territories (UTs), except for Lakshadweep, Andaman and Nicobar islands (Isloor *et al.*, 2019). The disease is endemic in the country and human and animal cases of rabies are reported from all over the country. Majority of the rabies cases are associated with dog bites. Cats, wolf, jackal, mongoose and monkeys are other important animals transmitting rabies in India. Bat rabies has not been conclusively reported from India. However, the real burden of rabies in India is unknown. According to WHO, India accounts for approximately 35 percent of rabies deaths annually. Over three quarters of cases in India occur in rural communities with limited access to diagnostic services and post-exposure prophylaxis, both of which are critical to disease prevention (Sudarshan *et al.*, 2007). More than 95 percent of cases are caused by dog bites, because of the approximately 60 million stray/free ranging dogs in the country (Gompper,

2013) with a dog to man ratio of approximately 1:36. Many cases of human rabies go undetected or misdiagnosed or are under reported (Mani and Willoughby 2017). Although rabies affects people of all age groups, children are the most vulnerable which constitutes 40 percent of the total human exposures. The National Rabies Control Programme (NRCP) has formulated the "National Action Plan for Dog Mediated Rabies Elimination from India by 2030" which is a multisectoral collaborative plan involving various stakeholders. In addition, few states in India are working towards the rabies elimination goals.

Virion Properties

The rabies virus is a negative-sense, non-segmented, bullet shaped single-stranded RNA (11-15 kb) virus measuring 100-300 nm in length and 75 nm in diameter (Figure 17.2). It is composed of an internal protein core or nucleocapsid, containing the nucleic acid, and an outer envelope, a lipid-containing bilayer covered with transmembrane glycoprotein spikes. The rabies virus genome encodes five different viral proteins, namely nucleoprotein (N), phosphoprotein (P), matrix protein (M), glycoprotein (G) and RNA-dependent RNA polymerase (L) in sequence 3′ -5′. The RNA exists as a ribonucleoprotein (RNP) with the N protein closely surrounding the RNA and is of helical structure. The N protein is present in abundance and is an important structural component of viral ribonucleoprotein core that is essential for virus propagation. Further N protein is the primary target for the diagnosis of rabies. Phosphoprotein is a cofactor component of viral polymerase. The M protein facilitates virion budding by binding the nucleocapsid to the lipid envelope that contains the G (glyco) protein. Glycoprotein is the major surface protein and peplomers or spikes are visible on the surface of the virion. These form approximately 400 trimer spikes or peplomers of 6-7 nm in length which is closely aligned on the viral envelope. G protein is highly antigenic and has epitopes for vaccine-induced

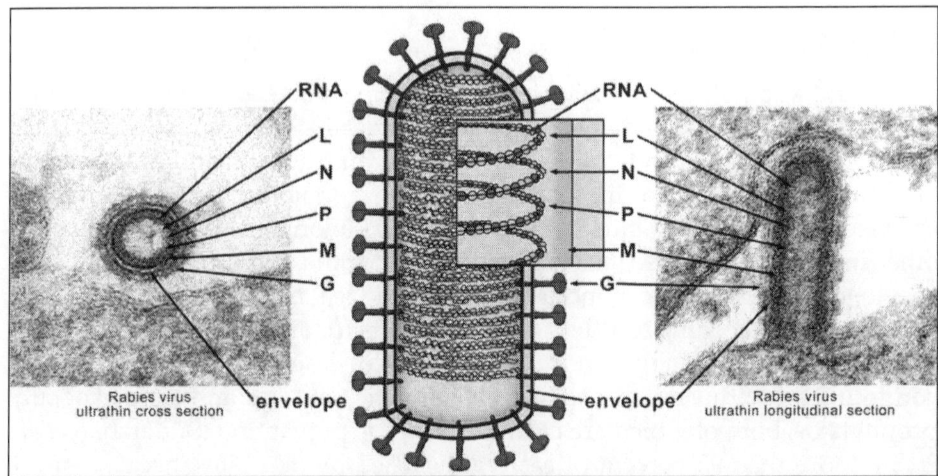

Figure 17.2: Rabies Virus Structure.

neutralizing antibodies and anti-rabies immunoglobulins (RIG). The G spike protein controls major aspects of host cell infection such as receptor binding, antigenicity and host adaptation. In addition, it is involved in trans-synaptic propagation in the central nervous system. The nucleoprotein includes epitopes involved in cell-mediated immunity. Polymerase aids in viral transcription and replication (ICTV, 2017).

The lipid component of virions is derived from the cell membranes of the host. The virus contains approximately 67 percent proteins, 26 percent lipid, 4 percent RNA and 3 percent carbohydrate and has a buoyant density of 1.19-1.20 g/cm3 in caesium chloride and 1.17-1.19 g/cm in sucrose gradients. The sedimentation value ranges between 500 and 1000S (Sarma, 2009; Chandra *et al.*, 2015). It is heat-labile and inactivated by heating at 56° C for 30 min. The virus is ether sensitive and readily inactivated by exposure to sunlight or UV radiation, formalin (1 per cent), cresol (3 per cent), beta-propiolactone (0.1 per cent), mercuric chloride (0.1 per cent), aqueous solutions of household bleach, quaternary ammonium compounds and hospital disinfectants. The virus persists in the infected brain tissue for up to 10 days at room temperature and several weeks at 4° C but is relatively susceptible to disinfection. It can be preserved indefinitely at -70°C and by freeze-drying. The virus is susceptible to pH below 7.0 and above 10 (Debbie and Trimarchi, 1992; Zee and MacLachlan, 2005; Chandra *et al.*, 2015). Laboratory adapted strains haemagglutinates goose erythrocytes at 0-4 °C and pH 6.8 (Sarma, 2009).

Replication of the Virus

The replication of rabies virus is believed to be similar to that of other negative-stranded RNA viruses. The rabies virus infects and replicates in all warm-blooded animals. The virus can thus be propagated in chicken embryos or duck embryos or in various cell cultures systems such as baby hamster kidney cells (BHK21), human diploid cells (WI-26), mouse neuroblastoma (MNA) cells (Zee and MacLachlan, 2005). At no stage of rabies viral replication, it enters into the nucleus of the infected cell. The virus attaches to the host cell membranes via the G protein, penetrates the cytoplasm by fusion or pinocytosis. The fusion of the viral membrane with the endosomal membrane is pH dependent after it is endocytosed by a process known as viropexis. This results in uncoating and the release of ribonucleoprotein (RNP) and virion associated RNA dependent RNA polymerase (RDRP) carried by the virus itself in the cytoplasm of the infected cell. The RDRP polymerises the positive sense RNA using the negative sense RNA of rabies virus. This is followed by polymerisation of several copies of progeny negative sense RNA using the positive sense intermediate RNA on one side. On the other side, the intermediate positive sense RNA of rabies virus functions like the mRNA, which undergoes translation to generate viral structural (N, P, M and G) and non structural proteins (L). The G protein is embedded into the cellular membranes, where as the negative sense viral RNA is encapsidated into the capsid. The nucleocapsid buds out from that part of the cellular membrane where the G protein is embedded. Then virions are released by budding through the plasma membrane of the infected neurons or the salivary glands or epithelial cells. The rabies virus replicates slowly and does not induce cytopathic effect (CPE) as it does not shut down the host cell machinery. However, it results in the formation of prominent inclusion bodies that are of diagnostic importance.

Clinical Signs of Rabies

The incubation period (the time the virus spreads from the peripheral nerves near the site of the bite via the spinal cord to the brain) ranges in general between 2 and 3 month (2 weeks to 6 years are reported) depending on the site of bite/ infection, the amount of virus and the virus strain. Due to its neurotropism, all known lyssaviruses cause severe neurological symptoms as a result of an acute encephalitis. Therefore, clinical signs in humans and animals are very similar.

The infection progresses in a predictable manner, from the initial prodromal phase to manifest as furious or paralytic forms of rabies; the outcome is almost always fatal. Initially, the infected animals show nonspecific signs like anorexia and lethargy. Furthermore, due to rapid progression of disease, the animal may become vicious or begin to show signs of paralysis. Clinical signs of rabies may vary from one animal to another. Once the animal shows severe signs they die within 5 to 6 days.

Dogs

The incubation period in dogs ranges from 10 days to 2 months, and clinical signs attributable to the CNS are predominant in dogs. Rabies in dogs is classified as dumb (predominantly paralytic manifestation with docile behavior of animal) or furious (mainly convulsions and aggressive behavior with greatly exaggerated biting tendencies). It can be hyperexcitability or lethargy, paralysis of the pharynx with frothing of saliva, posterior paresis or paralysis, sudden coma and death. Behavioral changes are common during the early stages of the disease, when the dog behaves abnormally, hides in dark corners, shows unusual excitement, becomes restless. Fever, dilated pupils, and photophobia are sometimes present. The rabid form follows the prodromal phase and affected dogs may exhibit aggressive biting (bite without any provocation) at anything. It can bite and cause serious injury. Some dogs only show the paralytic stage with the characteristic drooping jaw and incoordination. Progressive paralysis begins in the muscles of the head and neck. The tone of the bark changes due to partial paralysis of vocal cords. Convulsions are seen in the terminal phase followed by incoordination and posterior paresis. Once clinical signs appear, the disease rapidly progresses to death of the animal due to respiratory failure, usually within 3-8 days. It's during this clinical period and up to 5 days before the recognition of clinical signs that the virus may be present in saliva. This mandates a 14-day detention and observation of a dog that has bitten humans or other animals (Zee and MacLachlan, 2005). Virus shedding in the saliva of infected dogs is intermittent and variable (Hemachudha *et al.*, 2013).

Cats

Clinical signs in cats are fierce and similar to dogs, but affected cats have a greater tendency to hide in secluded areas and are more vicious than dogs. A cat exhibits great sensitivity to touch/noise, profuse salivation and might strike in air with its front paws. After 2-4 days of the excitation phase, the paralysis of posterior third of the body follows.

Cattle, Buffalo, Sheep and Goat

The incubation period may vary from 2 weeks to many months. In cattle, the prodromal signs may manifest as an animal being off feed and water and a drop in milk production which is of little diagnostic value. Lactation ceases abruptly, grinding of teeth, salivation or pharyngeal paralysis is often misdiagnosed as a choke. Cattle become aggressive with rabies infrequently. In furious form of the disease, the cattle loses its placid expression and becomes alert and restless. Eyes and the ears of the affected animal follow the sound and movement, and thus the animal butts moving objects, attack nearby people and animals. Salivation, choking, lack of rumination, rectal tension and hindquarter paralysis are noted. There may be sexual excitement and the animal begins to bellow abnormally due to vocal paralysis and intermittently until it dies within 12 to 24 hours. Once the clinical signs are evident, the disease progresses rapidly to the death of the animal generally within 5-7 days. During this period and up to 5 days before the detection of clinical symptoms the virus can be excreted through saliva.

Horses and Mules

In horses, the signs are similar to tetanus. Initially, there is a weakness or lameness. The infected horses and mules appear distressed and agitated. They start rolling, which is normally confused with colic. The animal may bite or strike viciously and becomes unmanageable causing self-inflicted wounds. Tremors and spasms are noted in specific muscles. Difficulty in swallowing, progressive paralysis, stiffening of the hindquarters, ataxia and eventual death within 2-4 days.

Pigs

The symptoms are characterized by excitement, irritation, rooting up the ground or rubbing at the surface, aggressiveness, biting of hard objects, other animals and man followed by paralysis and death in 2–4 days.

Humans

After (bite) exposure with a rabid animal, initial symptoms of rabies are of non-specific nature and may be very similar to those of other infectious diseases including general weakness or discomfort, fever, or headache. A first specific clinical sign is neuropathic pain or a prickling or itching sensation at the site of the bite. Signs rapidly progress and include cerebral dysfunction, anxiety, confusion, agitation, delirium, abnormal behavior, hallucinations, and insomnia. Once clinical signs of rabies appear, without intensive care, death occurs within 2 weeks after onset of symptoms. There is no treatment available.

Transmission and Pathogenesis

Herbivores do not transmit the disease. In most cases the disease is transmitted via the bite of rabid animals which shed infectious virus in their saliva. The virus enters the body through transdermal inoculation (*i.e.* wounds) or through direct contact of infectious material (*i.e.* saliva, cerebrospinal liquid, nerve tissue) to mucous membranes or skin lesions. Rabies virus can also gain entry by contaminated saliva through licking on pre existing breach on the skin. In bat infested caves, inhalation of aerosols containing the virus can result in rabies. Livestock are vulnerable victims of rabid carnivores and wild animals. Man is dead end host (since there is no biting tendency) and hence does not play any role in transmission to new hosts. The virus cannot penetrate intact skin, and thus the disease transmission is most commonly through a bite. The susceptibility of the virus is less by oral route than the intramuscular route. Thus ingestion is not a common mode of infection.

The rabies virus has an affinity to the nervous system (NS) (Wunner, 1987) and reaching the CNS is critical for the virus to establish the infection. Various mechanisms are used by the virus to strategically evade the immune system of the host, but the detailed mechanism of evading the host and its pathogenesis during the early stages is not well understood. The glycoprotein and phosphoprotein have a major role in axonal transport. In the affected animals, the incubation period depends on various factors like the location of bite, depth of the wound, load and virulence of the virus, susceptibility of the animal species that has been bitten, it's immune status all play a major role in the pathogenesis and course of the disease.

The virus enters the host through a bite or scratch of an infected animal. Following exposure, the virus is deposited and persists in the local muscle tissue for hours or days (3-9 weeks). After entry, the virus binds to cell receptors. Viruses may replicate within striated muscle cells or epithelial cells in the lower layers of the epidermis or directly infect nerve cells. Since the replication is minimum at the site of bite, no detectable immune response is seen during the incubation period. A viremic phase is not seen in rabies since the virus does not move through the blood or lymphatics. In the host, the binding of the virus to the neurons occurs at the postsynaptic muscle membrane to the nicotinic acetylcholine receptor (nAchR) which enriches the virus at the neuromuscular junction or synaptic cleft thus enabling it's entry to the motor neurons. Further, the virus enters neurons through neural cell adhesion molecule (NCAM) and p75 neurotrophic receptor and is transported to the cell body through the axon in vesicles. The virus then travels via retrograde axoplasmatic transport mechanisms to the central nervous system. Both motor and sensory fibres may be involved depending on the animal infected.

Once it has reached the CNS, rapid virus replication takes place, causing pathologic effects on nerve cell physiology. The virus then moves from the CNS via anterograde axoplasmic flow within peripheral nerves, leading to infection of some of the adjacent non-nervous tissues, for example, secretory tissues of salivary glands (Figure 17.3). The virus is widely disseminated throughout the body at the time of clinical onset. With shedding of infectious virus in saliva the infection cycle of rabies is completed. The virus is excreted in all secretions and excretions.

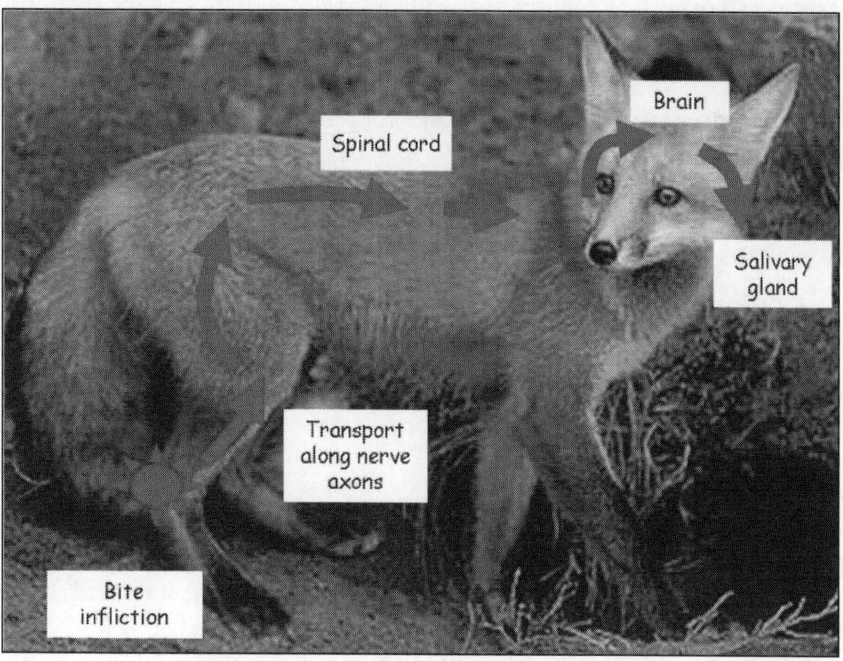

Figure 17.3: Rabies Virus Transmission.

Pathology

In general, gross pathological lesions are not visible in rabid dogs except fresh bite wounds and signs of self-mutilation. Microscopic changes are limited to CNS. The histopathology of brain tissue reveals moderate neuronal damage with encephalomyelitis and perivascular cuffing of lymphocytes, mononuclear infiltration and polymorphonuclear cells. Cytoplasmic eosinophilia, cytoplasmic vacuolation, pyknosis and karyorrhexis are more commonly seen in freshly fixed and adequately treated brain tissue. Acidophilic round or oval inclusions in the cytoplasm of infected neurons are seen with a clear halo around it which are referred to as Negri bodies. These are found in the pyramidal cells of Ammon's horns, Purkinje cells of the cerebellum and brain stem. A large number of small inclusions can also be demonstrated in the smears of brain tissue by immunofluorescence or immunoenzyme methods. Experimental inoculation of mice with fixed rabies virus may also show numerous virus particles without the formation of Negri body (Chandra *et al.,* 2015).

Diagnosis of Rabies

Even with signs of the disease quite characteristic for rabies, like changes in behaviour or difficulties in swallowing, the clinical observation and examination cannot rule out rabies nor confirm the diagnosis and hence, may only lead to a suspicion of rabies. Detection of the virus or some of its specific components using WHO and WOAH recommended standard laboratory tests is therefore, the only way to undertake a reliable diagnosis of rabies.

Brain tissue is the preferred specimen for post-mortem diagnosis in animals. While intra-vitam diagnosis of rabies in animals is strongly discouraged, it is the only way of confirming infection in a patient suspected of having the disease. Intra-vitam diagnosis in suspect cases is based on detecting virus or viral RNA in saliva, in innervated skin biopsy samples or hair follicles. All samples from suspected Rabies cases should be considered potentially infectious. All specimens should be collected in a primary leak-proof container and must be securely sealed before transport to the designated laboratory. However, due to intermittent shedding of virus in saliva and variable diagnostic sensitivity of the methods applied, only positive results are valid.

Detection of Rabies Antigen

The most widely used primary diagnostic test for rabies in animals is the Direct fluorescent antibody test (DFA). This test is based on antigen detection and is recommended by both WHO and WOAH as the gold standard for rabies diagnosis. In brief, chilled acetone (70 per cent) fixed brain tissue in -20°C deep freezer(preferably brain stem or cerebellum) impressions are air dried and then treated with anti N protein Monoclonal antibody conjugated with fluorescein isothiocyanate (FITC), washed and visualized under Fluorescent microscope at 490-492 nm wavelength. Specific aggregates of rabies viral inclusions are detected by their apple green fluorescence using a reflected light (incident light) in a fluorescence microscope. Using Evan's blue (along with the FITC conjugate) as counter stain provides background red color to the tissue and provides better contrast in case of rabies positive brain impression (Figure 17.4), whereas no apple green colored rabies viral inclusions are seen in case of rabies negative brain impression (Figure 17.5). The DFA is accurate, sensitive and rapid, *i.e.* results can often be obtained within 1 to 2 hours of receipt of the specimen.

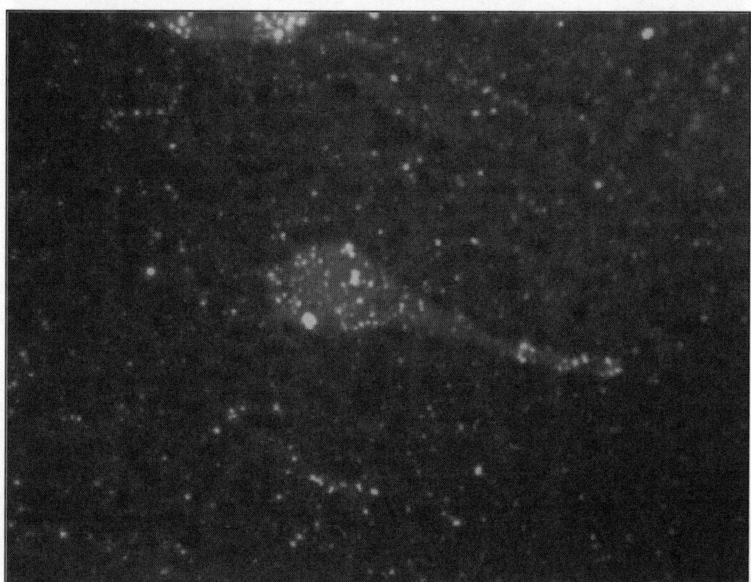

Figure 17.4: Rabid Animal Brain Impression Stained with the Rabies Virus Antinucleocapsid IgG-FITC Conjugate (DFA) (40X objective).

Figure 17.5: Non Rabid Animal Brain Impression Stained with the Rabies Virus Antinucleocapsid IgG-FITC Conjugate Tested by DFA) (40X objective).

Alternatively, particularly where fluorescence microscopy, incubators and -20° C deep freezers are not available, a direct rapid immunohistochemistry test (dRIT) was developed with a similar sensitivity and specificity as the

DFA(Isloor *et al.*, 2017). The principle is similar to the DFA except that the dRIT uses biotinylated anti N protein Monoclonal antibody conjugate and streptavidin conjugated HRP system followed by Amino Ethyl Carbizole as the chromogen and 3 percent H_2O_2 as the substrate. Gill's hematoxylene is used as the counter stain which provides blue color to the brain tissue cells in the impression. The rabies positivity is declared in the impression when the brick red colored viral inclusions are seen under the light microscope (Figure 17.6). The healthy or absence of brick red colored inclusions indicates negativity for rabies (Figure 17.7) The major advantage of dRIT is that it is performed at the room temperature itself and is independent of Fluorescent microscope, deep freezer and incubator.

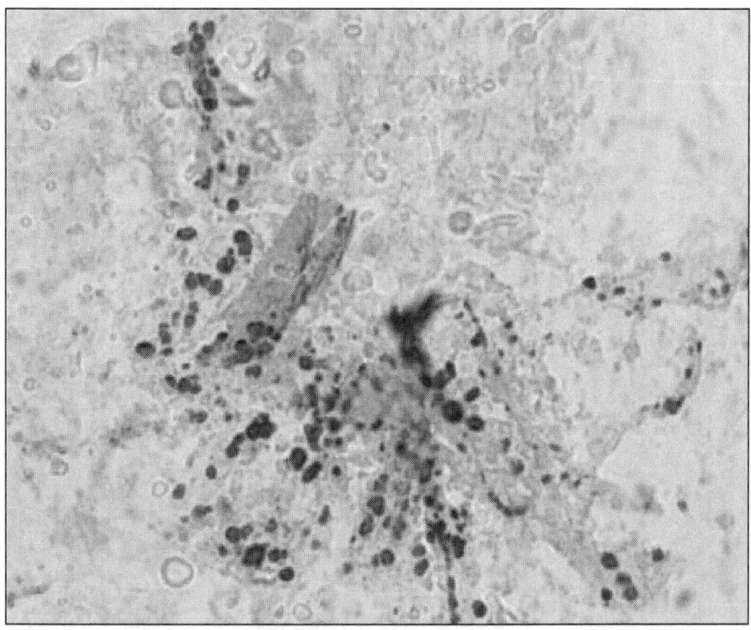

Figure 17.6: Rabid Animal Brain Impression, Revealing Brick Red Colored Inclusions Tested by dRIT (40X objective).

Recently, the immunochromatography based lateral flow assays (LFA) for rapid detection of rabies virus antigen under field conditions have been developed and kits are available commercially. This is a preliminary assay and found user friendly by the filed veterinarians. The development of colored lines at positions 'C' and 'T' indicate rabies positivity and development of colored line only at the C line indicates rabies negativity. Irrespective of presence or absence of rabies viral antigen in the tested brain sample, the colored line at position 'C' must develop to ensure the quality of the test (Figure 17.8). Adequate validation according to international standards is still required so that they can be useful in surveillance situations in which the resources for the use of recommended testing are lacking.

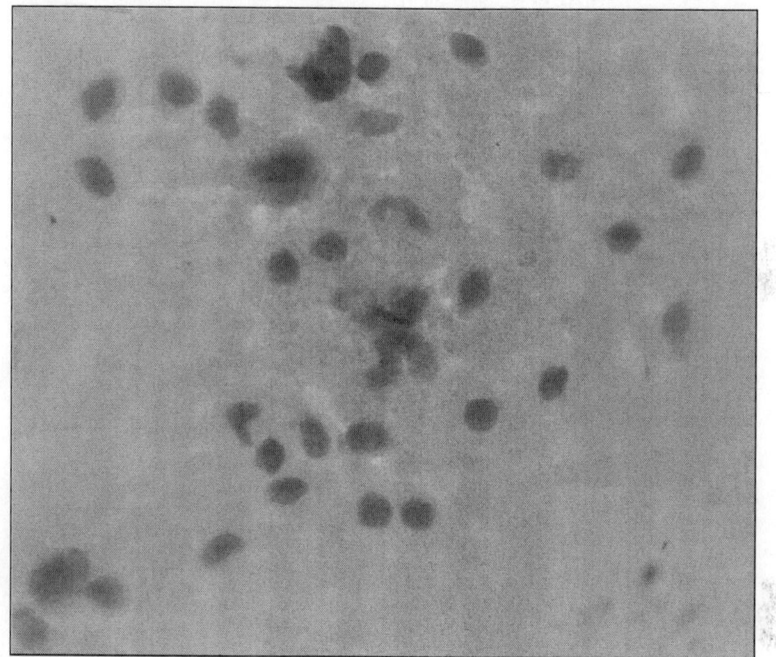

Figure 17.7: Absence of Brick Red Colored Inclusions in Non-Rabid Animal Brain Impression Tested by dRIT (40X objective).

Detection of Rabies Virus RNA

Molecular technique with reference to Polymerase Chain Reaction (PCR) is commonly used for molecular detection of rabies viral RNA for diagnostic purposes as well as molecular epidemiology. Both conventional Reverse Transcription PCR as well as Real Time PCR, mostly targeting N and G genes (complete or partial) are employed. The real time PCR has been developed to increase sensitivity and to obtain results faster. Furthermore, real-time PCR is used as a confirmatory test and is particularly useful for intra vitam diagnosis. Yet another advantage of PCR is its ability to detect positivity even in partially putrified brain samples from rabid animals.

Virus Isolation

Virus isolation is carried out especially for epidemiological purposes to characterize the viruses in a particular geographic area. Mice inoculation test or Rapid Tissue culture infection test (RTCT) is performed for isolation.

In Mouse inculation test, the suspected brain material is homogenized and inoculated intracerebrally to 2 day old mice or 3-4 wks old mice and observed for 28 days. Further, after 5-7 days, the mice exhibit characteristic signs of rabies like hunching of back, ruffling of hair, dragging of hind limbs followed by forelimbs. Confirmatory diagnosis is by extracting the brain tissue from these mice and subjecting it to DFA.

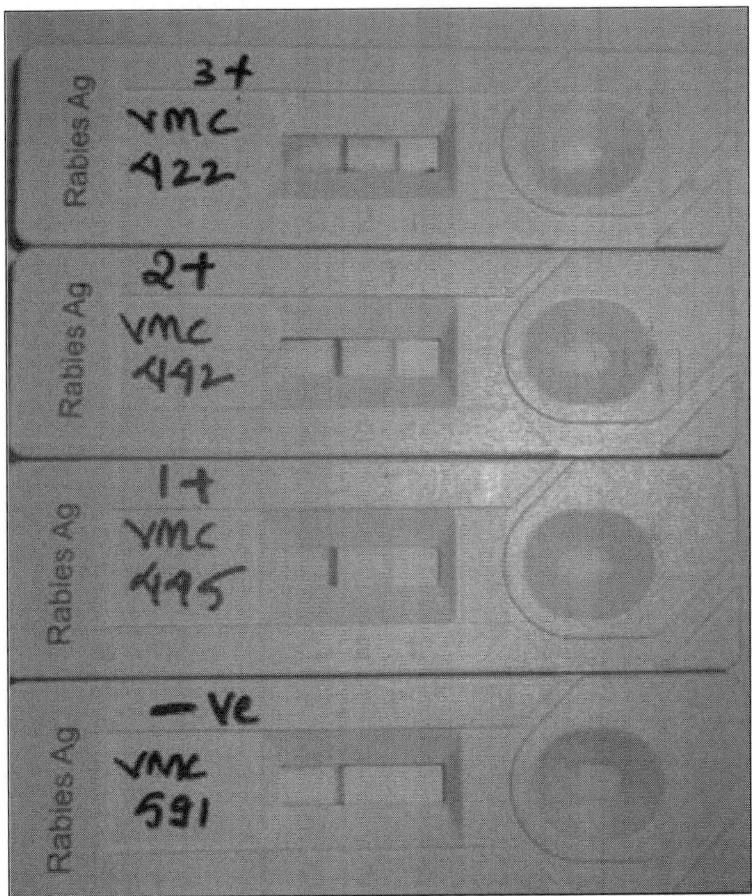

Figure 17.8: Lateral Flow Assay of Rabies Positive and Negative Brain Sample Suspension.

For RTCIT, viruses are isolated in cell culture. This method is faster and yields results in 48 hours. The neural origin cell lines are more suitable and murine neuroblastoma (MNA) cell line is commonly used. Baby Hamster Kidney (BHK 21) cell line and Human embryonic kidney (HEK) cell line can also be used for isolation. Here the cells are grown in tissue culture plates and then the suspected sample is added and incubated for 24-48 hrs. Further, these tissue culture plates are subjected to DFA for confirmation.

Other less common antigen detection assays used are the rapid sandwich ELISA for detecting lyssaviruses belonging to all seven genotypes. A dot blot immunoassay and an enzyme immunoassay is also employed for rapid diagnosis from suspected brain samples of humans and animals (Mashusudana *et al.*, 2004)

Serological Tests

Serological assays are not suitable for diagnosis of rabies infections in humans and animals as virus-specific antibodies in serum tend to appear only relatively late after the onset of clinical signs if at all. They are mainly used to evaluate the immune response to human and animal rabies vaccines.

The gold standard is the virus neutralisation test, virus neutralising antibody titres directly correspond to the level of immune response. Virus neutralisation assays are also the referred tests for international trade and travel of pets. Both FAVN (Fluorescent Antibody Virus Neutralization Test) and RFFIT (Rapid Fluorescent Focus Inhibition Test) are approved for determination of the titre of virus neutralizing antibodies.

ELISAs provide a rapid serological test that avoids the requirement to handle live rabies virus. These tests detect antibodies that can specifically bind to rabies virus antigens, primarily the rabies virus glycoprotein and nucleoprotein.

Control of Rabies

Rabies deaths are 100 percent preventable through prompt and appropriate medical care. It is a vaccine preventable disease. Especially companion animals such as dogs and cats are potential vectors for rabies transmission to humans. Canine or dog-mediated rabies contributes to more than 99 percent of all human rabies cases; half of the global human population especially in the developing world lives in canine rabies-endemic areas and is considered at risk of contracting rabies. Vaccinating dogs is the cost effective strategy for preventing rabies in humans and livestock. Currently, inactivated cell culture based anti rabies vaccines are used for Pre exposure prophylaxis (PrEP) in both animals as well as risk group of human beings such as veterinarians, laboratory personnel and slaughter house workers. Whereas, post exposure anti-rabies vaccination (PEP) is commonly practised in human beings wherein thorough washing of wounds, anti rabies vaccination (intra dermal or intra muscular) and administration of Rabies Immunoglobulins (RIGs) at the site of bite have to be strictly employed. There are many effective animal rabies vaccines available that provide a considerable duration of immunity. In countries endemic with dog-mediated rabies, mass parenteral vaccination programmes in domestic, pet dogs as well as free roaming dogs remain the mainstay of rabies control along with effective surveillance capability. Disease surveillance include ongoing sample collection, application of suggested diagnostic methods, and interpretation of findings and communication of information to relevant public and veterinary health authorities for the purpose of implementing targeted control measures. The United against Consortium (UAR) comprising of WHO, WOAH (World Organization for animal Health), FAO (Food and Agriculture Organization) and GARC (Global Alliance for Rabies Control) have jointly decided to eliminate "Dog mediated human rabies" by 2030. Accordingly, several countries including India have launched National Action Plan for Dog Mediated Rabies Elimination by 2030". This approach is a multisectoral collaborative plan involving multi stakeholders.

As for the control of wild animal rabies is concerned, conventionally, Oral rabies vaccines (ORVs) have been used since 1978 across Europe and the USA. However, there is need for the use of ORVs in free-roaming dogs also to control dog-transmitted rabies. This approach is necessary in accessing a large proportion of the free roaming dog population in the endemic countries as parenteral methods is a considerable challenge due to the large, evasive stray dog population in these countries with special reference to Asia and Africa. The limitation of the current parenteral approaches for free roaming dogs is the need of large skilled dog-catching teams to reach these dogs, for which present financial, operational and logistical limitations are the barriers to achieve 70 percent dog vaccination coverage in a short period (Gowri *et al.*, 2022).

Conclusion

The target of 'Elimination of Canine mediated human rabies by 2030' is set by the international agencies WHO, OIE, FAO and GARC for Africa, Asia, and Latin America. The main objective of this rabies elimination program in human beings include control, prevention, and subsequent elimination of rabies in dogs by successive annual mass vaccination covering at least 70 percent of the dog population. In addition, various measures should be taken up to control rabies. People should be made aware through Public education campaigns, particularly in rural areas. There has to be regular supply of vaccines and anti-rabies immunoglobulin in all hospitals and primary-health-care centres. Primary health care personals should be well trained to administer appropriate prophylaxis, including intradermal anti rabies vaccination. The primary school curriculum shall cover rabies awareness among students. Ensure that all dogs are vaccinated and stray dogs sterilized. Rabies should be incorporated into a "One health programme" at all levels. Further, a significant role is being played by the non - Governmental Organizations (NGOs) in addressing rabies. The initiatives of World Rabies Day, by the GARC, throws light on the need to systematically address rabies worldwide, and provide impetus for local governments to act. Success stories from various countries/region can be the guidance for local initiatives and in turn achieve the global goal of "Dog mediated human rabies elimination by 2030.

REFERENCES

Belotto, A., Leanes, L.F., Schneider, M.C., Tamayo, H. and Correa, E. (2005). Overview of rabies in the Americas. Virus Res, 111(1): 5-12.

Birhane, M.G., Cleaton, J.M., Monroe, B.P., Wadhwa, A., Orciari, L.A., Yager, P., Blanton, J., Velasco-Villa, A., Petersen, B.W. and Wallace, R.M. (2017). Rabies surveillance in the United States during 2015. J Am Vet Med Assoc,

250(10): 1117-1130. doi: 10.2460/javma.250.10.1117. Erratum in: J Am Vet Med Assoc, 2017 Jul 15; 251(2):146. doi: 10.2460/javma.251.2.146.

Bourhy, H., Dacheux, L., Strady, C. and Mailles, A. (2005). Rabies in Europe in 2005. Euro Surveill, 10(11): 213-216.

Chandra, R., Dutta, T.K. and Roychoudhury, P. (2015). Rhabdoviridae. In: Systematic veterinary virology. Kalyani Publishers, New Delhi, pp 412-417.

De Serres, G., Dallaire, F., Côte, M. and Skowronski, D.M. (2008). Bat rabies in the United States and Canada from 1950 through 2007: human cases with and without bat contact. Clin Infect Dis, 46(9): 1329-1337. https://doi.org/10.1086/586745

Debbie, J.G. and Trimarchi, C.V. (1992). Rabies. In: Castro AE, Heuschele WP (eds) Veterinary diagnostic virology. Mosby, St. Louis, pp 116-117.

Everard, C.O.R. and Everard, J.D. (1988). Mongoose rabies. Rev Infect Dis, 10(4): S610-S614.

Favi, M., de Mattos, C.A., Yung, V., Chala, E., Lopez, L.R. and de Mattos, C.C. (2002). First case of human tabbies in Chile caused by an insectivorous bat virus variant. Emerg Infect Dis, 8(1): 79-81.

Yale, G., Lopes, M., Isloor, S., Head, J.R., Mazeri, S., Gamble, L., Dukpa, K., Gongal, G. and Gibson, A.D. (2022). Review of oral rabies vaccination of dogs and its application in India. Viruses, 14(1): 155. doi: 10.3390/v14010155.

Gould, A.R., Kattenbelt, J.A., Gumley, A.G. and Lunt, R.A. (2002). Characterisation of an Australian bat lyssavirus variant isolated from an insectivorous bat. Virus Res, 89(1): 1-28.

Gompper, M.E. (2013). Free-ranging dogs and wildlife conservation. Oxford University Press, pp. 1-69.

Greene, C.E. and Rupprecht, C. (2006). Rabies and other lyssavirus infections. Infectious diseases of the dog and cat: pp.167-183.

Hemachudha, T., Ugolini, G., Wacharapluesadee, S., Sungkarat, W., Shuangshoti, S. and Laothamatas, J. (2013). Human rabies: neuropathogenesis, diagnosis, and management. Lancet Neurol, 12(5): 498-513. doi: 10.1016/S1474-4422(13)70038-3.

Isloor, S., Fooks, A.K., Santosh, A.K., Rathnamma, D. and Abdul Rahman, S. (2017). Advances in diagnosis of rabies in animals and its strengthening under OIE Initiatives in India. Paper presented in XV Annual Conference of Indian Association of Veterinary Public Health Specialists, College of Veterinary Science, Tirupati, Andhra Pradesh.

Isloor, S., Marissen, W.E., Veeresh, B.H., Nithin Prabhu, K., Kuzmin, I.V., Rupprecht, C.E., Satyanarayana, M.L., Deepti, B.R., Sharada, R., Neelufer, M.S., Yathiraj, S. and Abdul Rahman, S. (2014). First case report of rabies in a wolf (*Canis lupus pallipes*) from India. J Vet Med Res, 1(3): 1012.

Isloor, S., Mani, R. and Mahendra, B.J. (2019). Assessing rabies free status of Andaman, Nicobar and Lakshadweep islands, India. Indian J Pub Health, 63(5): 48-50.

Kobayashi, Y., Sato, G., Shoji, Y., Sato, T., Itou, T., Cunha, E.M., Samar, S.I., Carvalho, A.A., Nociti, D.P., Ito, F.H. and Sakai, T. (2005). Molecular epidemiological analysis of bat rabies viruses in Brazil. J Vet Med Sci, 67: 647-652.

Madhusudana, S.N., Shamsundar, R. and Seetharaman, S. (2004). In vitro inactivation of the rabies virus by ascorbic acid. Int J Infect Dis, 8(1): 21-25.

Mani, R.S. and Willoughby, R.E. (2017). Human rabies in south Asia. Negl Trop Dis South Asia, 349-371.

Nadin-Davis, S.A., Huang, W., Armstrong, J., Casey, G.A., Bahloul, C., Tordo, N. and Wandeler, A.I. (2001). Antigenic and genetic divergence of rabies viruses from bat species indigenous to Canada. Virus Res, 74: 139-156.

Nadin-Davis, S.A. and Loza Rubio, E. (2006). The molecular epidemiology of rabies associated with Chiropteran hosts in Mexico. Virus Res, 117(2): 215-226.

Nel, L.H. and Rupprecht, C.E. (2007). Emergence of lyssaviruses in the Old World: the case of Africa. Curr Top Microbiol Immunol, 315: 161-193.

OIE. (2013). World Organisation for animal health, terrestrial animal health code, 22nd edn, vol II, pp 440.

Rupprecht, C.E. *et al.* (2017). In: Plotkin, S.A., Orenstein, W.A., Offit, P.A. (eds), Vaccines, 7th edn. Elsevier Saunders, Philadelphia, pp 918-942.

Rupprecht, C.E., Hanlon, C.A. and Hemachudha, T. (2002). Rabies re-examined. Lancet Infect Dis, 2(6): 327-343.

Sarma, D.K. (2009). A textbook of veterinary virology and viral diseases, 2nd edn. Kalyani Publishers, Ludhiana, pp 229-234.

Sudarshan, M.K., Madhusudana, S.N, Mahendra, B J., Rao, N.S.N., Narayana, D.A., Rahman, S.A., Meslin, F.X., Lobo, D. and Ravikumar, K. (2007). Assessing the burden of human rabies in India: results of a national multi-center epidemiological survey. Int J Infect Dis, 11(1): 29-35.

Swanepoel, R. (2004). Rabies. In: Coetzer, J.A.W., Tustin, R.C. (eds) Infectious diseases of livestock, 2nd edn. Oxford University Press, Cape Town, pp 1123-1182.

Warrilow, D. (2005). Australian bat lyssavirus: a recently discovered new rhabdovirus. Curr Top Microbiol Immunol, 292: 25-44.

WHO. (2018). Expert consultation on rabies third report. In WHO technical series report no. 1012, Geneva, ISBN 978-92-4-121021-8.

Wunner, W.H. (1987). Rabies viruses - pathogenesis and immunity. In: Wagner, R.R. (ed) The rhabdoviruses. Plenum Press, New York, pp 361-426.

Zee, Y.C. and MacLachlan, N.J. (2005). Rhabdoviridae. In: Hirsh, D.C., MacLachlan, N.J., Walker, R.L. (eds) Veterinary microbiology, 2nd edn. Blackwell Publishing, Ames, pp 377-380.

Chapter 18

Malignant Catarrhal Fever

Sandeep Bhatia* and S. Kalaiyarasu

ICAR-National Institute of High Security Animal Diseases, Anand Nagar, Bhopal – 462 022, Madhya Pradesh, India

****e-mail: sbhatia1967@gmail.com***

ABSTRACT

Malignant catarrhal fever (MCF) is a highly contagious and often fatal viral disease that primarily affects cattle and other ruminants. It is caused by two distinct but related viruses: Ovine herpesvirus 2 (OvHV-2) and Alcelaphine herpesvirus 2 (AlHV-2). These two viruses are responsible for distinct forms of MCF, and their respective natural reservoirs are crucial to understanding the disease's epidemiology. OvHV-2 is primarily associated with sheep and is prevalent throughout the world, while AlHV-2 is primarily associated with wildebeest and is mainly found in Africa. Sheep, as the reservoir host for OvHV-2, play a central role in the transmission of the virus to other susceptible species, including cattle and various other ruminants. These reservoir animals often carry the virus without displaying clinical symptoms themselves, making them a critical source of infection for other species. The global presence of OvHV-2 in sheep populations poses a significant challenge for livestock industries worldwide, as the virus can easily be transmitted to domestic cattle, causing severe clinical effects. Conversely, AlHV-2, the causative agent associated with wildebeest, is primarily confined to Africa. Wildebeest are the natural hosts for this virus, and they often carry AlHV-2 without suffering from apparent illness. However, when the virus is transmitted to cattle or other ruminants, it can lead to clinical manifestations of MCF. This geographical limitation means that regions with wildebeest populations are at a higher risk of MCF outbreaks, whereas other parts of the world are less susceptible to AlHV-2-related MCF. In this chapter, we will explore the unique characteristics of OvHV-2 and AlHV-2 and other related viruses, their natural reservoirs, and the impact of MCF on cattle and other ruminants. Understanding the complex dynamics of these causative agents and their reservoir hosts is essential for effective disease management and prevention strategies in affected regions.

Keywords: *Malignant catarrhal fever, Ovine herpesvirus-2, Alcelaphine herpesvirus-1.*

Introduction

Malignant Catarrhal Fever (MCF) is a severe and lethal systemic illness that impacts a range of ungulate species. The disease has primarily been observed in animals belonging to the subfamily *Bovinae* and family *Cervidae*. However, it has also been documented in domestic pigs, as well as giraffes and species of antelope belonging to the subfamily *Tragelaphinae*. The causative agents responsible for malignant catarrhal fever (MCF) are gammaherpesviruses classified under the *Macavirus* genus (Li *et al.*, 2011). The two primary viruses responsible for generating malignant catarrhal fever (MCF) are Alcelaphine herpesvirus 1 (AlHV-1) and Ovine herpesvirus 2 (OvHV-2). AlHV-1 is predominantly linked to wildebeest and serves as the etiological factor for wildebeest associated Malignant Catarrhal Fever (WA-MCF) in cattle and deer. Conversely, OvHV-2 is harboured by sheep (Russell *et al.*, 2009) and acts as the causative agent for sheep associated Malignant Catarrhal Fever (SA-MCF) in cattle, deer, bison, pig, and giraffe.

Etiology

AlHV-1 and OvHV-2 are the two most commonly found etiological agents among the ten viruses associated with MCF as documented in Table 18.1 (Li *et al.*, 2011). The ten viruses responsible for causing MCF can be classified within the taxonomic genus *Macavirus*, which falls within the subfamily *Gammaherpesvirinae* within the larger family *Herpesviridae* (Figure 18.1). There are six viruses that are distinctly linked to clinical disease. These are AlHV-1, OvHV-2, caprine herpesvirus-2 (CpHV-2), MCFV of unknown origin which causes disease in white-tailed deer (MCFV-WTD), ibex MCFV (MCFV-ibex) found in ibex, and an AlHV-2-like virus found in Jackson hartebeest. The four remaining viruses present in roan antelope (Reid *et al.*, 1991), oryx, muskox (Li *et al.*, 2003), and aoudad (Li *et al.*, 2005) have not yet been linked to any pathological conditions (Li *et al.*, 2011). AlHV-1 is prevalent in wildebeest populations, where it triggers sub-clinical infection and leads to the development of WA-MCF in susceptible ruminant species (Plowright *et al.*, 1960). Ovine herpesvirus-2 (OvHV-2) is prevalent globally in domestic sheep (Baxter *et al.*, 1993) and is responsible for causing SA-MCF when transmitted to a diverse range of domestic and wild ruminants, including pigs (Løken *et al.*, 1998; Syrjala *et al.*, 2006; Vikøren *et al.*, 2006). AlHV-2 has been found to induce Malignant Catarrhal Fever (MCF) in hartebeest (*Alcelaphus buselaphus P*) and topi (*Damaliscus korrigum O*) (Mushi *et*

Table 18.1: MCF-Causing Viruses with their Reservoir and Susceptible Hosts

Sl.No.	MCF-Causing Virus	Reservoir Host	Susceptible Host
1.	Alcelaphine herpesvirus 1 (AlHV-1)	Wildebeest	Cattle, deer
2.	Alcelaphine herpesvirus 2 (AlHV-2)	Hartebeest Topi	Barbary red deer, Bison
3.	Hippotragine herpesvirus 1 (HiHV-1)	Roan antelope	Not reported
4.	Oryx-MCFV	Oryx	Not reported
5.	Ovine herpesvirus 2	Sheep	Cattle, deer, bison, pig, giraffe
6.	Caprine herpesvirus 2	Goat	Sika deer, white-tailed deer
7.	MCFV–white-tailed deer/Caprine herpesvirus 3	Goat	White-tailed deer, red brocket deer, reindeer
8.	Ibex-MCFV	Ibex	Bongo, anoa, pronghorn
9.	Muskox-MCFV	Muskox	Not reported
10.	Aoudad-MCFV	Aoudad	Not reported

al., 1981; Klieforth *et al.*, 2002). Hippotragine herpesvirus 1 (HiHV-1) has been identified in the roan antelope (*Hippotragus eguinus D*) and oryx (*Oryx gazella*) (Reid *et al.*, 1987; Li *et al.*, 2003). Caprine herpesvirus 2 (CpHV-2) is prevalent in domestic goats (Chmielewicz *et al.*, 2001; Li *et al.*, 2001) and has the potential to induce malignant catarrhal fever (MCF) in other ruminant species, such as

Order	*Herpesvirales*	
Family	*Orthoherpesviridae*	
Subfamily	*Gammaherpesvirinae*	
Genus	Macavirus	
Species	**Alcelaphine herpesvirus 1 (AlHV-1) Type species** Alcelaphine herpesvirus 2 (AlHV-2) Hippotragine herpesvirus 1 (HiHV-1)	**Alcelaphinae/Hippotragine group** (also includes MCFV-oryx)
	Ovine herpesvirus 2 (OvHV-2) Type species Caprineherpesvirus 2 (CpHV-2)	**Caprinae group** (also includes MCFV-WTD, MCFV-ibex, MCFV-muskox and MCFV-aoudad)
	Suid herpesvirus 3 Suid herpesvirus 4 Suid herpesvirus 5	

Figure 18.1: Classification of MCF Causing Viruses (ICTV 2022 Master Species List 38).

sika deer (*Cervus nippon T.*), white-tailed deer (*Odocolieus virginianus Z.*), moose (*Alcesalces L.*), and roe deer (*Capreolus capreolus L.*) (Crawford *et al.*, 2002; Keel *et al.*, 2003; Li *et al.*, 2003; Vikøren *et al.*, 2006).

It has been observed that OvHV-2, present in the nasal secretions of sheep, can infect naive sheep as well as bison and cattle to cause MCF in both the susceptible species. However, cattle have a lower susceptibility to OvHV-2 infection and may require much higher viral doses (up to 1000-fold) through intranasal nebulization to contract the virus as compared to sheep and bison. Despite being susceptible to OvHV-2, cattle and bison are usually considered dead-end hosts since they do not transmit the virus to other animals. The reason for the lack of natural transmission of OvHV-2 among cattle is attributed to the fact that OvHV-2 is bound to cells in infected cattle, and cattle are not capable of generating and discharging enough cell-free OvHV-2 to infect other cattle. Some researchers have hypothesized that OvHV-2 is cell-bound because the viral DNA gets integrated into the cellular genome or exists as an episome within infected cells (Metzler, 1991).

AlHV-1 in a cell-free state has been identified in the nasal secretions of wildebeest calves (Mushi *et al.*, 1980; Barnard *et al.*, 1989) This virus has the potential to infect cattle through the inhalation of aerosolized droplets that are produced by the nasal secretions of young wildebeest, particularly those that are less than 3 months old. Thus, it appears that young wildebeest are the primary carriers of AlHV-1.

Epidemiology

MCF is a global problem, with cases reported across Africa, North America, Europe, Asia, and Oceania (Li *et al.*, 2011). The disease is endemic in regions where the reservoir host populations are high, such as Africa with wildebeest and sheep in other parts of the world. The etiological agent of the sheep-associated MCF (SA-MCF), Ovine herpesvirus-2 (OvHV-2), exhibits a worldwide distribution, while Wildebeest-associated herpesvirus-1 (AlHV-1), responsible for wildebeest-associated MCF (WA-MCF), is predominantly prevalent in natural habitats of African antelopes and zoological collections housing mixed populations of Artiodactyla, including wildebeest. Although reports of OvHV-2 infection in South Africa exist, CpHV-2 infections in buffaloes in Switzerland are also reported. Serological and molecular studies conducted globally on various domestic and wild animals have provided insights into the prevalence of MCF viruses. A high frequency of seropositivity, exceeding 50 per cent, has been observed in certain species, such as sheep, goats, muskoxen, and oryxen. In contrast, MCF-susceptible species, such as cattle, bison, deer, caribou, elk, llama, reindeer, moose, pronghorn antelope, and various deer species, exhibited low degrees of seropositivity, less than 50 per cent.

The first confirmed report of MCF in India, based on histopathology, is from Punjab, where 31 buffaloes and cattle reportedly died from the disease in the year 1975 (Parihar *et al.,* 1975). About three decades later, the disease was reported in cows in Kashmir (Wani *et al.,* 2004), followed by the genomic detection of OvHV-2 in sheep from the same region (Wani *et al.,* 2006). Since then sporadic cases of SA-MCF have been reported from southern region of India. Genomic and serological evidence of OvHV-2 infection and clinical cases of MCF in India have been documented in various studies from 2008 to 2021 (Banumathi *et al.,* 2008; Kumar *et al.,* 2014; Sood *et al.,* 2014; Premkrishnan *et al.,* 2015; Sood *et al.,* 2017; Kishna *et al.,* 2019, Kumar *et al.,* 2021). OvHV-2 infection has been detected in captive bison in Karnataka state of India (Sood *et al.,* 2012).

Clinical Features

Malignant catarrhal fever (MCF) is a multisystemic syndrome that is generally characterized into four forms: (1) peracute form, (2) alimentary tract form, (3) common 'head and eye' form, and (4) mild form. The gradations of these forms are primarily based on the appearance of clinical signs, which are more prominent in the chronic form of the disease (Radostits *et al.,* 2010).

Peracute form characterized by rapid onset of depression and high fever, followed by the development of diarrhoea, which may be haemorrhagic, followed by death in 12-24 h.

Alimentary tract form characterized by marked diarrhoea in SA-MCF, whereas there is constipation in WD-MCF with only minor eye changes consisting of conjunctivitis rather than ophthalmia as seen in SA-MCF.

'Head and eye' form is the most common form of MCF, characterized by sudden onset of symptoms, such as extreme dejection, anorexia, agalactia, high fever, rapid pulse rate, and nasal discharge, which can later become mucopurulent. Dyspnea, stertor and foul odor accompany the nasal discharge, and ocular discharge and variable degrees of eyelid edema occur. The sclera becomes reddened due to congestion, and the cornea progressively becomes opaque, leading to blindness. The progression of corneal oedema correlates well with disease outcome and that non-improvement of uveitis is a bad prognostic sign. Lymph node swelling occurs in the head and neck region, and superficial necrosis is seen in the anterior nasal and buccal mucosa. Lesions may also occur at the skin-hoof junction of the feet, especially at the back of the pastern, and the skin of the teats, vulva, and scrotum may slough off entirely or become covered with dry, tenacious scabs in acute cases.

Mild form The mild form is generally observed in experimental animals, characterized by transient fever and mild erosions on the oral and nasal mucosae. The affected animals may completely recover from the disease.

Prevention and Control Measures

Currently, the only effective method to control MCF is to separate susceptible hosts from reservoir hosts. If sheep and cattle are being raised together, it is recommended to keep them in separate grazing areas, and define different grazing hours for them. For instance, if sheep graze in the morning, cattle can graze in the evening. This interval between grazing will help to inactivate labile MCFVs in sunlight. It is critical to segregate cattle from potential reservoir hosts such as sheep, goats, and wildebeest, especially during lambing, kidding or calving seasons, respectively. Removing lambs from infected sheep before they reach 3 months of age will keep them from being infected and allow them to be raised virus-free. In the event of an outbreak, it is essential to promptly separate cattle or susceptible hosts from sheep and goats (in the case of SA-MCF) or alcelaphine or wild ruminants (in the case of WD-MCF). Additionally, minimizing stress can aid in preventing this disease in sub-clinically or mildly affected animals.

Nucleic acid-based detection of MCF is the currently accepted method for diagnosis worldwide. The hemi-nested PCR test, which is an OIE-approved diagnostic test (OIE, 2018) for the detection of OvHV-2 infection, was developed by Baxter and co-workers (Baxter *et al.*, 1993).

Vaccinating reservoir hosts such as sheep or wildebeest is unlikely to be commercially viable, nor is it suitable for the majority of cattle herds at risk of sporadic SA-MCF. However, in farmed species that are particularly susceptible or exposed to MCF, vaccination could be a considered as a viable strategy. Despite several attempts to develop a protective vaccine against AlHV-1-induced MCF, the results have been unsatisfactory. However, trials focusing on inducing high titers of neutralizing antibodies in nasal secretions of cattle have shown promising results (Russell *et al.*, 2012). A live attenuated vaccine was successful in providing protection against intranasal experimental challenge with pathogenic AlHV-1, and this protection was found to persist for at least 6 months. Further research, including field trials are required to establish the utility of the vaccine.

Conclusions

Tackling MCF remains a challenge for field veterinarians due to a number of factors. The absence of a reliable vaccine, the existence of multiple reservoir hosts, and the inherent complexity of diagnosing this disease, characterized by its non-specific clinical manifestations, together compound the difficulties in its management. MCF has often been overshadowed by its sporadic occurrence and its symptoms' resemblance to those of other ailments, which has, resulted in insufficient attention from the veterinary community, researchers, and policymakers. Consequently, many regions around the world lack comprehensive national strategies for MCF control. There are no definitive treatments or vaccines are currently available for MCF, and the global inability to isolate the OvHV-2 infection agent has hindered the development of essential diagnostic tools and vaccines. Nevertheless, recent advancements in PCR-based assays have made the diagnosis more accessible, with confirmatory testing now possible in specialized laboratories. To enhance our understanding of MCF epidemiology and quantify the economic losses it incurs, nationwide surveys are imperative. Additionally, molecular characterization of the virus genome will undoubtedly provide crucial insights, paving the way for the formulation of effective control strategies to mitigate the impact of this challenging disease.

REFERENCES

Banumathi, N., Sood, R., Patil, S.S., Subramanian, M. and Pradhan, H.K. (2008). Genomic detection of Ovine herpes virus-2 in sheep and goat in states of southern India. Ind J Anim Sci, 78: 13-16.

Barnard, B.J., Bengis, R.G., Griessel, M.D. and de Vos, V. (1989). Excretion of Alcelaphine Herpesvirus-1 by captive and free-living wildebeest (*Connochaetes taurinus*) Onderstepoort J Vet Res, 56: 131-134.

Baxter, S.I., Pow, I., Bridgen, A. and Reid, H.W. (1993). PCR detection of the sheep-associated agent of malignant catarrhal fever. Arch Virol, 132: 145-159.

Chmielewicz, B., Goltz, M. and Ehlers, B. (2001). Detection and multigenic characterization of a novel gammaherpesvirus in goats. Virus Res, 75: 87-94.

Crawford, T.B., Li, H., Rosenburg, S.R., Norhausen, R.W. and Garner, M.M. (2002). Mural folliculitis and alopecia caused by infection with goat-associated malignant catarrhal fever virus in two sika deer. J Am Vet Med Assoc, 221: 843-847.

Keel, M.K., Patterson, J.G., Noon, T.H., Bradley, G.A. and Collins, J.K. (2003). Caprine herpesvirus-2 in association with naturally occurring malignant catarrhal fever in captive sika deer *Cervus nippon*. J Vet Diagn Invest, 15: 179-183.

Klieforth, R., Maalouf, G., Stalis, I., Terio, K., Janssen, D. and Schrenzel, M. (2002). Malignant catarrhal fever-like disease in Barbary red deer *Cervus elaphus barbarus* naturally infected with a virus resembling alcelaphine herpesvirus 2. J ClinMicrobiol, 40: 3381-3390.

Krishna Kumar, S., Sood, R., Selvaraj, P., Jayalakshmi, K. and Ranjith Kumar, M. (2019). Cohort study of bovine malignant catarrhal fever in Cavery delta districts of Tamil Nadu. Ann Rev Res, 5(2): 56-58.

Kumar, P., Nagarajan, N., Saikumar, G., Arya, R.S., Somvanshi, R., and Maan, S. (2014). Detection of malignant catarrhal fever viral DNA in cattle and buffaloes by polymerase chain reaction in India. J Anim Res, 4(2): 153-159.

Kumar, N., Sood, R., Pateriya, A.K., Venkatesakumar, E., Ramprabhu, R., Dixit, R., Bhatia, S. and Singh, V.P. (2021). First molecular evidence and genetic characterization of ovine herpesvirus 2 in multiple animal species in India. Front Vet Sci, 8: 610178.

Li, H., Cunha, C.W. and Taus, N.S. (2011). Malignant catarrhal fever: understanding molecular diagnostics in context of epidemiology. Int J Mol Sci, 12: 6881-6893.

Li, H., Gailbreath, K., Bender, L.C., West, K., Keller, J. and Crawford, T.B. (2003). Evidence of three new members of malignant catarrhal fever virus group in muskox (*Ovibos moschatus*) Nubian ibex (*Capra nubiana*) and gemsbok (*Oryx gazelle*). J Wildlife Dis, 39: 875-880.

Li, H., Gailbreath, K., Flach, E.J., Taus, N.S., Cooley, J., Keller, J., Russell, G.C., Knowles, D.P., Haig, D.M., Oaks, J.L., Traul, D.L. and Crawford, T.B. (2005). A novel subgroup of rhadinoviruses in ruminants. J Gen Virol, 86: 3021-3026.

Li, H., Keller, J., Knowles, D.P. and Crawford, T.B. (2001). Recognition of another member of the malignant catarrhal fever virus group: an endemic gammaherpesvirus in domestic goats. J Gen Virol, 82: 227-232.

Li, H., Wunschmann, A., Keller, J., Hall, D.G. and Crawford, T.B. (2003). Caprine herpesvirus-2-associated malignant catarrhal fever in white-tailed deer *Odocoileus virginianus*. J Vet Diagn Invest, 15: 46-49.

Løken, T., Aleksandersen, M., Reid, H. and Pow, I. (1998). Malignant catarrhal fever caused by ovine herpesvirus-2 in pigs in Norway. Vet Rec, 143: 464-467.

Metzler, A.E. (1991). The malignant catarrhal fever complex. Comp Immunol Microbiol Infect Dis, 14: 107-124.

Mushi, E.Z., Karstad, L. and Jessett, D.M. (1980). Isolation of bovine malignant catarrhal fever virus from ocular and nasal secretions of wildebeest calves. Res Vet Sci, 29(2): 168-171.

Mushi, E.Z., Rossiter, P.B., Jessett, D. and Karstad, L. (1981). Isolation and characterization of a herpesvirus from topi (*Damaliscus korrigum, Ogilby*). J Comp Pathol, 91: 63-68.

OIE, World Organization for Animal Health. (2018). Manual of Diagnostic Tests and Vaccines for Terrestrial Animals [online]. Paris: OIE; 2013a. Malignant catarrhal fever. Available at: http://www.oie.int/fileadmin/Home/fr/Health_standards/tahm/2.04.15_MCF.pdf.

Parihar, N.S., Rajya, B.S. and Gill, B.S. (1975). Occurrence of malignant catarrhal fever in India. Ind Vet J, 52: 857-859.

Plowright, W., Ferris, R.D. and Scott, G.R. (1960). Blue wildebeest and the aetiological agent of bovine malignant catarrhal fever. Nature, 188: 1167-1169.

Premkrishnan, G.N., Sood, R., Hemadri, D., Chanu, KhV., Khandia, R., Bhat, S., Dimri, U. and Bhatia, S. (2015). Cross-sectional study indicates nearly a quarter of sheep population in Karnataka state of India is infected with ovine herpesvirus 2. Virus Dis, 26(3): 180-188.

Radostits, O.M., Gay, C.C., Hinchcliff, K.W. *et al.* (2010). Veterinary Medicine: A text book of diseases of cattle, sheep goat pigs and horses, 10th edn. Saunders, Philadelphia, pp 1245.

Reid, H.W., Buxton, D., McKelvey, W.A., Milne, J.A. and Appleyard, W.T. (1987). Malignant catarrhal fever in Pe'reDavids deer. Vet Rec., 121: 276-277.

Reid, H.W. and Bridgen, A. (1991). Recovery of a herpesvirus from a roan antelope *Hippotragus equinus*. Vet Microbiol, 28: 269-278.

Russell, G.C., Benavides, J., Grant, D., Todd, H., Deane, D., Percival, A., Thomson, J., Connelly, M., and Haig, D.M. (2012). Duration of protective immunity and antibody responses in cattle immunised against Alcelaphine herpesvirus-1-induced malignant catarrhal fever. Vet Res, 43(1): 51. https://doi.org/10.1186/1297-9716-43-51.

Russell, G.C., Stewart, J.P. and Haig, D.M. (2009). Malignant catarrhal fever: a review. Vet J, 179: 324-335.

Syrjala, P., Saarinen, H., Laine, T., Kokkonen, T. and Veijalainen, P. (2006). Malignant catarrhal fever in pigs and a genetic comparison of porcine and ruminant virus isolates in Finland. Vet Rec, 159: 406-409.

Sood, R., Khandia, R., Bhatia, S., Hemadri, D., Kumar, M., Patil, S.S., Pateriya, A.K., Siddiqui, A., Kumar, M.S., Venkatesha, M.D. and Kulkarni, D.D. (2014). Detection and molecular characterization of naturally transmitted sheep associated malignant catarrhal fever in cattle in India. Trop Anim Health Prod, 46(6): 1037-1043.

Sood, R., Kumar, M., Bhatia, S., Pateriya, A.K., Khandia, R., Siddiqui, A., Kulkarni, D.D., Kumar, S. and Venkatesha, M.D. (2012). Transboundary Diseases: ovine herpesvirus type 2 infection in captive bison in India. Vet Rec, 170: 654.

Sood, R., Kumar, N. and Bhatia, S. (2017). Malignant Catarrhal Fever. In: Bayry, J. (eds) Emerging and Re-emerging Infectious Diseases of Livestock. Springer, Cham. https://doi.org/10.1007/978-3-319-47426-7_16

Vikøren, T., Li, H., Lillehaug, A., Jonassen, C.M., Bo¨ckerman, I. and Handeland, K. (2006). Malignant catarrhal fever in free-ranging cervids associated with OvHV-2 and CpHV-2 DNA. J Wildlife Dis, 42: 797-807.

Wani, S.A., Bhat, M.A., Samanta, I., Buchoo, B.A., Ishaq, S.M., Pandit, F., Shah, Gu.D. and Buchh, A.S. (2004). Clinical serological and molecular evidence of sheep-associated malignant catarrhal fever in India. Vet Rec, 155: 242-244.

Wani, S.A., Samanta, I., Pandit, F., Buchoo, B.A., Peer, F. and Bhat, M.A. (2006). Molecular epidemiology of ovine herpesvirus type 2 infection in Kashmir India. Vet Rec, 159: 587-590.

Chapter 19

Beyond Humans: Deciphering the Diversity of Rotavirus in the Animal Kingdom

Shubham Saini, Shriya Rawat and Z.B. Dubal*

Division of Veterinary Public Health, ICAR-Indian Veterinary Research Institute Izatnagar (Bareilly), Uttar Pradesh
****e-mail: drzunjar@yahoo.co.in***

ABSTRACT

Rotavirus, a highly contagious and diverse virus, poses a significant threat to the health of both humans and animals. Rotaviruses are classified into nine distinct species (A-I), with Rotavirus A (RVA) being the primary culprit for human infections. Within RVA, a complex system of genotyping based on VP7 and VP4 proteins further categorizes strains into numerous genotypes. This remarkable diversity, fuelled by point mutations and genetic reassortment, poses challenges for vaccine development and surveillance. Rotavirus spreads through the fecal-oral route, and environmental persistence and resistance to disinfectants contribute to its spread. Rotavirus infects mature enterocytes in the small intestine, causing cell damage and disrupting absorption, ultimately leading to acute diarrhoea, vomiting, abdominal pain, and dehydration. Laboratory diagnosis relies on electron microscopy, immunoassays, RNA-based methods, and genotyping techniques. Treatment focuses on supportive care, including fluid and electrolyte replacement, with specific antiviral therapy currently unavailable. Rotavirus vaccines are highly effective in preventing severe diarrhoea and inactivated or attenuated vaccines are employed to protect young animals and confer maternal immunity. Genetic reassortment between human and animal rotaviruses poses a significant public health concern, potentially leading to novel and more virulent strains. Therefore, understanding interspecies transmission and zoonotic potential through continual surveillance and research is crucial for comprehensive disease control and prevention strategies including the development of improved vaccines. Furthermore, it necessitates a holistic one-health approach to mitigate its global impact and any potential public health risk(s).

Keywords: *Enteric virus, Genotypes, Interspecies transmission, Rotavirus, VP6 protein, Zoonoses.*

Introduction

Diarrheal diseases remain a critical global health burden, ranking among the top ten contributors to disability-adjusted life years (DALYs) despite recent advancements (Behera and Mishra, 2022). Within low- and middle-income countries, diarrhoea poses a significant public health threat and economic challenge. Annually, an estimated 1.3 million individuals of all ages succumb to diarrheal diseases, disproportionately impacting children under 5 years old (IHME, 2020). Notably, rotaviruses reign as a primary culprit for gastroenteritis across a vast spectrum of vertebrates, including humans, animals, and avian species.

Rotaviruses, members of the *Sedoreoviridae* family, exhibit a broad host range, infecting humans and animals across all age groups (Matthijnssens *et al.*, 2022). However, their pathogenic potential is amplified in children under 5 and young animals, leading to severe acute diarrhoea with a risk of dehydration-related fatalities, particularly in resource-limited settings. Globally, rotavirus is estimated to be responsible for approximately 60 percent of paediatric hospital admissions due to acute diarrhoea (Tate *et al.*, 2016). In India alone, Giri *et al.* (2019) reported an annual burden of 11.37 million rotavirus gastroenteritis episodes, translating to 3.27 million outpatient visits, 872,000 hospitalizations, and an estimated direct cost of Rs 10.37 billion. Beyond its human impact, rotavirus reigns as the premier viral culprit in animal gastroenteritis. The virus has been isolated from an array of domesticated and wild mammals, notably pigs, calves, goats, lambs, and horses, as well as avian species (Otto *et al.*, 2015, Simsek *et al.*, 2021; Althof *et al.*, 2023). These infections inflict significant economic burdens on livestock industries, particularly cattle, swine, and horses, due to increased treatment costs and weight loss in affected animals, ultimately hampering their production potential. The above staggering information reveals rotavirus-induced diarrhoea as a significant public health and socioeconomic challenge, with substantial ramifications for global development.

Historical Overview of RVs

Rotavirus offers a captivating case study in zoonotic viruses – pathogens initially identified in animals later found to impact humans. Notably, the link between diarrhea and rotavirus was established in calves long before its recognition in humans (Parashar *et al.*, 1998). Pioneering work by Mebus and colleagues not only demonstrated rotavirus as the culprit behind bovine diarrhea but also established a critical cell culture system for virus propagation (Mebus *et al.*, 1969; 1971). Furthermore, research in calves shed light on the crucial role of local immunity and colostrum-derived antibodies in combating rotavirus infection (Woode *et al.*, 1975; Snodgrass and Wells, 1976). These discoveries significantly paved the way for understanding and preventing rotavirus illnesses in humans.

Turning to the human story, in May 1973, Ruth Bishop and colleagues employed electron microscopy to reveal an abundance of viral particles within the intestinal epithelial cells of children suffering from acute gastroenteritis (Bishop *et al.*, 1973). Initial characterization suggested a reovirus-like/orbivirus-like morphology, resembling that of the established diarrheal agents in neonatal mice (Adams and Kraft, 1963), vervet monkey (Malherbe and Strickland-Cholmley, 1967), and calves (Mebus *et al.*, 1969). This 70-nm virus acquired various temporary names, including reovirus-like, orbivirus-like, duovirus, infantile gastroenteritis virus, and even "a new virus." Ultimately, owing to the characteristic wheel-like appearance observed in electron microscopy, the name rotavirus (rota meaning wheel in Latin) was unanimously adopted (Flewett *et al.*, 1973).

Structure and Genomic Organization of RVs

Rotavirus, a non-enveloped member of the genus Rotavirus within the family *Sedoreoviridae*, boasts a complex architecture. Its mature virion, roughly 100 nm in diameter with protruding VP4 spikes, resembles a triple-layered icosahedral protein shell (Settembre *et al.*, 2011). This capsid comprises three concentric layers: an outermost capsid layer composed of VP7 and VP4, a middle layer formed solely by VP6, and an innermost core where VP2 encapsulates the viral genome replication machinery - VP1 (RNA-dependent RNA polymerase) along with VP3 (viral capping enzyme). The Rotavirus genome, approximately 18.5 kilobases in size, is a mosaic of 11 double-stranded RNA (dsRNA) segments. These segments encode six structural proteins (VP1-4, VP6, and VP7) and six non-structural proteins (NSP1-NSP6). Notably, VP1 and VP3, along with NSP1-NSP5, except NSP4 (which is a viral enterotoxin that induces diarrhoea in the host), drive viral genome replication and orchestrate interactions with cellular proteins, facilitating virus propagation (Crawford *et al.*, 2017). Individual segments range from 667 to 3302 base pairs in length, being monocistronic except for segment 11, which exhibits bicistronic behavior and, in certain strains, encodes for two proteins - NSP5 and NSP6 (Trask *et al.*, 2012).

RV Classification

Rotaviruses are classified into nine distinct species (A, B, C, D, F, G, H, I, and J) based on the immunogenicity of their group antigen - VP6 protein, and hence constitute a diverse viral landscape (Matthijnssens *et al.*, 2022). While Groups A, B, C, and H are known to infect both humans and animals, Groups D, F, and G are capable of infecting animals and birds. Groups H-J are detected from animals; H from humans as well as pigs, I from dogs and cats, and J from bats (Matthijnssens *et al.*, 2009; Marthaler *et al.*, 2014; Bányai *et al.*, 2017). Among them, rotavirus A (RVA) reigns supreme, posing a significant global health threat due to its widespread prevalence and medical complications from both medical as well as veterinary perspectives (Crawford *et al.*, 2017). Moreover, beyond the established nine species of rotavirus, two tentative new species have emerged in recent years. In 2019, Johneand colleagues identified rotavirus K (RVK) and rotavirus L (RVL) in common shrews (*Sorex araneus*) from Germany. The classification of RVL as a distinct species within the RVB-like clade has been solidified by the recent determination of a complete reference genome sequence (Johne *et al.*, 2022). In addition, with the near-complete genome of an RVK strain sequenced, its status as a distinct rotavirus species with a close relation with its RVA-like clade cousin, RVC stands firmly established (Johne *et al.*, 2023).

Beyond broad species classification, rotaviruses exhibit remarkable diversity within. RVs are classified into various serotypes and genotypes, that are determined by genetic as well as antigenic variations in the two crucial surface antigens - VP7 and VP4. VP7, a glycosylated protein, defines the G-type, while VP4, a protease-sensitive spike protein, dictates the P-type specificity. Beyond their structural roles, these proteins act as potent serum-specific antigens, eliciting neutralizing antibodies and informing serotype determination (Sadiq *et al.*, 2018). Currently, genotyping and serotyping classification based on VP7 and VP4 remain the most common typing method for rotavirus. VP7 exhibits remarkable consistency in both genotype and serotype identification, elegantly expressed with a unified G notation. In contrast, discerning the VP4 serotype proves more challenging, necessitating separate designations - P and P[n] for serotype and genotype, respectively (Estes and Cohen, 1989; Estes and Kapikian, 2007). This meticulous classification system has yielded an impressive inventory of rotavirus diversity, currently encompassing at least 42 G and 58 P genotypes in humans and animals across the globe (https://rega.kuleuven.be/cev/viralmetagenomics/virus-classification/rcwg).

Fueled by the ever-expanding trove of rotavirus sequence data beyond VP7 and VP4, genotyping has evolved to encompass all 11 rotaviral genes. This intricate nomenclature, also known as Matthijnssens classification system, employs the format Gx-P[x]-Ix-Rx-Cx-Mx-Ax-Nx-Tx-ExHx, where each letter symbolizes the genotype of a specific gene segment (VP7-VP4-VP6-VP1-VP2-VP3-NSP1-NSP2-NSP3-NSP4-NSP5 respectively), with 'x' replaced by the corresponding genotype number. To date, a staggering array of genotypes has been identified: at least 42 G, 58 P, 32 I, 28 R, 24 C, 24 M, 39 A, 28 N, 28 T, 32 E, and 28 H (https://rega.kuleuven.be/cev/viralmetagenomics/virus-classification/rcwg). This revolutionary classification scheme, coupled with gene-specific phylogenetic analyses, has illuminated far beyond mere genotype identification. It has not only revealed prevalent genotype constellations in humans and diverse animal hosts but also unlocked new insights into the origin and evolution of enigmatic rotavirus strains.

The phylogenetic analysis of rotaviruses reveals a dynamic diversification, resulting in the classification of these viruses into two distinct evolutionary lineages, referred to as clades. The RVA-like clade encompasses RVA, RVC, RVD, and RVF, while the RVB-like clade includes RVB and the RVG-RVJ quartet (Kindler *et al.,* 2013; Johne *et al.,* 2022). The structural and functional distinctions among viral proteins across various clades are indicative of the underlying phylogenetic relationships, as previously exemplified in the case of VP1 (Ogden *et al.,* 2012).

Epidemiology of RVs

Epidemiological investigations conducted across diverse regions globally reveal a plethora of rotavirus strains prevalent in nature. RVA stands as the primary etiological agent responsible for acute gastroenteritis in infants and young children under the age of five(Crawford *et al.,* 2017). It represents a major contributor to child mortality in numerous countries, with infection rates ranging from 30 percent to 50 percent (Troeger *et al.,* 2018). In 2019, rotavirus infections were estimated to be attributed to 151,714 deaths in children under five, with 90 percent of these cases concentrated in developing or low-income countries. This disparity is likely attributed to limited access to healthcare services, unavailability/insufficiency of hydration therapy, and malnutrition (Janko *et al.,* 2022). In general, the prevalence of strains G1P[8], G2P[4], G3P[8], G4P[8], G8P[8], G9P[8], and G12P[8] is noteworthy, as these collectively constitute over 70 percent of the total RVA strains circulating worldwide (Rojas *et al.,* 2019).

In contrast to human RVs, investigations into animal RVs are sparse and restricted. Presently, there is a scarcity of global data regarding RV-related fatalities in various animal species (Ghosh and Kobayashi, 2014). Nevertheless, RVA exhibits the capability to infect a diverse array of animal species, including pigs, cows, horses, dogs, and cats (Lestari *et al.,* 2020). In pigs, twelve G genotypes (G1 to G6, G8 to G12, and G26) and seventeen P genotypes (P[1] to P[8], P[11], P[13], P[19], P[23], P[26], P[27], P[32], P[34], and P[49]) have been identified globally (Vlasova *et al.,* 2017). Predominant G genotypes in pigs encompass G3, G4, G5, G9, and G11, often associated with P[6], P[7], P[13], P[19], and P[23] (Matthijnssens *et al.,* 2011; Dubal *et al.,* 2013). Furthermore, the most prevalent genotypes in pigs are G5 (46 per cent), G3 (11 per cent), and G4 (10 per cent), coupled with P[7] (47 per cent), P[6] (16 per cent), and P[13] (3 per cent). Among these, G5P[7] is particularly significant, which alone constitutes 37 percent of the identified combinations (Papp *et al.,* 2013a). In cattle, at least 12 G types (G1–G3, G5, G6, G8, G10, G11, G15, G17, G21, and G24) and 11 P types (P[1], P[3], P[5-7], P[11], P[14], P[17], P[21], P[29], and P[33]) have been documented, with the most prevalent genotypes in calves being G6, G8, and G10, as well as P[1], P[5], and P[11] (Doro *et al.*, 2015). The commonly prevailing genotypes of bovine RVA include G6 (57 per cent), G10 (21 per cent), and G8 (3 per cent), along with P[5] (26 per cent), P[11] (21 per cent), and P[1] (2 per cent). To date, approximately 20 distinct G and P combinations have been documented, with G6P[5], G6P[11], and G10P[11] emerging as predominant combinations, collectively accounting for ~40 percent prevalence across the globe (Papp *et al.*, 2013a). In the case of horses, molecular epidemiological studies frequently report the prevalence of genotypes G3P[12] and G14P[12] worldwide(Papp *et al.*, 2013b). Notably, the frequently prevailing RVA genotype in canines is G3P[3] and, while in felines G3P[3], G3P[9] and G6P[9] are the most common ones (Papp *et al.*, 2015; German *et al.*, 2015).

In contrast to domesticated animals, the circulating RVA genotypes in wild animals remain largely uncharacterized. While instances have been documented, such as the detection of G3P[9] in masked palm civets and raccoons in Japan, G1P[7] in giant pandas in China, G25P[6] in fruit bats (*Eidolon helvum*) in Kenya, G9P[23], G4P[23], G9P[13], and G4P[6] in wild boars in Japan, G3P[3] in lesser horseshoe bats (*Rhinolophus hipposideros*) in China, G8P[14] in vicunas in Argentina, and G10P[11] in a giraffe at the Dublin Zoo in Ireland (LuchsandTimenetsky, 2016), it remains challenging to ascertain whether these genotypes commonly infect these animals or if infections result from interspecies transmissions and/or gene reassortments.

A crucial epidemiological feature of RVA is its distinct seasonal pattern. In temperate climates, RVA infection exhibits peaks during the coldest and driest months, specifically in winter, earning it the designation of 'winter diarrhea'. Conversely, in tropical climates, the seasonality is less pronounced, with rates tending to be evenly distributed throughout the year (Levy *et al.*, 2009).

Transmission of RVs

Rotaviruses exhibit a high level of contagion and robust resistance to environmental conditions (Estes *et al.*, 1979). In typical environmental settings, characterized by ambient temperature, humidity, and sunlight, viral particles can retain infectiousness for up to seven months. This prolonged viability poses a potential risk for soil, and various crops, and can serve as a source of infection for both humans and animals. Furthermore, these viruses can remain infectious in raw foods and water for over 14 days, occasionally leading to food or water-borne disease outbreaks (Hurst and Gerba, 1980; Köroδlu *et al.*, 2011). Eliminating the virus presents challenges, as rotaviruses can maintain their infectivity even after exposure to various disinfectants, ultraviolet irradiation, or temperature treatments. Notably, only disinfectants containing ≥95 percent ethanol have demonstrated efficacy against the virus (Steele *et al.*, 2004; Li *et al.*, 2009).

Transmission primarily occurs through the fecal-oral and oral routes, with a remarkably low infectious particle threshold of around 10 particles (Ward *et al.*, 1986). The incubation period is 2 to 3 days, and a substantial number, approximately 100 billion rotavirus particles, are shed per gram of feces (Kirkwood *et al.*, 2019). Transmission can also take place through drinking contaminated water, ingestion of contaminated food, and contact with contaminated hands or surfaces/fomites. Nosocomial transmission of rotavirus in healthcare settings acts as a reservoir, contributing to an elevated rate of infections and significant economic losses due to prolonged hospitalizations. In developed countries, approximately 11 percent to 32 percent of rotavirus infections are attributed to hospital-acquired cases, indicating a substantial burden of disease (Johansen *et al.*, 2008). While transmission via the respiratory route has been suggested, conclusive evidence remains insufficient (Prince *et al.*, 1986).

Pathogenesis of RVs

Rotavirus exhibits a tropism for apical cells lining the villi of the small intestine, specifically targeting mature enterocytes. In the course of disseminating within enterocytes, rotavirus induces cell desquamation, resulting in the destruction of enterocytes. This leads to an accelerated migration of cells from the crypt to the villi, causing a transient loss of the intestine's absorptive capacity and subsequent onset of diarrhea (Ramig, 2004). Following the cytolytic replication of rotavirus in mature enterocytes, newly produced viral particles can infect the distal segments of the small intestine and/or be excreted through stools. The NSP4 protein plays a pivotal role in inducing diarrhea by acting as an enterotoxin (Estes and Kapikian, 2007). Beyond acute gastroenteritis, rotavirus infection has been associated with various complications, including hepatitis, cholestasis, respiratory illnesses, myocarditis, type-1 diabetes, renal failure, and thrombocytopenia. Recent findings also suggest that rotavirus infection can disseminate systemically, leading to systemic infection (Medici *et al.*, 2011). Neurological manifestations linked to rotavirus infection have been reported in approximately 2 to 5 percent of cases, ranging from benign seizures to fatal encephalitis. However, it remains unclear whether the rotavirus remains active and replicates in extraintestinal sites or the virus is passively transferred through the bloodstream (Dian *et al.*, 2021). Collectively, these data indicate that the pathogenesis of rotaviruses may be more intricate than currently understood.

Clinical Signs of RV Infection

Clinical manifestations of rotavirus infections vary from asymptomatic cases to severe infections leading to fatality (Saif and Jiang, 1994; Martela *et al.*, 2010). The severity of infection is influenced by multiple factors such as the age of the host, its immune status, the inoculum dose, variances in virulence among the virus strains, the occurrence of mixed infections, environmental stressors (such as weather conditions, housing, and overcrowding), and nutritional factors (Bridger *et al.*, 1992). The disease typically emerges abruptly after a brief incubation period, with an acute course. In calves, lambs, and kids, severe infection characterized by diarrhea occurs, predominantly in animals younger than 10 days, with mortality rates ranging from 1 to 50 per cent. Mortality may result from the direct pathogenic effects of the virus or secondary bacterial infections, depending on the virulence determinants of the viral strain and individual factors in each affected animal, including its immunological competence (Munoz *et al.*, 1996). Hyper-acute cases can present as sudden death (Woode, 1978). Piglets experience diarrhoea as the main clinical finding of the RV infection, with high morbidity and low mortality (Bohl *et al.*, 1984). Furthermore, RV infections are a primary cause of diarrhea in foals under three months of age (Browning and Begg, 1996). In all these aforementioned species, diarrhea is a prominent but not exclusive sign of the disease, with feces appearing white, yellow, or, in severe cases, blood-tinted or frank hemorrhagic. Additional clinical signs include vomiting, anorexia, acute abdominal pain, and dullness/depression. Death may result from severe dehydration and loss of proteins and electrolytes as an outcome of persistent diarrhoea. Recovery, typically within 3-9 days post-onset of clinical signs, is anticipated with appropriate supportive treatment, although growth retardation may follow in some cases, which ultimately hampers their production potential and hence results in economic loss to the farmers. In puppies and kittens, RV infections often manifest as self-limiting diarrhea whereas young rabbits (particularly of 1-3 months of age) are notably susceptible, and the disease manifests as an acute diarrheic syndrome with an elevated mortality rate (Thouless *et al.*, 1988; Squires, 2003). Unlike human rotavirus infections, animal cases generally do not exhibit clinical signs beyond those related to the gastrointestinal tract. Nevertheless, as viral particles have been identified in various organs of the RV-infected animals (Crawford *et al.*, 2006), potential incidences of extra-intestinal signs warrant further investigation.

Laboratory Diagnosis of RVs

The laboratory diagnosis of rotavirus infections encompasses diverse methods, including electron microscopy (EM) for the identification of characteristic morphology. While EM does not differentiate between various rotavirus groups and genotypes, this distinction can be achieved using group-specific antisera and immune EM. Culturing rotaviruses in foetal African green monkey kidney cells (MA104 cell line) is feasible but necessitates media containing trypsin for optimal growth, as trypsin is required to cleave the protease-sensitive protein, VP4, crucial for virus attachment (Tihova *et al.*, 2001). Although primary isolation of rotavirus in primary or secondary rhesus monkey kidney (RMK) cells is less efficient, it is commonly employed for preparing rotavirus antigens.

The detection of rotavirus VP6 protein, the most abundant virus protein which is highly conserved among mammalian RVAs is the preferred target for diagnostic assays (Nelsen *et al.*, 2022). Techniques such as indirect immunofluorescence, latex agglutination (LAT), in-house or commercial enzyme-linked immunosorbent assays (ELISA), and more recently, immunochromatographic near-patient test devices have gained widespread use in diagnosing rotavirus infections. Additionally, polyacrylamide gel electrophoresis (RNA-PAGE) is utilized to detect the rotavirus genome and differentiate among different groups based on the distinct migration patterns of the 11 segments of rotaviral dsRNA (also known as electropherogram).

Molecular amplification methods like RT-PCR, while highly sensitive, are not commonly employed for fecal rotavirus detection due to their capability to identify even asymptomatic infections, particularly in infants and young children. Instead, molecular methods are reserved for characterizing RV G and P types, investigating zoonotic transmissions, and uncovering the virus in low-viral-load extra-intestinal settings like the central nervous system.

In cases where conventional genotyping methods face challenges, nucleic acid sequence analysis becomes imperative. Genotyping failures may result from point mutations at primer-binding positions (Iturriza-Gomara *et al.*, 2004) or the emergence of novel types lacking specific primers (Banerjee *et al.*, 2007). In such cases, gene-specific consensus primers can be employed to generate amplicons for sequencing, facilitating genotype assignment and enabling potential modifications of existing primers or the inclusion of new primers to accommodate novel or emerging genotypes (Iturriza-Gomara *et al.*, 2003). The emergence of novel RV strains in the human population is attributed to their zoonotic transmission. Utilizing sequence analysis of NSP4 and VP6 genes proves instrumental in surveilling interspecies transmission, as these genes exhibit a tendency to co-segregate during reassortment and demonstrate distinct segregation patterns corresponding to the species of origin (Gray and Iturriza-Gómara, 2010).

Treatment of RV Infections

No specific treatment exists for rotaviral infections. The approach to treatment revolves around offering supportive care and addressing clinical manifestations and potential complications. In both livestock and companion animals, the essential component of treatment involves fluid administration to compensate for losses due to diarrhoea or vomiting, correct acidosis, and restore electrolyte balance. An effective rehydration solution is characterized by optimal sodium concentration and appropriate glucose-to-sodium ratios (Lorenz *et al.*, 2011). For young animals, fluid administration can be achieved using an esophageal catheter, while in older animals, intravenous administration is preferable. Supplemental administration of probiotics has demonstrated efficacy in facilitating rapid recovery, although the underlying mechanisms of action remain unclear (Munoz *et al.*, 2011). In instances of secondary bacterial infections, antimicrobial agents should be prescribed.

Prevention of RV Infections

Vaccination against rotavirus is of paramount importance in mitigating the severe gastroenteritis resulting from the emergence and re-emergence of new or wild RV strains, predominantly in humans. Unusual strains can directly impact vaccine efficacy, imposing a substantial financial burden on developing economies. Hence, continuous surveillance for rotavirus genotypes in humans, as well as animals, is crucial for the comprehensive understanding of the genomic diversity of RVs and ultimately for the effective development of vaccines for them. Four oral, live-attenuated RVA vaccines, namely Rotarix®, RotaTeq®, Rotavac®, and RotaSiil®, have been prequalified by the World Health Organization (WHO) and licensed to be used globally. Additionally, two more vaccines - Rotavin-M1 and Lanzhou lamb rotavirus vaccine are available for use in humans but are exclusively available in Vietnam and China respectively (WHO, 2021). In 2016, the introduction of the rotavirus vaccine (Rotavac®) in India, integrated into the Universal Immunisation Programme (UIP) in a phased manner, resulted in a documented decline in the prevalence and hospitalization due to RV infection in children under 5 years of age (Malik *et al.*, 2019; Varghese *et al.*, 2021).

In the veterinary context, vaccines for the prevention of rotavirus disease in calves, piglets, and foals have been available for many years. The prevention strategy for rotavirus disease in these animals involves the parenteral administration of either inactivated or attenuated vaccines. To confer robust maternal immunity, vaccines are administered to mothers in the late stages of pregnancy. The concept of passive immunization relies on maternal antibodies transferable through the placenta or secreted in colostrum, providing transient protective immunity to offspring against RVA infection (Dhama *et al.*, 2009).

For preventing neonatal calf diarrhea, although vaccination is not routinely performed (Papp *et al.*, 2013a), currently available vaccines include Guardian1®, containing G6P[1] and G6P[5] strains, Scourguard 3® composed of RVA/Cow-tc/USA/NCDV-Lincoln/967/G6P[1], Scourguard 4KC® composed of serotypes G6 and G10, and Trivaction 6®. For preventing diarrhea in piglets, the ProSystem® ROTA (Merck), a polyvalent live oral RVA vaccine containing strains RVA/Pig-tc/USA/OSU/1975/G5P[7], RVA/Pig-tc/USA/Gottfried/1975/G4P[6], and RVA/Pig-tc/USA/A2/19XX/G9P[7] is commercially available (Papp *et al.*, 2013a). In foals, inactivated vaccines have been developed to prevent diarrhea caused by RVA (Papp *et al.*, 2013b). The USA, United Kingdom, and Ireland currently use the Fort Dodge Animal Health and Zoetis vaccines, containing an RVA/Horse-wt/GBR/H-2/1976/G3P[12] strain. In Argentina, a vaccine named ROTAMIX EQUIN® is available which is a polyvalent vaccine containing strains RVA/Horse-wt/GBR/H-2/1976/G3P[12], RVA/Simian-tc/ZAF/SA11-H96/1958/G3P[2], and RVA/Cow-tc/USA/NCDV-Lincoln/967/G6P[1]. In Japan, an inactivated HRV vaccine, containing RVA/Horse-tc/JPN/HO-5/1982/G3BP[12], is available for use.

Unfortunately, for our feathered friends (avian species), the battle against avian RVA remains vaccine-less to date (Patzina-Mehling *et al.*, 2020).

Interspecies Transmission and Zoonotic Potential of RVs

RNA viruses exhibit higher genetic and evolutionary diversity due to their inferior RNA polymerase proofreading activity when compared to DNA viruses (Mandary *et al.*, 2019). Similar to other RNA viruses, the rotavirus genome is highly flexible, with observed point mutation rates at 5×10^{-5} per nucleotide, which is 1000 times higher than those for most DNA viruses (Blackhall *et al.*, 1996). Moreover, genome rearrangement and reassortment events in dsRNA viruses (RVs included) result from their high frequency of genetic diversity and immune evasion (Oki *et al.*, 2022).

The genetic diversity of rotavirus is propelled by interspecies transmission and genetic reassortment events, serving as crucial mechanisms for rotavirus evolution (Bányai and Pitzer, 2016). Genotype diversity among rotavirus strains is generated through genetic drift, marked by the continuous accumulation of point mutations driven by the high error rate of the RV RNA polymerase. This leads to genetic lineages within genotypes and monotypes within serotypes, displaying altered epitopes and specific antibody recognition patterns (Coulson and Kirkwood, 1991). Furthermore, the segmented nature of the rotavirus genome facilitates gene reassortment during co-infection with multiple strains, contributing to additional diversity of co-circulating strains. The widespread presence of rotaviruses across the animal kingdom constitutes a substantial reservoir of rotavirus strains, and interspecies transmission combined with reassortment can lead to the emergence of novel or unusual strains with a potential for global dissemination.

While specific G/P combinations of RVA genotypes are typically associated with distinct host species, zoonotic transmissions of rotaviruses between different hosts have been frequently documented (Martella *et al.*, 2010; Díaz Alarcón *et al.*, 2022). Humans can be infected by rotaviruses of animal origin through direct transmission or through the exchange of one or more genome segments among human and animal rotaviruses (Rojas *et al.*, 2019). Numerous reports have detailed interspecies transmission resulting in sporadic cases of human disease with rotaviruses originating from various animal species (Luchs*et al.*, 2012; Mukherjee *et al.*, 2013; Doro *et al.*, 2015; Komoto *et al.*, 2021; Abass *et al.*, 2022; Kuniæ *et al.*, 2023). Strains such as G3 (commonly found in pigs, horses, cats, dogs, rabbits, monkeys, and mice), G5 (pigs and horses), G6 and G8 (cattle), G9 (pigs and lambs), and G10 (cattle) have been isolated from the human population worldwide (Papp *et al.*, 2013a; 2013b; Ghosh and Kobayashi, 2014). The emergence of epidemiologically important strains such as G9P[8] globally, G10P[11] in India, and G8P[4] in Africa, Europe, and the USA is postulated to result from reassortment with animal strains leading to host adaptation and spread (De Donno*et al.*, 2009; Nyaga*et al.*, 2013). Additionally, some uncommon rotavirus strains detected in humans are derived from viruses transmitted between humans and animals (Bányai, and Pitzer, 2016). In the last decade, bats have been considered a potential source for emerging rotaviruses in humans and are deemed reservoirs for many other emerging viruses (Simsek *et al.*, 2021). Consequently, the increasing interaction between humans and wild animals, such as bats, heightens the risk of zoonotic rotavirus transmission.

Despite accumulating molecular evidence about interspecies transmission of RVAs, epidemiological studies describing natural interspecies transmission events are lacking. Few studies report limited evidence of indirect contact or cohabitation of rotavirus-positive patients with domestic animals (Zeller *et al.*, 2012; Doan *et al.*, 2013). Despite these reports, very little is known about the modes and frequencies of interspecies rotavirus transmission. Given that strain surveillance primarily focuses on clinically ill patients and animal rotavirus strains tend to be less virulent in humans, it is reasonable to speculate that documented cases of independent interspecies transmission events likely represent the tip of the iceberg.

Therefore, continuous surveillance of the distribution of RVA strains in the animal world is crucial for preventing outbreaks and monitoring circulating rotavirus strains. This contributes to a better understanding of interspecies transmission and aids in the design of new vaccines.

Conclusion

Investigating the genomic diversity of animal rotaviruses (RVAs) through comprehensive analysis of their genomic constellations (strain compositions) holds immense potential. This deeper understanding of RVA evolution and ecology will not only illuminate fundamental biological processes but also pave the way for improved vaccine development. Specifically, such insights can guide the selection of more suitable vaccine strains:

a) Enhancing the breadth and potency of cross-immunity: by identifying conserved antigenic elements amidst the diverse RVA landscape, vaccines can be designed to elicit broader protection against a wider range of circulating strains.

b) Optimizing vaccine efficacy through strain matching: analyzing co-circulating RVA strains in specific regions allows for the selection of vaccine strains that closely mirror the local viral landscape, maximizing vaccine effectiveness.

c) Ultimately, effective control of RVA in animals through advanced vaccines has the potential to prevent subsequent epidemic outbreaks in humans. This can be achieved through:

d) Reduced zoonotic transmission: By minimizing RVA prevalence and circulation in animal populations, the opportunities for spillover infections to humans are significantly diminished. b) Indirect herd immunity: Widespread vaccination of animals creates a protective barrier, indirectly shielding humans from exposure to circulating RVA strains.

e) Therefore, delving into the intriguing world of animal RVA genomic constellations can offer us a promising avenue for not only safeguarding animal health but also protecting human populations from the threat of zoonotic rotavirus epidemics, which is what the concept of one-health encompasses.

REFERENCES

Abass, G., Dubal, Z.B., Rajak, K.K., Kale, B.M., Raorane, A., Dudhe, N., Malla, B.A., Desai, D., Sinha, D.K., Vinodh Kumar,O.R. and Malik, Y.P.S. (2022). Molecular characterization of porcine rotavirus A from India revealing zooanthroponotic transmission. Ani Biotech, 33(6): 1073-1085.https://doi.org/10.1080/10495398.2020.1868486

Adams, W.R. and Kraft, L.M. (1963). Epizootic diarrhea of infant mice: identification of the etiologic agent. Science, 141(3578): 359-360. doi: 10.1126/science.141.3578.359

Althof, N., Trojnar, E. and Johne, R. (2023). Rotaviruses in Wild Ungulates from Germany, 2019–2022. Microorganisms, 11(3): 566. doi: https://doi.org/10.3390/microorganisms11030566

Banerjee, I., Ramani, S., Primrose, B., IturrizaGomara, M., Gray, J. J., Brown, D. W. and Kang, G. (2007). Modification of rotavirus multiplex RTPCR for the detection of G12 strains based on characterization of emerging G12 rotavirus strains from South India. J Med Virol, 79(9): 1413-1421.doi: https://doi.org/10.1002/jmv.20872

Bányai, K. and Pitzer, V.E. (2016). Molecular epidemiology and evolution of rotaviruses. In Viral Gastroenteritis, Academic Press, pp. 279-299. doi: https://doi.org/10.1016/B978-0-12-802241-2.00014-6

Bányai, K., Kemenesi, G., Budinski, I., Földes, F., Zana, B., Marton, S., Varga-Kugler, R., Oldal, M., Kurucz, K. and Jakab, F. (2017). Candidate new rotavirus species in Schreiber's bats, Serbia. Infect Genet Evol, 48: 19-26. doi: 10.1016/j.meegid.2016.12.002

Behera, D.K., and Mishra, S. (2022). The burden of diarrhea, etiologies, and risk factors in India from 1990 to 2019: evidence from the global burden of disease study. BMC Public Health, 22(1): 1-9. doi: https://doi.org/10.1186/s12889-022-12515-3

Bishop, R., Davidson, G.P., Holmes, I.H., and Ruck, B.J. (1973). Virus particles in epithelial cells of duodenal mucosa from children with acute non-bacterial gastroenteritis. Lancet, 302(7841): 1281-1283. doi: 10.1016/s0140-6736(73)92867-5

Blackhall, J., Fuentes, A. and Magnusson, G. (1996). Genetic stability of a porcine rotavirus RNA segment during repeated plaque isolation. Virol, 225(1): 181-190.doi: https://doi.org/10.1006/viro.1996.0586

Bohl, E.H., Theil, K.W., and Saif, L.J. (1984). Isolation and serotyping of porcine rotaviruses and antigenic comparison with other rotaviruses. J Clin Microbiol, 19(2): 105-111.doi: https://doi.org/10.1128/jcm.19.2.105-111.1984

Bridger, J.C., Hall, G.A. and Parsons, K.R. (1992). A study of the basis of virulence variation of bovine rotaviruses. Vet Microbiol, 33(1-4): 169-174.doi: https://doi.org/10.1016/0378-1135(92)90044-T

Browning, G.F. and Begg, A.P. (1996). Prevalence of G and P serotypes among equine rotaviruses in the faeces of diarrhoeic foals. Arch Virol, 141: 1077-1089. doi: https://doi.org/10.1007/BF01718611

Coulson, B.S. and Kirkwood, C. (1991). Relation of VP7 amino acid sequence to monoclonal antibody neutralization of rotavirus and rotavirus monotype. J Virol, 65(11): 5968-5974.doi: https://doi.org/10.1128/jvi.65.11.5968-5974.1991

Crawford, S.E., Ramani, S., Tate, J.E., Parashar, U.D., Svensson, L., Hagbom, M., Franco, M.A., Greenberg, H.B., O'Ryan, M., Kang, G. and Desselberger, U. (2017). Rotavirus infection. Nat Rev Dis Primers, 3(1): 1-16. doi: https://doi.org/10.1038/nrdp.2017.83

Crawford, S.E., Patel, D.G., Cheng, E., Berkova, Z., Hyser, J.M., Ciarlet, M., Finegold, M.J., Conner, M.E. and Estes, M.K. (2006). Rotavirus viremia and extraintestinal viral infection in the neonatal rat model. J Virol, 80(10): 4820-4832.doi: https://doi.org/10.1128/jvi.80.10.4820-4832.2006

De Donno, A., Grassi, T., Bagordo, F., Idolo, A., Cavallaro, A., Gabutti, G. and Collaborative Group for the surveillance of Rotavirus Infection (2009). Emergence of unusual human rotavirus strains in Salento, Italy, during 2006-2007. BMC Infect Dis, 9(43): 1-7. doi: https://doi.org/10.1186/1471-2334-9-43

Dhama, K., Chauhan, R.S., Mahendran, M. and Malik, S.V.S. (2009). Rotavirus diarrhea in bovines and other domestic animals. Vet Res Commun, 33(1): 1-23. doi: 10.1007/s11259-008-9070-x

Dian, Z., Sun, Y., Zhang, G., Xu, Y., Fan, X., Yang, X., Pan, Q., Peppelenbosch, M. and Miao, Z. (2021). Rotavirus-related systemic diseases: clinical manifestation, evidence and pathogenesis. Crit Rev Microbiol, 47(5): 580-595. doi: 10.1080/1040841X.2021.1907738

Díaz Alarcón, R.G., Liotta, D.J. and Miño, S. (2022). Zoonotic RVA: state of the art and distribution in the animal world. Viruses, 14(11): 2554. doi: https://doi.org/10.3390/v14112554

Doan, Y.H., Nakagomi, T., Aboudy, Y., Silberstein, I., Behar-Novat, E., Nakagomi, O. and Shulman, L.M. (2013). Identification by full-genome analysis of a bovine rotavirus transmitted directly to and causing diarrhea in a human child. J Clin Microbiol, 51(1): 182-189.doi: https://doi.org/10.1128/jcm.02062-12

Doro, R., Farkas, S.L., Martella, V. and Banyai, K. (2015). Zoonotic transmission of rotavirus: surveillance and control. Expert Rev Anti Infect Ther, 1337-1350. doi: 10.1586/14787210.2015.1089171

Dubal, Z.B., Bhilegaonkar, K.N., Barbuddhe, S.B., Kolhe, R.P., Simranpreet Kaur, Shriya Rawat, Prejit Nambiar andKarunakaran, M. (2013). Prevalence and genotypic (G and P) determination of porcine group A rotaviruses from different regions of India. Trop Ani Health Prod, 45(2): 609-615. DOI 10.1007/s11250-012-0267-1

Estes M.K. and Kapikian A.Z. (2007). Rotaviruses (5th ed.). In: D.M. Knipe, P.M. Howley (Eds.), Fields Virology, vol. 2, Lippincott Williams and Wilkins, Philadelphia, PA, pp. 1917-1958.

Estes, M.K., and Cohen, J. (1989). Rotavirus gene structure and function. Microbiol Rev, 53(4): 410-449.doi: https://doi.org/10.1128/mr.53.4.410-449.1989

Estes, M.K., Graham, D.Y., Smith, E.M., and Gerba, C.P. (1979). Rotavirus stability and inactivation. J Gen Virol, 43(2): 403-409.doi: https://doi.org/10.1099/0022-1317-43-2-403

Flewett, T.H., Bryden, A.S. and Davies, H. (1973). Virus particles in gastroenteritis. Virus particles in gastroenteritis. Lancet, 2(7844): 1497. doi: 10.1016/s0140-6736(73)92760-8

German, A.C., Iturriza-Gómara, M., Dove, W., Sandrasegaram, M., Nakagomi, T., Nakagomi, O., Cunliffe, N., Radford, A.D. and Morgan, K.L. (2015). Molecular epidemiology of rotavirus in cats in the United Kingdom. J Clin Microbiol, 53(2): 455-464.doi: https://doi.org/10.1128/jcm.02266-14

Ghosh, S. and Kobayashi, N. (2014). Exotic rotaviruses in animals and rotaviruses in exotic animals. Virus Dis, 25: 158-172. doi: https://doi.org/10.1007/s13337-014-0194-z

Giri, S., Nair, N.P., Mathew, A., Manohar, B., Simon, A., Singh, T., Suresh Kumar, S., Mathew, M.A., Babji, S., Arora, R. and Girish Kumar, C.P. (2019). Rotavirus gastroenteritis in Indian children < 5 years hospitalized for diarrhoea, 2012 to 2016. BMC Public Health, 19(1): 1-10. doi: https://doi.org/10.1186/s12889-019-6406-0

Gray, J. and Iturriza-Gómara, M. (2010). Rotaviruses. In: Stephenson, J., Warnes, A. (eds). Diagnostic Virology Protocols. Methods in Molecular Biology, vol 665. Humana Press, Totowa, NJ, pp. 325–355. DOI: https://

doi.org/10.1007/978-1-60761-817-1_18. https://rega.kuleuven.be/cev/viralmetagenomics/virus-classification/rcwg. Accessed on 05-01-2024.

Hurst, C.J. and Gerba, C.P. (1980). Stability of simian rotavirus in fresh and estuarine water. Appl Env Microbiol, 39(1): 1-5.doi: https://doi.org/10.1128/aem.39.1.1-5.1980

IHME. (2020). Global Burden of Disease Study 2019 (GBD 2019) Results. Global Burden of Disease Collaborative Network, vol. 2020. Institute for Health Metrics and Evaluation (IHME), *University of Washington,* United States. DOI: *http://ghdx.healthdata.org/gbd-results-tool*

Iturriza-Gomara, M., Anderton, E., Kang, G., Gallimore, C., Phillips, W., Desselberger, U., and Gray, J. (2003). Evidence for genetic linkage between the gene segments encoding NSP4 and VP6 proteins in common and reassortant human rotavirus strains. J Clin Microbiol, 41(8): 3566-3573.doi: https://doi.org/10.1128/jcm.41.8.3566-3573.2003

Iturriza-Gómara, M., Kang, G. And Gray, J. (2004). Rotavirus genotyping: keeping up with an evolving population of human rotaviruses. J Clin Virol, 31(4): 259-265. doi: https://doi.org/10.1016/j.jcv.2004.04.009

Janko, M.M., Joffe, J., Michael, D., Earl, L., Rosettie, K.L., Sparks, G.W., Albertson, S.B., Compton, K., Velandia, P.P., Stafford, L. and Zheng, P. (2022). Cost-effectiveness of rotavirus vaccination in children under five years of age in 195 countries: A meta-regression analysis. Vaccine, 40(28): 3903-3917. doi: https://doi.org/10.1016/j.vaccine.2022.05.042

Johansen, K., Hedlund, K.O., Zweygberg-Wirgart, B. and Bennet, R. (2008). Complications attributable to rotavirus-induced diarrhoea in a Swedish paediatric population: report from an 11-year surveillance. Scand J Infect Dis, 40(11-12): 958-964. doi: 10.1080/00365540802415509

Johne, R., Schilling-Loeffler, K., Ulrich, R.G. and Tausch, S.H. (2022). Whole genome sequence analysis of a prototype strain of the novel putative rotavirus species L. Viruses, 14(3): 462. doi: https://doi.org/10.3390/v14030462

Johne, R., Tausch, S.H., Ulrich, R.G., and Schilling-Loeffler, K. (2023). Genome analysis of the novel putative rotavirus species K. Virus Res, 334: 199171. doi: https://doi.org/10.1016/j.virusres.2023.199171

Kindler, E., Trojnar, E., Heckel, G., Otto, P.H., and Johne, R. (2013). Analysis of rotavirus species diversity and evolution including the newly determined full-length genome sequences of rotavirus F and G. Infect Genet Evol, 14: 58-67.doi: https://doi.org/10.1016/j.meegid.2012.11.015

Kirkwood, C.D., Ma, L.F., Carey, M.E. and Steele, A.D. (2019). The rotavirus vaccine development pipeline. Vaccine, 37(50): 7328-7335. doi: https://doi.org/10.1016/j.vaccine.2017.03.076

Komoto, S., Tacharoenmuang, R., Guntapong, R., Upachai, S., Singchai, P., Ide, T., Fukuda, S., Hatazawa, R., Sutthiwarakom, K., Kongjorn, S. and Onvimala, N. (2021). Genomic characterization of a novel G3P [10] rotavirus strain from a diarrheic child in Thailand: Evidence for bat-to-human zoonotic transmission. Infect Genet Evol, 87: 104667. doi: https://doi.org/10.1016/j.meegid.2020.104667

Koroglu, M., Yakupogullari, Y., Otlu, B., Ozturk, S., Ozden, M., Ozer, A. and Sener, K., Durmaz, R.A. (2011). Waterborne outbreak of epidemic diarrhea due to group A rotavirus in Malatya, Turkey. New Microbiol, 34(1): 17-24.

Kuniæ, V., Mikuletiè, T., Kogoj, R., Koritnik, T., Steyer, A., Šoprek, S., Tešoviæ, G., Konjik, V., RoksandiæKrižan, I., Prišlin, M. and Jemeršiæ, L. (2023). Interspecies transmission of porcine-originated G4P [6] rotavirus A between pigs and humans: a synchronized spatiotemporal approach. Front Microbiol, 14: 1194764. doi: https://doi.org/10.3389/fmicb.2023.1194764

Lestari, F.B., Vongpunsawad, S., Wanlapakorn, N. and Poovorawan, Y. (2020). Rotavirus infection in children in Southeast Asia 2008-2018: disease burden, genotype distribution, seasonality, and vaccination. J Biomed Scie, 27: 1-19. doi: https://doi.org/10.1186/s12929-020-00649-8

Levy, K., Hubbard, A.E., and Eisenberg, J.N. (2009). Seasonality of rotavirus disease in the tropics: a systematic review and meta-analysis. Int J Epidemiol, 38: 1487-1496. doi: 10.1093/ije/dyn260

Li, D., Gu, A.Z., He, M., Shi, H.C., and Yang, W. (2009). UV inactivation and resistance of rotavirus evaluated by integrated cell culture and real-time RT-PCR assay. Water Res, 43(13): 3261-3269.doi: https://doi.org/10.1016/j.watres.2009.03.044

Lorenz, I., Fagan, J. and More, S.J. (2011). Calf health from birth to weaning. II. Management of diarrhoea in pre-weaned calves. Irish Vet J, 64(1): 1-6. doi: 10.1186/2046-0481-64-9

Luchs, A. and Timenetsky, M.D.C.S.T. (2016). Group A rotavirus gastroenteritis: post-vaccine era, genotypes and zoonotic transmission. Einstein (Sao Paulo), 14: 278-287. doi: 10.1590/S1679-45082016RB3582

Luchs, A., Cilli, A., Morillo, S.G., Carmona, R.D.C.C. and Timenetsky, M.D.C.S.T. (2012). Rare G3P [3] rotavirus strain detected in Brazil: Possible human–canine interspecies transmission. J Clin Virol, 54(1): 89-92.doi: https://doi.org/10.1016/j.jcv.2012.01.025

Malherbe, H.H. and Strickland-Cholmley, M. (1967). Simian virus SA11 and the related 0 agent. Arch Gesamte Virusforsch, 22: 235-245. doi: https://doi.org/10.1007/BF01240518

Malik A., Haldar P., Ray A., Shet A., Kapuria B., Bhadana S., Santosham M., Ghosh R.S., Steinglass R. and Kumar R. (2019). Introducing rotavirus vaccine in the Universal Immunization Programme in India: From evidence to policy to implementation. Vaccine, 37(39): 5817-5824. doi: 10.1016/j.vaccine.2019.07.104

Mandary, M.B., Masomian, M., and Poh, C.L. (2019). Impact of RNA virus evolution on quasispecies formation and virulence. Int J Mol Sci, 20(18): 4657.DOI: https://doi.org/10.3390/ijms20184657

Martella, V., Bányai, K., Matthijnssens, J., Buonavoglia, C. and Ciarlet, M. (2010). Zoonotic aspects of rotaviruses. Vet Microbiol, 140(3-4): 246-255.doi:https://doi.org/10.1016/j.vetmic.2009.08.028

Marthaler, D., Rossow, K., Culhane, M., Goyal, S., Collins, J., Matthijnssens, J., Nelson, M., and Ciarlet, M. (2014). Widespread rotavirus H in commercially raised pigs, United States. Emerg Infect Dis, 20(7): 1203. doi: 10.3201/eid2007.140034

Matthijnssens, J., Attoui, H., Bányai, K., Brussaard, C.P., Danthi, P., Del Vas, M., Dermody, T.S., Duncan, R., Fâng, Q., Johne, R. and Mertens, P.P. (2022). ICTV virus taxonomy profile: Sedoreoviridae 2022. J Gen Virol, 103(10): 001782. doi: https://doi.org/10.1099/jgv.0.001782

Matthijnssens, J., Bilcke, J., Ciarlet, M., Martella, V., Bányai, K., Rahman, M., Zeller, M., Beutels, P., Van Damme, P. and Van Ranst, M. (2009). Rotavirus disease and vaccination: impact on genotype diversity. Future Microbiol, 4(10): 1303-1316. doi: 10.2217/fmb.09.96

Matthijnssens, J., Ciarlet, M., McDonald, S.M., Attoui, H., Bányai, K., Brister, J.R., Buesa, J., Esona, M.D., Estes, M.K., Gentsch, J.R. and Iturriza-Gómara, M. (2011). Uniformity of rotavirus strain nomenclature proposed by the Rotavirus Classification Working Group (RCWG). Arch Virol, 156: 1397-1413. doi: https://doi.org/10.1007/s00705-011-1006-z

Mebus, C.A., Kono, M., Underdahl, N.R. and Twiehaus, M.J. (1971). Cell culture propagation of neonatal calf diarrhea (scours) virus. Can Vet J, 12(3): 69.

Mebus, C.A., Underdahl, N.R., Rhodes, M.B. and Twiehaus, M.J. (1969) *Calf diarrhea (Scours): Reproduced with a virus from a field outbreak* (Research Bulletin: Bulletin of the Agricultural Experiment Station of Nebraska No. 233). Available at: https://digitalcommons.unl.edu/ardhistrb/69/

Medici, M.C., Abelli, L.A., Guerra, P., Dodi, I., Dettori, G., and Chezzi, C. (2011). Case report: detection of rotavirus RNA in the cerebrospinal fluid of a child with rotavirus gastroenteritis and meningism. J Med Virol, 83(9): 1637-1640. doi:https://doi.org/10.1002/jmv.22156

Moreno Muñoz, J.A., Chenoll, E., Casinos, B., Bataller, E., Ramón, D., Genovés, S., Montava, R., Ribes, J.M., Buesa, J., Fabrega, J. and Rivero, M. (2011). Novel probiotic Bifidobacteriumlongum subsp. infantis CECT 7210 strain active against rotavirus infections. Appl Env Microbiol, 77(24): 8775-8783. doi: https://doi.org/10.1128/AEM.05548-11

Mukherjee, A., Mullick, S., Deb, A. K., Panda, S., and ChawlaSarkar, M. (2013). First report of human rotavirus G8P [4] gastroenteritis in India: Evidence of ruminantstohuman zoonotic transmission. J Med Virol, 85(3): 537-545. doi: https://doi.org/10.1002/jmv.23483

Munoz, M., Alvarez, M., Lanza, I. And Carmenes, P. (1996). Role of enteric pathogens in the aetiology of neonatal diarrhoea in lambs and goat kids in Spain. Epidemiol Infect, 117(1): 203-211.doi: https://doi.org/10.1017/S0950268800001321

Nelsen, A., Lager, K.M., Stasko, J., Nelson, E., Lin, C.M. and Hause, B.M. (2022). Identification of pulmonary infections with porcine rotavirus a in pigs with respiratory disease. Front Vet Scie, 9: 918736. doi: https://doi.org/10.3389/fvets.2022.918736

Nyaga, M.M., Jere, K.C., Peenze, I., Mlera, L., van Dijk, A.A., Seheri, M.L. and Mphahlele, M.J. (2013). Sequence analysis of the whole genomes of five African human G9 rotavirus strains. Infect Genet Evol, 16: 62-77. doi: https://doi.org/10.1016/j.meegid.2013.01.005

Ogden, K.M., Johne, R. and Patton, J.T. (2012). Rotavirus RNA polymerases resolve into two phylogenetically distinct classes that differ in their mechanism of template recognition. Virol, 431(1-2): 50-57.doi: https://doi.org/10.1016/j.virol.2012.05.011

Oki, H., Masuda, T., Hayashi-Miyamoto, M., Kawai, M., Ito, M., Madarame, H., Fukase, Y., Takemae, H., Sakaguchi, S., Furuya, T. and Mizutani, T. (2022). Genomic diversity and intragenic recombination of species C rotaviruses. J Gen Virol, 103(2): 001703.doi: https://doi.org/10.1099/jgv.0.001703

Otto, P.H., Rosenhain, S., Elschner, M.C., Hotzel, H., Machnowska, P., Trojnar, E., Hoffmann, K. and Johne, R. (2015). Detection of rotavirus species A, B and C in domestic mammalian animals with diarrhoea and genotyping of bovine species A rotavirus strains. Vet Microbiol, 179(3-4): 168-176. doi: https://doi.org/10.1016/j.vetmic.2015.07.021

Papp, H., László, B., Jakab, F., Ganesh, B., De Grazia, S., Matthijnssens, J., Ciarlet, M., Martella, V. and Bányai, K. (2013a). Review of group A rotavirus strains reported in swine and cattle. Vet Microbiol, 165(3-4): 190-199.doi: https://doi.org/10.1016/j.vetmic.2013.03.020

Papp, H., Matthijnssens, J., Martella, V., Ciarlet, M. and Bányai, K. (2013b). Global distribution of group A rotavirus strains in horses: a systematic review. Vaccine, 31(48): 5627-5633.doi: https://doi.org/10.1016/j.vaccine.2013.08.045

Papp, H., Mihalov-Kovács, E., Dóró, R., Marton, S., Farkas, S.L., Giammanco, G.M., De Grazia, S., Martella, V. and Bányai, K. (2015). Full-genome sequencing of a Hungarian canine G3P [3] Rotavirus A strain reveals high genetic relatedness with a historic Italian human strain. Virus Genes, 50: 310-315. doi: https://doi.org/10.1007/s11262-014-1163-8

Parashar, U.D., Bresee, J.S., Gentsch, J.R., and Glass, R.I. (1998). Rotavirus. Emerg Infect Dis, 4(4): 561. doi: 10.3201/eid0404.980406

Patzina-Mehling, C., Falkenhagen, A., Gadicherla, A.K., Grützke, J., Tausch, S.H. and Johne, R. (2020). Whole genome sequence analysis of cell culture-adapted rotavirus A strains from chicken. Infect Genet Evol, 81: 104275. doi: https://doi.org/10.1016/j.meegid.2020.104275

Prince, D.S., Astry, C., Vonderfecht, S., Jakab, G., Shen, F.M. and Yolken, R.H. (1986). Aerosol transmission of experimental rotavirus infection. Pediatr Infect Dis J, 5(2): 218-222.

Ramig, R.F. (2004). Pathogenesis of intestinal and systemic rotavirus infection. J Virol, 78(19): 10213-10220.doi: https://doi.org/10.1128/jvi.78.19.10213-10220.2004

Rojas, M., Dias, H.G., Gonçalves, J.L.S., Manchego, A., Rosadio, R., Pezo, D. and Santos, N. (2019). Genetic diversity and zoonotic potential of rotavirus A strains in the southern Andean highlands, Peru. Transbound Emerg Dis, 66(4): 1718-1726. doi: https://doi.org/10.1111/tbed.13207

Sadiq, A., Bostan, N., Yinda, K.C., Naseem, S. and Sattar, S. (2018). Rotavirus: Genetics, pathogenesis and vaccine advances. Rev Med Virol, 28(6): e2003. doi: 10.1002/rmv.2003

Saif, L.J. and Jiang, B. (1994). Nongroup A rotaviruses of humans and animals. Curr Top Microbiol Immunol, 185: 339-371.

Settembre, E.C., Chen, J.Z., Dormitzer, P.R., Grigorieff, N. and Harrison, S.C. (2011). Atomic model of an infectious rotavirus particle. EMBO J, 30(2): 408-416. doi: https://doi.org/10.1038/emboj.2010.322

Simsek, C., Corman, V.M., Everling, H.U., Lukashev, A.N., Rasche, A., Maganga, G.D., Binger, T., Jansen, D., Beller, L., Deboutte, W., Gloza-Rausch, F., Seebens-Hoyer, A., Yordanov, S., Sylverken, A., Oppong, S., Sarkodie, Y.A., Vallo, P., Leroy, E.M., Bourgarel, M., Yinda, K.C., Van Ranst, M., Drosten, C., Drexler, J.F. and Matthijnssens, J. (2021). At least seven distinct rotavirus genotype constellations in bats with evidence of reassortment and zoonotic transmissions. mBio, 12(1): e02755-20. doi: 10.1128/mBio.02755-20.

Snodgrass, D.R. and Wells, P.W. (1976). Rotavirus infection in lambs: studies on passive protection. Arch Virol, 52(3): 201-205. doi: https://doi.org/10.1007/BF01348017

Squires, R.A. (2003). An update on aspects of viral gastrointestinal diseases of dogs and cats. New Zealand Vet J, 51(6): 252-261. doi: 10.1080/00480169.2003.36379

Steele, A.D., Geyer, A. and Gerdes, G.H. (2004). Rotavirus infections. In: Coetzer, J.A.W. and Tustin, R.C. (eds), Infectious diseases of Livestock, Oxford University Press, Southern Africa, pp.1256-1264.

Tate, J.E., Burton, A.H., Boschi-Pinto, C., Parashar, U.D., World Health Organization–Coordinated Global Rotavirus Surveillance Network, Agocs, M., Serhan, F., de Oliveira, L., Mwenda, J.M., Mihigo, R. and RanjanWijesinghe, P. (2016). Global, regional, and national estimates of rotavirus mortality in children < 5 years of age, 2000-2013. Clin Infect Dis, 62(suppl_2): S96-S105. doi: https://doi.org/10.1093/cid/civ1013

Thouless, M.E., DiGiacomo, R.F., Deeb, B.J. and Howard, H. (1988). Pathogenicity of rotavirus in rabbits. J Clin Microbiol, 26(5): 943-947. doi: https://doi.org/10.1128/jcm.26.5.943-947.1988

Tihova, M., Dryden, K.A., Bellamy, A.R., Greenberg, H.B. and Yeager, M. (2001). Localization of membrane permeabilization and receptor binding sites on the VP4 hemagglutinin of rotavirus: implications for cell entry. J Mol Biol, 314(5): 985-992. doi: 10.1006/jmbi.2001.5238

Trask, S.D., McDonald, S.M. and Patton, J.T. (2012). Structural insights into the coupling of virion assembly and rotavirus replication. Nat Rev Microbiol, 10(3): 165-177. doi: 10.1038/nrmicro2673

Troeger, C., Blacker, B.F., Khalil, I.A., Rao, P.C., Cao, S., Zimsen, S.R., Albertson, S.B., Stanaway, J.D., Deshpande, A., Abebe, Z. and Alvis-Guzman, N. (2018). Estimates of the global, regional, and national morbidity, mortality, and aetiologies of diarrhoea in 195 countries: a systematic analysis for the Global Burden of Disease Study 2016. Lancet Infect Dis, 18(11): 1211-1228. doi:https://doi.org/10.1016/S1473-3099(18)30362-1

Varghese, T., AlokitKhakha, S., Giri, S., Nair, N.P., Badur, M., Gathwala, G., Chaudhury, S., Kaushik, S., Dash, M., Mohakud, N.K. and Ray, R.K. (2021). Rotavirus strain distribution before and after introducing rotavirus vaccine in India. Pathogens, 10(4): 416.doi: https://doi.org/10.3390/pathogens10040416

Vlasova, A.N., Amimo, J.O. and Saif, L.J. (2017). Porcine rotaviruses: epidemiology, immune responses and control strategies. Viruses, 9(3): 48. doi: https://doi.org/10.3390/v9030048

Ward R.L., Bernstein D.I., Young E.C., Sherwood J.R., Knowlton D.R. and Schiff G.M. (1986). Human rotavirus studies in volunteers: determination of infectious dose and serological response to infection. J Infect Dis, 154(5): 871880. doi: https://doi.org/10.2307/30106704

WHO (2021). Rotavirus vaccines: WHO position paper - July 2021. Wkly Epidemiol Rec, 96(28): 301-320.

Woode, G.N. (1978). Epizootiology of bovine rotavirus infection. Vet Rec, 103(3): 44-46.doi: https://doi.org/10.1136/vr.103.3.44

Woode, G.N., Jones, J. and Bridger, J. (1975). Levels of colostral antibodies against neonatal calf diaahoea virus. Vet Rec, 97(8): 148-149. doi: 10.1136/vr.97.8.148

Zeller, M., Heylen, E., De Coster, S., Van Ranst, M. and Matthijnssens, J. (2012). Full genome characterization of a porcine-like human G9P [6] rotavirus strain isolated from an infant in Belgium. Infect Genet Evol, 12(7): 1492-1500. doi: https://doi.org/10.1016/j.meegid.2012.03.002

Chapter 20

Zika Virus: An Emerging Global Threat

Ekta Gupta, Shashi Sharma, Jyoti S. Kumar and Paban K. Dash*

Defence Research and Development Establishment, Jhansi Road, Gwalior – 474 002, Madhya Pradesh, India
****e-mail: pabandash.drde@gov.in***

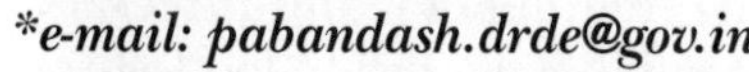

ABSTRACT

Zika infection has been a worldwide mosquito-borne disease since 1947. It is spread by Aedes mosquitoes in tropical and subtropical regions. Mild Zika infections cause symptoms such as fever and rash, while serious infections can cause various neurological and reproductive problems. As the population grows, more and more people come into contact with livestock or animals, increasing the possibility of zoonotic diseases being transmitted from animals to humans. Zika outbreaks have been observed before. This chapter describes all important aspects of Zika virus, including its structural organization, life cycle, transmission, symptoms, diagnosis, and treatment. This will help better understand Zika infection and inform future strategies for use of better tools and methods to monitor Zika virus.

Keywords: *Zika virus, Zoonotic diseases, Mosquito-borne.*

Introduction

Zika virus (ZIKV) is a mosquito borne virus, transmitted by bite of *Aedes* mosquitoes *i.e. Aedes aegypti* or *Aedes albopictus.* It belongs to genus *Flavivirus* and family *Flaviviridae*. Zika infection is closely related to other diseases like Dengue (DENV), Yellow Fever (YFV), West Nile (WNV), Japanese Encephalitis (JEV), *etc.* that are potentially able to cause epidemic. It was named ZIKA as it was first time isolated from the serum of rhesus monkey of Zika Forest in 1947 in Uganda. It was second time isolated from wild caught *Aedes africanus* mosquitoes of Zika forest in an attempt to isolate yellow fever virus in 1948 (Dick *et al.*, 1952(I)). Neutralising antibodies against ZIKV confirmed from the first human case in Uganda and Union Republic of Tanzania in 1952 (https://www.who.int/news-room/fact-sheets/detail/zika-virus). Other cases were reported in Eastern Nigeria during the widespread jaundice infection in 1954 (Macnamara *et al.*, 1954). Initially, the disease was believed to be confined to Africa and Asia but the first outbreak of the disease occurred in 2007 on Yap Island, Micronesia, where the patients were misdiagnosed with dengue virus (Duffy *et al.*, 2009). In 2013, another outbreak happened in French Polynesia, territory in South Pacific (Cao-Lormeau *et al.*, 2013). Another epidemic was reported in 2015 in North-eastern Brazil (Lowe *et al.*, 2018). Cases of autochthonous infection of ZIKV were also reported in America since 2014. The degree of threat connected to ZIKV was assessed by World Health Organization (WHO) and designated Zika as a worldwide global health emergency on 1st February 2016. The disease was associated with microcephaly and Guillain-Barré Syndrome (http://www.who.int/mediacentre/news/statements/2016/emergency-committee-zika microcephaly/en/). Genetic studies revealed that ZIKV evolution led to three distinct lineages – West African, East African and Asian. It is believed that the virus had originated from the East Africa and then was spread to West Africa and Asia later on.

Viral Structural Organization

In 2015, the virus was isolated from the patient in São Paulo state, Brazil and the first complete genome of ZIKV was deciphered in 2016 using Ion Torrent platform (Ion Personal Genome Machine Sequencer, ThermoFisher) (Cunha *et al.*, 2016). ZIKV consists of a positive sense, single stranded RNA with 10,807 nucleotides length. ZIKV RNA was found with an unusual nucleotide composition with rich in purine and poor in pyrimidine. This biasness towards purine is a characteristic feature of mosquito borne viruses (van Hemert and Berkhout, 2016). It encodes for a single open reading frame (ORF) which is flanked by un-translated regions (UTR) on 5′ and 3′ end. 5′ end of ZIKV RNA starts with a Type 1 capping, then non coding region (NCR) around 106 nucleotides, proceeded with open reading frame of 10,807 nucleotides and non-coding region (NCR) at 3′ end of around 428 nucleotides. ZIKV RNA has absence of poly A tail at 3′ end. The ORF gets translated into a single polyprotein of 3,423 amino acids that undergoes co or post translational modifications in which it is cleaved by host and viral proteases into ten proteins *viz.*, three structural proteins and seven non-structural proteins. Structural proteins are capsid C, pre-membrane/membrane prM/M and envelope E and non-structural proteins are NS1, NS2A, NS2B, NS3, NS4A, NS4B, and NS5 (Yun *et al.*, 2017). The C protein also called core protein is 11 kDa and forms nucleocapsid layer, E protein is largest structural protein of 50 kDa called envelope glycoprotein or fusion protein, M protein is present in immature virion as prM which is 26kDa and when it is cleaved by furin, it turns out 8 kDa resulting in mature virion (Singh *et al.*, 2016). M is a structural proteins help in the viral assembly whereas non-structural proteins help in significant functions including RNA replication, RNA translation, polyprotein modifications and evasion of host immune responses.

Cryo-EM Structure

Cryo-electron microscopy was used to discover 3.8Å resolution structure of ZIKV. The structure was quite similar to other flaviviruses. The dissimilarity existed of approximately 10 amino acids in the region of N154 glycan site. Zika virus like other flaviviruses is an enveloped protein with RNA enclosed within the capsid protein which is further surrounded by 180 copies of each Envelope (E) glycoproteins encoded by 500 amino acids and Membrane (M) protein encoded by 75 amino acids or Precursor Membrane (prM) protein encoded by 165 amino acid approximately in lipid membrane (Sirohi *et al.*, 2016).

E and M protein exist in heterodimeric form with 60 repeating units in icosahedral symmetry. E protein consists of four domains. E protein is bound to the viral membrane by two anti-parallel transmembrane domain (TMD) and has three other domains called ectodomains, E-DI, E-DII and E-DIII (Pierson *et al.*, 2018). E-D1 has a glycosylation site at Asn154 (N154) that serves as an attachment site of ZIKV to host cell by interaction with dendritic cell-specific intercellular adhesion molecule 3-grabbingnon-integrin (DC-SIGN) and the mannose receptor. E-DII forms a fusion loop that is highly conserved. It is responsible for interaction with the host membrane at the time of fusion (Shi *et al.*, 2017). ZIKV genome is enclosed within prM, E, and C protein forming immature virion. 'Pr' from the precursor membrane (prM) is cleaved by furin protease to form mature virion that can attach to the host cell to cause infection.

Life Cycle

When the mature virion attaches to the host cell and it is engulfed inside the cell by endocytosis. Clathrin mediated endocytosis occur where change in pH of endosome causes conformational changes of E protein that fuses host endosomal membrane with envelope of virion releasing the genome in cytoplasm of host cell (Abrams *et al.*, 2017). NS1 protein is responsible for replication of RNA, NS2 is a serine protease and has helicase activity with cofactor involved NS2B. NS4A and NS4B participate in the formation of complex for replication. Replication is further facilitated by RNA dependent RNA polymerase (RdRp), which is located on the C-terminal of the NS5 protein. The positive single stranded (+) ssRNA is transcribed by RdRp into complementary negative sense single strand RNA (-) ssRNA. This results in a double stranded RNA (dsRNA), which is unwound by the helicase activity of NS2, producing freshly generated (-) ssRNA and original (+) ssRNA. The (-) ssRNA strand subsequently serves as a template for transcription by NS5 RdRp, producing new (+) ssRNA and another dsRNA that is released when NS2 helicase unwinds it, separating the two RNA strands (+) and (-). Numerous (+) ssRNA strands are created in this manner. Post-transcriptional alteration of the (+) ssRNA occurs by capping by the methyl transferase enzyme, which is located on the N terminus of the NS5 protein. At the endoplasmic reticulum (ER), the modified (+) ssRNA is assembled with the help of the prM, C, and E proteins. This immature virion enters the Golgi network after leaving the ER. Furin splits and separates "Pr" from prM in the trans-golgi network, forming a mature virion with C, E, and M proteins encasing (+) ssRNA that is exocytosed and expelled from the cell. Since the E-prM heterodimer's repeating units provide immature virion its dimension of 60 nm, they contain C protein. The trans golgi network's low pH causes the E-prM heterodimer to reorient, cleaving Pr from prM to form mature virion with a diameter of 50 nm by host protease furin (Abrams *et al.*, 2017; Lin *et al.*, 2017).

Table 20.1: Structural and Non-structural Proteins and Function (Javed *et al.*, 2018; Hilgenfeld *et al.*, 2018)

Protein	*Function*
STRUCTURAL PROTEINS	
Capsid (C)	Formation of nucleocapsid layer outside viral genome
Pre membrane (prM)	It facilitates fusion with host cell.
Envelop (E)	It binds to host receptor, causes viral fusion and viral entry
NONSTRUCTURAL PROTEINS	
NS1 protein (NS1)	Tightly conserved protein, causes emmision, virulence and replication
NS2 protein (NS2A)	Involved in the replication of RNA
NS2 protein (NS2B)	Involved in replication complex formation, collaborates with NS3 for serine protease polyportein cleavage
NS3 protein (NS3)	Serine protease, helicase and triphosphatase activity present.
NS4 protein (NS4A)	Responsible for pathogenesis of virus, RNA Replication
NS4 protein (NS4B)	Involved in RNA Replication complex and localization of NS3 protein in membrane
NS5 protein (NS5)	RNA-dependent RNA polymerase (RdRp) and methyltransferase (MTase) activity.

Transmission

The transmission of ZIKV occurs through the bite of an infected female mosquito during its blood meal. It is primarily circulated by *Aedes* spp. (*Aedes aegypti* or *Aedes albopictus*). Apart from *Aedes africanus*, lots of other species of mosquitoes were found for the presence of the virus like *Ae. africanus, Ae. furcifer, Ae. luteocephalus, Ae. vittatus, Ae. dalzieli, Ae. hirsutus, Ae. metalicus Ae. taylori, Ae. aegypti, Ae. unilineatus, Anopheles coustani, Culex perfuscus,* and *Mansoniauniformis* in Africa. The virus is found to be circulating in the sylavtic cycle or enzootic cycle where the interaction is between mosquitoes and non-human primates *e.g.* monkeys. Spillover of infected mosquitoes from sylvatic habitats causes transmission of virus to humans, thus the virus entering the urban cycle. Healthy mosquitoes bite infected humans for blood meal and then transmitting infection to another healthy human. Additional modes of transmission include perinatal transmission *i.e.* virus transmitted from mother to child. One study showed that during the ZIKV outbreak in French Polynesia in 2013, sera collected from two mothers and their infants in less than four days after delivery were tested positive for ZIKV by RT-PCR. It was suggested that transmission could be transplacental or during parturition (Besnard *et al.*, 2014). Cases of the presence of ZIKV in breast milk are also seen in three mothers that were infected with ZIKV. Out of three newborns of these three infected mothers who were in taking breast milk, two were tested positive for ZIKV (Colt *et al.*, 2017). Potential risk of sexual transmission of ZIKV was also found where semen of a patient from Tahiti, French Polynesia was tested positive for ZIKV by rRT-PCR suggesting transmission of ZIKV by sexual intercourse (Musso et al., 2015a). During the investigation of arboviruses (DENV, JEV, ZIKV *etc.*) in blood donors in French Polynesia. Sera of blood donors were found positive for DENV, ZIKV, JEV, WNV. Thus, blood transfusion from asymptomatic people can also be responsible for disease transmission and sero-surveillance of blood donors can prevent the risk of virus spread (Aubry *et al.*, 2015). ZIKV RNA was also detected in urine sample of patients (Gourinat *et al.*, 2015; Da Conceicao *et al.*, 2022). One study suggested that larvae and pupae of mosquitoes can get ZIKV infection from urine or sewage contaminated water habitats which become the breeding sites for field mosquitoes leading to vertical transmission (Du *et al.*, 2019). Detection of ZIKV from saliva samples is also seen (Musso et al., 2015b).

Clinical Manifestation

Most cases of ZIKV infection are asymptomatic. Symptoms are very mild if prevail like fever, rashes, arthralgia, conjunctivitis, arthritis, edema. In some cases, headache, myalgia, low back pain, digestive disorders, cervical lymphadenopathy and vomiting is also seen. Cases where patients with co-morbidities have severe infection have been reported (McArthur *et al.*, 2017). Neurological symptoms are also seen in ZIKV infection. Guillain-Barre syndrome (GBS) is the most common neurological disorder in the symptomatic cases of ZIKV. It attributes to flaccid paralysis that contributes to peripheral nerve damage. About 48 cases were affected with GBS and more than 80 percent cases were found to suffer from symptomatic ZIKV at the time of French Polynesia outbreak (Rajah *et al.*, 2016). Other neurological disorders like meningoencephalitis, microcephaly, *etc.* are also been reported. In Rio de

Janeiro, pregnant women were tested for ZIKV using RT-PCR between 2015-2016. Abnormalities in foetus were checked by sonography. It was found that there was high percentage of microcephaly in foetuses of ZIKV positive mothers (Brasil *et al.,* 2016). ZIKV RNA was found in the amniotic fluids, placental tissue, foetal brain tissue, thus contributing in miscarriages. It may also affect neural progenitor cells (NPS) in uterus causing foetal malformations or severe microcephaly (Martines *et al.*, 2016; Mlakar *et al.,* 2016; Brady *et al.,* 2019).

Diagnosis

With the association of various other complications like GBS, microcephaly with ZIKV and the spread of virus due to variety of transmission routes, elevates the importance of development of diagnostic paradigm. It acts as a tool for surveillance of the disease especially in case of pregnant women or asymptomatic patients who might have disease history or are prevailing in an infected country. ZIKV diagnosis is challenging due to quite similar symptomatology and cross reactivity with other members of the family *Flaviviridae*. Most common approaches incorporated in detection of ZIKV comprises of viral isolation, detection of viral RNA from blood or other biological matrices (blood, urine, saliva *etc.*) and serological assays. RT-PCR or qRT-PCR assays are always considered as the most frequent used technique to detect ZIKV in blood of suspected patients due to its high sensitivity. It is advisable to collect and test samples in the first five days after the onset of symptoms in patients as the viremia is short lived (Faye *et al.,* 2013). Centers for Disease Control and Prevention (CDC) developed RT-PCR duplex assay targeting two genes prM and E gene. No cross reactivity was observed while testing samples (Lanciotti *et al.,* 2008). Serological tests can also be performed like enzyme-linked immunosorbent assays (ELISA), immunofluorescence assays *etc.* which can detect presence of IgM and IgG antibodies against ZIKV. There are no commercially approved FDA diagnostic kits. Zika IgM Antibody Capture Enzyme-Linked Immunosorbent Assay (Zika MAC-ELISA) was developed by CDC which has been approved under emergency use by FDA (https://www.cdc.gov/zika/laboratories/types-of-tests.html). High cross reactivity was observed in cases that have a history of dengue infection or other flavivirus infection, plaque reduction neutralization test (PRNT) aid in more specific result by using neutralizing antibodies specific for ZIKV (Charrel *et al.,* 2016). CDC has a mentioned diagnostic guideline that comprises different detection test as per sample type (https://www.cdc.gov/zika/laboratories/index.html).

Treatment and Therapeutics

As a primary treatment, adequate rest and fluid consumption are advised to avoid dehydration. Patients with dengue-like symptoms who may have ZIKV infection are not allowed to take aspirin or other non-steroidal anti-inflammatory medicines (NSAIDS) because they can cause haemorrhage in dengue virus infection. Fever is usually treated with acetaminophen (paracetamol) or dipyrone, and rashes can be treated with antihistamines. There are currently no medications available to treat or prevent ZIKV infection. To treat ZIKV infection, existing drugs were repurposed.

Merimepodib, an enzyme that inhibits synthesis of guanine nucleotides when used alongside the antiviral medicine ribavirin or T-705 (favipiravir) can block ZIKV RNA replication (Gorshkov *et al.,* 2019). In one study many compounds were tested like ribavirin and polymerase inhibitor 7-deaza-2′-C-methyladenosine (7DMA). AG129 mice was infected with ZIKV and administration of 7DMA was found to inhibit viral replication of ZIKV in mice and delaying viral disease thus exhibiting a potent anti ZIKV therapeutic candidate (Zmurko *et al.,* 2016). Another research where 774 compounds approved by FDA were tested for anti ZIKV activity. These were screened in their *in-vitro* assay where the compounds had to inhibit infection of ZIKV in human HuH-7 cells. Out of which, compounds like Auranofin, Azathioprine, Bortezomib, Clofazimine, Cyclosporine A, Dactinomycin, Daptomycin, Deferasirox, Digoxin, Fingolimod, Gemcitabine-HCl, Ivermectin, Mebendazole, Mefloquine-HCl, Mercaptopurine Hydrate, Methoxsalen, Micafungin, Mycophenolate Mofetil, Mycophenolic Acid, Palonosetron·HCl, Pyrimethamine, Sertraline-HCl, Sorafenib Tosylate and Thioguanine showed inhibition of viral infection (Barrows *et al.,* 2016). Many anti-malarial drugs like chloroquine and hepatitis C virus anti-viral like sofosbuvir were seen to suppress ZIKV replication and hence infection (McArthur *et al.,* 2017). Sofosbuvir blocks ZIKV RdRp and also protects NPCs. Flaviviruses, such as ZIKV, require low pH, or the acidic composition of endosomal vesicles, to fuse. In one study, Obatoclax was found to rapidly neutralise the acidic character of endolysosomal vesicles, making them unfavourable for viral fusion and thereby limiting ZIKV infection (Varghese *et al.,* 2017). At the time of ZIKV infection, a compound named cholesterol-25-hydroxylase (CH25H) is released in form of a immune response. This compound was studied and synthetic cholesterol-25-hydroxylase (CH25H) tested *in-vitro* revealed that it blocks the viral entry, reduces viremia thus a potent antiviral candidate to hinder ZIKV infection (Li C *et al.,* 2017). In another investigation, three medicines, temoporfin, niclosamide, and nitazoxanide, were discovered to affect the NS2B-NS3 interaction out of a total of 2816 substances examined. Tempoporfin was found to protect NPCs and the placenta while also reducing viremia and mortality in mice. Tempoporfin was also potent antiviral for other flaviviruses as well (Li Z *et al.,* 2017). In response to viral infection, Type 1 interferon elicits expression of genes that encodes protein Viperin (virus inhibitory protein, endoplasmic reticulum-associated, interferon-inducible, also known as RSAD2), inhibits ZIKV infection by degrading NS3 protein (Panayiotou *et al.,* 2018). Analog of adenosine BCX4430 was revealed to effectively treat ZIKV infected mice in chronic stage of viremia. It was useful in preventing congential infection and infection by sexual transmission (Gorshkov *et al.,* 2019).

Vaccines

There are currently no vaccines for prevention of ZIKV. Vaccine is the only possible method available to provide immunity to humans or animals against such viral diseases. They need to be cost effective and also should not have any adverse effect on pregnant women and infants. Due to increasing outbreaks and spread of diseases, many developers are trying to overcome this challenge of vaccine development.

Developers are working on inactivated whole organisms vaccines against ZIKV after its success against Japanese encephalitis virus (JEV) and tick borne encephalitis virus (TBEV). Inactivated whole organism vaccine provides multiple antigenic sites and non-replicative virus. ZIKV strain of Puerto Rican was used to develop purified inactivated virus (PIV). This PIV is used in many studies with or without adjuvants. In one study, PIV was inoculated with alum by intramuscular or subcutaneous route in Balb/c mice. ZIKV specific neutralising antibodies were formed and successful protection against ZIKV infection was seen in mice with intramuscular route of inoculation (Larocca *et al.*, 2016). In another study, PIV with alum or sham alone was inoculated subcutaneously in 16 rhesus monkeys. Monkeys inoculated with PIV induced both E protein specific antibodies and neutralising antibodies and no virus was detected by RT-PCR in body fluids. However, monkeys inoculated with sham had virus detected in body fluids (Abbink *et al.*, 2016).

DNA vaccines where cloning of selected gene segment takes place in plasmid. Then injecting the plasmid such that the antigen presenting cells that presents our desired sequence to elicit immune response. Many DNA vaccines against ZIKV are in phase 1 clinical trials. The prM/E genes of ZIKV are exploited for DNA vaccine. E protein elicitsneutralising antibodies and are effective in providing protection. Also these vaccines can lead to production of CD4+ and CD8+ T lymphocyte responses against E protein. Researchers are also working on RNA Vaccines as they are safer than DNA vaccines. A modified mRNA with gene of our interest like prM/E gene from ZIKV enclosed within a lipid nanoparticle. This was tested and revealed that it provided protection to AG129 mice as well as C57BL/6 mice leading to both production of neutralising antibody and CD8+ lymphocyte response. Recombinant viral vector vaccine where rhesus adenovirus serotype 52 (RhAd52) vector used in vaccine was administered to rhesus monkey. This vaccine led to production of ZIKV specific neutralising antibodies against ZIKV E protein. Recombinant viral vector vaccines are in pre clinical stages of development (McArthur *et al.*, 2017).

Prevention

As ZIKV prevention techniques, a variety of approaches are being used. One of these is supportive therapy, in which patients with dengue-like symptoms in endemic countries are treated by avoiding aspirin or NSAIDS. The vector control strategy is then utilised to eliminate mosquito breeding places by removing water collecting in plant pots, rubber tyres, gutters, and so on, especially during the monsoon season. Mosquito repellents containing 10-30 percent N,N-Diethyl-meta-toluamide (DEET), wearing full sleeves and pants, utilising bed nets and remaining indoors with window and door screens can all help to prevent mosquito bites. As a result, mosquito viruses' vectorial ability is reduced. Another option being explored is the incorporation of genetically modified mosquitoes into the population to limit the emergence of new adult mosquitoes. Oxitec, a UK biotechnology corporation, created OX513A, a transgenic male mosquito in which the OX513A gene was inserted into the male mosquito genome. This male mosquito mates with a normal female mosquito, and the larva develops normally but does not mature into an adult (Gulland *et al.*, 2016). This technique was 95 percent effective in lowering the population of *Aedes aegypti* in Bahia, Brazil (Carvalho *et al.*, 2015). Another strategy where bacteria, *Wolbachia* is exploited which is presented naturally in arthropods. It reduces the life span of mosquitoes thus reducing the number of viruses. It reduces ZIKV proliferation and also circulation of viruses to the salivary glands thus affecting vector competence. This, in turn, prevents transmission of virus to humans (Valentine *et al.*, 2016).

Pregnant women that are residing in ZIKV endemic countries should go for regular sonography to check fetal health. Also, pregnant women who are not residing in affected countries should restrain themselves from traveling to affected country (Plourde *et al.*, 2016).

Role of Animals in Spread of Zika Virus

First time ZIKV was isolated from a rhesus monkey with mild pyrexia in 1947 from Zika Forest in Uganda. Monkey was suffering from fever and was evaluated on third day by collecting the blood sample. The acquired sample was used to inoculate Swiss mice and additional rhesus monkeys. The mice all had the same disease, and samples were taken from their brains for neutralisation research. The inoculated rhesus monkey showed no signs of illness and exhibited asymptomatic disease behaviour (Gubler *et al.*, 2017). This sheds light on the function of animals in the propagation of the ZIKV virus.

Many studies were conducted to determine the presence of ZIKV antibodies in livestock like rodents, sheep and goats and primates were detected. Sera isolated from 172 domestic animals (Cow, buffalo, sheep, goat), 157 wild rodents (*Tatera indica, Merioneshuwianae, Rattus rattus, Rattus noruegicus, Rattus (Millardin) meltada, Nesokia indica, Bandicota bengalensis, Golundaellioti, Funambulus pennanti*) and 43 humans from Pakistan were examined for togaviridae viruses by complement fixation (CF) tests, out of which 6 rodents, 2 domestic animals and 1 human was tested positive for ZIKV antibodies (Darwish *et al.*, 1983). Another study was conducted where domestic animals like sheep, goat, cattle and vertebrates were examined for arboviruses by haemagglutination-inhibition tests. In Kenya, domestic animals, birds, small mammals and reptiles showed positive results for ZIKV (Johnson *et al.*, 1977). Real time PCR assays also revealed positive results in detection of ZIKV in sera and oral swabs from 15 marmosets (*Callithrix jacchus*) and 9 monkeys (*Sapajus libidinosus*) in Brazil. Sequencing was also undertaken, which demonstrated that the strain's sequence was 100 percent comparable to the South American ZIKV strain (Favoretto *et al.*, 2016). In-vitro cell culture investigations were also carried out, with cell lines derived from several animal species grown and inoculated with ZIKV. On day 7 after infection, 17 cell lines exhibited cytopathic effects. Domestic animals, old world primates, and wild animals were used to create these cell lines (Barr *et al.*, 2016). The increased potential of wild animals engaging in the sylvatic cycle raises the chance of ZIKV circulation in the people, posing a global danger of outbreaks. Thus, surveillance and regular monitoring may contribute in the eradication of these dangers, while vector control and personal protective measures are also taken into account (Althouse *et al.*, 2016).

Generally, animal models are used to study viral pathogenesis, viral propagation, immune response in adults, pregnant women, foetuses, vaccinations, and treatments. Rodent and nonhuman primate (NHP) models are the most commonly used for ZIKV infection. The first ZIKV isolate (ZIKV MR 66) was isolated from a rhesus monkey in Uganda in 1947 and was serially passaged numerous times in mice. Inoculation by the intracerebral or intracranial method resulted in neurological illness in mice. However, peripheral route had no sign of illness in immunocompetent mice (Dick *et al.*, 1952(II)). After a few years, biochemical analysis indicated that ZIKV with NS5 protein causes STAT2 to degrade, which is a transcription factor that facilitates transcription of Type I interferon receptor (IFNAR), hence suppressing Type I interferon signalling response. ZIKV NS5 protein is unable to degrade STAT2 in mice and therefore, immunocompetent mice show resistance to ZIKV infection (Grant *et al.*, 2016). Other route of inoculation like subcutaneous, intraperitoneal or intravenous showed sign of critical disease with paralysis, hind limb weakness when mice like A129 and C57BL/6 with IFNAR1 gene deficiency or mice without transcription factors like Irf3, Irf5, Irf7 in a triple knockout state ($Irf3^{-1}/Irf5^{-1}/Irf7^{-1}$) were used. Another work with same result of severe illness was seen where wild type (WT) C57BL/6 mice inoculated by intraperitoneal route, had been served with anti-IFNAR1 monoclonal antibody (MAR1-5A3) as it blocks the IFN-α/β receptor (Smith *et al.*, 2017). Mice with Type I IFN and Type II IFN receptor deficiency exhibit greater form of severe illness as both are involved in Janus kinases-signal transducers and activators of transcription pathway (Jak-STAT). Thus, innate immune response mediated by IFN play major role in forming resistance for ZIKV infection. ZIKV also causes conjunctivitis. In one study rhesus macaques was infected subcutaneously with ZIKV and ZIKV RNA was detected from urine, saliva, lacrimal fluid. Mice ($Ifnar^{-1}$) when infected with ZIKV suffered from conjunctivitis and panuveitis. Male reproductive system was affected by ZIKV infection which was correlated with another study where male WT C57BL/6 mice infected by subcutaneous route after administration of anti-IFNAR monoclonal antibody exhibited signs of infection in reproductive cells like spermatogonia, sertoli cells, spermatocytes, sperm *etc.* (Govero *et al.*, 2016). Mice administered to immunosuppressant drugs like Dexamethasone and then infected with ZIKV led to weight loss, high viremia, disseminated infection to other body tissue like spleen, kidney, lungs, brain *etc.* Neonatal mice were also utilised to study development with ZIKV infection. From one day old to 7 to 8 days old WT C57BL/6 mice were examined with infection by subcutaneous or intraperitoneal or intracranial route displayed signs of ZIKV associated disease like tremors, CD4+ and CD8+ cells infiltered brain tissue, degradation of cerebellum, ataxia, paralysis *etc.* (Fernandes *et al.*, 2017). ZIKV infected pregnant women can give rise to microcephaly in fetus or other congenital disorders or abnormalities, affecting fetal health. Pregnant mice or developing fetal brain when inoculated with ZIKV resulted in cortical malformations, brain injury, neuronal cell death. Mouse models have also aided in the studies of antivirals or therapeutics and vaccine development against ZIKV. In one study, human monoclonal antibody (Mab ZIKV-117) when administered to WT pregnant dams in addition with anti-IFNAR1 Mab showed considerable reduction of ZIKV infection in placenta and fetal brain (Sapparapu *et al.*, 2016). There few limitations that exist with mouse model like structural and immunological difference in human and mice placenta, ZIKV is not very competent to replicate in mice due to species specific immune evasion mechanism, some clinical manifestation like GBS is not observed in mice and absence of neonatal Fc receptor (FcRn) on trophoblast cells that may affect the transfer of IgG in mice thus a false analysis of a drug or vaccine would result.

Nonhuman primate (NHP) models, such as macaque monkeys, have similarities with humans in terms of gestation and foetal development, making studies including NHP models more credible. After 6-7 days of infection, viral RNA in plasma could be detected in the NHP model, exactly as it can in humans. In the NHP model, an increase in anti-ZIKV neutralising antibodies and a decrease in viral RNA in plasma can be co-related. ZIKV-specific natural killer cells, CD4+ and CD8+ T cell production, and IFN-γ was also observed (Javed *et al.*, 2018).

Other models where different doses of ZIKV strain MEX1-44 were administered to chicken amniotic membrane. Magnetic resonance imagining (MRI) of ZIKV-infected chicken embryos exhibited brain abnormalities, slowed brain growth, cortex malformations.

There are few limitations with NHP model as well. Although the structure of placenta of NHP and humans is almost alike. The duration of gestation period in NHP is much longer which results in increased time of experiment thus it is time consuming. As the experiments would require animal biosafety level 2 or 3 thus increasing the cost of experiment. Professional or skilled person is required for such experiments which very less in number with such expertise (Morrison *et al.*, 2017).

REFERENCES

Abbink, P., Larocca, R.A., De La Barrera, R.A., Bricault, C.A., Moseley, E.T., Boyd, M.and Barouch, D.H. (2016). Protective efficacy of multiple vaccine platforms against Zika virus challenge in rhesus monkeys. Science, 353(6304): 1129-1132. https://doi.org/10.1126/science.aah6157.

Abrams, R.P., Solis, J.and Nath, A. (2017). Therapeutic approaches for Zika virus infection of the nervous system. Neurotherapeutics, 14(4): 1027-1048. https://doi.org/10.1007/s13311-017- 0575-2.

Althouse, B.M., Vasilakis, N., Sall, A.A., Diallo, M., Weaver, S C.and Hanley, K.A. (2016). Potential for Zika virus to establish a sylvatic transmission cycle in the Americas. PLoSNegl Trop Dis, 10(12): e0005055. https://doi.org/10.1371/journal.pntd.0005055.

Aubry, M., Finke, J., Teissier, A., Roche, C., Broult, J., Paulous, S. and Musso, D. (2015). Seroprevalence of arboviruses among blood donors in French Polynesia, 2011-2013. Int J Infect Dis, 41: 11-12. https://doi.org/10.1016/j.ijid.2015.10.005.

Barr, K.L., Anderson, B.D., Prakoso, D.and Long, M.T. (2016). Working with Zika and Usutu viruses in vitro. PLoSNegl Trop Dis, 10(8): e0004931. https://doi.org/10.1371/journal.pntd.0004931.

Barrows, N.J., Campos, R.K., Powell, S.T., Prasanth, K.R., Schott-Lerner, G., Soto-Acosta, R., and Garcia-Blanco, M.A. (2016). A screen of FDA-approved drugs for inhibitors of Zika virus infection. Cell Host Microbe, 20(2): 259-270. http://dx.doi.org/10.1016/j.chom.2016.07.004.

Besnard, M., Lastere, S., Teissier, A., Cao-Lormeau, V.M., and Musso, D. (2014). Evidence of perinatal transmission of Zika virus, French Polynesia, December 2013 and February 2014. Eurosurveill, 19(13). https://doi.org/10.2807/1560-7917.ES2014.19.13.20751.

Brady, O.J., Osgood-Zimmerman, A., Kassebaum, N.J., Ray, S.E., de Araújo, V.E.M., da Nóbrega, A.A., Frutuoso, L.C.V., Lecca, R.C.R., Stevens, A., Zoca de Oliveira, B., de Lima, J.M.Jr, Bogoch, I.I., Mayaud, P., Jaenisch, T., Mokdad, A.H., Murray, C.J.L., Hay, S.I., Reiner, R.C.Jr. and Marinho, F. (2019). The association between Zika virus infection and microcephaly in Brazil 2015-2017: An observational analysis of over 4 million births. PLoS Med, 16(3): e1002755. doi: 10.1371/journal.pmed.1002755.

Brasil, P., Pereira Jr, J.P., Moreira, M.E., Ribeiro Nogueira, R.M., Damasceno, L., Wakimoto, M.and Nielsen-Saines, K. (2016). Zika virus infection in pregnant women in Rio de Janeiro. N Engl J Med, 375(24), 2321-2334. https://doi.org/10.1056/NEJMoa1602412.

Cao-Lormeau, V.M., Roche, C., Teissier, A., Robin, E., Berry, A.L., Mallet, H.P., Sall, A.A. and Musso, D. (2014). Zika virus, French polynesia, South pacific, 2013. Emerg Infect Dis, 20(6): 1085-1086. doi: 10.3201/eid2006.140138.

Carvalho, D.O., McKemey, A.R., Garziera, L., Lacroix, R., Donnelly, C.A., Alphey, L., Malavasi, A. and Capurro, M.L. (2015). Suppression of a field population of *Aedes aegypti* in Brazil by sustained release of transgenic male mosquitoes. PLoSNegl Trop Dis, 9(7): e0003864. doi: 10.1371/journal.pntd.0003864.

Charrel, R.N., Leparc-Goffart, I., Pas, S., de Lamballerie, X., Koopmans, M.andReusken, C. (2016). Background review for diagnostic test development for Zika virus infection. Bull World Health Organ, 94(8): 574. https://doi.org/10.2471/BLT.16.171207.

Colt, S., Garcia-Casal, M.N., Peña-Rosas, J.P., Finkelstein, J.L., Rayco-Solon, P., Weise Prinzo, Z.C.and Mehta, S. (2017). Transmission of Zika virus through breast milk and other breastfeeding-related bodily-fluids: A systematic review. PLoSNegl Trop Dis, 11(4): e0005528. https://doi.org/10.1371/journal.pntd.0005528.

Cunha, M.S., Esposito, D.L., Rocco, I.M., Maeda, A.Y., Vasami, F.G., Nogueira, J.S., de Souza, R.P., Suzuki, A., Addas-Carvalho, M., Barjas-Castro, Mde. L., Resende, M.R., Stucchi, R.S., Boin Ide, F., Katz, G., Angerami, R.N. and da Fonseca, B.A. (2016). First complete genome sequence of Zika Virus (Flaviviridae, Flavivirus) from an autochthonous transmission in Brazil. Genome Announc, 4(2): e00032-16. doi: 10.1128/genomeA.00032-16.

da Conceição, P.J.P., de Carvalho, L.R., de Godoy, B.L.V., Nogueira, M.L., Terzian, A.C.B., de Godoy, M.F., Calmon, M.F., Bittar, C. and Rahal, P. (2022). Detection of Zika virus in urine from randomly tested individuals in Mirassol, Brazil. Infection. 50(1): 149-156. doi: 10.1007/s15010-021-01667-w.

Darwish, M.A., Hoogstraal, H., Roberts, T.J., Ahmed, I.P.and Omar, F. (1983). A seroepidemiological survey for certain arboviruses (Togaviridae) in Pakistan. Trans R Soc Trop Med Hyg, 77(4): 442-445. https://doi.org/10.1016/0035-9203(83)90106-2.

Dick, G.W., Kitchen, S.F.and Haddow, A.J. (1952). Zika virus (I). Isolations and serological specificity. Trans R Soc Trop Med Hyg, 46(5): 509-520. https://doi.org/10.1016/0035-9203(52)90042-4.

Dick, G.W., Kitchen, S.F.and Haddow, A.J. (1952). Zika virus (II). Pathogenicity and physical properties. Trans R Soc Trop Med Hyg, 46(5): https://doi.org/10.1016/0035-9203(52)90043-6.

Duffy, M.R., Chen, T.H., Hancock, W.T., Powers, A.M., Kool, J.L., Lanciotti, R.S., Pretrick, M., Marfel, M., Holzbauer, S., Dubray, C., Guillaumot, L., Griggs, A., Bel, M., Lambert, A.J., Laven, J., Kosoy, O., Panella, A., Biggerstaff, B.J., Fischer, M. and Hayes, E.B. (2009). Zika virus outbreak on Yap Island, Federated States of Micronesia. N Engl J Med, 360(24):2536-2543. doi: 10.1056/NEJMoa0805715. PMID: 19516034.

Du, S., Liu, Y., Liu, J., Zhao, J., Champagne, C., Tong, L., Zhang, R., Zhang, F., Qin, C.F., Ma, P., Chen, C.H., Liang, G., Liu, Q., Shi, P.Y., Cazelles, B., Wang, P., Tian, H. and Cheng, G. (2019). Aedes mosquitoes acquire and transmit Zika virus by breeding in contaminated aquatic environments. Nat Commun, 10(1): 1324. doi: 10.1038/s41467-019-09256-0.

Favoretto, S., Araújo, D., Oliveira, D., Duarte, N., Mesquita, F., Zanotto, P.andDurigon, E. (2016). First detection of Zika virus in neotropical primates in Brazil: a possible new reservoir. BioRxiv, 10: 049395. http://dx.doi.org/10.1101/049395.

Faye, O., Faye, O., Diallo, D., Diallo, M., Weidmann, M., and Sall, A.A. (2013). Quantitative realtime PCR detection of Zika virus and evaluation with field-caught mosquitoes. Virol J, 10: 1-8. https://doi.org/10.1186/1743-422X-10-311.

Fernandes, N.C., Nogueira, J.S., Réssio, R.A., Cirqueira, C.S., Kimura, L.M., Fernandes, K.R., Cunha, M.S., Souza, R.P. and Guerra, J.M. (2017). Experimental Zika virus infection induces spinal cord injury and encephalitis in newborn Swiss mice. Exp ToxicolPathol, 69(2): 63-71. doi: 10.1016/j.etp.2016.11.004.

Grant, A., Ponia, S.S., Tripathi, S., Balasubramaniam, V., Miorin, L., Sourisseau, M., Schwarz, M.C., Sánchez-Seco, M.P., Evans, M.J., Best, S.M. and García-Sastre, A. (2016). Zika virus targets human STAT2 to inhibit type I interferon signaling. Cell Host Microbe, 19(6): 882-890. doi: 10.1016/j.chom.2016.05.009.

Gorshkov, K., Shiryaev, S.A., Fertel, S., Lin, Y.W., Huang, C.T., Pinto, A., Farhy, C., Strongin, A.Y., Zheng, W. and Terskikh, A.V. (2019). Zika Virus: origins, pathological action, and treatment strategies. Front Microbiol, 9: 3252. doi: 10.3389/fmicb.2018.03252.

Gourinat, A.C., O'Connor, O., Calvez, E., Goarant, C.and Dupont-Rouzeyrol, M. (2015). Detection of Zika virus in urine. Emerg Infect Dis, 21(1): 84. https://doi.org/10.3201/eid2101.140894.

Govero, J., Esakky, P., Scheaffer, S.M., Fernandez, E., Drury, A., Platt, D.J., Gorman, M.J., Richner, J.M., Caine, E.A., Salazar, V., Moley, K.H. and Diamond, M.S. (2016). Zika virus infection damages the testes in mice. Nature, 540(7633):438-442. doi: 10.1038/nature20556.

Gubler, D.J., Vasilakis, N.and Musso, D. (2017). History and emergence of Zika virus. J Infect Dis, 216(suppl 10): S860-S867. https://doi.org/10.1093/infdis/jix451.

Gulland, A. (2016). Genetically modified mosquitos may be used in fight against Zika. Br Med J (Online), 352. https://doi.org/10.1136/bmj.i1086.

van Hemert, F. and Berkhout, B.(2016). Nucleotide composition of the Zika virus RNA genome and its codon usage. Virol J, 13: 95. doi: 10.1186/s12985-016-0551-1.

Hilgenfeld, R., Lei, J. and Zhang, L. (2018). The structure of the Zika virus protease, NS2B/NS3 pro. Dengue and Zika: control and antiviral treatment strategies, 131-145. https://doi.org/10.1007/978-981-10-8727-1_10.

Javed, F., Manzoor, K.N., Ali, M., Haq, I.U., Khan, A.A., Zaib, A.and Manzoor, S. (2018). Zika virus: what we need to know?. Journal Basic Microbiol, 58(1): 3-16. https://doi.org/10.1002/jobm.201700398.

Johnson, B.K., Chanas, A.C., Shockley, P., Squires, E.J., Gardner, P., Wallace, C., Simpson, D.I., Bowen, E.T., Platt, G.S., Way, H., Parsons, J. and Grainger, W.E. (1977). Arbovirus isolations from, and serological studies on, wild and domestic vertebrates from Kano Plain, Kenya. Trans R Soc Trop Med Hyg, 71(6): 512-7. doi: 10.1016/0035-9203(77)90146-8.

Lanciotti, R.S., Kosoy O.L., Laven, J.J., Velez, J.O., Lambert, A.J., Johnson, A.J., Stanfield, S.M. and Duffy, M.R. (2008). Genetic and serologic properties of Zika virus associated with an epidemic, Yap State, Micronesia, 2007. Emerg Infect Dis, 14(8):1232123-9. doi: 10.3201/eid1408.080287.

Larocca, R.A., Abbink, P., Peron, J.P., Zanotto, P.M., Iampietro, M.J., Badamchi-Zadeh, A., Boyd, M., Ng'ang'a, D., Kirilova, M., Nityanandam, R., Mercado, N.B., Li, Z., Moseley, E.T., Bricault, C.A., Borducchi, E.N., Giglio, P.B., Jetton, D., Neubauer, G., Nkolola, J.P., Maxfield, L.F., De La Barrera, R.A., Jarman, R.G., Eckels, K.H., Michael, N.L., Thomas, S.J. and Barouch, D.H. (2016). Vaccine protection against Zika virus from Brazil. Nature, 536(7617): 474-478. doi: 10.1038/nature18952.

Li, C., Deng, Y.Q., Wang, S., Ma, F., Aliyari, R., Huang, X.Y., Zhang, N.N., Watanabe, M., Dong, H.L., Liu, P., Li, X.F., Ye, Q., Tian, M., Hong, S., Fan, J., Zhao, H., Li, L., Vishlaghi, N., Buth, J.E., Au, C., Liu, Y., Lu, N., Du, P., Qin, F.X., Zhang, B., Gong, D., Dai, X., Sun, R., Novitch, B.G., Xu, Z., Qin, C.F. and Cheng, G. (2017). 25-Hydroxycholesterol Protects Host against Zika Virus Infection and Its Associated Microcephaly in a Mouse Model. Immunity, 46(3): 446-456. doi: 10.1016/j.immuni.2017.02.012.

Li, Z., Brecher, M., Deng, Y.Q., Zhang, J., Sakamuru, S., Liu, B., Huang, R., Koetzner, C.A., Allen, C.A., Jones, S.A., Chen, H., Zhang, N.N., Tian, M., Gao, F., Lin, Q., Banavali, N., Zhou, J., Boles, N., Xia, M., Kramer, L.D., Qin, C.F. and Li, H. (2017). Existing drugs as broad-spectrum and potent inhibitors for Zika virus by targeting NS2B-NS3 interaction. Cell Res, 27(8): 1046-1064. doi: 10.1038/cr.2017.88.

Lin, H H., Huang, L.M. and Wu, S.C. (2017). Zika virus molecular biology and perspectives for vaccine development: a review. J Nurs Res, 25(1): 3-6. https://doi.org/10.1097/jnr.0000000000000205.

Lowe, R., Barcellos, C., Brasil, P., Cruz, O.G., Honório, N.A., Kuper, H. and Carvalho, M.S. (2018). The Zika virus epidemic in Brazil: from discovery to future implications. Int. J. Environ. Res. Public Health, 15(1): 96. https://doi.org/10.3390/ijerph15010096.

Macnamara, F.N. (1954). Zika virus: a report on three cases of human infection during an epidemic of jaundice in Nigeria. Trans R Soc Trop Med Hyg, 48(2): 139-145. https://doi.org/10.1016/0035-9203(54)90006-1.

Martines, R.B., Bhatnagar, J., Keating, M.K., Silva-Flannery, L., Muehlenbachs, A., Gary, J., Goldsmith, C., Hale, G., Ritter, J., Rollin, D., Shieh, W.J., Luz, K.G., Ramos, A.M., Davi, H.P., Kleber de Oliveria, W., Lanciotti, R., Lambert, A. and Zaki, S. (2016). Notes from the field: evidence of Zika virus infection in brain and placental tissues from two congenitally infected newborns and two fetal losses-Brazil, 2015. MMWR Morb Mortal Wkly Rep, 65(6): 159-160. doi: 10.15585/mmwr.mm6506e1.

McArthur, M.A. (2017). Zika virus: recent advances towards the development of vaccines and therapeutics. Viruses, 9(6): 143. https://doi.org/10.3390/v9060143.

Mlakar, J., Korva, M., Tul, N., Popoviæ, M., Poljšak-Prijatelj, M., Mraz, J., Kolenc, M., Resman Rus, K., VesnaverVipotnik, T., Fabjan Vodušek, V., Vizjak, A., Pižem, J., Petrovec, M. and AvšièŽupanc, T. (2016). Zika Virus Associated with microcephaly. N Engl J Med, 374(10): 951-958. doi: 10.1056/NEJMoa1600651.

Morrison, T.E. and Diamond, M.S. (2017). Animal models of Zika virus infection, pathogenesis, and immunity. JVirol, 91(8):e00009-17. doi: 10.1128/JVI.00009-17.

Musso, D., Roche, C., Robin, E., Nhan, T., Teissier, A. and Cao-Lormeau, V.M. (2015a). Potential sexual transmission of Zika virus. Emerg Infect Dis, 21(2): 359. https://doi.org/10.3201/eid2102.141363.

Musso, D., Roche, C., Nhan, T.X., Robin, E., Teissier, A.and Cao-Lormeau, V.M. (2015b). Detection of Zika virus in saliva. J Clin Virol, 68: 53-55. https://doi.org/10.1016/j.jcv.2015.04.021.

Panayiotou, C., Lindqvist, R., Kurhade, C., Vonderstein, K., Pasto, J., Edlund, K., Upadhyay, A.S. andÖverby, A.K. (2018). Viperinrestricts Zika virus and tick-borne encephalitis virus replication by targeting NS3 for proteasomal degradation. J Virol, 92(7):e02054-17. doi: 10.1128/JVI.02054-17. Erratum in: J Virol, 2018 May 29;92(12):e00501-18. doi: 10.1128/JVI.00501-18.

Pierson, T.C. and Diamond, M.S. (2018). The emergence of Zika virus and its new clinical syndromes. Nature, 560(7720): 573-581. https://doi.org/10.1038/s41586-018-0446-y.

Plourde, A.R. and Bloch, E.M. (2016). A literature review of Zika virus. Emerg Infect Dis, 22(7):1185. https://doi.org/10.3201/eid2207.151990.

Sapparapu, G., Fernandez, E., Kose, N., Bin Cao, Fox, J.M., Bombardi, R.G., Zhao, H., Nelson, C.A., Bryan, A.L., Barnes, T., Davidson, E., Mysorekar, I.U., Fremont, D.H., Doranz, B.J., Diamond, M.S. and Crowe, J.E. (2016).

Neutralizing human antibodies prevent Zika virus replication and fetal disease in mice. Nature, 540(7633):443-447. doi: 10.1038/nature20564.

Shi, Y. and Gao, G.F. (2017). Structural biology of the Zika virus. Trends Biochem Sci, 42(6): 443-456. https://doi.org/10.1016/j.tibs.2017.02.009.

Sirohi, D., Chen, Z., Sun, L., Klose, T., Pierson, T.C., Rossmann, M.G. and Kuhn, R.J. (2016). The 3.8 Å resolution cryo-EM structure of Zika virus. Science, 352(6284): 467-470. https://doi.org/10.1126/science.aaf5316.

Singh, R.K., Dhama, K., Malik, Y.S., Ramakrishnan, M.A., Karthik, K., Tiwari, R., Saurabh, S., Sachan, S. and Joshi, S.K. (2016). Zika virus - emergence, evolution, pathology, diagnosis, and control: current global scenario and future perspectives - a comprehensive review. Vet Q, 36(3):150-175. doi: 10.1080/01652176.2016.1188333.

Smith, D.R., Hollidge, B., Daye, S., Zeng, X., Blancett, C., Kuszpit, K., Bocan, T., Koehler, J.W., Coyne, S., Minogue, T., Kenny, T., Chi, X., Yim, S., Miller, L., Schmaljohn, C., Bavari, S. and Golden, J.W. (2017). Neuropathogenesis of Zika virus in a highly susceptible immunocompetent mouse model after antibody blockade of type I interferon. PLoSNegl Trop Dis, 11(1):e0005296. doi: 10.1371/journal.pntd.0005296.

Rajah, M.M., Pardy, R.D., Condotta, S.A., Richer, M.J. and Sagan, S.M. (2016). Zika virus: emergence, phylogenetics, challenges, and opportunities. ACS infectious diseases, 2(11): 763-772. https://doi.org/10.1021/acsinfecdis.6b00161.

Valentine, G., Marquez, L. and Pammi, M. (2016). Zika virus epidemic: an update. Expert Rev Anti Infect Ther, 14(12): 1127-1138. https://doi.org/10.1080/14787210.2016.1245614.

Varghese, F.S., Rausalu, K., Hakanen, M., Saul, S., Kümmerer, B.M., Susi, P., Merits, A. and Ahola, T. (2017). Obatoclax inhibits Alphavirus membrane fusion by neutralizing the acidic environment of endocytic compartments. Antimicrob Agents Chemother, 61(3):e02227-16. doi: 10.1128/AAC.02227-16.

Yun SangIm, Y.S. and Lee YoungMin, L.Y. (2017). Zika virus: an emerging flavivirus. https://doi.org/10.1007/s12275-017-7063-6.

Zmurko, J., Marques, R.E., Schols, D., Verbeken, E., Kaptein, S.J. andNeyts, J. (2016). The viral polymerase inhibitor 7-Deaza-2′-C-methyladenosine is a potent inhibitor of in vitro Zika virus replication and delays disease progression in a robust mouse infection model. PLoSNegl Trop Dis, 10(5):e0004695. doi: 10.1371/journal.pntd.0004695.

Chapter 21

Monkeypox: An Emerging Threat to Global Population

Anshuman Behera, Sushil Kumar Sharma, Ambuj Shrivastava and Paban Kumar Dash*

Defence Research and Development Establishment, Jhansi Road, Gwalior – 474 002, Madhya Pradesh, India
****e-mail: pabandash.drde@gov.in***

ABSTRACT

Mpox, earlier known as Monkeypox, has recently emerged as an important Orthopox virus infection in many parts of the world including India. More than 90,000 cases of Mpox infection were reported worldwide in 2022-23. Mpox, being a zoonotic virus, has the ability to transmit from rodents, primates to humans. Mpox virus (MPXV) is a close relative of Orthopoxviruses like smallpox, cowpox, buffalopox, horsepox and camelpox viruses that are known to infect livestock and other wild animals. Due to factors including surging human-animal interaction, wide diversity of Mpox cases has been observed across the world. MPXV was initially endemic to the Central and West African countries but it has presently spread to over a hundred countries. The relatively easy transmission routes among humans through direct contact and respiratory droplets have the potential to cause Mpox epidemic in the future. This warrants for an urgent requirement for development of suitable medical countermeasures against Mpox. Various detection strategies have been developed for Mpox including rapid, specific and sensitive detection techniques using conventional PCR, real-time PCR, LAMP as well as RPA based detection. A plethora of ongoing research is currently focussed on the development of specific vaccine and antiviral compounds exclusively targeting Mpox. Vaccines against smallpox like ACAM 2000, JYNNEOS and antivirals like Brincidofovir and Tecovirimat have been proposed for use against Mpox. Considering the prospective capability of the virus to develop into a pandemic, there is an urgent need to comprehensively study the genotypic and phenotypic characteristics of the circulating Mpox viruses across the globe.

Keywords: *Mpox, Orthopox virus (OPV), Epidemic, Vaccine, Antivirals.*

Introduction

Monkeypox (Mpox) is a zoonotic virus that causes disease in humans comparable to that of smallpox disease with slightly less severity. Mpox virus disease is caused by infection with MPXV – a dsDNA virus belonging to the Orthopoxvirus genus and shares a close relation with several other Pox viruses. Diseases associated with this genus include smallpox, cowpox (CPXV), buffalopox (BPXV), horsepox (HSPV) and camelpox (CMLV). The re-emergence of Mpox virus in non-endemic countries has elevated this previously obscure pathogen to the forefront of global public health. In 2022, thirteen cases were reported from India, a non-endemic country for Mpox. The isolation of Mpox virus was also conducted with detailed phenotypic and genotypic characterization (Sharma *et al.*, 2023). Although the virus was initially endemic to the tropical rainforest regions of Central and West Africa, it has gradually progressed to several other countries mostly through human-to-human transmissions (Huang *et al.*, 2022). Due to common transmission routes, as seen in the SARS-CoV-2 pandemic, such as direct/indirect contact, respiratory droplets and body fluids, Mpox disease has the potential to develop into an epidemic. In 2022, the Mpox outbreak was categorized by the WHO as a Public Health Emergency of International Concern (PHEIC) (World Health Organization, WHO, 2022). With fairly inadequate development of treatment and vaccination strategies against Mpox up till now, there is an urgent requirement towards comprehensive research on the virus and establishment of appropriate mitigation strategies against MPXV.

History

MPXV was first identified in Copenhagen, Denmark (1958) in Cynomolgus monkeys (Magnus *et al.*, 1959). In the 1960s, several Mpox outbreaks were documented in non-human primates like monkeys and chimpanzees in the USA and the Netherlands (Arita *et al.*, 1968). The first case of Monkeypox virus infection in humans was reported in a nine-month-old boy in the Democratic Republic of the Congo, Africa in 1970 (Ladnyj et al., 1972). Mpox diseases remained endemic to the African continent, with a few occasional cases reported in the West and Central African regions until 2003, when for the first time MPXV infections were reported outside Africa (Beer *et al.*, 2019). The MPXV genome is categorized into two distinct genetic clades - clade I (earlier referred as the Congo Basin clade) and clade II (West African clade), that is further segregated into 2 distinct phylogenetic subclades, referred to as clade IIa and IIb. The Mpox viruses belonging to clade I are reported as more virulent as they have a higher case fatality rate of around 10 per cent, while Clade IIb has been found more prevalently in the recent global outbreak (Ulaeto *et al.*, 2023). As of 01st December 2023, the number of Mpox infection cases has gone up over 90,000 along with 170 deaths being reported worldwide, most of which have been traced to non-endemic countries.

Structure and Genome Organization

MPXV virion particles are brick or ovoid-shaped (200–250 nm) with a complex internal structure consisting of a dsDNA genome and related enzymes (Shchelkunov *et al.*, 2001; Kugelman *et al.*, 2014). Electron microscopy (EM) technique has enabled the observation and identification of virus particles after isolation in cell culture-based systems.

The Mpox virus genome size is approximately 196kb encoding about 200 proteins. The MPXV nucleic acid is comprised of double-stranded DNA along with covalently closed hairpin ends (without 5′ and 3′ free ends). Further, it also consists of 10kb inverted terminal repeats (ITR) flanking both the ends of the genome. The Mpox genes are relatively closely packed to each other with rare intergenic regions greater than 100bp. The central conserved region of MPXV genome encodes for "housekeeping" genes that are primarily associated with replication, transcription and virion assembly stages. The terminal region genes vary in comparison to other poxviruses as they encode proteins involved in pathogenesis and host range determination.

Transmission Cycle

Mpox is a zoonosis and spreads from animals to humans. The animal reservoirs of MPXV are rodents such as squirrels, pouched rats, mice, prairie dogs, hedgehogs, and non-human primates such as monkeys found in Africa where Mpox has previously been reported. However, the recent outbreak was predominantly disseminated by human-to-human transmission through fomites, respiratory droplets and indirect as well as direct contact with MPXV infected individuals (Alakunle *et al.*, 2020). The Mpox viral loads are found to be high in body fluids, including saliva, urine, feces, and semen, as well as in swabs collected from rectum and oropharynx. This further suggests that sexual transmission is a key component responsible for the disease spread (Peiró-Mestres *et al.*, 2022).

Disease Pathology

Mpox viral entry generally occurs through nasopharynx, oropharynx, or intradermal routes, followed by replication of the virus and consequent spreading towards the lymph nodes. The initial viremia further leads to systemic spread of the Mpox virus and infection of other organs. This stage lasts for 7 to

14 days (maximum about 21 days) and is considered as the incubation period (Moore *et al.*, 2021).

Subsequently, it is followed by secondary viremia stage which effectively leads to prodromal symptoms like lymphadenopathy and fever before lesions start to appear. During this stage, infected patients may or may not be contagious. Concurrently, lesions in the oropharynx and skin begin to develop (Hutson *et al.*, 2015).

The primary clinical manifestations comprise of skin lesions (umbilicated, vesicular, and pseudo-pustular) accompanied by fever, weakness, headache, and regional lymphadenopathy. Lesion clustering in the perianal or genital area is also very frequently associated with MPXV disease progression, which is most likely due to the sexual nature of transmission (Orviz *et al.*, 2022).

The onset of fatigue, fever, myalgia, headache, and lymphadenopathy marks the beginning of Mpox disease symptoms. Mucosal and skin lesions in the mouth and face, palms and soles of the infected individual are characteristic manifestations of Mpox disease. The overall number of lesions in the infected body varies from a small amount to thousands and skin rash sometimes spreads to the entire body (McCollum and Damon, 2014). The lesions advance in 2-day increments through macular, vesicular, pustular and papular phases within the next 4 weeks. These lesions can be observed as firm 2-10 mm sized pustules and remain in this phase for about 5-7 days followed by crust formation. Crusts are formed and peeled off within the next 2 weeks and the disease condition usually improves after around 4 weeks of symptom onset in most cases. Only after all the crusts have fallen off from the body, the patients are no longer considered infectious (Weaver *et al.*, 2008).

Life Cycle

The Mpox virus replicates in the cytoplasm of the host cell. Host cell entry involves 3 steps including virus attachment, hemifusion and core entry (Buller *et al.*, 1991; Moss, 2006). A unique feature of MPXV and other Orthopoxviruses is the production of two types of infectious particles: mature virions (MV) with only one membrane and extracellular virions (EV) with an additional membrane consisting of varying protein composition (Smith *et al.*,2002). Both these particles express distinct surface epitopes and hence, depending on their infectious forms, penetrate the host cells through specialized mechanisms. Several glycosaminoglycans including heparin sulfates, chondroitin, and laminin, contribute to their attachment to cells (Schmidt *et al.*,2012; Moss, 2012). The viral proteins including D8L, A34R, A26L, A27L, and H3L are vital for binding to the cell surface (Wolffe *et al.*, 1997; Chung *et al.*, 1998; Hsiao *et al.*, 1999; Lin *et al.*, 2000; Chiu *et al.*, 2007; Berhanu *et al.*, 2008). The virion binds and fuses with the host cell membrane after attachment which leads to its core being released into the host cytoplasm. The core encompasses various factors and enzymes that are responsible for transcription initiation. The viral envelope proteins L1, F9, A28, A27, L5, H2 are vital for facilitating entry to the host cell (Townsley *et al.*, 2005a; Townsley *et al.*, 2005b; Senkevich and Moss, 2005; Brown *et al.*, 2006; Foo *et al.*, 2009). The virus-encoded DdRP or DNA-dependent RNA polymerase promotes transcription, which is followed by translation of early, intermediate, and late proteins mediated by host cell ribosome. These proteins are consequently involved in the rapid proliferation and packaging of complete virions in the host cell (Rosales *et al.*, 1994a; Rosales *et al.*, 1994b; Wright *et al.*, 1998; McFadden *et al.*, 2005). Most of the viral particles remain in the host cytoplasm as IMVs or intracellular mature virions. The remaining virions, intracellular enveloped virions or IEVs, are attached to the cell membrane after acquiring a second envelope (Roberts and Smith, 2008). These cell surface-associated enveloped virions contribute towards cell-to-cell transmission of MPXV, while extracellular enveloped virions (EEV) are responsible for systemic transmission (Pauli *et al.*, 2010).

Diagnosis

Polymerase Chain Reaction (PCR)

A few important genes most commonly employed in conventional PCR testing include hemagglutinin, the ATI gene, and the crmB gene. In comparison to other diagnosis techniques, real-time PCR provides high throughput, sensitive and fast results. Important viral genes such as envelope protein gene B6R, B7R, complement binding protein C3L, DNA polymerase gene E9L, DdRP subunit gene RPO18, N3R and F3L have been targeted for MPXV detection (Kulesh *et al.*, 2004; Li *et al.*, 2006; Shchelkunov *et al.*, 2011; Orba *et al.*, 2015; Peiró-Mestres *et al.*, 2022). Distinct target sites in the MPXV genome leads to varying degrees of sensitivity and limits of detection for Mpox.

Isothermal Amplification Assay (RPA)

During the recent Mpox outbreak, the use of RPA has evidenced it to be a viable alternative to qPCR. The MPXV-RPA detection method has demonstrated a high sensitivity with a limit of detection value of only 16 DNA copies/µL (Davi *et al.*, 2019). Multiple DNA-based detection assays have been studied, optimized and validated for the detection of Mpox, such as recombinase aided amplification (RAA) and loop-mediated isothermal amplification (LAMP) (Jiang *et al.*, 2022). The use of isothermal amplification assays has significantly enhanced Point of Care testing facilities for Mpox diagnosis even in remote locations.

Serological Testing

Other diagnostic approaches for MPXV include ELISA, immunohistochemistry as well as western blot (Sejvar *et al.*, 2004; Hammarlund *et al.*, 2005; de Marco *et al.*, 2010; Dubois *et al.*, 2012). It is quite imperative to employ these serological techniques to diagnose Mpox virus in cases where there is no virological sample. In such instances, serological testing of IgG and IgM antibodies are usually deployed. However, due to the immunologic cross-reactivity between Mpox and other Orthopoxviruses, the specificity of the assay is low (Weinstein *et al.*, 2005). Electron microscopic observation, though a more tedious process, has also been used as an auxiliary method for Orthopox virus detection (Stagles *et al.*,1985).

Table 21.1: Summary of Limit of Detection (LoD) Values for Various MPXV Diagnosis Techniques

Detection Technique	*Gene Target*	*Detection Limit*	*Sample*	*References*
Conventional PCR	ATI	100 copies/ reaction	Rodent	(Neubauer *et al.*,1998)
Real-time PCR	B6R	10 copies/reaction (2 fg)	Human	(Li *et al.*, 2006)
	F3L	20 copies/reaction	Rodent	(Kulesh *et al.*, 2004)
	G2R	8.2 copies/ reaction (1.7 fg)	Human	(Li *et al.*, 2010)
LAMP	D14L	~250copies/ reaction	Human	(Iizuka *et al.*, 2009)
	ATI	1000 copies/ reaction	Human	(Iizuka *et al.*, 2009)
RAA	A27L	10 copies/reaction	Human	(Cui *et al.*, 2023)
RPA	G2R	16 copies/μL	Human	(Davi *et al.*, 2019)

Virus Isolation

Virus isolation, being considered as the gold standard for the laboratory diagnosis of virus infections, has also been used as the confirmatory step for the diagnosis of Mpox virus (Schnurr, 2006; Sharma *et al.*, 2023). However, with MPXV being a Risk Group 3 (RG-3) agent, it can only be performed in the laboratory with bio-safety level - 3 (BSL3) facility.

Treatment and Vaccination

Clinical management of Mpox disease involves isolation and careful monitoring of the temperature and symptoms of Mpox infected individuals and their close contacts twice daily for at least 21 days (upper limit of MPXV incubation period). Prolonged contact with mucous membranes, broken skin, scabs, respiratory droplets or body fluids is considered as "high risk" exposure warranting the immediate need for therapeutics and vaccination. The infected individuals must wear surgical mask, stay in isolation, and their lesions must stay covered as much as feasible until the natural falling off of lesion crusts and formation of new skin layer.

Vaccine Development

Although no specific vaccine candidate exclusively for MPXV has been designed yet, the live-attenuated vaccines ACAM 2000 and JYNNEOS have been in restricted use among the MPXV-infected individuals. The replication-deficient modified vaccinia vaccine is a two-shot vaccine and has a better safety profile in comparison to 1st and 2nd-gen smallpox vaccination. Administration of the modified vaccinia Ankara vaccine does not create lesions and diminishes the risk of local or disseminated spreading of the virus (McCollum and Damon, 2014). Furthermore, clinical trials have also illustrated that the modified vaccinia Ankara can stimulate antibody production in patients accompanied with weak or immune-compromised systems. As suggested by the CDC (Centers for Disease Control and prevention), prompt vaccination may aid in reducing disease severity as well as prevent disease onset.

Antivirals against Mpox

At the present scenario, no particular clinically proven antiviral treatments for Mpox infection exist. There are, however, countermeasures for prevention of Mpox outbreak. For severe cases of Mpox infection, compounds with demonstrated efficacy against other Orthopoxviruses may be utilized. Antivirals like intravenous vaccinia immune globulin (VIG), brincidofovir - the DNA polymerase inhibitor and tecovirimat - the intracellular viral release inhibitor have unknown efficacy against Mpox virus (McCollum and Damon, 2014). Tecovirimat inhibits VP37, a viral envelope protein that leads to blocking of viral maturation preventing the release of the virus from infected cells. Brincidofovir is an inhibitor of MPXV DNA polymerase and has been approved for the treatment of smallpox and other poxviruses in the USA (Rizk *et al.*, 2022).

Table 21.2: Current Medical Counter Measures against Mpox Virus Infection

		Mode of action	*FDA Approval*
Antiviral drugs	Tecovirimat	Prevents formation of EEVs that are responsible for the spread of the virus.	Yes
	Brincidofovir	Prodrug, selective inhibition of viral DNA synthesis.	Yes
	Vaccinia Immune Globulin Intravenous (VIGIV)	Purified IgG fraction of human plasma containing vaccinia virus antibodies	Yes
Vaccine	ACAM 2000	Live attenuated non replicating vaccine, illicits humoral and cellular immunity againsts OPVs	Yes
	JYNNEOS	Smallpox vaccine that provides limited resistance against other OPVs	Yes

Conclusion

The rate of Mpox infection in people is swiftly going up all over the world. A steady increase in human–wildlife contacts due to various factors like deforestation, climate change, travel industry boom and use of exotic species as a source of food consumption are the likely causes of the vast disease spread. All these contribute significantly towards the risk of viral zoonosis. Additionally, due to the cessation of smallpox vaccination in 1980, Mpox virus, being closely related to smallpox, has become more prevalent. With MPXV sharing several common genes with other related Orthopoxviruses like CPXV, BPXV, HSPV and CMLV, there is also a probability of genetic variations arising in the Mpox genome that can lead to host range modification and ultimately enhance the virus' ability to infect livestock and other animals.

Global public health has been critically challenged during the COVID-19 pandemic in the last few years. As confirmed cases of Mpox infection keep on rising, we might face a similar or far worse epidemic scenario in the near future, which may result in serious negative repercussions on the global economy. With Mpox being announced as a PHEIC by the WHO, it is imperative to develop sufficient diagnostic evaluation capability to limit further spread of the virus. Hence there is a critical necessity to improve our in depth understanding of the Mpox virus as well as the management and prevention/control of Mpox disease, especially among public health personnel. Finally, a collective collaboration of the global scientific community to conduct clinical investigations testing the safety and efficacy of Mpox vaccination and antiviral drug strategies is the need of the hour.

REFERENCES

Alakunle, E., Moens, U., Nchinda, G. and Okeke, M.I. (2020). Monkeypox virus in Nigeria: infection biology, epidemiology, and evolution. Viruses, 12(11): 1257. https://doi.org/10.3390/v12111257.

Arita, I. and Henderson, D.A. (1968). Smallpox and monkeypox in non-human primates. Bull World Health Organ, 39(2): 277. https://iris.who.int/handle/10665/266595

Berhanu, A., Wilson, R.L., Kirkwood-Watts, D.L., King, D.S., Warren, T.K., Lund, S.A., Brown, L.L., Krupkin, A.K., Vandermay, E., Weimers, W., Honeychurch, K.M., Grosenbach, D.W., Jones, K.F. and Hruby, D.E. (2008). Vaccination of BALB/c mice with *Escherichia coli*-expressed vaccinia virus proteins A27L, B5R, and D8L protects mice from lethal vaccinia virus challenge. J Virol, 82(7):3517-3529. doi: 10.1128/JVI.01854-07.

Brown, E., Senkevich, T.G. and Moss, B. (2006). Vaccinia virus F9 virion membrane protein is required for entry but not virus assembly, in contrast to the related L1 protein. J Virol, 80(19): 9455-9464. https://doi.org/10.1128/JVI.01149-06

Buller, R.M. and Palumbo, G.J. (1991). Poxvirus pathogenesis. MicrobiolRev, 55(1): 80-122. https://doi.org/10.1128/mr.55.1.80-122.1991

Chiu, W.L., Lin, C.L., Yang, M.H., Tzou, D.L.M. and Chang, W. (2007). Vaccinia virus 4c (A26L) protein on intracellular mature virus binds to the extracellular cellular matrix laminin. J Virol, 81(5): 2149-2157. https://doi.org/10.1128/JVI.02302-06

Chung, C.S., Hsiao, J.C., Chang, Y.S. and Chang, W. (1998). A27L protein mediates vaccinia virus interaction with cell surface heparan sulfate. JVirol, 72(2): 1577-1585. https://doi.org/10.1128/JVI.72.2.1577-1585.1998

Cui, X., Du, B., Feng, J., Feng, Y., Cui, J., Yan, C., Zhao, H., Gan, L., Fan, Z., Fu, T., Xu, Z., Zhang, R., Du, S., Zhou, Y., Tian, Z., Zhang, Q., Fu, H., Xue, G. and Yuan, J. (2023). Rapid detection of mpox virus using recombinase aided amplification assay. Front Cell Infect Microbiol, 13:1008783. doi: 10.3389/fcimb.2023.1008783.

Davi, S.D., Kissenkötter, J., Faye, M., Böhlken-Fascher, S., Stahl-Hennig, C., Faye, O., Faye, O., Sall, A.A., Weidmann, M., Ademowo, O.G., Hufert, F.T., Czerny, C.P. and AbdEl Wahed, A. (2019). Recombinase polymerase amplification assay for rapid detection of Monkeypox virus. DiagnMicrobiol Infect Dis, 95(1):41-45. doi: 10.1016/j.diagmicrobio.2019.03.015.

de Marco, M.D.M.F., Alejo, A., Hudson, P., Damon, I.K. and Alcami, A. (2010). The highly virulent variola and monkeypox viruses express secreted inhibitors of type I interferon. FASEB J, 24(5): 1479. https://doi.org/10.1096/fj.09-144733

Dubois, M.E., Hammarlund, E. and Slifka, M.K. (2012). Optimization of peptide-based ELISA for serological diagnostics: a retrospective study of human monkeypox infection. Vector Borne Zoonotic Dis, 12(5), 400-409. https://doi.org/10.1089/vbz.2011.0779

Foo, C.H., Lou, H., Whitbeck, J.C., Ponce-de-León, M., Atanasiu, D., Eisenberg, R.J. and Cohen, G.H. (2009). Vaccinia virus L1 binds to cell surfaces and blocks virus entry independently of glycosaminoglycans. Virol, 385(2): 368-382. https://doi.org/10.1016/j.virol.2008.12.019

Hammarlund, E., Lewis, M.W., Carter, S.V., Amanna, I., Hansen, S.G., Strelow, L.I., Wong, S.W., Yoshihara, P., Hanifin, J.M. and Slifka, M.K. (2005). Multiple diagnostic techniques identify previously vaccinated individuals with protective immunity against monkeypox. Nat Med, 11(9):1005-1011. doi: 10.1038/nm1273.

Hsiao, J.C., Chung, C.S. and Chang, W. (1999). Vaccinia virus envelope D8L protein binds to cell surface chondroitin sulfate and mediates the adsorption of intracellular mature virions to cells. J Virol, 73(10): 8750-8761. https://doi.org/10.1128/JVI.73.10.8750-8761.1999

Huang, Y., Mu, L. and Wang, W. (2022). Monkeypox: epidemiology, pathogenesis, treatment and prevention. Signal Transduct Target Ther, 7(1): 1-22. https://doi.org/10.1038/s41392-022-01215-4

Hutson, C.L., Carroll, D.S., Gallardo-Romero, N., Drew, C., Zaki, S.R., Nagy, T., Hughes, C., Olson, V.A., Sanders, J., Patel, N., Smith, S.K., Keckler, M.S., Karem, K. and Damon, I.K. (2015). Comparison of Monkeypox virus clade kinetics and pathology within the prairie dog animal model using a serial sacrifice study design. Biomed Res Int, 2015:965710. doi: 10.1155/2015/965710.

Iizuka, I., Saijo, M., Shiota, T., Ami, Y., Suzaki, Y., Nagata, N., Hasegawa, H., Sakai, K., Fukushi, S., Mizutani, T., Ogata, M., Nakauchi, M., Kurane, I., Mizuguchi, M. and Morikawa, S. (2009). Loop-mediated isothermal amplification-based diagnostic assay for monkeypox virus infections. J Med Virol, 81(6):1102-1108. doi: 10.1002/jmv.21494. PMID: 19382264.

Jiang, Z., Sun, J., Zhang, L., Yan, S., Li, D., Zhang, C., Lai, A. and Su, S. (2022). Laboratory diagnostics for monkeypox: An overview of sensitivities from various published tests. Travel Med Infect Dis, 49:102425. doi: 10.1016/j.tmaid.2022.102425.

Kugelman, J.R., Johnston, S.C., Mulembakani, P.M., Kisalu, N., Lee, M.S., Koroleva, G., McCarthy, S.E., Gestole, M.C., Wolfe, N.D., Fair, J.N., Schneider, B.S., Wright, L.L., Huggins, J., Whitehouse, C.A., Wemakoy, E.O., Muyembe-Tamfum, J.J., Hensley, L.E., Palacios, G.F. andRimoin, A.W. (2014). Genomic variability of monkeypox virus among humans, Democratic Republic of the Congo. Emerg Infect Dis, 20(2):232-239. doi: 10.3201/eid2002.130118.

Kulesh, D.A., Loveless, B.M., Norwood, D., Garrison, J., Whitehouse, C.A., Hartmann, C., Mucker, E., Miller, D., Wasieloski,L.P.Jr, Huggins, J., Huhn, G., Miser, L.L., Imig, C., Martinez, M., Larsen, T., Rossi, C.A. and Ludwig, G.V. (2004). Monkeypox virus detection in rodents using real-time 3′-minor groove binder TaqMan assays on the Roche LightCycler. Lab Invest, 84(9):1200-1208. doi: 10.1038/labinvest.3700143.

Ladnyj, I.D., Ziegler, P. and Kima, E.J.B.W.H.O. (1972). A human infection caused by monkeypox virus in Basankusu Territory, Democratic Republic of the Congo. Bull World Health Organ, 46(5): 593. PMID: 4340218

Li, Y., Olson, V.A., Laue, T., Laker, M.T. and Damon, I.K. (2006). Detection of monkeypox virus with real-time PCR assays. J Clin Virol, 36(3): 194-203. https://doi.org/10.1016/j.jcv.2006.03.012

Li, Y., Zhao, H., Wilkins, K., Hughes, C. and Damon, I.K. (2010). Real-time PCR assays for the specific detection of monkeypox virus West African and Congo Basin strain DNA. J VirolMethods, 169(1): 223-227. https://doi.org/10.1016/j.jviromet.2010.07.012

Lin, C.L., Chung, C.S., Heine, H.G. and Chang, W. (2000). Vaccinia virus envelope H3L protein binds to cell surface heparan sulfate and is important for intracellular mature virion morphogenesis and virus infection in vitro and in vivo. Journal Virol, 74(7): 3353-3365.https://doi.org/10.1128/jvi.74.7.3353-3365.2000

Magnus, P., Andersen, E., Petersen, K. and BirchAndersen, A. (1959). A Pox-like disease in Cynomolgus monkeys. Acta PatholMicrobiol Scand, 46: 156-176. https://doi.org/10.1111/j.1699-0463.1959.tb00328.x

McCollum, A.M. and Damon, I.K. (2014). Human monkeypox. Clin Infect Dis, 58(2): 260-267. https://doi.org/10.1093/cid/cit703

McFadden, G. (2005). Poxvirus tropism. Nat Rev Microbiol, 3(3): 201-213. https://doi.org/10.1038/nrmicro1099

Moore, M.J., Rathish, B. and Zahra, F. (2023). Mpox (Monkeypox). In: StatPearls [Internet]. Treasure Island (FL): StatPearls Publishing; 2024 Jan. PMID: 34662033.

Moss, B. (2006). Poxvirus entry and membrane fusion. Virol, 344(1): 48-54. https://doi.org/10.1016/j.virol.2005.09.037

Moss, B. (2012). Poxvirus cell entry: how many proteins does it take?. Viruses, 4(5): 688-707. https://doi.org/10.3390/v4050688

Neubauer, H., Reischl, U., Ropp, S., Esposito, J.J., Wolf, H. and Meyer, H. (1998). Specific detection of monkeypox virus by polymerase chain reaction. J VirolMethods, 74(2): 201-207. https://doi.org/10.1016/s0166-0934(98)00099-8

Orba, Y., Sasaki, M., Yamaguchi, H., Ishii, A., Thomas, Y., Ogawa, H., Hang'ombe, B.M., Mweene, A.S., Morikawa, S., Saijo, M. and Sawa, H.(2015). Orthopoxvirus infection among wildlife in Zambia. J Gen Virol, 96(Pt 2):390-394. doi: 10.1099/vir.0.070219-0.

Orviz, E., Negredo, A., Ayerdi, O., Vázquez, A., Muñoz-Gomez, A., Monzón, S., Clavo, P., Zaballos, A., Vera, M., Sánchez, P., Cabello, N., Jiménez, P., Pérez-García, J.A., Varona, S., Del Romero, J., Cuesta, I., Delgado-Iribarren, A., Torres, M., Sagastagoitia, I., Palacios, G., Estrada, V. and Sánchez-Seco, M.P.(2022). Grupo Viruela del Simio Madrid CNM/ISCIII/HCSC/Sandoval. Monkeypox outbreak in Madrid (Spain): Clinical and virological aspects. J Infect, 85(4):412-417. doi: 10.1016/j.jinf.2022.07.005.

Pauli, G., Blümel, J., Burger, R., Drosten, C., Gröner, A., Gürtler, L., Heiden, M., Hildebrandt, M., Jansen, B., Montag-Lessing, T., Offergeld, R., Seitz, R., Schlenkrich, U., Schottstedt, V., Strobel, J., Willkommen, H. and von König, C.H. (2010). Orthopoxviruses: Infections in humans. Transfus Med Hemother, 37(6):351-364. doi: 10.1159/000322101.

Peiró-Mestres, A., Fuertes, I., Camprubí-Ferrer, D., Marcos, M.Á., Vilella, A., Navarro, M., Rodriguez-Elena, L., Riera, J., Català, A., Martínez, M.J. and Blanco, J.L. (2022). Hospital Clinic de Barcelona Monkeypox Study Group. Frequent detection of monkeypox virus DNA in saliva, semen, and other clinical samples from 12 patients, Barcelona, Spain, May to June 2022. Euro Surveill, 27(28):2200503. doi: 10.2807/1560-7917.ES.2022.27.28.2200503.

Rizk, J.G., Lippi, G., Henry, B.M., Forthal, D.N. and Rizk, Y. (2022). Prevention and treatment of monkeypox. Drugs, 82(9):957-963. doi: 10.1007/s40265-022-01742-y. Erratum in: Drugs. 2022 Aug;82(12):1343. doi: 10.1007/s40265-022-01767-3.

Roberts, K.L. and Smith, G.L. (2008). Vaccinia virus morphogenesis and dissemination. TrendMicrobiol, 16(10): 472-479. https://doi.org/10.1016/j.tim.2008.07.009

Rosales, R., Sutter, G. and Moss, B.A. (1994a). A cellular factor is required for transcription of vaccinia viral intermediate-stage genes. Proc Natl Acad Sci U S A, 91(9):3794-3798. doi: 10.1073/pnas.91.9.3794.

Rosales, R., Harris, N., Ahn, B.Y. and Moss, B. (1994b). Purification and identification of a vaccinia virus-encoded intermediate stage promoter-specific transcription factor that has homology to eukaryotic transcription factor SII (TFIIS) and an additional role as a viral RNA polymerase subunit. JBiol Chem, 269(19): 14260-14267.https://doi.org/10.1016/S0021-9258(17)36783-2

Schmidt, F.I., Bleck, C.K.E. and Mercer, J. (2012). Poxvirus host cell entry. Curr OpinVirol, 2(1): 20-27. https://doi.org/10.1016/j.coviro.2011.11.007

Schnurr, D. (2006). Diagnosis of viral infections by viral isolation and identification or by direct detection. In: Congenital and Perinatal Infections: A Concise Guide to Diagnosis (pp. 23-35). Totowa, NJ: Humana Press. https://doi.org/10.1385/1-59259-965-6:023

Sejvar, J.J., Chowdary, Y., Schomogyi, M., Stevens, J., Patel, J., Karem, K., Fischer, M., Kuehnert, M.J., Zaki, S.R., Paddock, C.D.,Guarner, J., Shieh, W.J., Patton, J.L., Bernard, N., Li, Y., Olson, V.A., Kline, R.L., Loparev, V.N., Schmid, D.S., Beard, B., Regnery, R.R. and Damon, I.K. (2004). Human monkeypox infection: a family cluster in the midwestern United States. J Infect Dis, 190(10):1833-1840. doi: 10.1086/425039.

Senkevich, T.G. and Moss, B. (2005). Vaccinia virus H2 protein is an essential component of a complex involved in virus entry and cell-cell fusion. JVirol, 79(8): 4744-4754. https://doi.org/10.1128/JVI.79.8.4744-4754.2005

Sharma, S.K., Dash, P.K., Yadav, R.G., Shrivastava, A., Menon, R., Kumar, J.S., Sharma, S., Dhankher, S., Dhiman, S., Kumari, D., Shukla, M., Relhan, V., Kumar, S. andParida, M. (2023). Isolation and characterization of emerging Mpox virus from India. J Med Virol, 95(7):e28911. doi: 10.1002/jmv.28911.

Shchelkunov, S.N., Totmenin, A.V., Babkin, I.V., Safronov, P.F., Ryazankina, O.I., Petrov, N.A., Gutorov, V.V., Uvarova, E.A., Mikheev, M.V., Sisler, J.R., Esposito, J.J., Jahrling, P.B., Moss, B. andSandakhchiev, L.S. (2001). Human monkeypox and smallpox viruses: genomic comparison. FEBS Lett, 509(1):66-70. doi: 10.1016/s0014-5793(01)03144-1.

Shchelkunov, S.N., Shcherbakov, D.N., Maksyutov, R.A. and Gavrilova, E.V. (2011). Species-specific identification of variola, monkeypox, cowpox, and vaccinia viruses by multiplex real-time PCR assay. J VirolMethods, 175(2): 163-169. https://doi.org/10.1016/j.jviromet.2011.05.002

Smith, G.L., Vanderplasschen, A. and Law, M. (2002). The formation and function of extracellular enveloped vaccinia virus. JGen Virol, 83(12): 2915-2931.https://doi.org/10.1099/0022-1317-83-12-2915

Stagles, M.J., Watson, A.A., Boyd, J.F., More, I.A. andMcSeveney, D. (1985). The histopathology and electron microscopy of a human monkeypox lesion. Trans R Soc Trop Med Hyg, 79(2):192-202. doi: 10.1016/0035-9203(85)90333-5.

Townsley, A.C., Senkevich, T.G. and Moss, B. (2005). Vaccinia virus A21 virion membrane protein is required for cell entry and fusion. JVirol, 79(15): 9458-9469.https://doi.org/10.1128/JVI.79.15.9458-9469.2005

Townsley, A.C., Senkevich, T.G. and Moss, B. (2005). The product of the vaccinia virus L5R gene is a fourth membrane protein encoded by all poxviruses that is required for cell entry and cell-cell fusion. J Virol, 79(17): 10988-10998. https://doi.org/10.1128/jvi.79.17.10988-10998.2005

Ulaeto, D., Agafonov, A., Burchfield, J., Carter, L., Happi, C., Jakob, R., Krpelanova, E., Kuppalli, K., Lefkowitz, E.J., Mauldin, M.R., de Oliveira, T., Onoja, B., Otieno, J., Rambaut, A., Subissi, L., Yinka-Ogunleye, A. and Lewis, R.F. (2023). New nomenclature for mpox (monkeypox) and monkeypox virus clades. Lancet Infect Dis, 23(3):273-275. doi: 10.1016/S1473-3099(23)00055-5.

Weaver, J.R. and Isaacs, S.N. (2008). Monkeypox virus and insights into its immunomodulatory proteins. Immunol Rev, 225(1): 96-113. https://doi.org/10.1111/j.1600-065X.2008.00691.x

Weinstein, R.A., Nalca, A., Rimoin, A.W., Bavari, S. and Whitehouse, C.A. (2005). Reemergence of monkeypox: prevalence, diagnostics, and countermeasures. Clin Infect Dis, 41(12): 1765-1771. https://doi.org/10.1086/498155

WHO. (2022). WHO Director-General declares the ongoing Monkeypox outbreak a Public Health Emergency of International Concern. https://www.who.int/europe/news/item/23-07-2022-who-director-general-declares-the-ongoing-monkeypox-outbreak-a-public-health-event-of-international-concern

Wolffe, E.J., Katz, E., Weisberg, A. and Moss, B. (1997). The A34R glycoprotein gene is required for induction of specialized actin-containing microvilli and efficient cell-to-cell transmission of vaccinia virus. J Virol, 71(5): 3904-3915. https://doi.org/10.1128/jvi.71.5.3904-3915.1997

Wright, C.F., Hubbs, A.E., Gunasinghe, S.K. and Oswald, B.W. (1998). A vaccinia virus late transcription factor copurifies with a factor that binds to a viral late promoter and is complemented by extracts from uninfected HeLa cells. J Virol, 72(2): 1446-1451.https://doi.org/10.1128/JVI.72.2.1446-1451.1998

Index

D

E

F

G

H

I

W

Z